DATE DUE

GAYLORD			PRINTED IN U.S.A.

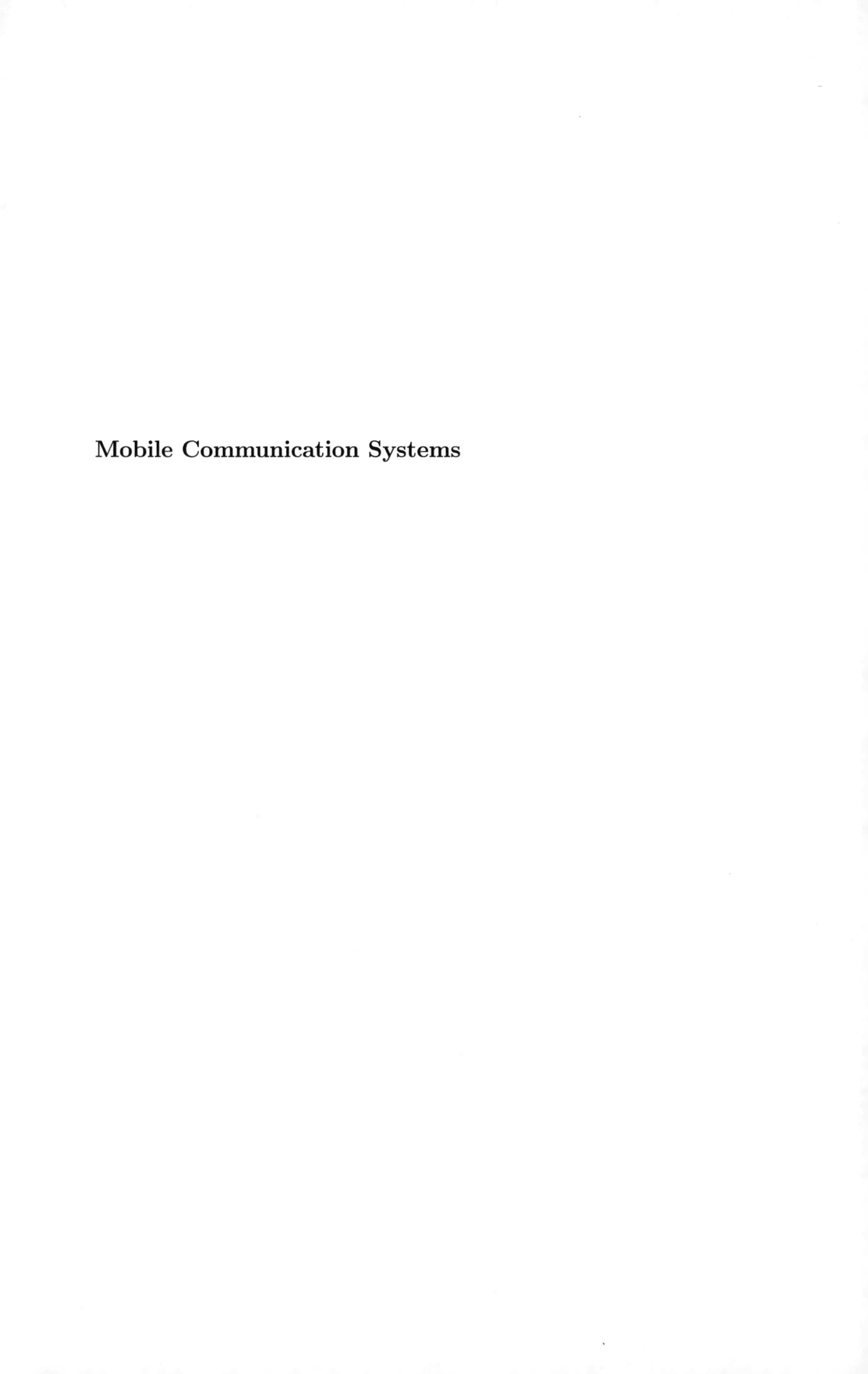

Mobile Communication Systems

To the memory of my Mother

Mobile Communication Systems

Krzysztof Wesołowski
Poznań University of Technology, Poland

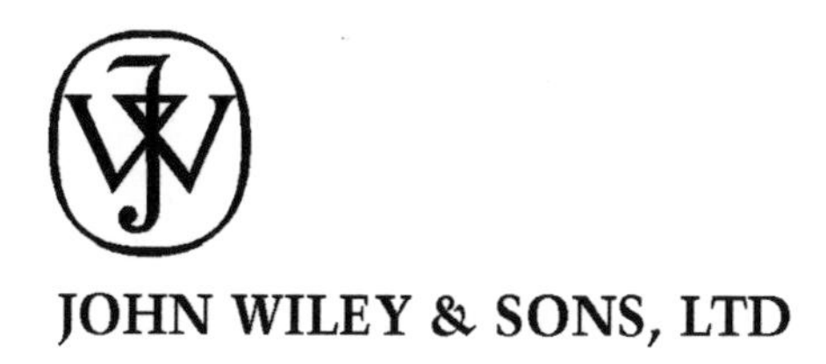

JOHN WILEY & SONS, LTD

Translation from the Polish language edition published by Wydawnictwa Komunikacji i Łączności sp z o.o., Warszawa

Baffins Lane, Chichester,
West Sussex, PO19 1UD, England

National 01243 779777
International (+44) 1243 779777

e-mail (for orders and customer service enquiries): cs-books@wiley.co.uk

Visit our Home Page on http://www.wiley.co.uk or http://www.wiley.com

Other Wiley Editorial Offices

John Wiley & Sons, Inc., 605 Third Avenue,
New York, NY 10158-0012, USA

WILEY-VCH Verlag GmbH
Pappelallee 3, D-69469 Weinheim, Germany

John Wiley & Sons Australia Ltd, 33 Park Road, Milton,
Queensland 4064, Australia

John Wiley & Sons (Canada) Ltd, 22 Worcester Road
Rexdale, Ontario, M9W 1L1, Canada

John Wiley & Sons (Asia) Pte Ltd, 2 Clementi Loop #02-01,
Jin Xing Distripark, Singapore 129809

British Library Cataloguing in Publication Data

A catalogue record for this book is available from the British Library

ISBN 0471 49837 8

Produced from PostScript files supplied by the author.
Printed and bound in Great Britain by Antony Rowe Ltd, Chippenham, Wiltshire.
This book is printed on acid-free paper responsibly manufactured from sustainable forestry, in which at least two trees are planted for each one used for paper production.

Contents

Preface

Mobile communication systems have become one of the hottest topics in communications. It has been forecast that within ten years half of the connections will be at least partially wireless. Mobile communications evolved from an upmarket toy for rich users towards a massive popular market. Rapid development of the Internet, its applications and new services created even higher challenges for further development of mobile communication systems. These systems in their theoretical and practical aspects have become a topic of academic lectures and have become an interesting subject for many engineers and technicians who work in the area of telecommunications. A long list of books available from leading publishers are devoted to several aspects of mobile communication systems. The book you are holding in your hand is trying to enter this list. It has evolved from the lecture notes prepared for telecommunication students at Poznań University of Technology, through the book published in Polish by WKiŁ, Warsaw in 1998 and 1999. This edition is a completely revised English translation, which takes into account new developments in mobile communications since 1999. Some of the details interesting for a Polish reader have been replaced by more general considerations. Due to the limited space, the book does not go into great detail when presenting examples of mobile communication systems. The readers interested in detailed information are asked to check the references included at the end of each chapter.

Chapter 1 is devoted to the basic theory of digital communication systems. Modern mobile communication systems are mostly digital and they share many aspects of transmission, multiple access and digital signal processing with other digital communication systems. The key material presented in this chapter is intended as reference and refreshment for those readers who completed their studies several years ago, or who have never studied digital communications before.

Chapter 1 presents a model of a digital communication system, the principles of speech coding, the basics of channel coding, an overview of digital modulations applied in mobile communication systems, the principles of spread spectrum systems, the multiple access and random access methods, the OSI reference model, basic information on X.25 and the Signalling System No. 7.

Chapter 2 characterizes the main types of mobile communication systems such as cellular telephony, cordless (wireless) telephony, trunking systems, wireless LANs and personal mobile satellite systems.

Chapter 3 presents the characteristics of a mobile communication channel. We start from the basic information on antennas and signal propagation in free space, then we consider the influence of the multipath effect on the received signal. Next we derive the channel model in the form of a transversal filter with time-varying tap coefficients and we present a model of the Doppler effect. We show the channel models which have been used in the GSM design. A substantial part of Chapter 3 is devoted to modeling the propagation loss and to the description of the most popular propagation models. After considering the influence of the channel properties on the transmitted narrowband and wideband signals, diversity reception is presented.

In Chapter 4 we consider the paging systems. After classification of paging networks we concentrate on the popular POCSAG protocol, the European ERMES paging system and the family of FLEX protocols.

The cellular system concept is the subject of Chapter 5. We describe a simplified design of a classical cellular system, we present the basic elements of the traffic theory applied in cellular system design and we consider the ways of increasing the system capacity such as cell sectorization, cell splitting and microcell zones. Next we present the rules of channel allocation in the cells and several strategies used in the practice.

Although first generation analog cellular systems slowly lose popularity and soon will have a historical meaning only, they deserve a general description which has been presented in Chapter 6. Two representative examples have been selected: the Scandinavian NMT system and American AMPS.

Chapter 7 is devoted to GSM which is the most important cellular system from the European point of view. We consider the GSM architecture and system aspects. We present basic radio transmission system parameters, and describe logical channels, GSM time hierarchy, burst structures and frame organization. We also describe the call set-up procedure, types of handover and the means of ensuring privacy and user authentication. Finally, we present modifications and derivatives of GSM.

Chapter 8 covers the GSM physical layer aspects. We consider the construction of a typical mobile station, and coding and decoding of a speech signal according to three possible algorithms. Next we describe the GMSK modulation applied in GSM and the principle of sequential detection applied in a GSM receiver.

In Chapter 9 we discuss the principles of data transmission in GSM. First, we consider typical data transmission based on the circuit-switched mode in which data rates of up to 14.4 kbit/s are achieved. We also sketch the rules of SMS transmission. Next we present the HSCSD system which works in the circuit-switched mode and substantially enhances data transmission capabilities. Finally, we describe the GPRS system based on the packet-switched mode. We present the GPRS architecture, its physical layer, transmission management, offered services and protocol architecture. At the end of the chapter we consider EDGE - *Enhanced Data Rate for Global Evolution*, mostly concentrating our attention on its physical layer.

Chapter 10 presents the basic knowledge necessary to understand the rules of operation of CDMA systems. We survey typical spreading sequences and basic transmitter and receiver schemes. We focus our attention on the RAKE receiver and we present basic rules of joint detection receivers. Finally, we consider basic properties of a CDMA system.

In Chapter 11 we describe the most popular CDMA system, i.e. IS-95. We concentrate mostly on the physical layer of forward and reverse (downlink and uplink) transmission and give some information on IS-95 enhancements.

Chapter 12 presents trunking systems. We start with the idea of trunking. Next we consider the MPT 1327 standard and we concentrate on TETRA, showing its general architecture, the offered services and the physical layer.

Chapter 13 describes basic types of digital cordless telephony. We concentrate on DECT, although the American PACS and the Japanese PHS systems are also briefly considered.

In Chapter 14 we present general rules of operation of wireless local loops, which have recently attracted a lot of interest due to the development of mobile communication technology.

Chapter 15 is devoted to personal mobile satellite systems. General classification of these systems is presented and a short description of INMARSAT systems follows. Most attention is given to Iridium, GLOBALSTAR, ICO and future broadband satellite communication systems such as Teledesic and Skybridge.

Chapter 16 presents basic properties and the rules of operation of wireless local loops (WLANs). We describe the most important WLAN standards such as HIPERLAN/1 and 2, and a few versions of IEEE 802.11. We also briefly consider Bluetooth, which has the opportunity to become a popular standard of wireless connections in everyday life.

Finally, in Chapter 17 we consider the third generation mobile communication systems. We present basic properties of the *Universal Mobile Telecommuncations System* (UMTS) and concentrate our attention on two types of air interface: WCDMA FDD and WCDMA TDD. We show the system architecture, the physical layer and some basic system procedures. Next we present a short description of the *cdma2000* system, which has evolved from IS-95. We end the chapter with the presentation of the idea of *Software Radio*, which can become quite useful in the world of many mutually incompatible air interface standards.

The final chapter of the book is devoted to a general description of smart antenna technology and its application to cellular telephony. This is a related subject because smart antennas are applied in the third generation mobile communication systems.

The above chapter description shows that this book attempts to tackle most of the important subjects in mobile communications. It is not possible to do it in great depth within the limited space of this book. Almost each chapter of this book could be, or already is, the subject of a separate book. We hope that the reader will understand this fact and, if needed, he/she will study other, more specialized books.

This book would not be in its present form if it had not been given attention and time by many people. First of all, I would like to thank Prof. Zdzisław Kachlicki and Dr. Tomasz Kosiło, the reviewers of the Polish edition, for their valuable comments. Similar thanks are directed to the anonymous reviewers of the English proposal. I am also grateful to Mark Hammond, the Senior Publishing Editor of John Wiley & Sons, Ltd., Sarah Hinton, the Assistant Editor and Zoë Pinnock, the Production Editor who were particularly patient and helpful. Someone who substantially influenced the final form of the book is Mrs Krystyna Ciesielska (M.A., M.Sc.) who was the language consultant and as an electrical engineer was a very critical reader of the English translation. Finally, the book would not have appeared if I did not have the warm support of my family, in particular of my wife Maria and my father Aleksander.

KRZYSZTOF WESOŁOWSKI

1

Elements of digital communication systems theory

1.1 INTRODUCTION

The following chapter contains the introduction to digital communication systems to the extent which is necessary to understand problems encountered in the mobile communication systems. The author assumes that only some readers have studied digital communications in recent years. The progress in this area is very fast. Many technical solutions now applied in modern mobile communication systems were no more than the subject of theoretical considerations several years ago and they seemed beyond technical implementation using then existing technical means. Today they can be found in popular mobile phones.

The aim of this chapter is to present an overview of basic issues of digital transmission over communication channels, methods of digital modulation and digital receiver structures applied in mobile communication systems. We will also briefly describe the basic rules of error detection and correction, as well as other methods aimed at improving the performance of digital transmission over dispersive channels. We will also introduce the Open System Interconnection (OSI) reference description model. We will use it for an ordered description of digital communication systems and networks. This model is applicable in defining the interfaces between the systems on the level of their several layers. We will also consider selected protocols for information exchange in higher layers of the OSI description model.

The author leaves the choice to the reader if he/she wishes to become acquainted with the theoretical material presented in this chapter or to refresh his/her knowledge or simply to skip this chapter and study the next chapters.

1.2 MODEL OF DIGITAL COMMUNICATION SYSTEM

Let us consider the basic model of a digital communication system presented in Figure 1.1.

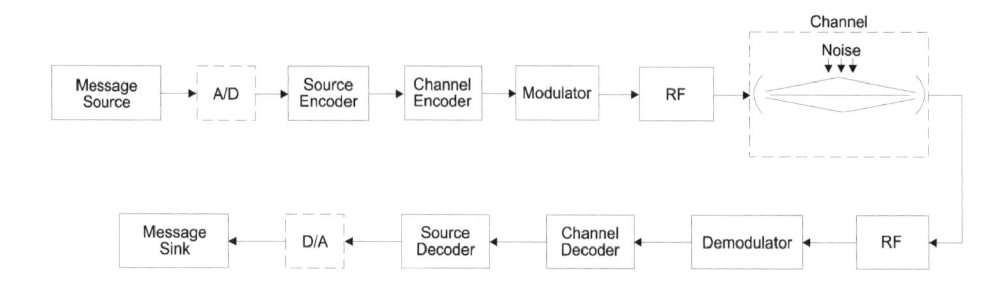

Figure 1.1 Model of a digital communication system

The *source* generates messages which have either the form of continuous functions of time or are streams of discrete symbols. An example of a message continuous in time is a waveform characterizing human speech. In this case, in order to send such messages over the digital communication system, the continuous speech signal has to be discretized in time and its samples have to be quantized. To fulfill this requirement, an analog-to-digital (AD) converter samples the analog signal at the *sampling frequency* f_s and a binary block determined by the sample amplitude is assigned to each sample. Such an operation is performed, for example, by the PCM encoder in standard telephony. The PCM encoder samples the analog signal from the microphone at the frequency $f_s = 8$ kHz and assigns 8-bit sequences to each sample, using a standardized nonlinear characteristic. Due to *quantization* performed with the selected accuracy, a certain part of information contained in the sampled signal is lost. As a result a *quantization noise* is observed. In case of speech signals, a nonlinear quantizer assigning binary sequences to the signal samples can be considered as a *source encoder*. Due to nonlinear characteristics applied in this process and taking into account the dynamic properties of the human ear, a single sample can be represented by an 8-bit word. If the AD conversion was linear, a comparable quality would be achieved with 12–13 bits.

Another example of a speech source encoder is the DPCM encoder. This encoder is based on strong correlation between subsequent speech samples. It encodes only the differences between them. More advanced methods of speech encoding will be briefly described in Section 1.3 and in the chapters describing the GSM system and wireless telephony.

An example of a source of discrete messages is a computer terminal. One can consider the alphanumeric characters generated by the terminal as the source messages. They are usually represented by 8-bit blocks in accordance with the widely used ASCII alphabet. Although very popular, the ASCII alphabet is not an efficient representation of the alphanumeric characters. Some characters occur very often, the others have low rate of occurrence. A good source encoder fits the length of the binary sequences to the statistical properties of the message source. More and more frequently, data compression is applied, which allows for efficient representation of messages generated by the source. This efficiency manifests itself by a low mean number of bits used for encoding of a single message.

Several physical phenomena occurring in communication channels result in errors made by the receivers. The errors are visible as the difference between the transmitted binary sequences and the binary sequences decided upon on the basis of the received signals. In order to correct the errors or at least to detect them, a *channel encoder* at the transmitter and a *channel decoder* at the receiver are applied. The information blocks are supplemented by a certain number of specially selected additional bits. These bits are the results of modulo-2 addition of information bits selected in such a way that algebraic interdependencies among them are created which allow for potential correction or at least detection of errors. If error detection is applied, a signal of the erroneous reception of the binary sequence is an indication that transmission of this sequence should be repeated. Both error correction and detection are frequently applied in mobile communication systems.

A *modulator* is a block which generates a sinusoidal signal (a *carrier*), whose parameters such as frequency, amplitude and/or phase are the functions of the digital sequence applied to its input. As a result of *modulation* the signal carrying the information is placed in the appropriate part of the spectrum and has properly shaped spectral properties. This is a very important feature in mobile communication systems. They should use the spectral resources assigned to them efficiently, so as not to distort the signals transmitted by neighboring spectrum users. They should also guarantee the largest possible number of their own users in the assigned frequency band. The electromagnetic spectrum is a valuable and limited resource. Recently, many systems have been put in operation, and new services have been offered on the market. As a result, new equipment and systems working in higher and higher frequency ranges are designed and more and more sophisticated technology is required. The distribution of the electromagnetic spectrum among radio systems is the subject of international negotiations and agreements.

Multiple access to the transmission medium is a subject closely related to the channel properties and the applied modulations. Multiple access can be realized by different methods. The first one is the division of the spectrum assigned to the system into a certain number of subbands which are used by different users (mostly, only if they are active). That kind of access to the channel is called *Frequency Division Multiple Access* (FDMA). In another approach, the users share the same spectrum but they divide time among themselves. This approach is called *Time Division Multiple Access* (TDMA). Users can also generate signals that occupy the whole system bandwidth and time; however, due to the application of specific signal sequences (code sequences) uniquely characterizing each user, the receiver is able to extract the signal of the selected user from the sum of signals emitted by different users. This access method is called *Code Division Multiple Access* (CDMA). Combinations of the three above mentioned multiple access methods are possible.

The *RF block* operates in the range of radio frequencies and amplifies the radio signal to the required level. The bandwidth of the signal depends on the selected modulation and the multiple access method applied. A frequently encountered limitation for the RF amplifier applied in a mobile communication system is its energy consumption. For example, a mobile phone should utilize as little energy as possible in order to lengthen the time between subsequent battery recharging. For this reason, the RF amplifier should have large dynamics and should work in the nonlinear range of its characteristics. This fact has serious implications for the choice of digital modulations applied in the mobile communication systems. So far, mostly digital modulations, characterized by a constant or low-dynamic envelope, have been applied in order to minimize the nonlinear distortions caused by the characteristics of the RF amplifier.

In mobile communication systems the transmitter emits the signal into space using the *antenna*. The channel properties strongly depend on the type of transmitter and receiver antennas, in particular on their directivity and gains. The antenna parameters determine the range of the system and its performance. Although important from the practical point of view, the antenna theory remains beyond the scope of this book. The reader can find books entirely devoted to antennas such as [1] and [2].

The processes performed in the receiver are matched by those taking place in the transmitter. After amplification and filtration in the RF front-end the received signal is demodulated. The last process strongly depends on the digital modulation applied and on the channel properties. The acceptable cost of the receiver also has an impact on the type of the demodulator applied. In general, the demodulator extracts the pulse sequence from the modulated signal received from the RF part. On the basis of these pulses, the *detector* makes the decisions upon transmitted data symbols and transforms them into binary sequences.

The *channel decoder*, using the redundant bits introduced by the channel encoder and sometimes applying additional information on the reliability of the received symbols, attempts to find the code sequence and, based on it, a binary information sequence. The latter is the subject of source decoding. Two examples of channel decoders are a decompression block recovering the original data from the received compressed data and a speech synthesizer generating speech samples. In the latter case, the source–decoded samples are D/A-converted and fed through the amplifier and a loudspeaker to the *message sink* - the user's ear.

In following sections we will describe the processes taking place during the transmission of digital signals over mobile channels.

1.3 SPEECH CODERS AND DECODERS

Transmission of speech is the most frequent communication service. In analog systems the signal representing human speech modulates one of the parameters of a sinusoidal signal such as amplitude or frequency. The second generation mobile communication systems transmit speech signals in digital form. For this reason the effective digital representation of a speech signal is of particular importance.

Pulse Code Modulation[1] (PCM) is the oldest method of digital coding of speech. Figure 1.2 presents the basic scheme of PCM encoder and decoder.

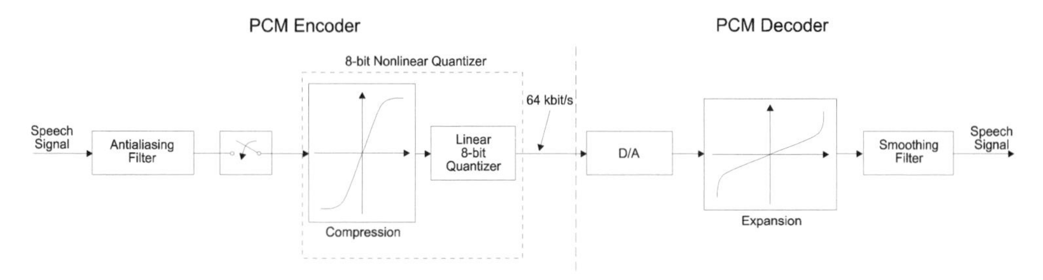

Figure 1.2 PCM encoder and decoder

The PCM encoder consists of an *anti-aliasing filter*, a sampler and a nonuniform quantizer. The anti-aliasing filter limits the bandwidth of the input signal to such frequency that is at most half the sampling frequency f_s. In practice, the anti-aliasing filter cuts off the bandwidth above 4 kHz, and the sampler picks up the samples of the input signal at the frequency of 8 kHz. The nonuniform quantizer can be theoretically decomposed into a nonlinear memoryless signal compression circuit and a linear quantizer. The characteristic of the nonlinear circuit is standardized by the ITU-T recommendations and is described by the formula

$$f(x) = \begin{cases} \dfrac{Ax}{1+\ln A} & \text{for} \quad 0 \leq x \leq \frac{1}{A} \\ \dfrac{1+\ln Ax}{1+\ln A} & \text{for} \quad \frac{1}{A} \leq x \leq 1 \end{cases} \tag{1.1}$$

where for the 8-bit quantizer the constant $A = 87.6$. For this quantizer sampling at the frequency equal to 8 kHz, the data rate of the resulting binary stream is 64 kbit/s. Let us note that for small signal amplitudes the characteristic is linear and for the amplitude higher than the threshold $1/A$ it is logarithmic. In the PCM decoder the applied characteristic is described by the function that is inverse to (1.1), so the cascade connection of both nonlinearities is linear. As a result, the shape of the processed signal is not changed. Thanks to the compression characteristic applied in the transmitter, a small amplitude signal is more strongly amplified than a large amplitude signal. As a result, the signal power to quantization noise power ratio is almost constant in the large range of the quantized signal amplitudes. In case of linear quantization that ratio would change linearly with the power of the quantized signal. The PCM speech coding is used in some early solutions of wireless links replacing a subscriber loop. It is otherwise seldom used because of low spectral efficiency of the speech signal representation in the form of a 64-kbit/s binary stream as compared with the speech signal represented by the

[1]This is a traditional term. PCM is not a modulation but a source coding procedure.

analog FM modulation. Recently PCM speech coding with logarithmic characteristic (1.1) or (1.2) has been applied as one of the speech coding methods in a new *de facto* standard called Bluetooth [20].

Another standard of the compression characteristic, similar to that described by formula (1.1), has been adopted in the USA and Canada. The normalized input–output magnitude characteristic is described by the formula

$$f(x) = \mathrm{sgn}(x)\frac{\ln(1+\mu|x|)}{\ln(1+\mu)} \tag{1.2}$$

where the value of $\mu = 255$.

The speech signal is characterized by strong autocorrelation. This means that two subsequent samples do not differ too much from each other. In consequence, if the differences between subsequent samples are encoded, the data rate of the binary stream representing the speech waveform can be decreased. Moreover, knowing a sequence of the recent samples and their correlation properties, one can predict the next sample. We take advantage of the observation that in a limited time span the speech signal is quasi-stationary. Thus, instead of coding the subsequent samples or even coding the differences among them, one can code the difference between the current sample and its predicted value calculated by the *predictor* on the basis of a few previous samples. Figure 1.3 presents the encoder and decoder of *Differential Pulse Code Modulation* (DPCM) exploiting this idea. The anti-aliasing filter in the transmitter and the smoothing filter in the receiver have been omitted for the clarity of presentation of the DPCM idea.

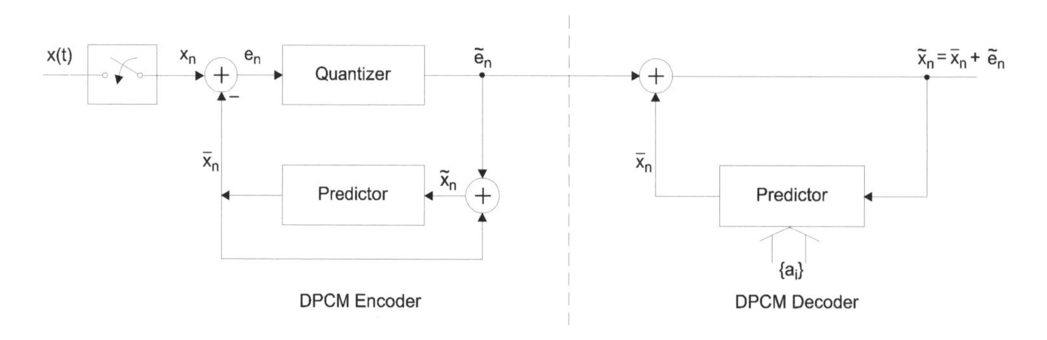

Figure 1.3 DPCM encoder and decoder

As we see in Figure 1.3, the subject of the quantization in the DPCM encoder is the difference between the input signal x_n and the predictor output $\overline{x}_n$, the latter being derived as the weighted sum of the recent p samples of which each is the sum of the predictor and quantizer outputs. This means that $e_n = x_n - \overline{x}_n$, where

$$\overline{x}_n = \sum_{i=1}^{p} a_i \widetilde{x}_{n-i} \quad \text{and} \quad \widetilde{x}_{n-i} = \overline{x}_{n-i} + e_{n-i} \tag{1.3}$$

Equation (1.3) describes the operation of the predictor and shows that this functional block can be realized in the form of a finite impulse response (FIR) digital filter. The quantized form $\widetilde{e}_n$ of the prediction error e_n is sent to the receiver.

The DPCM decoder derives the output signal as the sum of the predicted sample and the quantized prediction error which has been received from the DPCM encoder. Usually the predictor coefficients are selected in such a way that the mean square error between the signal sample and its predicted value is minimized. The prediction coefficients are the solution of the system of linear equations derived from minimization of the following expression

$$\begin{aligned} E[e_n^2] &= E\left[\left(x_n - \sum_{i=1}^{p} a_i x_{n-i}\right)^2\right] = \\ &= E[x_n^2] - 2\sum_{i=1}^{p} a_i E[x_n x_{n-i}] + \sum_{i=1}^{p}\sum_{j=1}^{p} a_i a_j E[x_{n-i} x_{n-j}] \end{aligned} \tag{1.4}$$

where $E[.]$ denotes the ensemble average. Assuming stationarity of the signal x_n and denoting the values of the autocorrelation function of the signal x_n as $r_{i-j} = E[x_{n-i}x_{n-j}]$, after calculating the first derivative of (1.4) and setting it to zero we end up with the following system of equations

$$\sum_{i=1}^{p} r_{i-j} a_i = r_j \quad \text{for } j = 1, ..., p \tag{1.5}$$

Let us stress that the assumption of signal stationarity is valid only in short time periods. The solution of the system (1.5) is the set of the predictor coefficients a_i, $(i = 1, ..., p)$ so the calculated coefficients get their averaged values. Periodical updating of the coefficients would be a certain improvement in the predictor's operation; however, the coefficients should be sent to the receiver or calculated adaptively on the basis of the received signals.

Adaptive Differential Pulse Code Modulation (ADPCM) is a meaningful improvement of the PCM and DPCM encoding. In 1984 this method was specified in the ITU-T Recommendation G.721, which was later replaced by Recommendation G.726. It is worth noting that this method of waveform encoding is applied in a few popular mobile communication systems, in particular in wireless telephony and wireless subscriber loops. Subjective speech quality when ADPCM coding is involved is comparable to that which is achieved when a typical PCM processing is performed.

ADPCM encoding combines two PCM improvements - the differential encoding described earlier and the adaptive quantization. The basic scheme of ADPCM encoder and decoder is shown in Figure 1.4. The analog signal is sampled and linearly processed in a 12-bit quantizer, resulting in the signal representation x_n. Subsequently, the difference $e_n = x_n - \overline{x}_n$ between the sample x_n and its prediction $\overline{x}_n$ is calculated. The resulting error signal e_n represented by a 12-bit word is the subject of processing in the quantizer having the base-2 logarithmic characteristic with 16 quantization thresholds. As a result, a 4-bit representation of the error sample is received. At the sampling

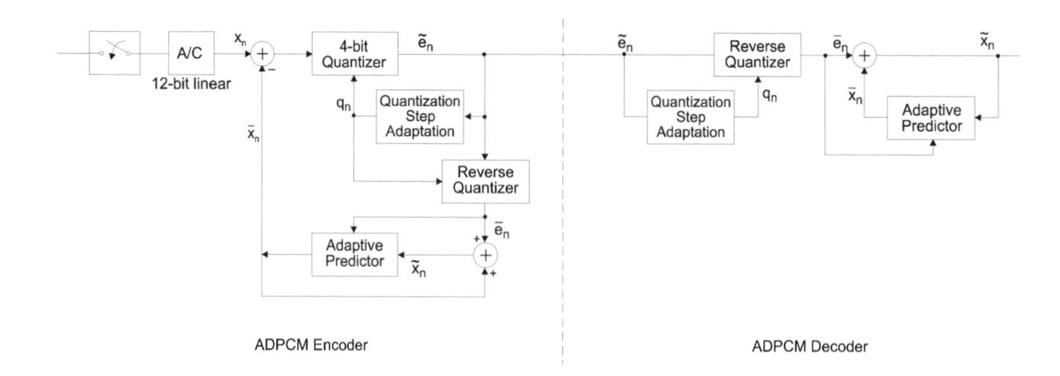

Figure 1.4 Scheme of ADPCM encoder and decoder

frequency of 8 kHz the resulting ADPCM data stream has a data rate of 32 kbit/s. The 4-bit error signal $\widetilde{e}_n$ determines the quantization threshold applied both in the adaptive quantizer and in the circuit recovering a linear quantized form of the error $\overline{e}_n$. The latter error is added to the signal $\overline{x}_n$ that is received at the output of the adaptive predictor. This sum constitutes the predictor's input signal for the next timing instant. The adjustment of the predictor's coefficients is performed on the basis of the output signal received from the adaptive inverse quantizer (see Figure 1.4).

Some blocks of the ADPCM decoder are identical to those applied in the encoder. The adaptation of the inverse quantizer is performed on the basis of the quantized error $\widetilde{e}_n$, as it is done in the encoder. The resulting linearly quantized error $\overline{e}_n$ is added to the adaptive predictor output signal, giving the approximation $\widetilde{x}_n$ of the ADPCM encoder's input signal. The adaptation processes are determined by the same signals in the transmitter and receiver so the results of adaptation performed in the transmitter do not need to be sent to the receiver. The adaptation algorithms are constructed in such a way that in case of transmission errors occurring in the encoded binary signal the algorithms return to their correct operation so their stability is ensured.

Detailed description of the ADPCM encoder and decoder can be found in the Recommendations G.721 and 726 [3] and the application note [4] presenting the ADPCM encoder/decoder implementation on a fixed point signal processor. Their more detailed analysis is beyond the scope of this chapter.

The next method of waveform encoding, which is worth mentioning, is *Delta Modulation* (DM). The DM can be considered as a special case of DPCM. In the DM the difference between the current and previous samples is calculated and quantized using a two-level quantizer. The price paid for a much simplified encoding scheme is the need to apply a much higher sampling frequency as compared with the minimum sampling frequency used in the PCM encoder. In the basic delta modulation encoder the sampling frequency is a compromise between the resulting encoder's output data rate and the tolerable level of the quantization errors. *Slope overloading* and *granulation noise* are two phenomena resulting in the particularly large values of the signal quantization errors. In the slope overloading the errors are a result of slow tracking of the steep input signal slope performed by the DM encoder which generates the linearly increasing quantized output signal. Granulation noise is a result of the quantization of the quasi-constant signal. A constant signal encoded using the DM method results in the alternating positive and negative binary pulses. The adaptation of the quantization step size performed in *Adaptive Delta Modulation* removes the above mentioned drawbacks at the price of the increased encoder and decoder complexity. Typically, the DM encoder generates a binary stream at the data rate of 16 kbit/s. Such a method of speech signal representation used to be applied in military equipment. Unfortunately it does not possess sufficient quality for commercial mobile communication systems.

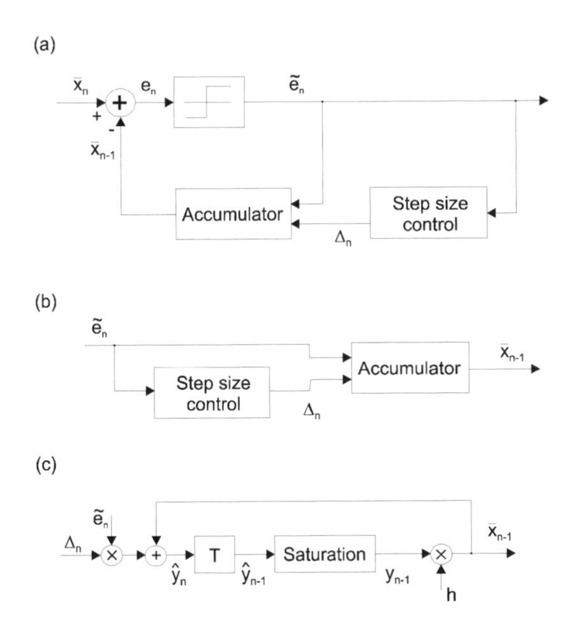

Figure 1.5 Scheme of the continuous variable slope delta encoder (a), decoder (b) and accumulator (c) used in the encoder and decoder

Much higher quality, even better than for regular PCM encoding, can be achieved by applying a version of adaptive delta modulation called *Continuous Variable Slope Delta Modulation* (CVSD). It has been selected as an alternative speech coding method in the Bluetooth standard [20]. The CVSD has been designed to reduce slope overload effects. Figure 1.5 shows the schemes of the CVSD encoder, decoder and of the accumulator used in both blocks. As in regular delta modulation, the object of two-level quantization is the difference between the input signal sample x_n and the sample $\overline{x}_{n-1}$ at the output of the accumulator, which approximates the previous sample x_{n-1}. Thanks to the factor h, which is slightly lower than one, the accumulator has a property of slow forgetting its contents (see Figure 1.5c). The CVSD modulation can be described by the following set of equations

$$\begin{aligned} \widetilde{e}_n &= \text{sgn}(e_n), \qquad e_n = x_n - \overline{x}_{n-1} \\ \overline{x}_{n-1} &= h \cdot y_{n-1} \quad y_{n-1} = \text{saturation}(\widehat{y}_{n-1}) \\ \widehat{y}_n &= \overline{x}_{n-1} + \Delta_n \cdot \widetilde{e}_n \end{aligned} \tag{1.6}$$

where saturation(.) is self-explanatory. The value of the step size Δ_n depends on the sequence of the most recent values of $\widetilde{e}_n$. Figure 1.6 shows a typical approximation of the continuous signal by a CVSD quantized signal and shows a binary representation of $\widetilde{e}_n$ [21]. The receiver is able to adjust its step size in the same way as the transmitter on the basis of the received values of $\widetilde{e}_n$ only.

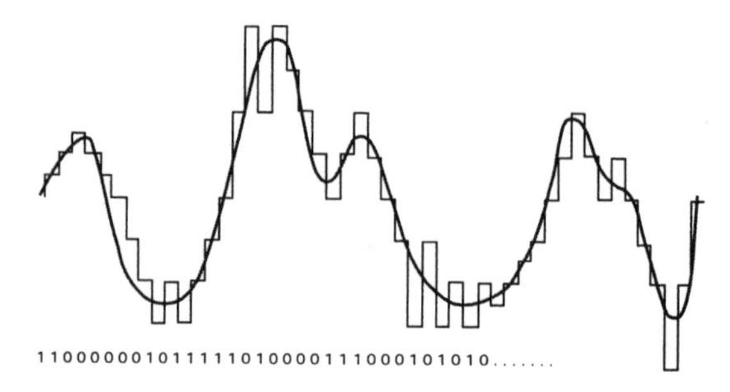

Figure 1.6 A typical approximation of the continuous signal by a CVSD quantized signal [18] (©Ericsson Review)

The speech encoding methods presented so far relied on the encoding of speech samples or their differences. More and more sophisticated methods leading to higher and higher binary stream reduction have been proposed. However, there are even more effective coding methods which are used in the mobile cellular systems. These methods rely on the *Linear Predictive Coding* (LPC) and *Vector Quantization* principles.

Let us start with the LPC method in its general form [5]. The speech signal encoding is a process of speech analysis in which the encoder parameters and the type of excitation signal are determined. These parameters are subsequently transmitted to the receiver where they are used by the decoder working as a speech synthesizer.

The vocal tract can be modeled using a filter with periodically updated coefficients. This filter is excited every few tens of milliseconds by a periodical signal or a noise-like signal. The speech encoder synthesizes the filter modeling the vocal tract, determines the type of excitation (noise or the pulse sequence) and its period. The criterion of the filter synthesis is usually the minimum of the mean square error, where the latter is understood as the weighted sum of squares of the differences between the input speech samples and the samples synthesized by the encoder with the given coefficients.

Figure 1.7 presents the general scheme of the LPC encoder [5]. The filter modeling the vocal tract is digital. It works on the input samples according to the equation

$$x_n = \sum_{i=1}^{p} a_i x_{n-k} + G v_n \tag{1.7}$$

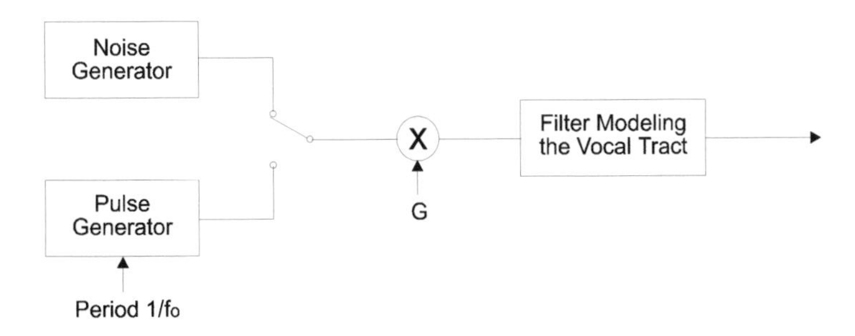

Figure 1.7 General scheme of the LPC encoder

where G is the gain of the excitation signal v_n and a_k are the filter coefficients. In practice the modeling filter is usually a *lattice filter* in which the filter coefficients a_k are replaced by the coefficients a_{kk} called the *reflection coefficients*. Thus, the speech synthesizer looks like that shown in Figure 1.8.

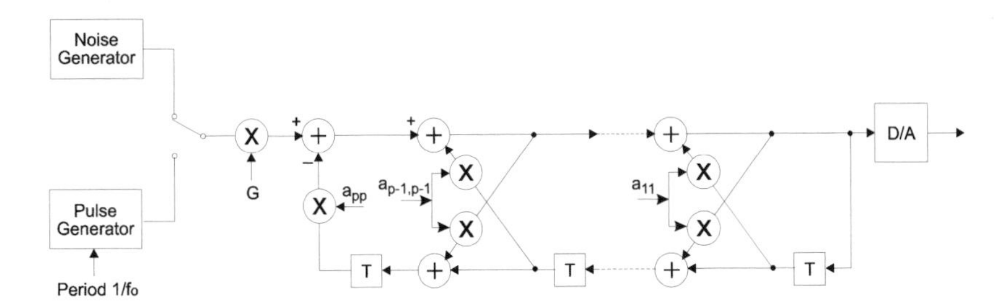

Figure 1.8 LPC encoder with the lattice filter

The LPC model parameters have to be renewed every 15–30 milliseconds due to the quasi-stationarity of the speech waveform, so this time period determines the speech encoder frame length. The frame contains the binary block from the LPC output. For example, [5], this block consists of a single bit determining the type of excitation signal, 6 bits determining the period of the excitation signal, and 5 bits describing the gain G in the logarithmic scale. The reflection coefficients a_{kk} of the modeling filter require a 6-bit representation for each coefficient. Up to 10 coefficients are applied. In consequence, a typical frame consists of 72 bits and the binary stream has the data rate of 2400 to 4800 bits/s depending on the frame period. The speech signal recovered by the receiver operating in accordance with the LPC model has a relatively low quality and sounds "synthetic". For this reason more advanced speech models are applied in commercial systems, resulting in much higher speech quality. Figure 1.9 presents a general scheme of an encoder using such a model [6], called the *analysis-by-synthesis* speech encoder.

The analysis-by-synthesis encoder consists of a speech synthesizer attempting to generate the signal similar to the speech signal being the subject of encoding. The weighted difference between both signals is a cost function used to adjust the parameters

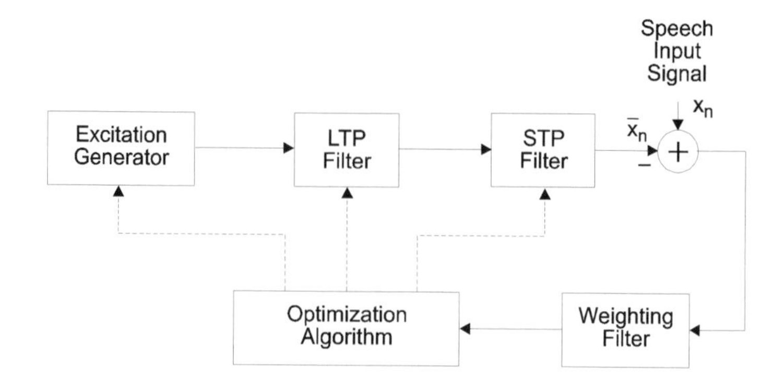

Figure 1.9 Scheme of the speech encoder applying "analysis-by-synthesis" rule

of the speech synthesizer. The synthesizer consists of the excitation generator, the *Long-Term Prediction* filter (LTP) and the *Short-Term Prediction* (STP) filter. The STP filter models a short-term correlation of the speech signal, or equivalently, its spectral envelope. The LTP filter reflects a long-term correlation or the precise speech signal spectral structure. As previously, the subjects of transmission are the STP and LTP filter parameters and the parameters of the excitation signal. The structure of the speech encoder applied in the GSM system is similar to that described above and its details are presented in the chapter dealing with the GSM physical layer.

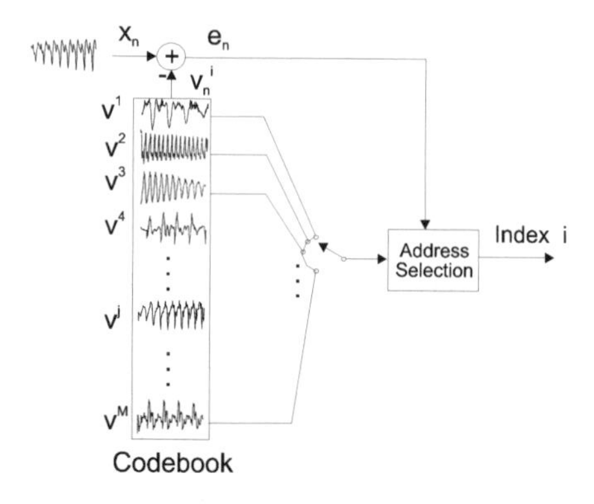

Figure 1.10 Vector quantization principle

In order to describe another important encoding method called *Code Excited Linear Prediction* (CELP), we have to present the idea of *vector quantization.* In this type of the quantization process, the object of processing is a block of N subsequent signal samples. The quantizer uses the table of appropriately selected signal sample representations ordered in the form of a *codebook.* The quantizer associates the block of N input signal samples with one of the *code words.* The criterion of the code word assignment is often the minimization of the mean square error between the input signal samples and the code word samples. The result of the vector quantization is the address of the selected code word that is transmitted to the receiver. The encoding process is shown in Figure 1.10. The decoder contains the same codebook so the received address of the code word allows for generation of the selected code word which best approximates the encoded sample sequence.

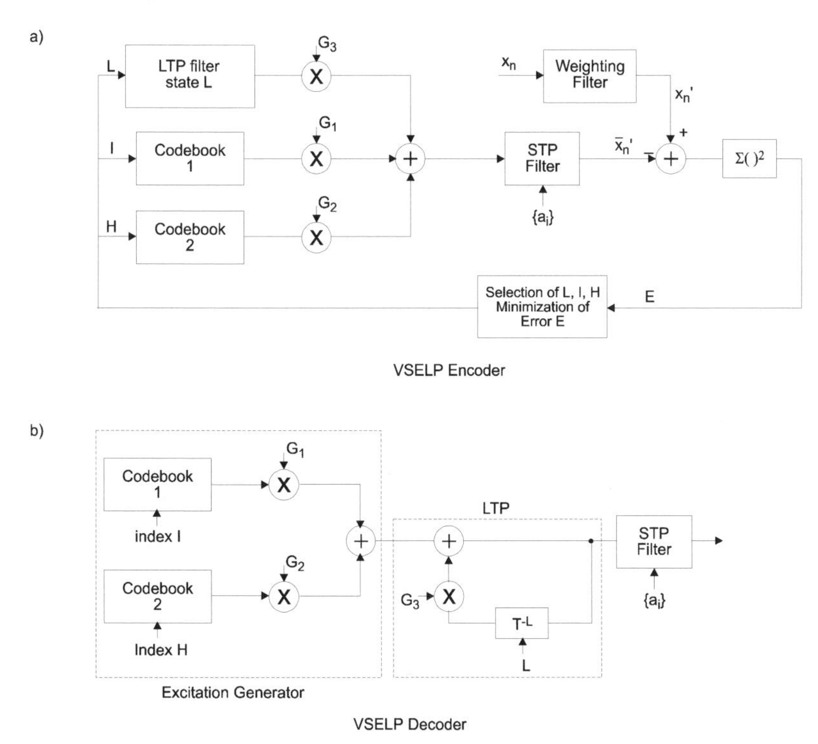

Figure 1.11 Scheme of the VSELP encoder (a) and decoder (b)

Figure 1.11 shows the CELP encoder and decoder in the version called *Vector Sum Excited Linear Prediction* (VSELP) applied in the second generation cellular system IS-54B/136 operating in the USA. As we see, the encoder is a modification of the "analysis-by-synthesis" circuit shown in Figure 1.9. The excitation generator is implemented on the basis of the selection of N-element words received from two codebooks, and their appropriate weighting. The LTP filter is realized as a two-tap filter in which the current sample is summed with the sample delayed by L sampling instants and weighted by the coefficient G_3. The weighting coefficients G_1, G_2 and G_3, the indices L, H, I and the set $\{a_i\}$ of the STP filter coefficients are selected so as to minimize the sum of the squared differences between the synthesized samples generated by the encoder and the weighted sequence of the input samples x_n. We have to stress once more that the whole encoding and decoding procedure is performed on the sample blocks collected in the time period of about 20 ms. The optimized encoder parameters are sent to the decoder and, since the codebooks in the encoder and decoder are identical, the transmitted parameters allow for the synthesis of the speech signal frame in the decoder. The binary stream generated by the described encoder has the data rate of about 8 kbit/s. Due to the type of calculations performed by the encoder and decoder, both blocks are usually implemented in software on a digital signal processor. Computational complexity of the CELP encoder and decoder does not exceed the capabilities of a typical DSP processor.

1.4 CHANNEL CODING

The application of the *Forward Error Correction* (FEC) or *Error Detection* coding is one of the most important means of ensuring the reliability of digital transmission. In this section we will present the basic rules of channel coding. First, we will consider simple channel models representing the whole system contained between the channel encoder and decoder.

1.4.1 Channel models for channel coding

Figure 1.12 presents a few basic versions of the channel models useful for analysis of the channel coding process. The simplest model version is known as a *binary symmetric memoryless channel* model (Figure 1.12a). The channel inputs and outputs are binary. The transmitted and received blocks are observed at the input and output of the channel model on the bit-after-bit basis. Every bit of the encoded sequence appears unchanged at the channel output with the probability $1-p$. With the probability p the transmitted bits are negated, which is equivalent to the bit errors. The decoder makes a decision

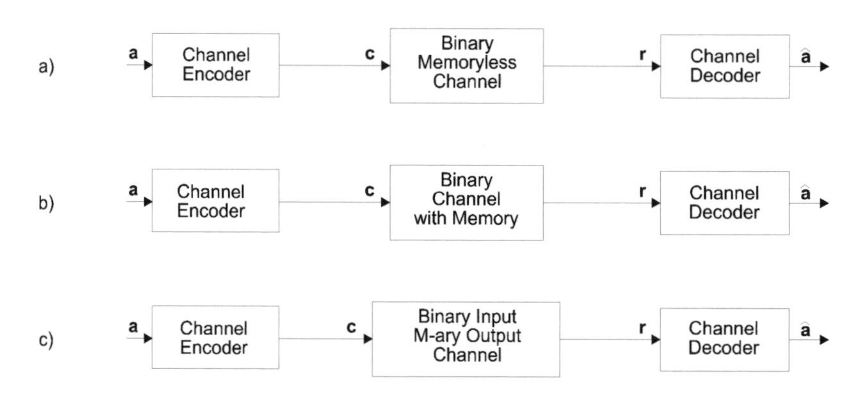

Figure 1.12 Channel models from the point of view of channel coding

about the transmitted coded sequence $\mathbf{c}$ on the basis of the received binary sequence $\mathbf{r}$. In the decision process it can apply only the algebraic interdependencies among particular bits of the transmitted sequence which have been implied by the coding rule. Due to the memoryless nature of the considered model, the occurring errors are mutually statistically independent, i.e. the occurrence of errors at previous moments does not have any influence on the error probability at the current moment. In reality only some transmission channels can be considered as memoryless. In most channels the errors occur in bursts. On the other hand there are many decoding algorithms which are designed for the correction of random errors, i.e. for memoryless channels. In order to ensure the error correction with the sufficient quality, additional means are undertaken in order to spread channel error bursts in the receiver. A widely applied method of destroying the error bursts is *interleaving*. It will be explained further in this chapter.

The second channel model reflects the bursty nature of errors occurring in the transmission channel. In this case the occurrence of a single error at one moment increases the probability of errors at the following moment. In this sense the channel has a memory of its previous states. There are special codes and decoding algorithms fitted to such a situation.

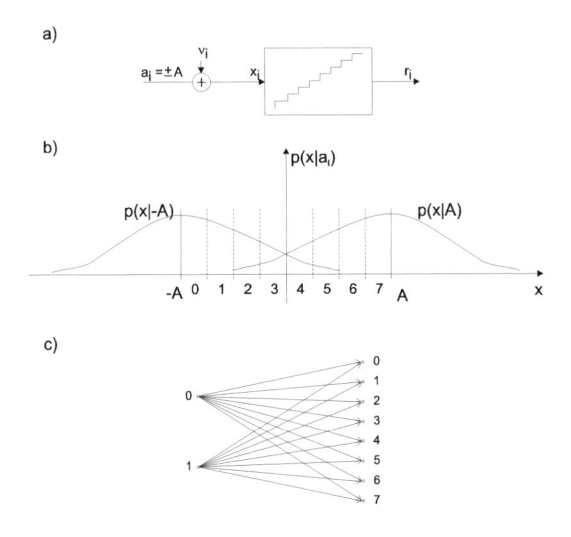

Figure 1.13 Example of use of the additional knowledge from the channel output for soft-decision decoding

The third model, similar to the first one, is also memoryless; however, it illustrates the case when more than binary information only is retrieved from the channel output. It means that the decoder uses not only algebraic interdependencies among particular bits in the coded sequence but also additional knowledge received from the channel, allowing for the improvement of the decoding process. Figure 1.13 illustrates a simple example of this case. Binary symbols are represented by bipolar pulses having values equal to $\pm A$. They are distorted by additive statistically independent Gaussian noise samples. Let the binary pulse $-A$ represent the binary symbol "0", whereas the pulse $+A$ represent binary "1". The sample x, being the sum of the pulse and the noise sample, has the probability density function conditioned on the transmitted symbol $+A$ or $-A$. In the receiver the sample x is quantized by an M-level quantizer, giving the output symbol r. Assigning a digit in the range from 0 to $M-1$ to each possible quantization level, we obtain a channel model with binary input and M-ary output. In the case of a binary quantizer this channel model is reduced to the binary symmetric memoryless channel model. As we see, in our channel model the channel output is measured much more precisely as compared with the binary channel model. This allows us to use that additional knowledge on the received symbols to improve the decoding quality, i.e. to decrease the probability of a false decision upon the received coded sequence. In Figure 1.13b the dashed lines indicate the subsequent quantization levels. The type of decoding in which additional channel knowledge is used is called *soft-decision decoding*, as opposed to *hard-decision decoding* when only binary symbols are used by the decoder. Most of the decoding algorithms applied in modern digital cellular telephony use soft decisions. The kind of knowledge used in the soft-decision decoding resulting from the M-level quantization is not the only one used to improve the decoding quality. There are other methods of measuring the bit reliability applied by soft-decision decoding. The power level of the signal carrying the information bit is one of them.

1.4.2 The essence of the redundant coding

As already mentioned, the channel coding relies on appending the information sequence by additional bits which constitute information redundancy. Let the subject of coding be a k-bit information sequence $\mathbf{a}$. Assume that the information source can generate any combination of bits in the k-bit block. Thus, 2^k different information sequences are possible. As a result of supplementing k-bit information blocks by $n-k$ additional bits we receive n-bit sequences. There are 2^n different binary sequences of length n, however, only 2^k sequences are selected from them. Each of them represents one of the possible information sequences $\mathbf{a}$. Let us call them *code words*. The n-bit sequences are selected in such a way that the sequences should differ from one another as much as possible. Thus, despite the erroneous reception of some bits, the decoder can assign with a high probability that coded sequence to the received sequence, which has been sent by the transmitter. Difference among the coded sequences can be measured by the number of positions on which the bits of any pair of two coded sequences are different. This number is called the *Hamming distance* between two sequences. One can show that if binary errors occur statistically independently of each other (which means that we represent the channel by the binary, symmetric memoryless channel model), then 2^k code words of length n should be selected in such a way that the minimum Hamming distance occurring between some pairs of them should be maximized. The optimum *maximum likelihood* decoder finds that sequence among 2^k of code words which is the closest to the received n-bit sequence in the sense of the Hamming distance. If the minimum Hamming distance $d_{\min}$ between the coding sequences is maximized, the coding sequence can be erroneous in no more than $t = \lfloor (d_{\min} - 1)/2 \rfloor$ positions[2] and the maximum likelihood decoder is still able to make a correct decision upon the received sequence.

[2] $\lfloor x \rfloor$ denotes the integer part of x.

Let the decoder have the unquantized channel output samples x_i ($i = 1, ..., n$) at its disposal. Assume that the additive noise samples are Gaussian and statistically independent. One can prove that in case of the channel model shown in Figures 1.13a and 1.12c, the optimum decoder finding the maximum likelihood code word should select the coding sequence $\mathbf{c} = (c_1, c_2, ..., c_n)$ which is the closest in the sense of the Euclidean distance to the received sequence $\mathbf{x} = (x_1, x_2, ..., x_n)$. It means that the decoder selects that code word which fulfills the criterion

$$\min_{\mathbf{c}} \sum_{i=1}^{n} (x_i - c_i)^2 \tag{1.8}$$

In practice, the decoder does not handle the ideal values of the samples x_i, but their quantized versions. Moreover, from the implementation point of view, it is much easier to calculate the distance between the received sequence and the code word in a sub-optimum way in form of the sum of modules of the differences between the elements of both sequences, i.e. the decoder has to search for such a coding sequence $\mathbf{c}$, for which

$$\min_{\mathbf{c}} \sum_{i=1}^{n} |r_i - c_i| \tag{1.9}$$

is fulfilled.

The decoder operating according to criterion (1.8) is the optimum maximum likelihood soft-decision decoder. In practice, slightly higher error probability is received when criterion (1.9) is applied.

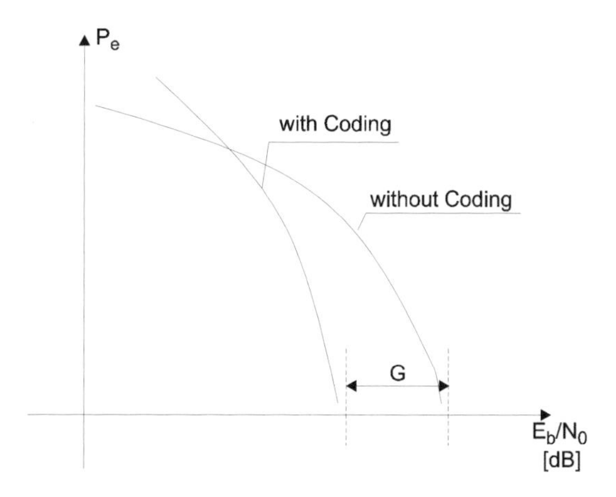

Figure 1.14 Illustration of the coding gain on the plot of block error probability versus E_b/N_0 with and without channel coding

The term *coding gain* is also strictly associated with the essence of coding. If we wish to compare the communication system with channel coding with the system without it, we have to assume that the time periods used for the transmission of sequences representing the same k-bit information block are equal in both cases. If the signal energy per bit in the system without coding is equal to E_b then the energy of a single bit in the system with coding has to be smaller due to the fact that redundant bits must be additionally transmitted. Thus, for the n-bit sequences representing the k-bit information block the energy per bit is equal to $\frac{k}{n}E_b$. In general, the probability of erroneous decoding of a code word is a function of the ratio of the energy per bit to the noise power density N_0. Despite the fact that the energy per bit in the system with coding is lower than that for the system without coding, the performance of the system with coding is higher if the energy per bit to the noise power density ratio exceeds a certain threshold value. Figure 1.14 shows that feature. For the raising value of E_b/N_0 the plots of the probability of the erroneous block decoding for coded and uncoded systems are becoming more and more parallel and asymptotically they are shifted one with respect to the other by G dB along the E_b/N_0 axis. The value G is called an *asymptotic coding gain.*

1.4.3 Classification of codes

There are several criteria of classification of channel codes. The first one is the function they perform. From this point of view we divide channel codes into *error correction* and *error detection* codes. The differences between these two code categories have already been explained.

The second criterion is the way the codes are created. Let us assume, similar to the previous section, that the binary information stream is divided into k-bit blocks $\mathbf{a}_j$, where j is the block number. If the code word $\mathbf{c}_j$ is a function exclusively of the current information block $\mathbf{a}_j$ for each j, the code is called a *block code.* If the code word is a function of the current information block $\mathbf{a}_j$ and a few previous information blocks $\mathbf{a}_{j-1}, \mathbf{a}_{j-2}, ..., \mathbf{a}_{j-i}$, it is called a *convolutional code.* From the point of view of the logical circuit theory a block code encoder can be implemented using only the combinatorial circuitry (logical gates), whereas the convolutional code encoder is an automaton and requires some memory cells. The term "convolutional code" originates from the observation that the binary sequence at the output of the encoder can be considered as a discrete convolution of the binary input stream with the *encoder impulse response.* The encoder impulse response is understood as the response of the encoder to a single "one" followed by a stream of zeros.

The basis of the next classification criterion is the number of different symbols of which code words are built. The symbols are mostly binary. A code in which binary symbols are used to compose code words is called a *binary code.* All operations performed on the elements of code words are realized in an algebraic field which consists of two elements: zero and one. In consequence, an additive operation is the addition[3] modulo-2, whereas a multiplication operation is a logical conjunction.

In some special applications *nonbinary codes* are used. The number of different symbols used for representation of the code words is a primary number or is its power.

[3] Recall that the modulo-M operation is the calculation of the remainder of division by M.

An example of a practical application of a nonbinary code is correction of a binary stream distorted by errors concentrated in bursts. A code word of the nonbinary code, in which the symbols selected from the set consisting of digits $\{0, ..., (2^m - 1)\}$ are used, is created in such a way that subsequent symbols of the code word are represented by m-bit blocks. The additive operation is then addition modulo-2^m, whereas the multiplication operation is multiplication modulo-2^m. If the error burst does not exceed m subsequent bits, it distorts at most two subsequent nonbinary code symbols. Then, in order to correct all m-bit long error bursts, it is sufficient to apply a nonbinary code which is able to correct at least two erroneous symbols.

According to another criterion of code classification, we divide codes into *systematic* and *nonsystematic*. In systematic codes the information blocks appear in the code words in the direct form and they are followed by parity bits. On the other hand, the symbols of the code word in a nonsystematic code are the sum of information symbols calculated in conformity with a selected coding rule and the information symbols do not appear in a direct form.

1.4.4 Block codes and their polynomial description

There are a few ways of describing block codes. The simplest way is to show the algebraic equations for the parity bits. For example, the expression

$$(a_1, a_2, a_3, a_4, (a_1 + a_2 + a_3), (a_1 + a_2 + a_4), (a_1 + a_3 + a_4)) \tag{1.10}$$

describes a block code of the code word length equal to $n = 7$ with the number of information bits $k = 4$. We symbolically denote such a code as (7,4). The symbol "+" denotes addition modulo-2. As we see in (1.10), the first four bits are independent, whereas each of the remaining three redundant bits is the sum of selected independent bits. These redundant bits are called *parity bits*. If we denote the parity bits as

$$a_5 = (a_1 + a_2 + a_3), \qquad a_6 = (a_1 + a_2 + a_4), \qquad a_7 = (a_1 + a_3 + a_4) \tag{1.11}$$

then the direct consequence of a modulo-2 operation is the set of equations

$$a_5 + a_1 + a_2 + a_3 = 0, \qquad a_6 + a_1 + a_2 + a_4 = 0, \qquad a_7 + a_1 + a_3 + a_4 = 0 \tag{1.12}$$

called *parity equations*. They are used in the decoder to check if the elements of the received block fulfill them and, in consequence, if this block is a code word. The above description of the block code is efficient only for codes using short code words.

In practice, for large n and k it is much more comfortable to describe the method of generating code words applying a polynomial notation. This is possible for a class of codes called *polynomial codes*. For such codes each code word can be represented by a polynomial, whose coefficients are elements of the code word. Strictly speaking, the polynomial

$$c(x) = c_{n-1}x^{n-1} + c_{n-2}x^{n-2} + \ldots + c_1 x + c_0 \tag{1.13}$$

describes the code word $(c_{n-1}, c_{n-2}, \ldots, c_1, c_0)$. For binary codes, among 2^n possible n-bit blocks and, what is equivalent, among different polynomials of the degree equal at most to $n-1$, the code (n, k) contains these code words, whose polynomials are divisible by a certain common polynomial $g(x)$ of degree $n-k$, called a *generator polynomial*. Division of polynomials is performed in a traditional manner, bearing in mind that the polynomial coefficients are equal to 0 or 1 and the additive operation is the addition modulo-2 and the multiplication operation is a logical conjunction. Each code word polynomial is divisible by the generator polynomial $g(x)$ so it can be represented as the product of two polynomials

$$c(x) = a(x)g(x) \tag{1.14}$$

where $g(x)$ is, as previously, the generator polynomial of degree $n-k$ and the degree of $a(x)$ is not higher than $k-1$. The form of the polynomial $a(x)$ depends on the information bits. We often wish to place k information bits at the beginning of the code word, which means that the code is systematic. The sequence of k information bits can be represented by the polynomial as

$$b(x) = b_{k-1}x^{k-1} + b_{k-2}x^{k-2} + \ldots + b_1 x + b_0 \tag{1.15}$$

In order to create a code word with the bits $(b_{k-1}, b_{k-2}, ..., b_1, b_0)$ in its highest positions, the polynomial $b(x)$ should be multiplied by x^{n-k} and the remaining $n-k$ bits should be calculated. This is equivalent to finding such a polynomial $p(x)$ of the degree equal at most $n-k-1$ that the sum of the polynomials is divisible by the generator polynomial $g(x)$. In other words

$$c(x) = x^{n-k}b(x) + p(x) = a(x)g(x) \tag{1.16}$$

In case of modulo-2 operations, after addition of $p(x)$ to both sides of expression (1.16) we obtain

$$x^{n-k}b(x) = a(x)g(x) + p(x) \tag{1.17}$$

As we see, the polynomial $p(x)$ of the degree lower than the degree of $g(x)$ is the remainder of division of the polynomial $x^{n-k}b(x)$ by the polynomial $g(x)$. Knowing that the calculation of the remainder of division by the polynomial $g(x)$ can be represented as a modulo-$g(x)$ operation we get

$$p(x) = \left[x^{n-k}b(x)\right] \bmod g(x) \tag{1.18}$$

For the code word polynomial the remainder of its division by the generator polynomial is equal to zero, which means that

$$c(x) \bmod g(x) = 0 \tag{1.19}$$

Let us note that the test if the given polynomial is divisible by $g(x)$ is checking if this polynomial describes a code word. This observation is utilized by many block code decoders which detect or correct errors in the code words.

Cyclic codes are an important subclass of the polynomial codes. For these codes, if the sequence $(c_1, c_2, \ldots, c_n)$ is a code word, then the sequence $(c_n, c_1, c_2, \ldots, c_{n-1})$ is a code word as well. One can prove that the generator polynomial of the cyclic codes is a dividend of the polynomial $x^n - 1$. In practice $n = 2^m - 1$. Among the cyclic codes the BCH[4] codes are particularly important due to their features. They are able to correct more than one error and their minimum distance among code words at the assumed word length n is larger as compared with other block codes of the same length. Detailed consideration of these codes is beyond the scope of this chapter. The reader interested in coding theory is advised to study the rich literature in particular [10] and [11]. Here we only sketch the general idea of BCH codes. In order to do it we have to introduce some algebra.

The operations performed on the polynomial coefficients considered so far have been made exclusively on two digits "0" and "1". This set of digits with the multiplicative and additive operations defined above creates an algebraic structure known as a *finite field* or *Galois field* $GF(2)$. One can show that finite fields exist for the sets of digits $\{0, 1, ..., p-1\}$, where p is a prime number. In this case the multiplicative and additive operations are simply multiplications and additions modulo-p.

Consider a polynomial description of the code words for $p = 2$. We know very well that each polynomial can be represented as a product of the polynomials of the degree lower than the original one. If the polynomial factorization has the form

$$f(x) = (x - \beta_1)(x - \beta_2) \cdot \ldots \cdot (x - \beta_k) \tag{1.20}$$

then $\beta_1, \beta_2, \ldots, \beta_k$ are the roots of the polynomial $f(x)$. However, in the same way as in the polynomial algebra over the field of real numbers, the roots do not always belong to the same field[5] as the polynomial coefficients. Some of them belong to the *extension field* $GF(p^m)$ of which the number of elements is a power of a prime number. The elements of $GF(p^m)$ are all possible polynomials of the degree lower than m. They constitute a set of all possible remainders obtained from the division of polynomials by a certain *irreducible polynomial* $p(x)$ of degree m. Irreducible polynomials are analogous to prime numbers and are tabulated. The additive and multiplicative operations are performed in $GF(p^m)$ on the polynomials modulo-$p(x)$, i.e. the result of the addition of two polynomials over $GF(p^m)$ is a remainder from the division of the sum of them by $p(x)$. In each extension field there is a single element denoted as α and called the *generator* or *primitive element*, such that every other nonzero element can be expressed as a power of this element. As a result, each β_i $(i = 1, ..., k)$ in (1.20) is a certain power of α.

Recall that polynomial codes are determined by the generator polynomial $g(x)$. Because the generator polynomial of degree $n - k$ of the polynomial code (n, k) can be factorized in the form

$$g(x) = (x - \beta_1)(x - \beta_2) \cdot \ldots (x - \beta_{n-k}) \tag{1.21}$$

this code is equivalently determined by the set of the roots $\{\beta_1, \beta_2, ..., \beta_{n-k}\}$. Because each code word polynomial is divisible by the generator polynomial, each code word

[4]The abbreviation BCH comes from the first names of the code inventors: Bose, Chaudhuri and Hockenghem.
[5]The roots of the polynomial with real coefficients can be complex numbers.

polynomial has the roots of which $\{\beta_1, \beta_2, ..., \beta_{n-k}\}$ is a subset. Instead of explicitly describing the generator polynomial, some codes are defined by selecting the roots of $g(x)$. The BCH codes belong to such codes. The formal definition of BCH codes is following

Definition 1 *A t-error correcting BCH code (n, k) with the code symbols belonging to $GF(p)$ is a block code of length n which has $\beta^{m_0}, \beta^{m_0+1}, \ldots, \beta^{m_0+2t-1}$ as roots of the generator polynomial $g(x)$. β is an element of $GF(p^m)$. If β is a primitive element α of $GF(p^m)$ then the length of the code words $n = p^m - 1$. Otherwise n is such a number for which $\beta^n = 1$ in $GF(p^m)$. m_0 is a parameter selected during the BCH code synthesis.*

A particularly important subclass of the BCH codes is the class with $m = m_0 = 1$. These codes are called the *Reed-Solomon codes*. The codes are defined over $GF(p)$, their block length is $n = p-1$, the roots of the generator polynomial are $\alpha, \alpha^2, \ldots, \alpha^{2t}$, so the generator polynomial is given by the formula

$$g(x) = (x - \alpha)(x - \alpha^2) \cdot \ldots \cdot (x - \alpha^{2t}) \tag{1.22}$$

where t is the number of correctable symbol errors. The codes are *non-binary* because p is obviously higher than 2 and is a power of a prime number. Thus, a Reed-Solomon code is determined in the extension field. Typically $p = 2^m$ which means that the code operates on 2^m–ary symbols. Each symbol is represented by a m-bit block, so the codes correct burst errors. For this reason they are very useful in two-level concatenated coding schemes which will be described later in this chapter.

We have already mentioned that the codes can be used for error correction or detection. Let us assume that during the transmission on n bits of the code word $\mathbf{c}$ some bits have been the subject of errors and instead of $\mathbf{c}$ the sequence $\mathbf{r}$ has been received. This sequence can be denoted as a modulo-2 addition of the code word $\mathbf{c}$ and the unknown error sequence $\mathbf{e}$, so $\mathbf{r} = \mathbf{c} + \mathbf{e}$. In the polynomial notation we have $r(x) = c(x) + e(x)$. Checking if $r(x)$ is a code polynomial results in $s(x)$, where

$$\begin{aligned} s(x) &= r(x) \bmod g(x) = [c(x) + e(x)] \bmod g(x) = \\ &= c(x) \bmod g(x) + e(x) \bmod g(x) = e(x) \bmod g(x) \end{aligned} \tag{1.23}$$

The remainder from the division of $r(x)$ by $g(x)$ denoted as $s(x)$ is called a *syndrome polynomial* and, as we see in (1.23), is determined only by the error polynomial $e(x)$. The result of calculations shown in (1.23) is a consequence of the following facts:

- calculation of the remainder from the division of the sum of two polynomials by another polynomial is a disjunctive operation, and
- the remainder from the division of the code word polynomial by the generator polynomial is zero (see (1.19)).

The idea of decoding very often relies on the calculation of the syndrome polynomial $s(x)$ and determining $e(x)$ on the basis of $s(x)$. This last step is the most difficult and

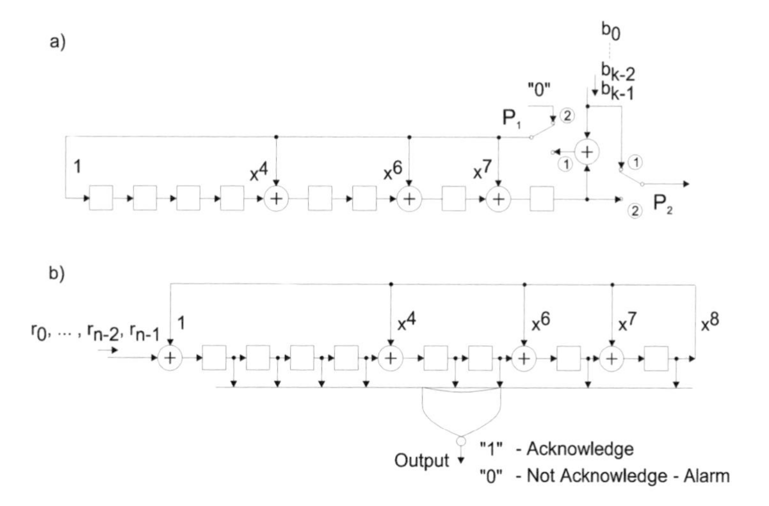

Figure 1.15 Example of the error detection encoder and decoder for the block code generated by the polynomial $g(x) = x^8 + x^7 + x^6 + x^4 + 1$

requires most of the decoder resources in form of the required memory, computational complexity, etc. After finding $e(x)$ the error polynomial is added to the polynomial $r(x)$ resulting in the most probable code word polynomial.

The operation of error detection system applying block coding relies on the calculation of the parity bits in the encoder according to expression (1.18). The decoder checks if the polynomial $r(x)$ describing the received sequence $\mathbf{r}$ is divisible by the generator polynomial $g(x)$. In other words, it checks if the syndrome $s(x) = r(x) \bmod g(x)$ is equal to zero. The parity bits calculated in the manner described above are often called CRC (*Cyclic Redundancy Check*) bits. Figure 1.15 presents an example of the circuit calculating the CRC bits (Figure 1.15a) and the circuit calculating the syndrome $s(x)$ (Figure 1.15b). Let us note that the configuration of both circuits depends only on the form of the generator polynomial and does not depend on the length of the code words. As a result, this error detection method can be applied for very long data sequences. Let us recall again that the symbol "+" in Figure 1.15a denotes modulo-2 addition.

The encoder of the (n, k) block code operates as follows. It is assumed that at the start of the operation all the memory cells contain zeros. During the first k clock cycles the switches P_1 and P_2 are in the first position so the information bits are transferred directly to the output and to the input of the circuit dividing by $g(x)$. One can check that after k clock cycles, the coefficients of the polynomial $p(x)$ are contained in the memory cells of the circuit implementing expression (1.18). Then the switches change their positions. The feedback in the circuit dividing by $g(x)$ is interrupted and the contents of the memory cells is gradually transferred to the output.

The error detecting decoder of the block code (n, k) divides the received sequence represented by the polynomial $r(x)$ by the generator polynomial $g(x)$. After n clock cycles the memory cells of this device contain the syndrome coefficients. If the received sequence is a code word, the contents of the memory cells is zero. Any non-zero contents is an indication that the received binary block has been corrupted. Such a state is detected by the logical circuit. Its output returns logical one if the sequence is a code word, otherwise it shows logical zero.

One can easily show [9], that such error detection is a very reliable method of detection of error bursts of any length b. The error burst of length b is a sequence of errors which starts and ends with a logical one and contains any sequence of zeros and ones between them. One can prove that if the degree of the generator polynomial $g(x)$ is equal to $n-k$ then all the error bursts of length not higher than $n-k$ will be detected, whereas the fraction of undetected error bursts of length higher than $n-k$ among all possible error bursts is equal to $2^{-(n-k)}$. Typically the length of the CRC block is 16 or 32 so the fraction of undetected error bursts among all errors bursts of length higher than 16 (or 32, respectively) is 2^{-16} or 2^{-32}, which are very small numbers.

1.4.5 Application of error detection in block coding – ARQ technique

Transmission of data sequences supplemented with CRC blocks for error detection, is often applied in data transmission systems, including mobile data systems. In this case it is necessary to introduce the feedback channel in which the information on acceptance or rejection of the transmitted block is sent. If the feedback channel cannot be created (e.g. due to an excessive loop delay) the only way to increase the transmission performance is to use a sufficiently strong feedforward error correction (FEC).

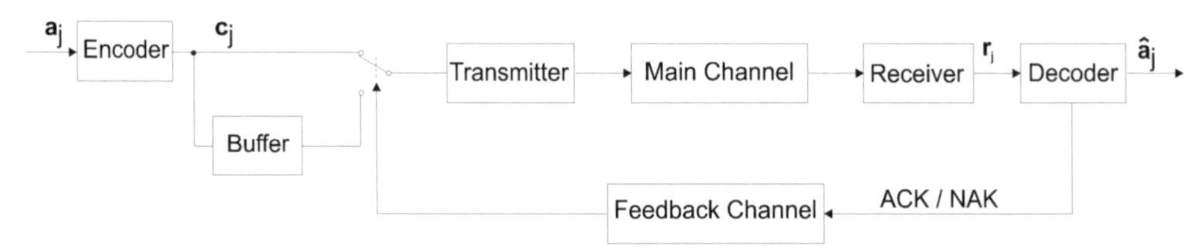

Figure 1.16 Transmission system with the block repetition and the feedback channel

Figure 1.16 presents a transmission scheme if a *feedback channel* is used. The main data stream flows from the transmitter to the receiver over the *main* channel. Each data block $\mathbf{a}_j$ supplemented with the CRC block constitutes a code word $\mathbf{c}_j$. In the receiver the syndrome of the received sequence $\mathbf{r}_j$ is calculated. In case of the zero syndrome the receiver sends a short block which acknowledges the positive reception of the data block. Such information is usually denoted as ACK (*Acknowledge*). In case of a nonzero syndrome the receiver sends to the transmitter the message NAK (*Not-Acknowledged*). In consequence, the block which has not been acknowledged is sent to the receiver once more. The procedure of information block exchange is realized automatically, so this technique is often denoted as ARQ (*Automatic-Repeat-Request*).

In general, the time necessary to transmit a data block using the ARQ technique is random and it depends on the number of block repetitions and the arrangement of the information block exchange. In return, the quality of the received and accepted data blocks is high and constant, as opposed to the FEC coding scheme. In this latter case the quality of the received data is variable because in case of breaking the error correction capabilities the decoder emits erroneous blocks. However, the data are sent at the constant delay which is an important feature for some applications. The system designer has to decide which solution is more advantageous for him.

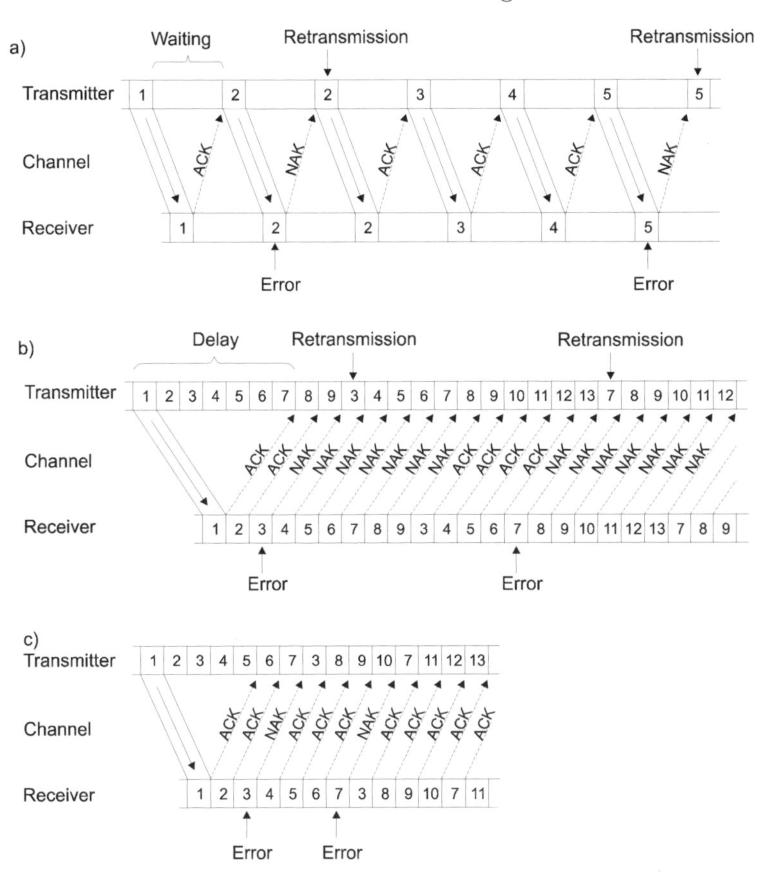

Figure 1.17 Examples of basic ARQ techniques

There exist many ARQ methods differing in circuit complexity and associated data transmission efficiency. In general they can be categorized into the following three groups [11] (see Figure 1.17):

- the stop-and-wait ARQ (*idle* ARQ, Figure 1.17a),
- continuous ARQ with *go-back-N* retransmission strategy (Figure 1.17b), and
- continuous ARQ with *selective repeat* strategy (Figure 1.17c).

The first kind of ARQ strategy is simple in implementation, however, the link utilization is often low. It is worth using in short links with the moderate data rates. The transmitter sends a data block and waits for its acknowledgement. If the receiver, on the basis of CRC block and syndrome calculation, decides that the received sequence is a code word, it sends the ACK message. Subsequently, the transmitter sends the next data block. In case of erroneous reception of the data block the NAK message is sent to the transmitter and the latter emits the last data block once more. The retransmission can occur a few times. If the subsequent attempts of the block retransmission are not successful, the link is considered unreliable. In general, in the time between transmission of subsequent code words the link remains unused, which causes a relatively low efficiency of the link utilization. It can be improved by lengthening the data block; however, the probability of the block error increases and, in consequence, the frequency of block repetition increases as well. Despite its inefficiency, the stop-and-wait ARQ procedure is applied in the widely used *bisync* (*Binary Synchronous Control*), the protocol originally developed by IBM.

The transmitter applying the continuous ARQ with go-back-N retransmission strategy sends subsequent data blocks continuously. It does not wait until the ACK message is received. However, the transmitted blocks are stored in the memory buffer until their acknowledgement is received. The size of the memory buffer depends on the size of data blocks and the maximum expected round-trip delay. After the reception of the NAK message the transmitter sends the whole sequence of blocks again, starting from the corrupted one. If this strategy is applied, the receiver does not need to be equipped with a buffer. The blocks following the corrupted one are marked as erroneous blocks until the positive acknowledgement of the erroneous block is received. This type of the procedure is applied in another famous communication protocol known as *Synchronous Data Link Control* - SDLC. Low efficiency of this procedure becomes noticeable when the delay loop is long and the data transmission rate is high. This inefficiency arises from the fact that the positively received blocks which are preceded by an erroneous block have to be retransmitted.

The continuous ARQ with selective retransmission avoids the above mentioned drawback. In this strategy, the receiver repeats only those blocks which have not been received correctly. Because the end-user should receive the data blocks in a correct sequence, the buffers in the receiver and transmitter are necessary. In the receiver buffer the received blocks are properly reordered to maintain the appropriate block sequence. Let us note that in the latter case data blocks must be equipped with block numbers which allow for block ordering.

More information on ARQ procedures can be found in [11] and [19].

1.4.6 Convolutional codes

Convolutional codes are an important class of error correction codes which are more and more frequently applied in digital communication systems. One of their features is simplicity of the coding process and a well-known decoding technique both in the hard- and soft-decision forms. In the sense of the theory of logical circuits the convolutional code encoder is an automaton. It features a certain number of states which are entered and left due to the input information bits which can be considered as the automaton's excitations. The output signal, which is a code word, is the result of the encoder's transition between two subsequent states.

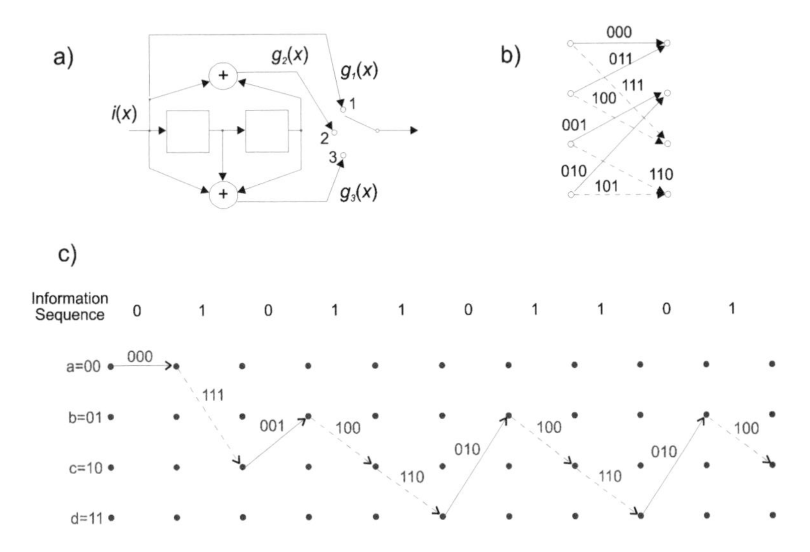

Figure 1.18 Example of the convolutional code encoder (a), related trellis diagram (b) and the path on the trellis diagram resulting from the encoder input sequence 0101101101 (c)

Let us consider a simple example of the convolutional code encoder, shown in Figure 1.18. As we see, the encoder has two memory cells. The encoder output signals are obtained by modulo-2 addition of the input bit and the selected bits contained in the memory cells. The switch periodically sends the bits from the subsequent outputs of the logical circuits directly to the encoder output. It selects each of the logical outputs in one-third of the input bit cycle.

A state diagram is one of the typical ways of automaton description. The encoder state is determined by the contents of the memory cells. In case of convolutional codes it is more advantageous to present the operation of the encoder using a particular form of the state diagram called a *trellis diagram*. This kind of diagram describes possible transitions from the states at the n-th moment to the states at the $n + 1$-st moment. The output bit blocks associated with particular transitions among states are placed above the arrows symbolizing these transitions. The excitation being the reason of a particular state transition is shown by a solid line in case of the zero input signal or by a dashed line for the input signal equal to logical one. Let us note that for the considered convolutional code encoder, for each input bit three bits are generated at the output of the encoder, so the code has the coding rate $R = k/n = 1/3$. Figure 1.18b presents possible state transitions for the n-th moment. One can easily imagine a whole sequence of transitions starting at the initial moment and ending at the current moment, so the operation of the encoder is equivalent to wandering between subsequent states along a particular path of the trellis diagram. As we know, state transitions are determined by the state of the encoder and its input excitation signal. Let us note that assuming that the initial state of the encoder at the zero moment (mostly characterized by zeros in the memory cells) is known to the decoder, the determination of the encoder path along the trellis diagram in the encoding process is equivalent to the determination of the sequence of information symbols. Concluding, the decoding algorithm can be based on finding the most likely path on the trellis diagram which has been travelled by the encoder.

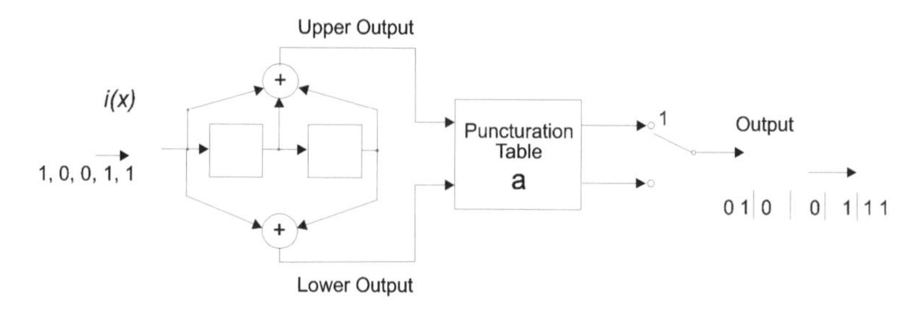

Figure 1.19 RCPC encoder with the coding rate $R = 4/5$

There are applications in which the coding rate of the form $R = 1/n$ is not appropriate. For example, a coding rate of 4/5 or 2/3 is desired. One of the ways of achieving this goal is elimination of selected output bits, called *puncturing.* Such codes are called *Rate Compatible Punctured Convolutional Codes* (RCPC). Figure 1.19 presents the example of generation of the RCPC code of the coding rate $R = 4/5$. The convolutional encoded bits are sent to the output, taking into account the so called *puncturing table.* Such a table applied in the encoder shown in Figure 1.19 has the form:

$$\mathbf{a} = \begin{bmatrix} 1 & 1 & 1 & 0 \\ 1 & 0 & 0 & 1 \end{bmatrix} \tag{1.24}$$

The table shows that for each four pairs of the encoder output puncturing results in elimination of the second and third bit from the lower encoder output and the fourth bit from its upper output.

As in a block code decoder, the decoder of a convolutional code optimal in a maximum likelihood sense selects the code word which is the closest to the received sequence. In case of the hard-decision decoding the measure of distance is the Hamming distance, whereas for soft-decision decoding the Euclidean distance is applied. As we have already mentioned, very often a suboptimal measure is used, leading to simplification of the decoder implementation.

An effective method of selection of the decoded code word was proposed in 1967 by Andrew Viterbi [12]. The *Viterbi algorithm* searches for the optimum code sequence which is associated with the 'shortest' path (in the sense of the selected distance measure) on the trellis diagram. Such a path is found by extending the paths starting from the known initial state to each possible trellis state at the current (n-th) moment. A specific cost is associated with each transition between trellis states. It is the distance between the code sequence associated with that particular transition and the received sequence. The key point of the algorithm results from the observation that the shortest path to the i-th state at the n-th moment consists of the transition from one of possible states (e.g. the k-th state) at the $(n-1)$-st moment, from which the i-th state is achievable, and the shortest path to the k-th state at the $(n-1)$-st moment. Thus, the selection of the shortest route to each state is a recursive procedure - we find it using the results of the search at the preceding moment. In case of a finite code sequence the algorithm makes a decision upon the transmitted code sequence by tracking the path to that state for which the distance measure is minimal. Sometimes, as it happens in the GSM system, the information sequence is supplemented with a few known data symbols, e.g. zeros. Thus, the decoder knows the number of the final state or the subset of the states in which the path along the trellis diagram can end. This knowledge can be utilized in the selection of the code sequence by the decoder.

In the case of a continuous transmission it is necessary to make a decision with a finite time delay. It is possible due to the fact that at the probability close to unity, the shortest paths to each state contain a common route up to D time instants back with respect to the current moment. It means that the route from the initial moment up to the $(n-D)$-th time instant is the same, independent of which state at the n-th moment features the lowest distance measure (has the shortest path). Thus, the decoder is able to generate the final decision upon the transmitted symbols delayed by at least D steps.

Figure 1.20 presents an example of the shortest path to each trellis state for the encoder shown in Figure 1.18 when a particular data sequence has been received. The code sequence from Figure 1.18c distorted by five binary errors has been selected for illustration. The lowest distance measure (the cost) of reaching each state at a given moment is denoted above each state. This cost is in fact the Hamming distance between the received binary sequence and the code sequence associated with the shortest path ending at the considered state. In Figure 1.20c the route featuring the shortest distance measure is bordered by a double dotted line. This is the route which would be selected by the decoder if the decision upon the transmitted binary sequence had to be made at that moment. Let us note that the path selected by the decoder is identical to that in Figure 1.18c, so the decoder has corrected all the errors contained in the received sequence.

Beside the Viterbi algorithm there are other methods of decoding the convolutional codes, e.g. the *Fano algorithm* or *algebraic decoding*; however, the Viterbi algorithm has gained a fundamental practical meaning and it has made the convolutional codes very attractive from the implementation point of view. It is applied in the main second and third generation mobile communication systems such as GSM, IS-54/136, IS-95, UMTS and cdma2000.

The Viterbi algorithm determining the code sequence on the basis of the minimum distance from the received sequence has been introduced following the assumption of statistically independent errors occurring during the transmission, i.e. the assumption that the transmission path can be modeled as a memoryless channel. In the case of mobile communication systems this assumption is often not fulfilled due to the channel properties. The errors frequently arise in bursts which substantially decreases the performance of the Viterbi algorithm. The remedy lies in application of *interleaving*.

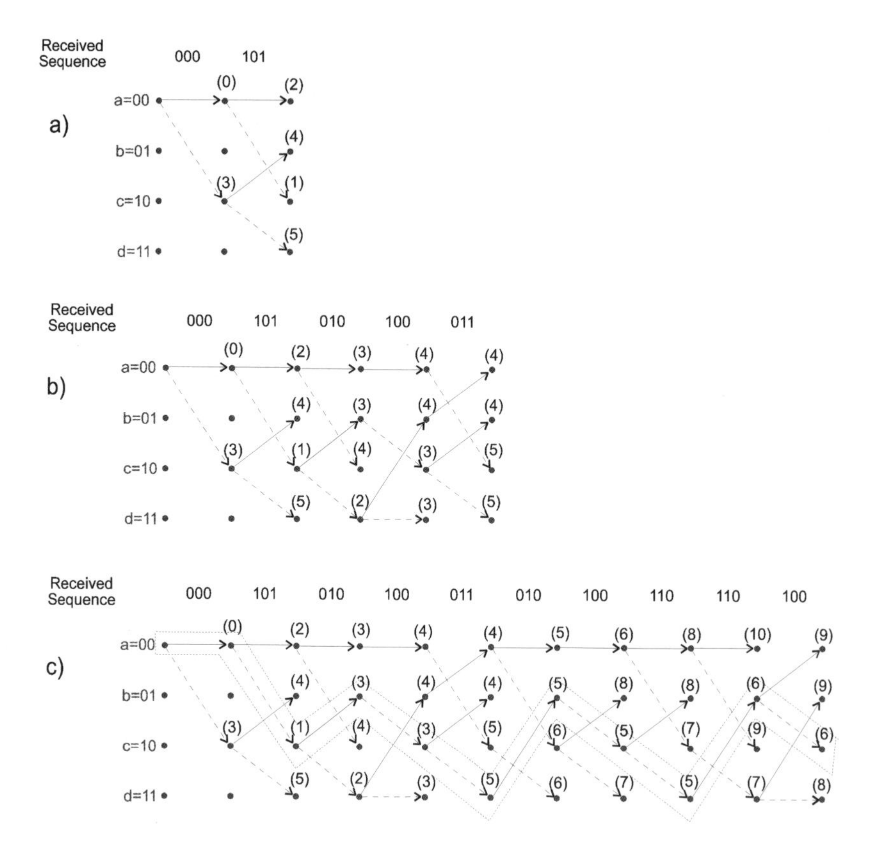

Figure 1.20 Illustration of the shortest path search for each trellis stage at the n-th moment: after two clock cycles (a), after five clock cycles (b), after 10 clock cycles (c)

1.4.7 Error spreading – application of interleaving

Application of interleaving in the transmitter at the output of the FEC encoder and *deinterleaving* at the receiver in front of the decoder improves the operation of most of the decoding algorithms for block and convolutional codes. Subsequent bits of the code sequence are transmitted over the channel in a modified order as compared with the order of generation. Such a modified sequence may be distorted by burst errors caused by transmission conditions in the mobile channel. At the receiver the received bits are reordered so that in case of the error absence the original code sequence could be recovered. As we see, the operations of interleaving and deinterleaving are complementary. In practice, the deinterleaving causes tearing apart the burst errors and spreading them over the whole range of the processed block. Thus, at the output of the deinterleaver the errors are quasi statistically independent. There are two basic categories of the interleavers: *block interleaver* and *convolutional interleaver.*

In case of the block interleaver the bits from the encoder are written in a two-dimensional matrix in a prescribed sequence. The simplest method is writing them into subsequent rows. If the whole matrix is filled with the input bits, the phase of reading begins, in which the bits are read out of the matrix in a different order, usually along subsequent columns. In the receiver the incoming bits are stored in an identical two-dimensional matrix in the order in which they have been read out in the transmitter. When the incoming bits fill up the matrix they are read out in the order of writing them into the interleaver matrix. Writing in rows and reading in columns is the simplest way of spreading the burst errors. One can define another sequence of writings and readings to/from the memory of the interleaver and deinterleaver. Both processes have to be complementary. The sequence of addresses and the size of the matrix determining so-called *depth of interleaving* have to be selected in such a way that the burst errors occurring in the channel are dispersed to that extent that they look like random errors. Figure 1.21 presents the simplest example of interleaving with writing in the rows and reading in columns on the transmitter side.

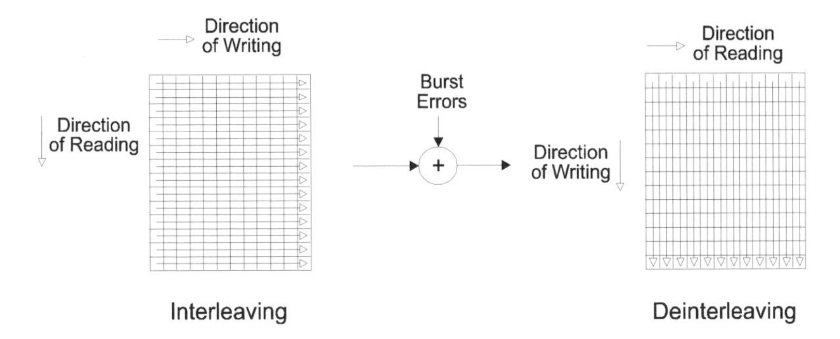

Figure 1.21 Example of block interleaving and deinterleaving

We have to stress the crucial meaning of synchronization for the correct operation of the interleaver and deinterleaver. If the frames of the transmitter and receiver are shifted in time with respect to each other, the ordering of the bits in the deinterleaver does not give appropriate results. Therefore the transmitted sequence often starts with a short synchronization word. Because writing into the matrix and reading from it have to be performed at the same time, the number of matrices is doubled. One of them is used for storing the input bits whereas the other one is in the phase of reading of the bits written to it in the previous phase. After reading/writing from/to the whole single matrix the matrices reverse their functions.

As already mentioned, the second way of spreading the burst errors is the application of the convolutional interleaver. Its operation is shown in Figure 1.22.

The binary sequence is sent by the commutator on the bit-by-bit basis to the inputs of B parallel registers. The i-th register ($i = 1, \dots, B$) delays the input bits by $(i-1)M$ cycles. The output signals of each delay register are fed serially to the interleaver output through the second commutator. The deinterleaver operates exactly in the same way as the interleaver. The only difference is the ordering of the delay register blocks according to the decreasing delay. Therefore the upper register delays the input sequence by $(B-1)M$ cycles, whereas the lowest branch of the deinterleaver does not

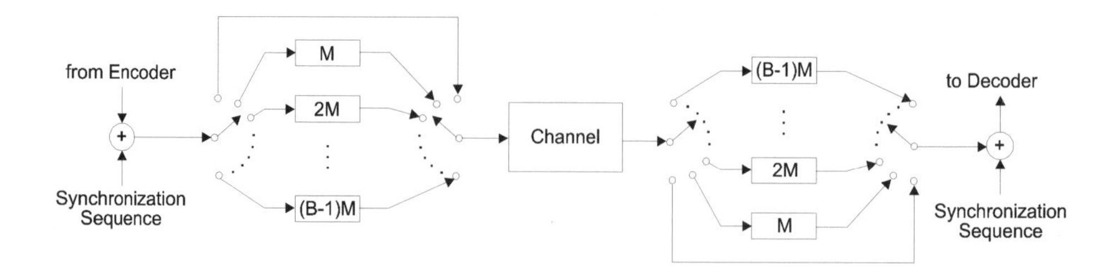

Figure 1.22 Basic scheme of convolutional interleaving and deinterleaving

introduce any delay. As in case of the block interleaver, the synchronous operation of the interleaver's and deinterleaver's commutators has a fundamental meaning. Thus, the delay introduced by the cascade of interleaver and deinterleaver in each parallel branch is the same and equal to $(B-1)M$ cycles.

As already mentioned, the quality of error spreading depends on the interleaving depth. However, we have to note that the interleaver/deinterleaver pair introduces a substantial delay, which in some communication systems can be tolerated to a limited extent.

1.4.8 The concept of concatenated coding

Transmission of digitized speech over mobile channels sets a moderate requirement on the binary error rate which can be achieved mostly by application of convolutional coding and interleaving. However, other services such as data and multimedia transmission require much lower BER than that which is sufficient for speech signals. One way to achieve a very low BER is the application of *concatenated coding*.

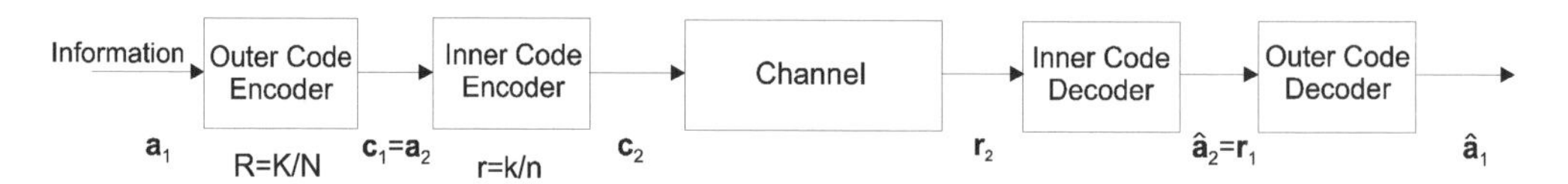

Figure 1.23 Principle of concatenated coding

The idea of concatenated coding was introduced by Forney [22] in 1966 and is shown in Figure 1.23. The channel coding at the transmitter side is realized by two encoders. The first one is called an *outer code* encoder. Let its coding rate be $R = K/N$, where K is the number of information bits and N is the code word length. The code words are treated as the information bits by the *inner code* encoder with the coding rate equal to $r = k/n$. One can easily show that the overall coding rate is $rR = kK/nN$. In the receiver the data stream appearing at the output of the channel is the subject of decoding in the *inner decoder* and the information bits at its output are the input bits for the *outer decoder.* Several arrangements of the inner and outer codes have been investigated. In many applications the inner code is a convolutional code, whereas the outer code is a nonbinary Reed-Solomon code. The task of a strong inner code is to correct as many errors caused by the channel as possible. However, some error patterns can be beyond the capabilities of error correction and cause error bursts at the output of the convolutional code decoder. Thus, the outer code decoder corrects the remaining burst errors.

Such a configuration in which the outer code is a Reed-Solomon code and the inner code is a convolutional code is often called *classical*. In another widely-known concatenated coding scheme two parallel convolutional codes with the interleaver between the outer and inner encoders are applied. Such a scheme is called *turbo-coding* and is the subject of our considerations below.

1.4.9 The turbo-coding principle

The idea of turbo-coding was first presented in 1993 [23]. An example of a *turbo-code encoder* is shown in Figure 1.24, whereas a respective *turbo-code decoder* is depicted in Figure 1.25.The turbo-code encoder shown in Figure 1.24 consists of two *Recursive Systematic Code* (RSC) encoders, an interleaver and a multiplexing and puncturing device. The recursive systematic code is a kind of a convolutional code where in the

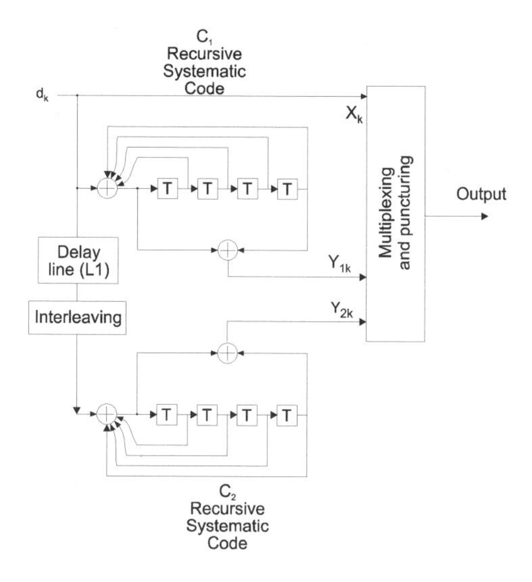

Figure 1.24 Example of turbo-code encoder (based on [23], © 1993 IEEE)

encoder the input information bits are directly transferred to the output, and redundancy bits are generated in the logical circuit containing a feedback shift register (see Figure 1.24). It turns out that the application of two parallel RSC encoders with the interleaver preceding the second one results in very good properties of the generated code words, such as a large Hamming distance between code words. The construction of the interleaver is crucial for the code performance. Usually, non–uniform (pseudorandom,

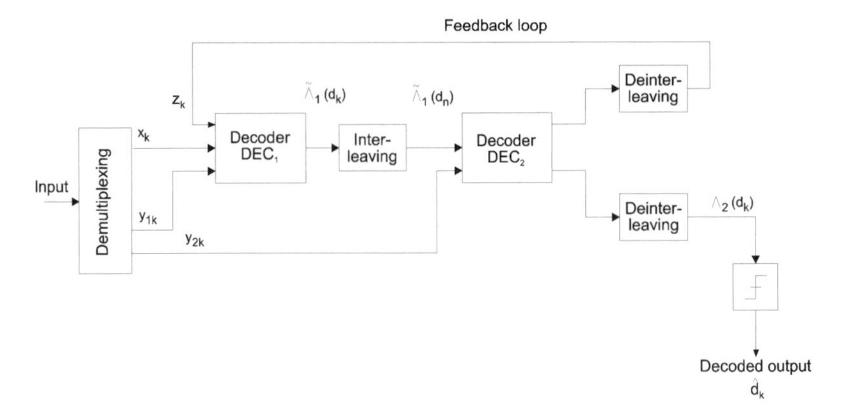

Figure 1.25 Turbo-code decoder for the code shown in Figure 1.24 [23] (© 1993 IEEE)

as opposed to block or convolutional) interleavers are applied. In such an interleaver, reading to and writing from memory are performed in a pseudorandom fashion. As in other types of interleavers, the processes of reading and writing are complementary. In order to achieve high code performance, the size of the interleaver memory has to be large, leading to a substantial delay introduced by the coding and decoding processes. The application of two RSC encoders determines the lowest coding rate $R = 1/3$ (three output bits are generated as a response to each information bit). The coding rate can be increased by applying puncturing.

The turbo-code decoder consists of two soft input/soft output decoders[6] separated by an interleaver. The output signal of the second decoder is fed back to the input of the first decoder. The decoding process is performed iteratively in the feedback loop. Both decoders use channel state information consisting of instantaneous signal amplitudes and noise variance. Both decoders process the information signals, redundancy signals and the LLR values.[7]

The performance of the turbo-code decoder, expressed as the BER versus E_b/N_0, improves with the number of iterations performed in the decoding process. Figure 1.26 shows the gradual improvement of the performance with the increasing number of iterations for the code presented in Figure 1.24 applied for transmission over additive white Gaussian noise (AWGN) channel. Let us note that the greatest improvement is

[6]Soft-decision decoding is performed in the decoders.

[7]The LLR means the Logarithm of Likelihood Ratio and is denoted in Figure 1.25 as $\Lambda(d_k)$. It is determined by formula

$$\Lambda(d_k) = \log \frac{\Pr\{d_k = 1|\text{observation}\}}{\Pr\{d_k = 0|\text{observation}\}}$$

where $\Pr\{d_k = 1|\text{observation}\}$ is the *a posteriori* probability determining the probability that the data symbol $d_k = 1$ has been transmitted knowing the observation signal at the receiver. The variables $\tilde{\Lambda}(d_k)$ shown in Figure 1.25 denote extrinsic information generated by the decoder. Details can be found in [23] or [29].

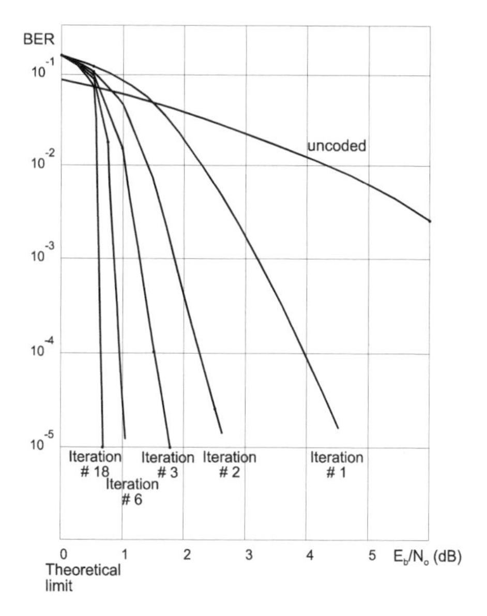

Figure 1.26 Performance of the turbo–code decoder in function of the number of iterations [23] (© 1993 IEEE)

achieved at the beginning of the iterative process and the final performance is close to the Shannon limit. In this sense turbo-coding is a considerable achievement in the coding theory. Turbo-codes can find applications in services requiring very low BER values, which can tolerate high delay introduced by the decoder.

1.5 DIGITAL MODULATIONS APPLIED IN MOBILE COMMUNICATION SYSTEMS

Let us consider the next block of the digital communication system model shown in Figure 1.1. The binary stream, having been protected against possible transmission errors, is fed to the modulator, which, depending on the logical values of the binary stream, modifies one or more parameters of the output sinusoidal signal, such as phase, frequency, or amplitude.

Let us consider a very general model of operations performed in the modulator. In fact this model characterizes all modulation types and is given by the formula

$$s(t) = x^I(t)\cos 2\pi f_c t - x^Q(t)\sin 2\pi f_c t = \mathrm{Re}\left\{x(t)\exp(j2\pi f_c t\right\} \tag{1.25}$$

where Re$\{.\}$ denotes the real part of the complex argument, and $x(t) = x^I(t) + jx^Q(t)$. Signals $x^I(t)$ and $x^Q(t)$ are the signals modulating the cosinusoidal and sinusoidal carrier of frequency f_c. The modulating signals are called the *in-phase* and *quadrature components*, respectively. By selecting these signals appropriately, we are able to describe any digital modulation. Due to the introduction of the complex signal $x(t)$, we can consider each modulation in the complex plane as a set of its characteristic points (so called *constellation points*) with the trajectory characterizing the movement in time of the signal point having the coordinates $(x^I(t), x^Q(t))$ in the complex plane. Signal $x(t)$ is called a *baseband equivalent signal.*

As we have already mentioned, one of the desired features of the modulations applied in the mobile communication systems is a constant envelope. This feature results from the necessity of obtaining a possibly high level of the signal at the output of the nonlinear power amplifier. A constant envelope is the attribute of phase (PM) or frequency (FM) modulations, which are generally described by the equations

$$x^I(t) = r \cos \varphi(t) \quad \text{and} \quad x^Q(t) = r \sin \varphi(t) \tag{1.26}$$

If

$$\varphi(t) = 2\pi k_{FM} \int_{-\infty}^{t} m(\tau)\mathrm{d}\tau, \quad \text{where} \quad |m(t)| \leq m_{\max} \tag{1.27}$$

and $m(t)$ is a continuous signal, then formulae (1.26) and (1.27) represent the analog frequency modulation. Factor k_{FM} is the *FM modulation index* given by the expression $k_{FM} = \Delta f / m_{\max}$. Δf is the *frequency deviation*, i.e. the maximum deviation of the instantaneous signal frequency from the carrier frequency. The analog frequency modulation has been applied in the first generation cellular and wireless telephony systems for transmission of analog speech signals. However, even in those systems the control signals are digital, so in this case the carrier is digitally modulated. In the case of the *Frequency Shift Keying* (FSK) the equation describing the instantaneous phase as a function of time and of the transmitted digital stream is given by the equation

$$\varphi(t) = 2\pi h \sum_{i=-\infty}^{n} a_i \int_{-\infty}^{t} g(\tau - iT)\mathrm{d}\tau \quad \text{for} \quad nT \leq t \leq (n+1)T \tag{1.28}$$

where a_i is the data symbol ($a_i = \pm 1$) transmitted in the i-th signalling period and $h = 2\Delta f T$ is the FSK *modulation index.* T is the *modulation period* and Δf is, as previously, the frequency deviation. The data symbols are mostly binary (equal to ± 1), although in some cases multilevel data symbols are applied. Function $g(t)$ is a *frequency pulse* and determines the frequency variation in time. Let us note that the instantaneous frequency can be derived from the formula

$$f(t) = \frac{1}{2\pi} \frac{\mathrm{d}\varphi(t)}{\mathrm{d}t} = h \sum_{i=-\infty}^{n} a_i g(t - iT) \quad \text{for} \quad nT \leq t \leq (n+1)T \tag{1.29}$$

In turn

$$q(t) = \int_{-\infty}^{t} g(\tau)\mathrm{d}\tau \tag{1.30}$$

is a *phase pulse* and is a phase response to the single unit data pulse. In the simplest case the frequency pulse $g(t)$ is a gate function of length T and of the height of $1/2T$. Then the instantaneous frequency with respect to the carrier frequency f_c equals $\pm\Delta f$. The frequency and phase pulses for this case are shown in Figure 1.27.

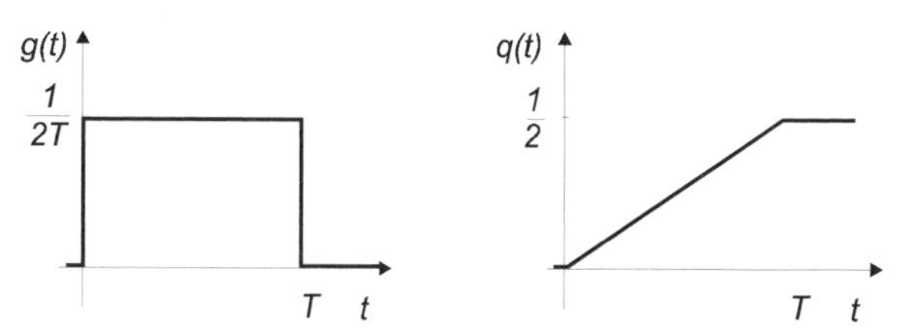

Figure 1.27 Frequency and phase pulse for continuous phase FSK

The signal described by equations (1.25) and (1.26) is characterized by the phase continuity for any integrable shape of the frequency pulse $g(t)$. This feature has a fundamental impact on the spectral properties of the modulated signal. In practice, the FSK signal generated by the modulator described by (1.25) and (1.26) with the frequency pulse shown in Figure 1.27 has a continuous phase. The example of a modulator generating a signal with a non-continuous phase would be an FSK modulator implemented in the form of two unsynchronized sinusoidal generators of nominal frequencies $f_c \pm \Delta f$ followed by the circuit switching between the generators' output signals, controlled by the current data symbol. The signal generated by such a modulator has poor spectral properties.

It was found relatively early that the choice of the frequency pulse, or equivalently, the phase pulse, and the value of the modulation index h have a fundamental influence on the spectral properties of the modulated signal. For the frequency pulse shown in Figure 1.27 and for $h = 1/2$ we have a special case of the FSK modulation, so called *Minimum Shift Keying* (MSK). One can show that the MSK modulation can be interpreted as a linear modulation[8] as opposed to FSK modulations with other than $h = 1/2$ values of the modulation index.

Several frequency pulses have been investigated with respect to their influence on the spectral and detection properties of the modulated signal. Beside the narrow signal spectrum, it is desired that the sequences of elementary signals determined by different input data sequences differ maximally from each other in the sense of the selected distance measure. This can be achieved by extending the frequency pulse beyond the modulation period T. Unfortunately, this leads to high complexity of the receiver, which has to apply the Viterbi algorithm for implementing the sequential detection.

One of the best binary modulations having excellent spectral properties results from the application of the frequency pulse given by the formula

$$g(t) = \frac{1}{\sqrt{2\pi}\sigma T} \exp\left(\frac{-t^2}{2\sigma^2 T^2}\right) * \operatorname{rect}\left(\frac{t}{T}\right) \tag{1.31}$$

[8] A modulation is linear if the rule of superposition with respect to the modulating signal is valid.

where $*$ denotes convolution and the function $\text{rect}(t/T)$ describes the rectangular pulse (gate function) of the unit height and lasting from $-1/2T$ to $1/2T$. The FSK modulation which uses the frequency pulse described by (1.31) and which has the modulation index $h = 1/2$ is called *Gaussian Minimum Shift Keying* (GMSK). In a GMSK modulator the rectangular frequency pulse is filtered by a Gaussian-shaped filter. In (1.31) the parameter $\sigma = \sqrt{\ln 2}/(2\pi BT)$, where B is a 3-dB bandwidth of the Gaussian filter. The pulse $g(t)$ usually lasts for a few modulation periods, so subsequent pulses which are the response to the data symbols interfere with each other and the Viterbi algorithm is useful in the data sequence detection. The GMSK modulation is described in detail in the chapter explaining the physical layer of the GSM system. GMSK is applied in the GSM and in a few other systems because of its particularly good spectral properties manifesting themselves as a very narrow main lobe and very low and sharply decreasing levels of sidelobes.

The constant envelope of the FSK (MSK, GMSK) modulated signals can be easily presented in the complex plane. One can conclude from formulae (1.25) and (1.26) that for the above listed modulations their signal envelope is determined by the expression

$$r(t) = \sqrt{(x^I(t))^2 + (x^Q(t))^2} = r\sqrt{\cos^2\varphi(t) + \sin^2\varphi(t)} = r = const \tag{1.32}$$

The steadiness of the envelope is ensured for any angle $\varphi(t)$. Figure 1.28 presents the envelope of the MSK modulated signal. Let us note that in this case, within a single modulation period the angle $\varphi(t)$ changes its value by $\pm\pi/2$ and the final value of the angle depends not only on the current data symbol but also on the phase determined by the data sequence transmitted before the current modulation period. In this sense the MSK modulator has a memory.

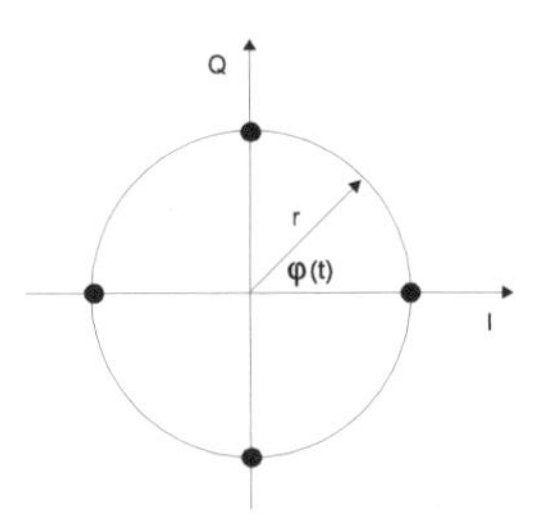

Figure 1.28 Envelope of the MSK modulated signal in the in-phase - quadrature plane

The steadiness of the envelope is usually not preserved if the signals modulating the in-phase and quadrature components result from the linear filtration of the data signals. Let us consider this case in detail (see Figure 1.29).

A binary data stream (mostly received from the output of the error correction or detection code encoder or interleaver) is fed to the input of the mapping block, which maps the blocks of binary data onto pairs of data symbols, d_n^I and d_n^Q. These data symbols constitute the input of the baseband transmit filters with the impulse responses $p(t)$ and $q(t)$. The signals modulating the in-phase and quadrature carriers are described

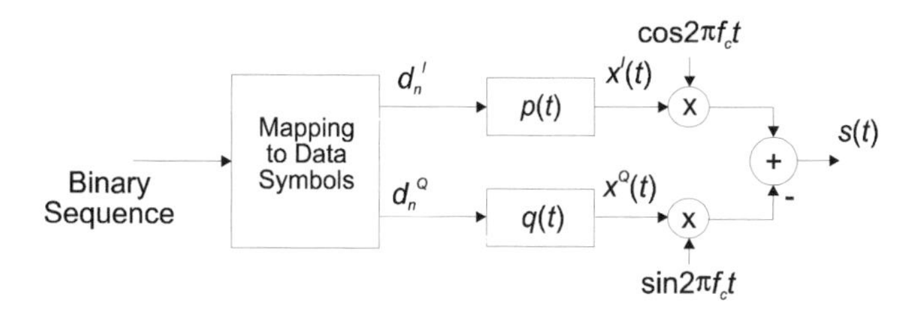

Figure 1.29 Linear modulator for a two-dimensional modulation

by the formulae

$$x^I(t) = \sum_{n=-\infty}^{\infty} d_n^I p(t - nT), \quad x^Q(t) = \sum_{n=-\infty}^{\infty} d_n^Q q(t - nT) \tag{1.33}$$

Various types of linear modulations can be described using (1.33). Assuming $q(t) = 0$, $d_n^I = \pm 1$ and $p(t) = \text{rect}(t/T)$, we obtain a two-level phase modulation called *Binary Phase Shift Keying* (BPSK). In turn, taking $d_n^I = d_n^Q = \pm 1$ and $p(t) = q(t) = \text{rect}(t/T)$ we obtain a four-level phase modulation called *Quadrature Phase Shift Keying* (QPSK). Higher level modulations such as *Quadrature Amplitude Modulations* (QAM) are obtained by selection of multilevel data symbols d_n^I and d_n^Q. Figure 1.30 illustrates constellations of a few most important digital modulations. It is worth noting that the QAM modulations have not been directly applied in the mobile communication systems so far, due to their variable envelope and the necessary precise gain control in the receiver. However, they are used for modulation of the subcarriers in a multicarrier modulation, which will be discussed in the next subsection.

The spectrum and envelope of the modulated signal can be determined by selection of the transmit filters $p(t)$ and $q(t)$. A typical transmit filter $p(t)$ applied in digital communication systems has a square-root raised cosine spectrum given by the formula

$$P(f) = \begin{cases} \sqrt{T} & \text{for} \quad 0 \leq |f| \leq \frac{1-\alpha}{2T} \\ \sqrt{\frac{T}{2}\left\{1 + \cos\left[\frac{\pi T}{\alpha}\left(|f| - \frac{1-\alpha}{2T}\right)\right]\right\}} & \text{for} \quad \frac{1-\alpha}{2T} \leq |f| \leq \frac{1+\alpha}{2T} \\ 0 & \text{for} \quad |f| \geq \frac{1+\alpha}{2T} \end{cases} \tag{1.34}$$

The parameter α is called a roll-off factor and its value falls within the range $0 \leq \alpha \leq 1$. In the receiver, the filter of the same characteristics is applied. The signal spectrum obtained by the application of such filters ensures low sidelobes and concentration of the signal energy in the main lobe. However, from the point of view of mobile communication systems such a signal has a substantial drawback. Its envelope is variable and can take instantaneous values close to zero. Figure 1.31a illustrates this phenomenon for the QPSK modulation with the square-root raised cosine transmit filter characteristics ($\alpha = 0.35$). Such a QPSK signal is not very immune from nonlinear distortion. A substantial improvement is achieved if the quadrature component is delayed by a half of the modulation period, i.e. if $q(t) = p(t - T/2)$. Such modified QPSK modulation

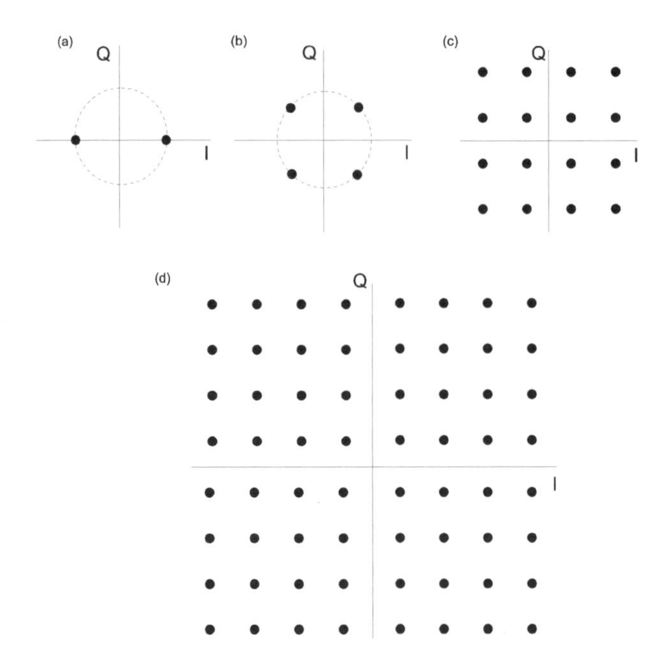

Figure 1.30 Examples of signal constellations: BPSK (a), QPSK (b), 16-QAM (c), 64-QAM (d)

is called *Offset Quarternary Phase Shift Keying* (OQPSK). This type of modulation is applied in the second generation cellular system IS-95, which will be considered in Chapter 11. The OQPSK signal envelope versus time in the complex plane is shown in Figure 1.31b.

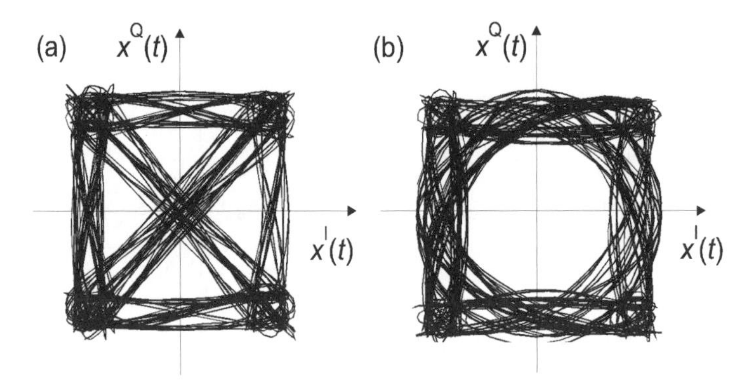

Figure 1.31 Envelope in the I-Q plane for QPSK (a) and OQPSK (b) modulations

Another method of decreasing the dynamics of the envelope is the addition of the $\pi/4$ phase shift to the differential quarternary phase shift keyed signal in each modulation period. Such a modulation, denoted as $\pi/4$-DQPSK, is applied in a few mobile communication systems, e.g. in IS-54/136 and TETRA (see appropriate chapters for the details).

At the present state of development of the mobile communication systems which apply single carrier modulations, the modulation level higher than 8 is not used.[9] Higher level QAM modulations are applied in some wireless LANs which use multicarrier modulations, e.g. in HyperLAN/2 [24].

The problem of demodulation is directly related to the modulation applied in a mobile communication system. This is the subject of many pages in academic textbooks devoted to communication systems. We will only sketch the most important demodulation methods applicable in mobile communication systems.

In general, the signal reception can be *coherent* (synchronous) or *non-coherent* (asynchronous). The coherent reception requires the knowledge of the beginning and end of the modulation period (*timing synchronization*) and of the carrier frequency and phase (*carrier synchronization*). In a coherent receiver, the whole knowledge of the received signal parameters is used so it is no surprise that the performance of a coherent demodulator, shown as the bit error rate (BER) versus the SNR, is better than that of other demodulators. Figure 1.32 presents a synchronous QPSK receiver (we assume that both transmit filters in the in-phase and quadrature branches are identical, i.e. $p(t) = q(t)$).

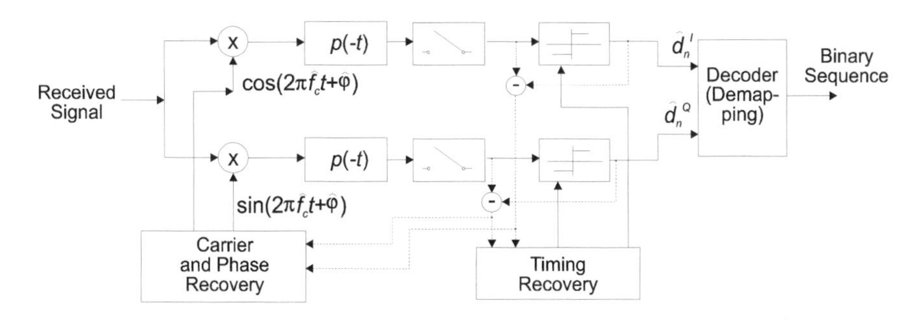

Figure 1.32 Scheme of the optimal synchronous receiver for QPSK signals

The received signal (usually converted down to the intermediate frequency) is put to the inputs of two synchronous demodulators consisting of mixers and lowpass filters. On the one hand these filters are the immanent part of the synchronous demodulator (they cut off the spectral components around double carrier frequency), on the other hand they serve as the filters matched to the transmit filters. One can prove that in the case of transmission over a nondistorting channel with an additive white Gaussian noise, the impulse response of the receive filter should be a mirrored response of the transmit filter. In the case of the filters with the characteristics given by formula (1.34) for which the impulse response is symmetric, the receive filter has an identical impulse response.

[9] 8Φ-PSK with appropriately spectral shaping has been proposed for EDGE GSM radio interface.

The carrier and phase recovery block shown in Figure 1.32 uses the difference between symbol decisions generated by a decision device in the in-phase and quadrature branches and the signals fed to the decision device input. This error is a measure of the phase difference between the received and reference signals and the rate of change of its value is a function of the frequency difference between both signals. This type of a carrier recovery feedback circuit is one of the possible solutions. There are other carrier recovery circuits, among them those based on the nonlinear processing of the received signal, resulting in the artificial generation of the spectral line equal to the carrier frequency.

The difference between decision device input and output signals is often used for adjusting the timing recovery circuit as well.

Distortions introduced by a typical transmission channel, such as a multipath, flat and selective fading, intersymbol interference and phase jitter (see Chapter 3), imply the application of the increased complexity synchronous receiver or even make such a receiver non-realizable. The latter occurs in particular when the channel properties change so quickly that the carrier recovery circuit is not able to track them and generates the estimates of the carrier frequency and phase of insufficient quality. In this case the solution is the application of the appropriately selected modulation with its non-coherent reception. We will illustrate this kind of reception with three examples: a non-coherent reception of a differential quarternary phase shift keying (DQPSK) and two methods of reception of FSK modulated signals.

First, let us consider a DQPSK non-coherent receiver. The DQPSK modulated signal can be described by formula (1.25), where the in-phase and quadrature baseband components are given by

$$x^I(t) = \sum_{n=-\infty}^{\infty} d_n^I p(t - nT) \quad x^Q(t) = \sum_{n=-\infty}^{\infty} d_n^Q p(t - nT) \tag{1.35}$$

This time a two-bit information block is not represented by the data symbols d_n^I and d_n^Q nor by their argument $\varphi_n = \arg(d_n^I + jd_n^Q)$ but by the phase difference between two subsequent modulation periods $\Delta\varphi_n = \varphi_n - \varphi_{n-1}$. The scheme of this receiver is shown in Figure 1.33.

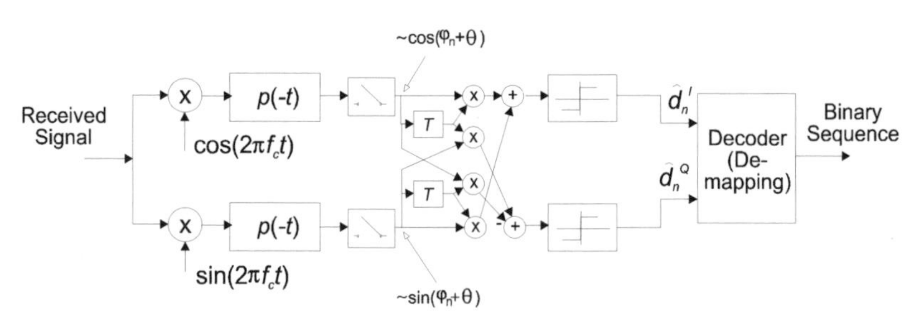

Figure 1.33 Non-coherent receiver for DQPSK signals

The samples from the outputs of in-phase and quadrature branches are taken once per data symbol and are proportional to the cosine and sine of the angle $\varphi_n + \theta$, respectively, where θ characterizes the difference between the carrier phase of the received signal and the phase of the reference carrier signal used in the demodulator. The receiver does not track the carrier phase of the received signal. The sine and cosine of the angle $\varphi_{n-1}+\theta$ received in the previous signaling period are stored in the memory cells, so the receiver can use the samples from the current and previous modulation periods. Let us note that for the DQPSK modulation it is sufficient to know in which quadrant of the in-phase/quadrature plane the angle $\Delta\varphi_n$ can be found. In order to check it, it is sufficient to determine the sign of the functions $\sin\Delta\varphi_n$ and $\cos\Delta\varphi_n$. The latter can be found from the well-known trigonometric dependencies

$$\begin{aligned}\cos\Delta\varphi_n &= \cos\left((\varphi_n+\theta)-(\varphi_{n-1}+\theta)\right) = \\ &= \cos(\varphi_n+\theta)\cos(\varphi_{n-1}+\theta)+\sin(\varphi_n+\theta)\sin(\varphi_{n-1}+\theta)\end{aligned} \tag{1.36}$$

$$\begin{aligned}\sin\Delta\varphi_n &= \sin\left((\varphi_n+\theta)-(\varphi_{n-1}+\theta)\right) = \\ &= \sin(\varphi_n+\theta)\cos(\varphi_{n-1}+\theta)-\cos(\varphi_n+\theta)\sin(\varphi_{n-1}+\theta)\end{aligned} \tag{1.37}$$

Let us note that the set of multipliers and adders in Figure 1.33 implements formulae (1.36) and (1.37). The receiver works well at the assumption that the angle θ hardly changes within the signaling period. That means the phase jitter is slow as compared with the signaling period.

Let us now illustrate the non-coherent reception of the frequency modulated signals. The first type of receiver realizing this kind of reception is shown in Figure 1.34. It is called a *non-coherent optimal* FSK *receiver.*

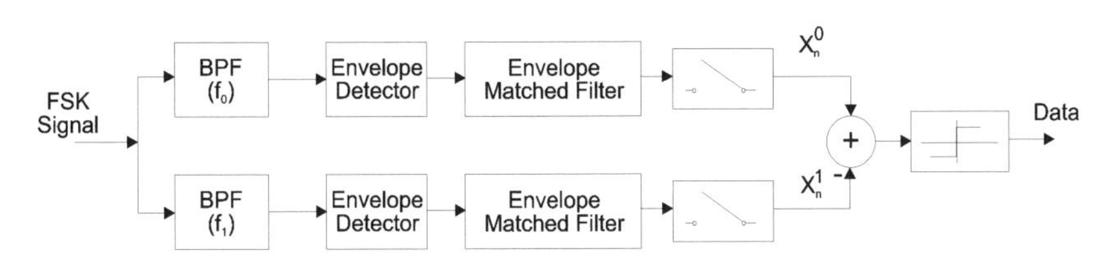

Figure 1.34 Non-coherent optimal FSK receiver

The received FSK signal is the subject of bandpass filtration in two filters whose center frequencies are equal to the nominal frequencies $f_c \pm \Delta f$ characterizing the logical data symbols equal "0" and "1". Thus, at the output of one of the bandpass filters a sinusoidal signal is received, whereas at the output of the second one only noise is observed. The envelope detectors calculate envelopes in both parallel branches. In order to maximize the signal to noise ratio, the outputs of the envelope detectors are fed to the matched filters working in the same way as in the synchronous receiver. The outputs of the matched filters are sampled once per modulation period and compared with each other. The larger sample indicates the more probable data signal transmitted.

Another, even simpler way of non-coherent detection of FSK signals is reception on the basis of a frequency discriminator. The FSK signal is treated as a regular

FM modulated signal for which the modulating signal is a stream of binary pulses representing a logical data sequence. Therefore the frequency discriminator converts the instantaneous frequency of the received signal into a signal of the amplitude proportional to this frequency. The next part of the receiver is realized in the baseband. A typical circuit realizing frequency discrimination often applied in mobile communications is a phase-locked loop (PLL). Figure 1.35 presents the block scheme of this receiver.

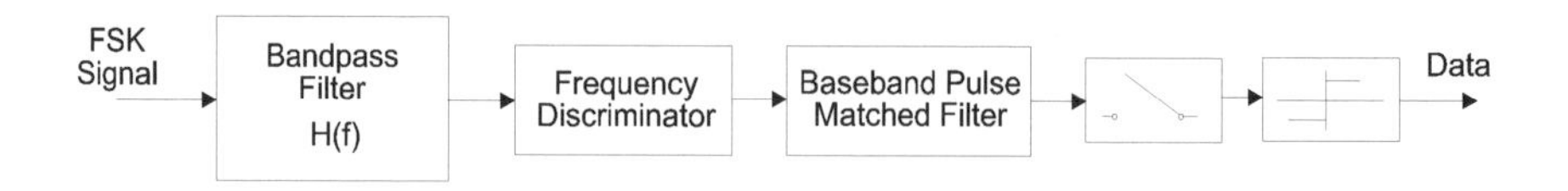

Figure 1.35 FSK receiver based on the frequency discriminator

An important task in the design of the FSK receiver with a frequency discriminator is choosing the bandpass filter which extracts the desired FSK signal. For a given modulation index h, determining the difference between two nominal FSK frequencies equal to $2\Delta f$ with respect to the modulation period T, the optimum characteristics of the bandpass filter can be found. It ensures the minimum error rate at the output of the non-coherent receiver.

Summarizing the signal reception methods, we have to stress that the detection quality of several types of receivers can be quite different. For a given signal-to-noise ratio, the synchronous reception using the whole knowledge of the received signal results in the lowest probability of error, the non-coherent reception with optimal envelope detection has worse performance, whereas the non-coherent reception with a frequency discriminator is characterized by the highest bit error rate. Nevertheless, the latter is often used in traditional systems due to its simplicity.

1.5.1 Multicarrier modulation

So far we have considered digital modulations of a single carrier. If the data symbol rate is comparable with the channel bandwidth, or if the channel time spread (see Chapter 3) is a considerable part of the modulation period, the intersymbol interference (ISI) arises which makes the reception much more complicated and decreases the system performance. The ISI countermeasures include adaptive channel equalization and sequential data detection. In the latter method the real or estimated channel impulse response is needed. The alternative to a single carrier modulation with a complicated receiving procedure is the *multicarrier modulation*.

Instead of transmitting a fast data stream on a single carrier serially, the stream is divided into a large number of much slower data streams. Each of them modulates a separate subcarrier. The data signaling on each subcarrier is so slow that the ISI lasts for a small portion of a data symbol only. Moreover, the subcarrier frequencies can be selected so densely that their spectra overlap. Despite this, the receiver is able to detect data symbols on each subcarrier by correlation of the multicarrier signal with the required reference tones. Below we will explain the operation of the transmitter and the receiver for the multicarrier modulation in detail. The reason for that is the increasing importance of that type of modulation, in particular in the wireless LANs.

For the n-th modulation period ($nT \leq t < (n+1)T$) the multicarrier (MC) signal is represented by the formula

$$x(t) = \sum_{k=0}^{N-1} [a_{k,n}p(t-nT)\cos 2\pi(f_c + k\Delta f)t - b_{k,n}p(t-nT)\sin 2\pi(f_c + k\Delta f)t] \tag{1.38}$$

As previously, $p(t)$ denotes the shape of the baseband data pulse, the pair $(a_{k,n}, b_{k,n})$ represents the data symbols modulating the in-phase and quadrature components of the k-th subcarrier and Δf is the subcarrier spacing. The set of the data symbol pair is determined by the type of modulation applied on each subcarrier. In actual systems using the MC modulation, the signal constellations range from BPSK to 64-QAM. Let us note that the designer is free to select the modulation individually on each subcarrier.

The choice of the subcarrier spacing strongly influences the operation of the system applying the MC modulation. As we have already mentioned, the parameters of the MC modulation are selected in such a way that the ISI caused by the channel lasts for a small fraction of the modulation period T. Let us divide T into two parts: the so-called guard time T_g and the orthogonality period T_{ort}, *i.e.* $T = T_g + T_{ort}$. The guard time covers at least that fraction of the modulation period in which the channel response to the transmitted data pulse has not yet reached a steady state. After that time the channel output becomes stable. This phenomenon is illustrated in Figure 1.36. The guard time is usually not more than 25 percent of the modulation period.

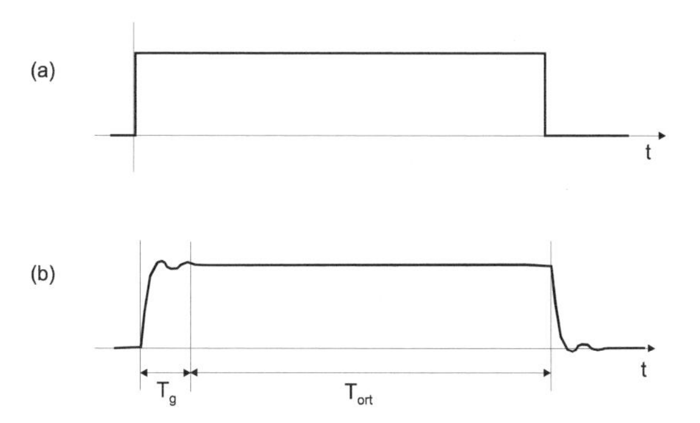

Figure 1.36 Baseband equivalent channel response (b) to a rectangular pulse (a) (illustration of the guard time and the orthogonality period)

If the subcarrier spacing Δf is selected to be equal to $1/T_{ort}$ then in the time period T_{ort} all the subcarriers are mutually orthogonal. Very often the MC system with the subcarrier spacing selected according to this principle is called *Orthogonal Frequency Division Multiplexing* (OFDM) system. The mutual orthogonality of subcarriers results from the fact that for any j and k

$$\int_{T_g}^{T} \cos(2\pi(f_c + k\frac{1}{T_{ort}})t)\cos(2\pi(f_c + j\frac{1}{T_{ort}})t)\mathrm{d}t = \begin{cases} T_{ort}/2 & \text{for } k = j \\ 0 & \text{for } k \neq j \end{cases} \quad (1.39)$$

$$\int_{T_g}^{T} \sin(2\pi(f_c + k\frac{1}{T_{ort}})t)\sin(2\pi(f_c + j\frac{1}{T_{ort}})t)\mathrm{d}t = \begin{cases} T_{ort}/2 & \text{for } k = j \\ 0 & \text{for } k \neq j \end{cases} \quad (1.40)$$

and

$$\int_{T_g}^{T} \cos(2\pi(f_c + k\frac{1}{T_{ort}})t)\sin(2\pi(f_c + j\frac{1}{T_{ort}})t)\mathrm{d}t = 0 \quad \text{for any } k \text{ and } j \quad (1.41)$$

Without loss of generality let us consider the operation of the receiver in the first modulation period ($n = 1$). The determination of the data symbols $(a_{k,1}, b_{k,1})$ on the k-th subcarrier can be performed based on the formulae

$$\int_{T_g}^{T} x(t)p(t)\cos(2\pi(f_c + k\Delta f)t)\mathrm{d}t = a_{k,1}\frac{1}{2}\int_{T_g}^{T} p^2(t)\mathrm{d}t \quad (1.42)$$

$$\int_{T_g}^{T} x(t)p(t)\sin(2\pi(f_c + k\Delta f)t)\mathrm{d}t = b_{k,1}\frac{1}{2}\int_{T_g}^{T} p^2(t)\mathrm{d}t \quad (1.43)$$

The right–hand sides of (1.42) and (1.43) are explained by the fact that the pulse shape $p(t)$ is often rectangular or at least constant in the integration period so the results of the correlation of the MC-modulated signal with the reference tones are proportional to the data symbols $a_{k,1}$ and $b_{k,1}$.

The realization of the MC transmitter and receiver based on (1.38), (1.42) and (1.43) respectively can be difficult if the number of subcarriers is large. Fortunately, the generation of the discrete samples of the signal described by (1.38) and the correlation of the received signal samples with the reference tones can be efficiently realized using Fast Fourier Transformation (FFT). Let the number of signal samples collected in the orthogonality period T_{ort} be denoted by N. Thus, the samples $x_i = x(i\frac{T_{ort}}{N})$, ($i = 0, \ldots, N-1$) can be calculated from the formula

$$\begin{aligned} x_i &= \mathrm{Re}\left\{\sum_{k=0}^{N-1}(a_{k,1} + jb_{k,1})\exp\left(j2\pi(f_c + \frac{k}{T_{ort}})i\frac{T_{ort}}{N}\right)\right\} = \\ &= \mathrm{Re}\left\{\exp\left(j2\pi f_c i\frac{T_{ort}}{N}\right)\sum_{k=0}^{N-1}(a_{k,1} + jb_{k,1})\exp\left(j2\pi\frac{ki}{N}\right)\right\} \end{aligned} \quad (1.44)$$

The factor $\exp\left(j2\pi f_c i \frac{T_{ort}}{N}\right)$ in (1.44) expresses the shifting of the MC signal to the carrier frequency f_c, whereas the second factor calculates the samples of the MC modulated signals in the baseband. Let us compare the latter with the well known formula for inverse discrete Fourier transform (IDFT)

$$x(i) = \sum_{k=0}^{N-1} X(k) \exp\left(j2\pi \frac{ki}{N}\right) \tag{1.45}$$

As we see, the samples of the MC modulated signal can be generated using the IDFT if we treat the data symbols $(a_{k,1} + jb_{k,1})$ modulating each subcarrier as spectral samples, i.e.

$$X(k) = (a_{k,1} + jb_{k,1}) \quad \text{for } k = 0, 1, \ldots, N-1 \tag{1.46}$$

An efficient way to implement the IDFT is to apply the Inverse Fast Fourier Transform (IFFT) algorithm. This is possible if the number of signal samples N is a power of 2, i.e., $N = 2^m$. Thus, even for a few hundred subcarriers the generation of the multicarrier modulated signal can be implemented in hardware or using a digital signal processor. Let us note that N is not only the number of samples in the time domain but it is also the number of spectral samples spaced by $1/T_{ort}$. Thus, the maximum number of subcarriers is equal to $N = 2^m$. However, mostly fewer than N subcarriers are applied. Some of them, in the spectral guard intervals on both edges of the signal spectrum, are left unused.

The above considerations deal with the calculation of the samples within the orthogonality period. The guard time is usually filled with the samples taken from the end of the orthogonality period. Such a set of samples is called a *cyclic prefix* and makes synchronization with the MC signal at the receiver much easier, in particular if the subcarriers arrive at the receiver with different delays.

Now let us consider the implementation of the receiver. Let the received signal be described by the expression

$$y(t) = x(t) * h(t) + n(t) \tag{1.47}$$

where $x(t)$ is the transmitted signal, $n(t)$ is the additive noise and $h(t)$ is the channel impulse response. We stress again that the channel impulse response is much shorter than the modulation period. The received signal has to be down-converted to the baseband. Let us denote the baseband signal as $w(t)$. Thus, in the remaining part of the receiver, the baseband signal is processed. The core of the baseband receiver is the set of the correlators operating according to (1.42) and (1.43) in the time period $[T_g, T]$. For $p(t)$ being constant within the orthogonality period and for digital realization of the correlators in which the samples $w(i) = w(i\frac{T_{ort}}{N})$ of the baseband signal $w(t)$ are processed, we have the formula

$$W(k) = \sum_{i=0}^{N-1} w(i) \exp\left(-j2\pi \frac{ik}{N}\right) \tag{1.48}$$

The samples $W(k)$ $(k = 0, \ldots, N-1)$ received at the output of each of N correlators are the samples of the channel output spectrum when the signal with the spectrum given by (1.45) is fed to their inputs. It is clearly seen from (1.48) that the samples $W(k)$ $(k = 0, ..., N-1)$ can be calculated from the time domain samples $w(i)$ $(i = 0, ..., N-1)$ using the discrete Fourier Transformation (DFT) implemented efficiently by FFT. Taking into account the baseband signals and the baseband equivalent channel the correlators outputs are described by the formula

$$W(k) = H(k)X(k) + N(k) \quad \text{for} \quad k = 0, \ldots, N-1 \tag{1.49}$$

where $N(k)$ is the noise sample on the output of the k-th correlator and $H(k)$ is the sample of the channel transfer function. Thanks to a long modulation period and the application of the cyclic prefix, the channel for each subcarrier can be interpreted as the channel with the gain $|H(k)|$ and shifting the phase of the k-th subcarrier by $\arg H(k)$. In order to make a decision on the transmitted data, each correlator's output has to be modified to compensate the gain and phase shift introduced by the transmission channel. This is done by multiplying the correlators' outputs by complex coefficients $C(k)$. The block performing this function is called an *equalizer*. Thus, the equalizer outputs are described by the expression

$$Z(k) = C(k)W(k) \quad \text{for} \quad k = 0, \ldots, N-1 \tag{1.50}$$

The decisions on the transmitted data can be made on the basis of samples $Z(k)$

$$\widehat{a}_{k,1} + j\widehat{b}_{k,1} = \text{dec}\,(Z(k)) \tag{1.51}$$

The scheme of the transmitter and receiver in the OFDM transmission system is presented in Figure 1.37. It summarizes our considerations of the OFDM transmitter and receiver. The OFDM transmission is highly flexible. As we have already mentioned, it is possible to select the type of modulation individually for each subcarrier and assign to each subcarrier an appropriate power level. Moreover, it is possible to switch off some deeply attenuated subcarriers. The necessary condition for the OFDM system optimization is the existence of the feedback between the OFDM receiver and OFDM transmitter. One can prove [5] that the optimum power allocation to each subcarrier ensuring the highest overall bit rate at the assumed error probability should take into account the channel characteristics and obey a "water pouring principle". This principle is illustrated in Figure 1.38.

The multicarrier modulation has gained a lot of attention in digital communication systems. It is the basis of high speed digital transmission in a subscriber loop in the *Asymmetric Digital Subscriber Line* (ADSL) mode [25], one of the alternatives in *Very High-Speed Digital Subscriber Lines* (VDSL) [26], the transmission method in European *Digital Video Broadcasting* (DVB) terrestrial segment [27], *Digital Audio Broadcasting* (DAB) [28]. It is also the basis of the radio access method in high speed wireless LANs of which HyperLAN/2 [24] is an example.

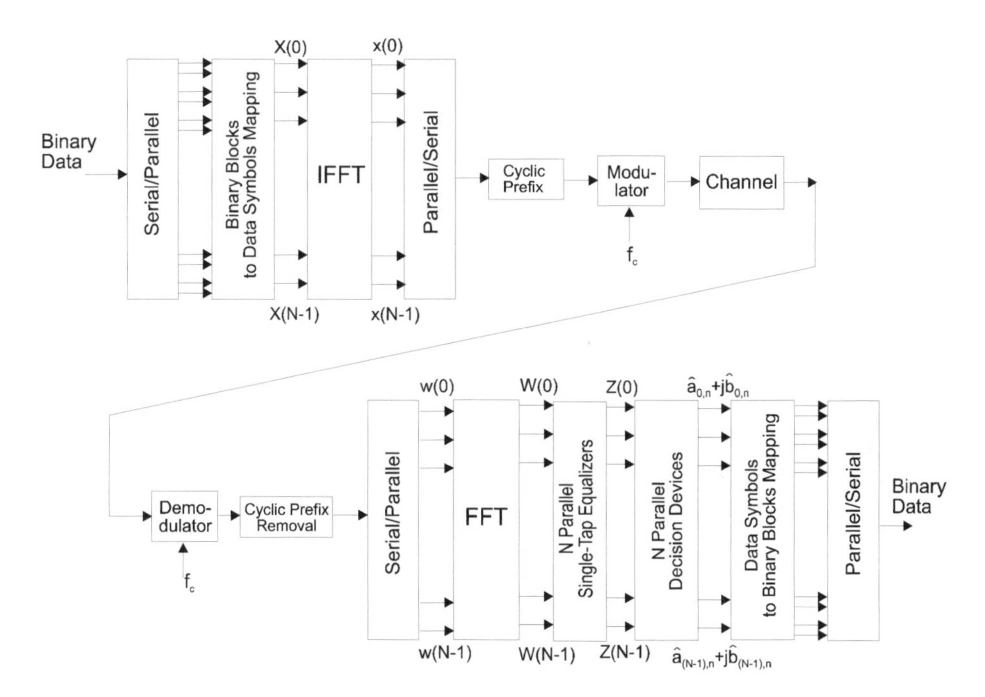

Figure 1.37 General scheme of the OFDM system

1.6 THE PRINCIPLES OF SPREAD SPECTRUM SYSTEMS OPERATION

Digital modulations described so far have been created to maximally utilize the limited bandwidth assigned to the given digital communication system. Claude Shannon, in his fundamental paper [13] giving the foundations of information theory, derived the formula for the capacity of the channel limited to W Hz in which the signal is distorted by the additive white Gaussian noise of the power density spectrum $N_0/2$. The formula has the following form

$$C = W \log_2 \left(1 + \frac{P_{av}}{W N_0} \right) \tag{1.52}$$

where P_{av} is the input signal average power. This formula is the result of the optimization procedure with respect to the input signal properties. It turns out that the amount of information which can be sent through the channel with additive white Gaussian noise achieves its upper bound called the *channel capacity* if the input signal is Gaussian. A multicarrier digitally modulated signal is Gaussian distributed if the number of subcarriers is large. Many other digitally modulated signals do not have this probability distribution. In traditional systems the achievable data rate is maximized and is getting closer to the channel capacity bound if the signal-to-noise ratio expressed by the ratio P_{av}/WN_0 in (1.52) is maximized. This can usually be achieved in a limited range using very sophisticated means such as trellis coding, equalization, etc.

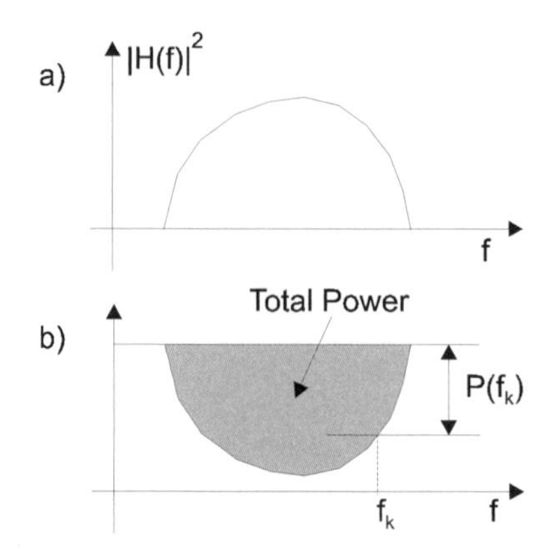

Figure 1.38 Water pouring principle: example of the channel characteristic (a), the signal power assignment along the frequency axis (b)

One can easily show that the same channel capacity can be reached by expanding the signal spectrum (provided expansion is possible from the point of view of spectrum management and implementation limitations) to such a width that the signal level falls even below the noise level. This observation has been utilized in spread spectrum systems.

Consider the most popular type of spread spectrum systems, denoted in the literature as DS-SS (*Direct Sequence Spread Spectrum*). In the DS-SS system the digital signal carrying information is spectrally spread by being directly multiplied by a pseudorandom sequence. Let T_b denote the duration of a data symbol. A binary sequence of length M is used to represent a single data symbol. Each element of the binary sequence, called a *chip*, lasts for $T_c = T_b/M$ second. The sequence is selected in such a way that the external observer sees it as a random sequence, so for him it has properties similar to the noise. Because the duration of a single chip is M times shorter than the duration of the information bit, the spectrum of the signal representing the data bits in form of the pseudorandom sequences is M times wider than the spectrum of the original data signal.

It is known from the communication systems theory that the optimal receiver of the signals distorted by a white Gaussian noise is a correlator. It correlates the distorted received signal with a known reference signal which is synchronized with respect to the received signal. In our case the reference signal is a pseudorandom signal used in the transmitter for representation of the data bit. Figure 1.39 shows the transmitter and receiver in a DS-SS system. We assume that the binary information signals shown in Figure 1.39 have bipolar representation, therefore controlling the polarization of the pseudorandom sequence by the information bits is equivalent to the multiplication of the pseudorandom sequence by -1 or $+1$.

Summarizing our consideration of the scheme shown in Figure 1.39, we conclude that the pseudorandom sequence is treated as an elementary signal which characterizes a single data bit, whereas the same sequence with the reversed polarity represents the

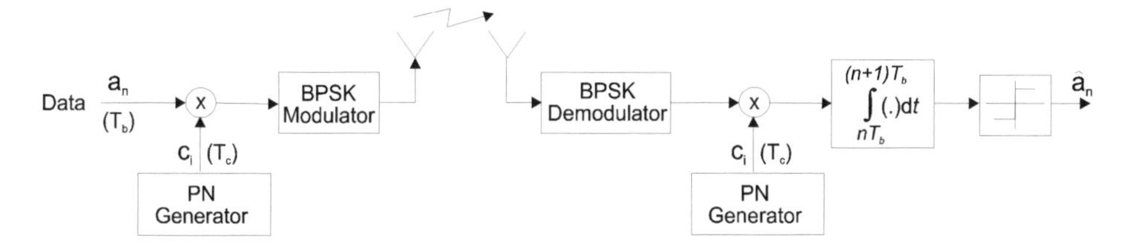

Figure 1.39 General scheme of the DS-SS system

data bit logically negated. The effect of spread spectrum is also achieved if the period of the pseudorandom sequence is longer than the duration of a single data bit.

The DS-SS system is an alternative to narrowband systems. One could wonder what it offers with respect to them. For many years the spread spectrum systems have had military applications mostly. Transmission of the pseudonoise signals close or below the noise level makes them difficult to detect. In order to make detection possible the receiver has to know the particular pseudorandom sequence applied in the transmitter and has to be synchronized with it. Pseudorandom sequences used in real systems have the period ranging from a few tens of bits to many thousand of bits. For the sequences having such long period their number becomes very large. The sequences are selected in such a way that their autocorrelation function is approximately equal to zero independent of the time shift between the sequence and its shifted replica, with the exception of the zero shift for which the autocorrelation function achieves its maximum. At the same time the mutual correlation function for different sequences of the same length should be equal to zero for any time shift between the correlated sequences. The zero autocorrelation function for the time shifts different from zero makes the system robust against the multipath. This robustness is possible if the chip duration is shorter than the smallest difference of the delays between different signal paths. The signal reaches the receiver in the form of a few replicas shifted in time. These time shift differences are usually larger than the chip duration. In consequence, the receiver is synchronized with the strongest component of the received signal. As a result of correlation all other replicas of the desired signal are eliminated. In the same way, thanks to the zero cross-correlation between any two different sequences, the other users' signals are eliminated as well. However, we have to stress that the elimination of all but the strongest component of the received signal is not the optimal strategy. It leads to losing information contained in the eliminated echoes. These echoes can be positively used after having been individually extracted and after their weighted summation has been performed in such a way that the energy of the signal sum is maximized. That operation is realized by a RAKE receiver, which will be explained in Chapter 11. The RAKE is the basic type of the receiver for multipath channels.

Directly from the cross-correlation properties of the applied pseudorandom sequences we can deduce the following feature of the spread spectrum systems. Because the correlators applied in the receiver eliminate all other sequences but their own one, the same spectrum can be shared by many users applying different pseudorandom sequences. This observation is the basis of the *Code Division Multiple Access* method.

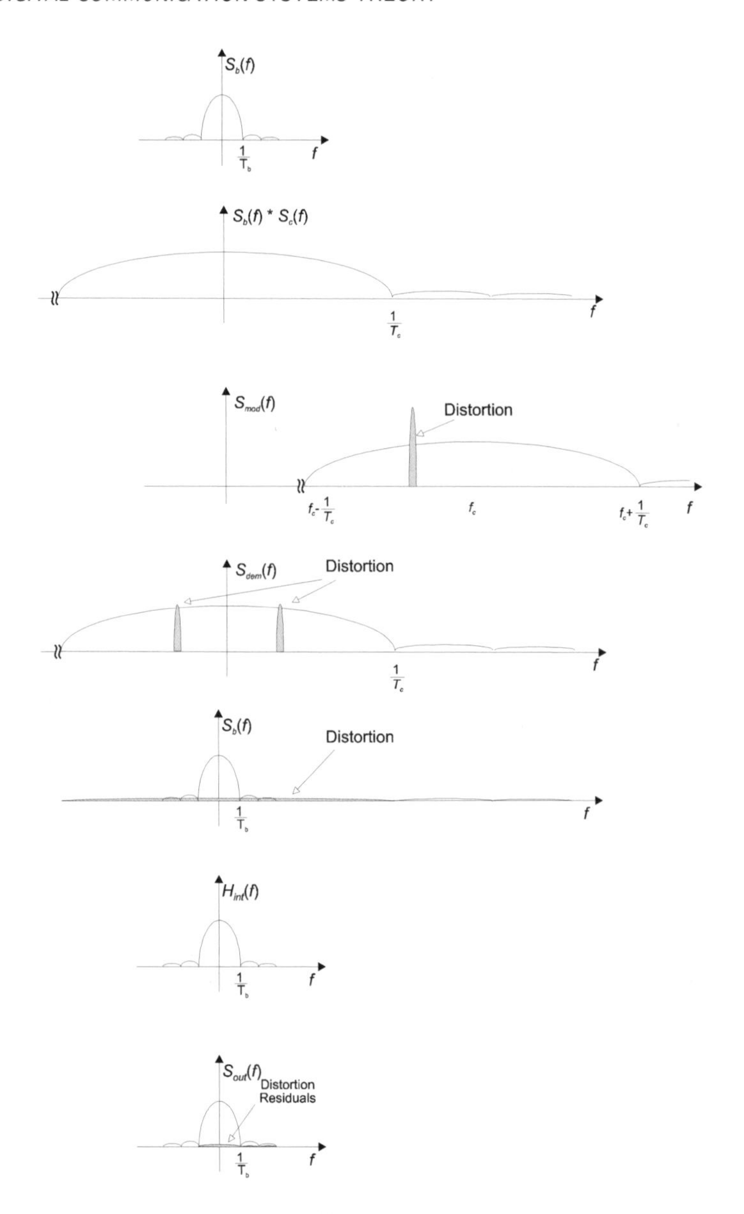

Figure 1.40 Spectra in particular locations of the DS-SS system in the presence of the narrowband distortion

Military system designers have been interested in spread spectrum systems because these systems show natural immunity to narrowband distortions. This feature is qualitatively illustrated in Figure 1.40. This figure also shows the operation of the DS-SS system in the frequency domain. The following notation has been introduced in it: $S_b(f)$ – spectrum of the information-carrying signal, $S_c(f)$ – spectrum of the spreading sequence, $S_{\rm mod}(f)$ – spectrum at the output of the BPSK modulator, $S_{\rm dem}(f)$ – spectrum at the output of BPSK demodulator, $H_{\rm int}(f)$ – the transfer function of the integrating circuit and $S_{\rm out}(f)$ – the output signal spectrum. As a result of correlation consisting of multiplication by a pseudorandom sequence, followed by integration, the spectrum of the narrowband distortion is spread. Recall that the spectrum of the pseudorandom signal is very wide and the multiplication of the distortion and the pseudorandom signal in the time domain is equivalent to the convolution of their spectra. Thus, only a small part of the distorting signal energy appears at the output of the integrator.

There are other versions of spread spectrum systems. However, the DS-SS system shown above is the one most frequently used in second and third generation cellular telephony, wireless subscriber loops and is also to be used in personal satellite communication systems. We will now briefly present two other system types.

As already mentioned, ensuring the synchronous reception, in particular the realization of timing recovery with the accuracy of a fraction of a chip, can be a very difficult task if the channel is time variant. In such case a possible solution for a spread spectrum system is the application of *frequency hopping*. The *Frequency Hopping Spread Spectrum* (FH-SS) system is shown in Figure 1.41.

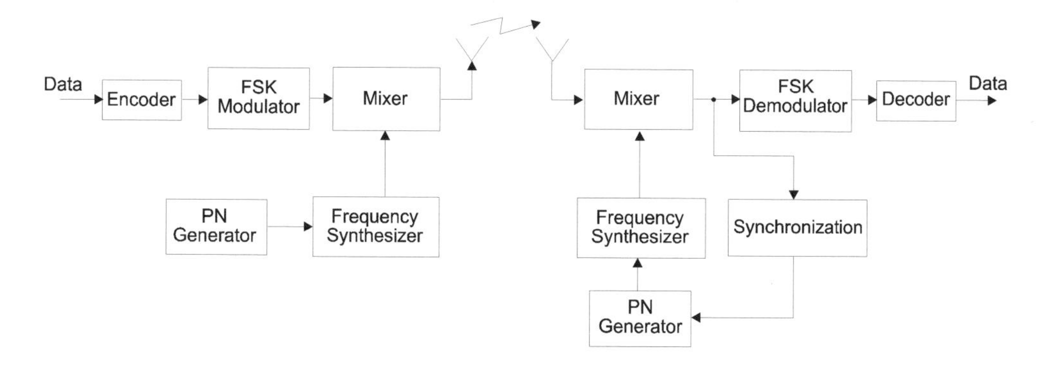

Figure 1.41 Example of the basic FH-SS system

The data bits, which are optionally FEC encoded, control the output of the FSK modulator. The FSK signal is shifted by the frequency interval determined by the pseudorandom generator controlling the frequency synthesizer. If the synthesizer is able to generate $2^m - 1$ different frequencies, then m subsequent bits of the pseudorandom generator determine the output frequency. Due to a large frequency range of generated signals it is very difficult to ensure phase synchronization between carriers selected in the subsequent frequency hops. Therefore a non-coherent FSK demodulator is used in the receiver. We have to stress that frequency hops occur many times within a single information bit period. The FSK modulation period T_b is divided into many short time intervals called *hop times* T_h. In this case the system is called *fast frequency hopping*. Figure 1.42 illustrates the operation of such a system for $T_b = 8T_h$. The change of

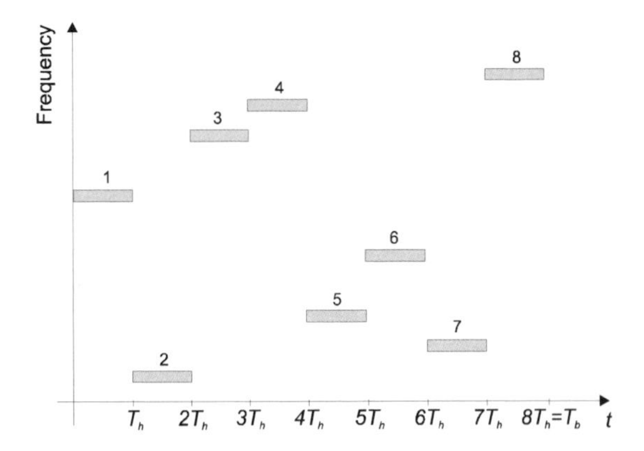

Figure 1.42 Example of the frequency hopping in the FH-SS system

carrier frequency can also take place once for a data block and such a system of carrier frequency changes is called *slow frequency hopping.*[10]

The FH-SS systems are often applied in military systems. They are robust against intentional jamming. The robustness is the result of a short occupancy time of a single carrier frequency, which does not allow the jammer to tune to this frequency. Slow frequency hopping performed jointly with FEC coding allows many users to share the common spectrum. It can also be a tool for increasing cellular system capacity.

The third type of the spread spectrum systems is called a *Time Hopping Spread Spectrum* (TH-SS) system. In this system the data bit period is divided into M_T time slots. The pseudorandom generator determines in which time slot the information signal is being transmitted. Figure 1.43 presents the TH-SS system. Its characteristic feature

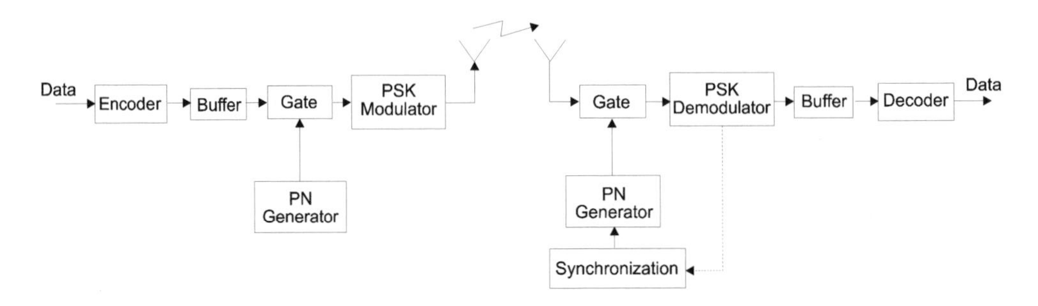

Figure 1.43 The basic scheme of TH-SS system

[10]In the GSM cellular system the term slow frequency hopping denotes the pseudorandom change of channel frequency which occurs once in a whole data slot containing a 148-bit burst.

is its burst nature – the signal is transmitted during $1/M_T$ fraction of the data signalling period. A natural choice of M_T is the number of the order of a thousand. However, such a number of slots creates serious synchronization problems which are much more difficult to cope with than those in the DS-SS systems. In order to ensure smooth transmission of information over the TH-SS system the transmitter and receiver have to be equipped with buffers.

Hybrid solutions of spread spectrum systems combining the three above mentioned system types are also considered in the literature. However, the DS-SS and FH-SS systems have the biggest practical meaning.

1.7 MULTIPLE ACCESS METHODS USED IN MOBILE COMMUNICATIONS

Multiple access is one of the most important issues in communications if more than two users share a common resource such as a common band of the electromagnetic spectrum, or a common wireline (metallic or optical fiber) channel. We briefly addressed this problem when describing the model of a typical digital communication system. Now we will consider it in more detail. We will concentrate on the peculiarity of the multiple access for wireless communication systems.

In wireless communications the common resource is the band assigned to the system by an administrative body such as FCC (*Federal Communications Commission*) in the USA or CEPT (*Conférence Européene des Postes et Télécommunications*) in the European Union. National administrative institutions also have a decisive influence on the spectrum assignment.

The oldest multiple access method is *Frequency Division Multiple Access* (FDMA). In this method the total band assigned to the system is divided into a certain number of frequency intervals which can be used for individual transmission between two users. They can also be used in the broadcasting mode. The frequency intervals are usually narrow enough to maximize the number of created channels and, consequently, the number of simultaneous users. On the other hand, they have to be broad enough to ensure the required transmission performance. If transmission is analog, the FDMA is the only possible multiple access method because it guarantees continuous access to the transmission medium required by analog signals. Figure 1.44 illustrates this method. The characteristic feature of the FDMA is the existence of the guard bands between neighboring channels which decreases the number of possible channels and the spectral efficiency of the system. Transmitters and receivers have to be equipped with high quality channel filters. On the transmitter side, they filter the transmitted signal to fit its spectrum into the channel bandwidth, whereas in the receiver they extract the signal received in the desired channel. A problem arises if the powers of two neighboring channels differ too much from each other. The attenuated sidelobes of the strong channel can have a similar level as the mainlobe of the weak channel, causing serious deterioration of the performance of the latter. Therefore, a certain level of power control of each transmitted signal is required to avoid this phenomenon.

Time Division Multiple Access (TDMA) is another multiple access method (Figure 1.45). Instead of access in a fraction of the band assigned to the system, users are allowed

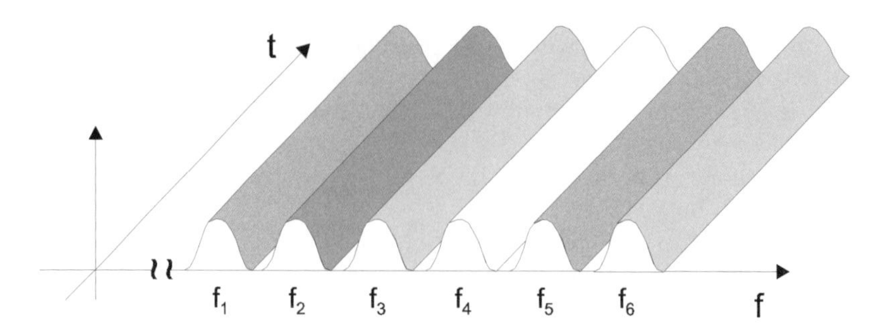

Figure 1.44 Illustration of the FDMA multiple access method

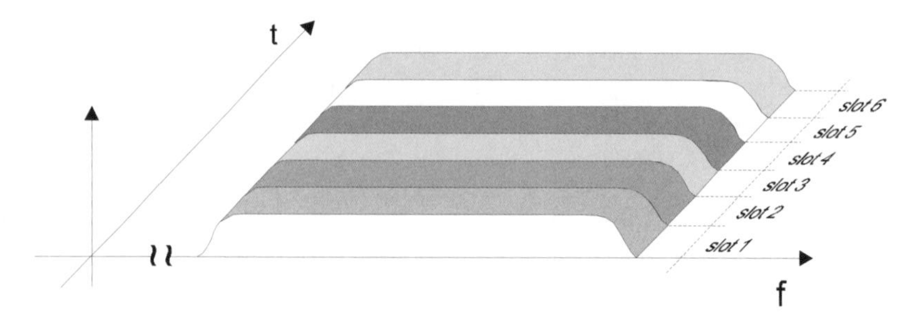

Figure 1.45 Illustration of TDMA method

to transmit their signals in the whole band but only in a fraction of time assigned to them periodically. The basic time unit is called a *frame*. The frame is divided into a number of *slots*. The maximum number of simultaneous users is equal to the number of slots in the frame. On average, however, the number of users is slightly lower because some slots are used for control, synchronization and maintenance purposes. For that reason the frames are often organized in higher order structures such as multiframes, superframes etc. The characteristic feature of the TDMA is the necessity to compress user's data stream into short blocks which fit into the assigned slots. If there are M slots in the frame, then the data rate within a slot must be M times higher than the rate of a single user. The consequence of that fact is the widening of the signal spectrum M times as compared with a continuous data stream. This is why the spectra in Figure 1.45 are much wider than those for FDMA shown in Figure 1.44. Another consequence of applying the TDMA is that the transmitted signals must be digital. Otherwise the time compression would not be possible. In the same way as for the FDMA, for which the frequency guard range between the neighboring channels is applied, the TDMA applied in the radio systems requires time guard periods between the data blocks occupying the neighboring slots. The reason for that is a finite time of switching the amplifiers on and off and possible propagation time differences between different users communicating with a common radio station. The guard time, in the same way as the guard band in the FDMA, decreases the spectral efficiency of the TDMA multiple access mode.

In practice, the hybrid TDMA/FDMA mode is very often applied. Thus, a number of frequency channels occupies the system spectrum and the time axis of each channel is divided into time slots. Such an approach has been applied in the GSM, IS-54/136 and in Japanese PDC cellular systems.

The third type of multiple access is *Code Division Multiple Access* (CDMA). We mentioned its existence when explaining the principles of spread spectrum systems. We noted that if the spreading codes are selected so that their cross-correlation is equal to zero, the signal of the desired user can be retrieved from the mixture of signals generated by many other users by correlating the received signal with the selected reference signal. Thus, all the users occupy the same bandwidth and transmit continuously their binary data in the form of pseudorandom sequences modulated by their information data. Figure 1.46 illustrates this type of multiple access. The CDMA allows extraction of

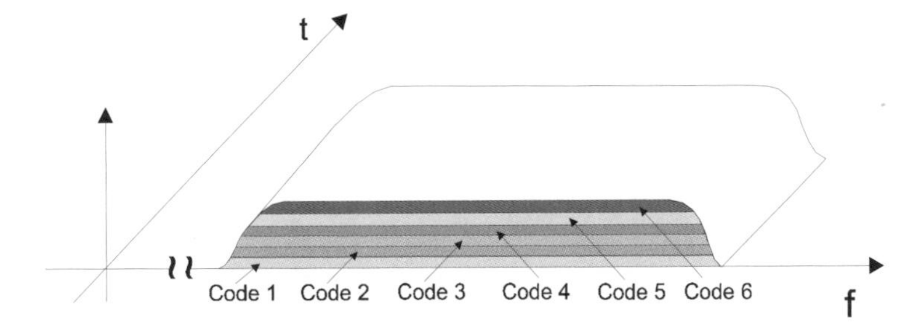

Figure 1.46 Illustration of the idea of CDMA

the user data ideally as long as the used codes are perfectly orthogonal i.e. their cross-correlation is zero. Unfortunately mobile channels mostly suffer from multipath (see Chapter 3) which causes the loss of the signal orthogonality at the receiver input. It is very difficult to find a large number of spreading sequences which are mutually orthogonal and at the same time orthogonal to their time shifts. As a result, the orthogonality of user spreading sequences is not perfect which has negative influence on the signal-to-noise ratio at the receiver and the receiver's error rate.

The CDMA can be used in a hybrid mode with other multiple access methods such as FDMA or TDMA. In the case of CDMA/FDMA the total spectrum assigned to the system is divided into a number of frequency bands and in each of them the CDMA is applied. In the hybrid TDMA/CDMA, spreading and CDMA take place in the assigned time slots leaving the remaining time slots to the other users.

In recent years another multiple access method supporting the other already described methods has been introduced. It is the *Space Division Multiple Access* (SDMA). SDMA relies on the application of antenna arrays which are able to generate highly directional, electronically steerable beams. Thus, if the users are sufficiently separated in angle, they are able to use the same frequency channels, time slots, codes or combinations thereof, depending on the main multiple access method used in the system. The application of SDMA has a serious influence on the overall system performance and will be discussed in Chapter 18.

1.8 METHODS OF DUPLEX TRANSMISSION

Unidirectional transmission is characteristic for broadcasting systems. Apart from these systems, it rarely appears in digital and mobile communications, as the information flow mostly takes place in two directions. A telephone call is its simplest example. Similarly, data transmission often requires a feedback channel (see the Section 1.4.5 on ARQ techniques) or it is simply bidirectional or *duplex*. In consequence, the problem of organization of two-directional information exchange has to be solved. This task is common for many communication systems so we will discuss it briefly.

In the first method of duplex transmission the total spectrum assigned to the system is divided between two opposite directions. Such an approach is called a *Frequency Division Duplex* (FDD). The frequency bands assigned to both directions have to be disjoint and sometimes other systems can use the spectrum between them. In many applications both bands have the same width. FDD is advantageous if independent transmission in both directions is desired.

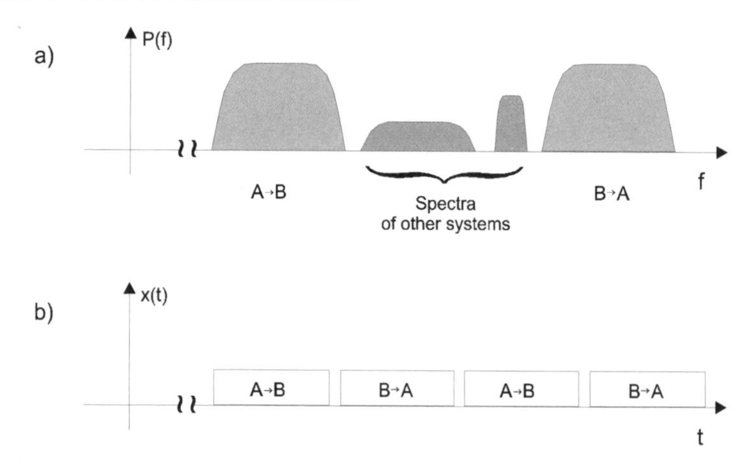

Figure 1.47 Examples of FDD and TDD duplex modes

Time Division Duplex (TDD) is another duplex transmission method. In this case the whole bandwidth is used in both directions. However, the time axis is divided between them. Similar to TDMA, the frame structure has to be defined. TDD transmission is useful if the propagation time is short with respect to the applied data blocks. Guard times for reversing the direction of the transmission have to be foreseen. In more advanced systems the time assigned to both transmission directions can be adjusted to the realized service. In this sense in some applications the TDD mode is more flexible and can result in a higher system throughput. Figure 1.47 illustrates both types of duplex transmission.

Frequency or time division duplex transmission is applied with one of the above described multiple access methods. In practice we find systems using TDMA/FDMA with FDD (GSM), TDMA/FDMA with TDD (DECT), CDMA with FDD (IS-95, UMTS WCDMA FDD), TDMA/CDMA with TDD (UMTS WCDMA TDD). The examples of these combinations will appear in the chapters describing the most important mobile communication systems.

1.9 COMPETING FOR CHANNEL ACCESS

Multiple access and duplex transmission methods are not sufficient to solve the problems occurring while using a common channel. Usually information exchange between two users lasts for a determined time period. During this time some system resources, particularly a channel meant as a frequency channel, a sequence of time slots and/or a code sequence are used by the pair of users. Before and after this period the channel can be used by other users or it can remain idle. If the total number of users is large, the situation is dynamic. It can happen that several users want to set a connection and use a particular channel. As a result, different users may compete for the channel and collisions can occur. In order to cope with them several arrangements are introduced. They will be briefly described below. From the communication system point of view the attempts at channel reservation by particular users look like a random process. The problem of the *random access* is particularly important if messages sent by users are in the form of short packets and setting a fixed connection would be a waste of network resources.

The simplest although the least effective arrangement of random access is called *pure* ALOHA protocol. It was developed at the University of Hawaii and put into operation in 1971. According to this protocol, users transmit packets encoded with an error detection code whenever they have something to send. Collisions among packets occur if at least two of them overlap in time. The reception of each packet has to be acknowledged. If the ACK packet does not reach the sender in the determined time, the user considers the packet lost in a collision and he/she sends it again with a randomly selected delay. In this way repeated collisions can be at least partially avoided.

Introducing some order in packet emission improves the properties of the ALOHA protocol. The simplest modification of the pure ALOHA is known as the *slotted* ALOHA protocol. It relies on the division of time into time slots equal to a packet transmission time. A user has to be synchronized to the time slots and is allowed to send his/her packets within the slot. Thus, the packets of different users cannot partially overlap, as happens in the pure ALOHA protocol; however, they can overlap, if two terminals tend to use the same time slot. Slotted ALOHA is often applied in wireless communications. It is used in the GSM system in the process of setting a connection and in some other procedures.

The users applying ALOHA protocols and wishing to send a packet do not take into account what other users are doing. If they first "listened" to the channel and then sent a packet, the number of packet collisions would decrease. The protocols which apply this rule are called *Carrier Sense Multiple Access* (CSMA) protocols. Selected protocols of this type are used in wireless packet data networks and wireless LANs.

The simplest protocol belonging to the CSMA group is called 1-*persistent* CSMA. The time remains unslotted. The terminal wishing to send a packet "listens" to the channel and sends it as soon as the channel is free. Subsequently, it waits for the packet acknowledgement. If ACK is not received in a specified time period, the terminal starts to "listen" to the channel after a random time period and sends the packet again as soon as the channel is available. One of the drawbacks of this protocol is the influence of the propagation delay on the overall protocol performance. If two terminals attempting to transmit a packet are separated in space, packet collisions can occur. The more distant terminal considers the channel available whereas the channel has already been occupied by another terminal; however, the detection of this fact is impossible due to the propagation delay.

Another version of the CSMA protocol is known as *nonpersistent* CSMA. In this protocol, if the terminal detects that the channel is busy, it waits for a random time interval before sensing the channel again. If the channel is available, the packet is sent immediately. In this way, the collisions occurring in the 1-persistent CSMA and resulting from possible start of packet transmission by more than one terminal are eliminated. On the other hand, if the number of users is low, the 1-persistent CSMA protocol is more efficient because the random waiting time before next channel sensing is a waste of time.

The *p-persistent* CSMA is an improvement of the 1-persistent protocol. Here time is divided into slots. Typically, the duration of the slot is equal to maximum propagation delay. The terminal which has a packet to send "listens" to the channel. If the channel is free, it transmits the packet with probability p. Thus, with probability $\overline{p} = 1 - p$ the current slot is not used for packet transmission. The action is repeated in the next slot and possibly in the next ones till the packet is finally sent. If the channel is busy, the terminals check its availability continuously. If the channel becomes free again, the whole procedure starts from the beginning.

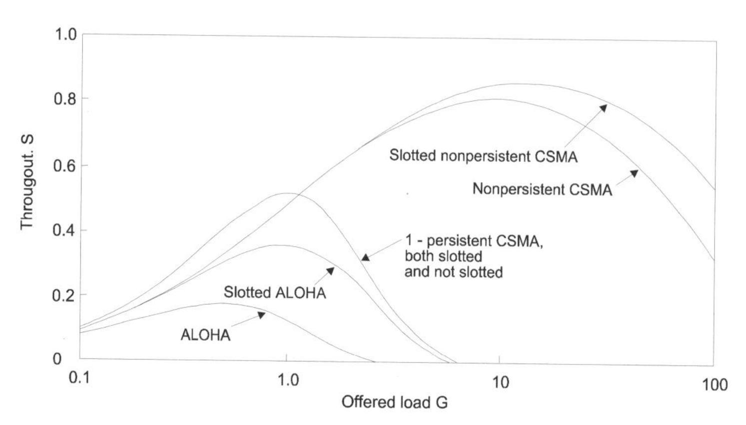

Figure 1.48 Comparison of throughput versus offered load for ALOHA and CSMA protocols

We conclude this short overview with the comparison of various random access protocols shown in Figure 1.48 [28]. The curves in Figure 1.48 were derived at the assumption that the propagation delay is one-tenth of the packet transmission time T_p. Figure 1.48 presents the throughput S versus the offered load G. The throughput S is defined as the average number of successful packet transmissions per packet transmission time T_p.

The offered load G is the number of packet transmissions attempted per packet time T_p, including both new packets and retransmitted ones.

As we see in Figure 1.48, the CSMA protocols substantially outperform the ALOHA protocols. Despite that, the slotted ALOHA protocol is frequently applied. The probable reason for that is its insensitivity to the values of the propagation delay as compared with the CSMA protocols.

1.10 THE OSI MODEL

So far we have concentrated on the transmission and reception methods of analog and digital signals. We have considered the basics of channel coding and associated information exchange protocols, as well as multiple access and duplexing methods. The organization of the digital communication systems has become so complex and the equipment and services markets have been so rich that setting standards on different levels of digital systems organization has become a necessity. In 1977 the *International Standardization Organization* (ISO) derived the *Open System Interconnection* (OSI) reference model which was aimed at setting in order the description of the information transmission procedures and simplifying the network design.

The description of the information transmission networks is organized in the form of a set of layers. A layer relates to the process or device within the system which performs a particular function [15]. The designers of a particular layer have to know the details of the system operation in their own layer. For system users and the designers of higher layers the process or device is a "black box" with the specified input and output signals and with interdependencies among them. In such a way a complicated architecture and system operation is decomposed into a set of smaller and better defined structures and operations.

In the OSI reference model particular layers communicate with each other through strictly defined interfaces. Each layer performs a number of basic functions using a set of functions defined in the OSI reference model in the layer located directly below it. Each layer offers a number of services to the layer located one level higher. The upper layer does not know the details of implementation of the services offered by the lower layer. In order to make the services possible, the interfaces between the layers have to be defined.

Figure 1.49 presents the OSI reference model, using as an example the interconnection of two systems. Each layer of the first system communicates with the same layer of the second system using a set of rules called *layer protocols*. On the level of each layer the respective processes cooperate with each other. Let us note that physical connection between both systems exists only on the physical layer level, whereas the communication between the layers of both systems has a virtual character. In reality it is realized through the subsequent layers located below the given layer in both systems.

We will shortly describe the meaning and functions of the layers in the OSI model.

The task of the *Physical Layer* is the realization of the physical connection ensuring the data bits transmission over the channel linking the terminal with the network node or connecting two network nodes. The physical layer is equipped with the physical

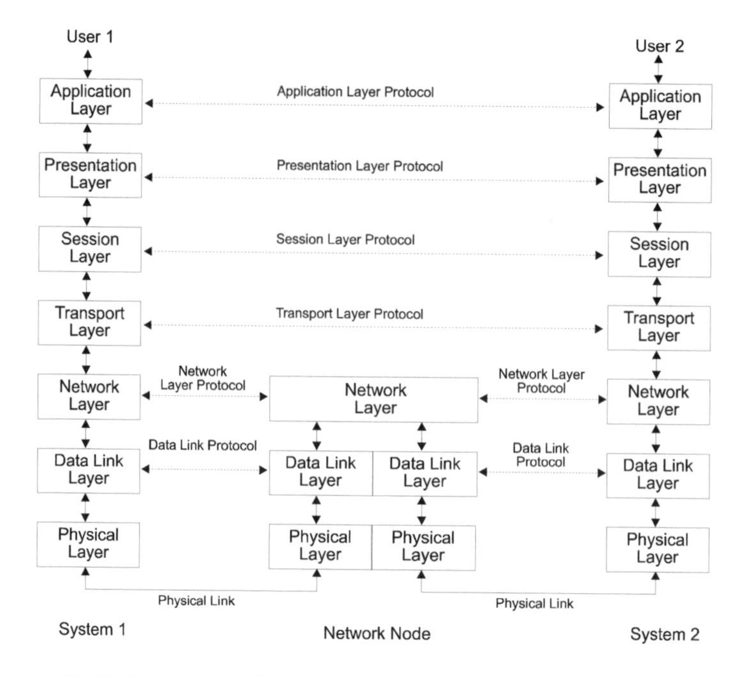

Figure 1.49 Reference model of the open systems interconnection (OSI model)

interface modules placed on each end of the physical link. On the transmit end, the task of the interface is to adapt the binary stream received from the *Data Link Control Layer* to the form in which it can be transmitted over the physical channel. On the receive end the interface module performs a function which is complementary to that on the transmit end. The typical functions of the physical interface are the modem functions. Problems associated with digital modulations, transmission over particular channels and reception of digital signals must be solved at this level.

The main task of the *Data Link Control Layer* is protection of the transmitted data stream against errors occurring during transmission between the network nodes. As a result, in the interface between the data link control layer and the *Network Layer* a binary stream of constant quality is received. The problems of error detection and correction, framing, as well as the ARQ methods are dealt with in the data link control layer.

The third layer of the OSI model is the *Network Layer*. The main tasks of this layer are the routing of the data packets and their flow control. Within the first task the network layer process operating in a network node decides to which link direct packets arriving to this node. In the destination node the packets are processed for transfer to the next layer – the *Transport Layer*. In turn, the task of the packet flow control is to prevent blocking and overloading of the buffers accumulating the data packets arriving nonrhythmically from the network. In order to avoid these events, the network layer process determines when the packets from the higher layer should be accepted and when to transmit the packet to another network or destination node. The network layer is believed to be one of the most complicated layers [15]. It has a key meaning for reliable functioning of the whole digital network.

The next remaining layers of the OSI model are so called higher level layers [17]. The protocols associated with these layers deal with the interconnections between systems and they are not directly associated with the functions of a particular communication system. Therefore their meaning for the description of the mobile communication systems is lower. However, for the completeness of the OSI model presentation we will shortly describe the higher layers as well.

The *Transport Layer* supports the reliable mechanism of data exchange between processes run in different systems. The transport layer ensures that the received data units do not contain errors, that they are ordered in the appropriate sequence, they are complete and do not contain duplicated elements. The transport layer is also associated with the optimization of the network services and ensuring their quality required on the level of the session layer by the network terminals.

The *Session Layer* assures the means for controlling the dialog between the applications. This layer assigns the resources to two application processes in order to establish a connection between them, called a *session*. After establishing a session between two users the processes of this layer manage the dialog between them.

The *Presentation Layer* establishes a common format to be applied between two terminals, using common rules for data representation. This layer process transforms input data to the format which allows selection of a set of services by the application layer. One of the transformations which can be realized by the presentation layer is data encryption.

The highest layer, the *Application Layer*, ensures that the end user has access to the OSI environment.

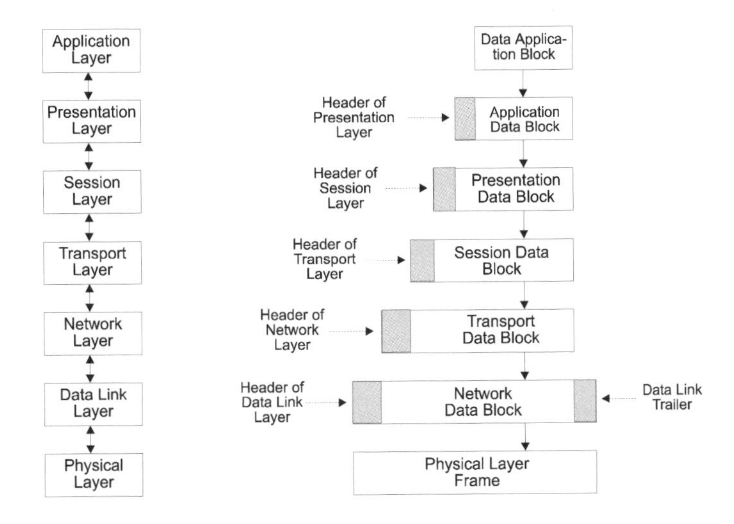

Figure 1.50 Packet structures of the particular layers in the OSI description model

Figure 1.50 illustrates the operation of the digital communication system from the perspective of the OSI reference model. Information exchange between two users is initiated in the highest layer. This layer receives a data block from the user, denoted as the application block. This block is transferred down the subsequent layers. On each layer level a header is added which contains the control information necessary to realize the protocol of the given layer. The whole block, supplemented by the header, is treated by the lower layer as a data information block. In the data link control layer usually not only the header but also a trailer is added. The task of the trailer is error detection and correction. For example, the trailer is a CRC binary block which we have described in the subsection on channel coding.

On the receiver side the dual processes take place. The blocks of a particular layer are unpacked; headers and trailers are removed and the received information is transferred to the higher layer.

1.11 X.25 – A PROTOCOL FOR A PACKET SWITCHING NETWORK

In general, networks can be divided into *channel switching* and *packet switching* networks. In case of a channel switching network, a physical connection is established for a period of time in which the information exchange between two users takes place. In case of radiocommunications a physical connection is the channel frequency (in FDMA mode), the channel frequency and the assigned time slot (in TDMA/FDMA mode) or the spreading sequence (in CDMA mode). A representative example of a channel switching network is a network based on the GSM standard.

In the channel switching networks some information, in particular control messages, has the form of packets. On the other hand there are networks of entirely packet type. During the information exchange between two network users there is no physical connection between them. Instead, the users send to the network the numbered data packets containing the destination addresses. Numbering assures the appropriate order of the packets on the receiver side. The packets are sent to the closest network node and from this node they are directed through a dynamically established route to the destination. In the case of the packet type of traffic, the network resources can be used much more efficiently in packet switching networks (assuming that there is no network overloading) than in channel switching networks. The network resources of a packet switching network are only used if there is a packet to transmit, while in channel switching networks the whole chain of devices on the path between the transmitter and receiver is reserved whether the transmitter has something to send or not.

One of the most frequently applied protocols in packet switching networks is the X.25 protocol. Its name is set by the number of ITU-T Recommendation describing it. This protocol is also applied in the data transmission offered by the GSM. It will be the subject of our consideration in Chapter 11. A short description of the X.25 is presented according to [15].

The X.25 protocol has a level structure which matches the three lowest layers of the OSI reference model. The level–to–layer relationship is shown in Figure 1.51. In the

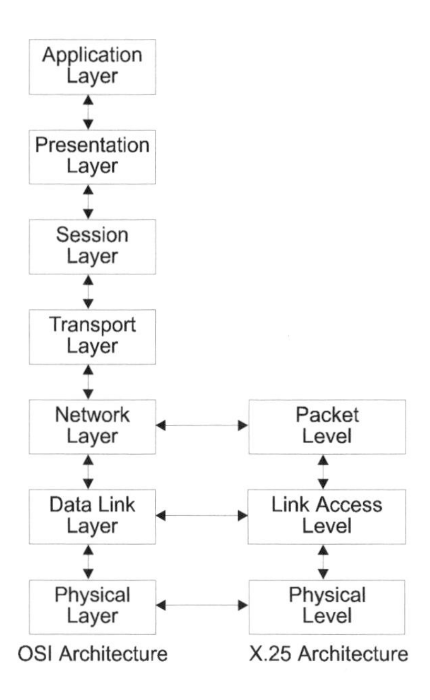

Figure 1.51 Relationship between the architecture of network OSI description and the architecture of X.25 protocol

X.25 terminology the subsequent levels are called: physical level, link access level and packet level.

The *Physical Level* determines the form of bit representation and how the connection and synchronization with the network node are established. This level defines the electrical, physical and procedural interfaces between the user's end device and the network node.

The *Link Access Level* ensures the synchronous operation of the receiver with respect to the transmitter and it also performs error detection and/or correction. On the *Packet Level* the services of the virtual channel established in the network are realized.

The functions of the link access level are realized using the LAP-B (*Link Access Protocol – Balanced*) protocol. This protocol determines how any number of information bits is transmitted through the network. The binary sequences are transmitted in sequenced frames whose structure is shown in Figure 1.52.

Number of bits	8	8	8	8n >0	16	8
	Flag	Address	Control	Packet	CRC	Flag

Figure 1.52 Frame structure in the LAP–B protocol

This protocol allows transmission of any number of bytes. The frame starts and ends with a specific 8–bit sequence of the form 01111110. This flag is used to synchronize the receiver with respect to the received frame (*frame synchronization*). The reception of a sequence of seven 1s indicates the link problems. The sequence of the length equal to at least fifteen 1s sets the channel in the idle state. In order not to detect the end of frame erroneously while observing the received data sequence or detecting the signal indicating network problems, so-called *bit stuffing* is applied in the transmitter which is realized by adding an additional 0 to the sequence of five 1s. In this way sequences denoting the start and end of frame do not appear within the frame except at the beginning and end of a frame. In the receiver the binary stream is continuously monitored. In case of detection of five subsequent 1s (let us note that the sequence of five 1s must has been preceded by a 0), the following sixth bit is checked. If it is a 0, it is removed because its value indicates that an additional bit has been stuffed in the transmitter. If the sixth bit is a 1, the receiver checks the value of the seventh one. If it is a 0, the end of the frame has been detected. If the seventh bit is equal to 1 and if an appropriate number of the following bits are also equal to 1, such a sequence indicates either the channel problems or the transfer of the channel to the idle state.

The address field of the LAP–B frame has a size of a single byte. It contains the address of the station transmitting or receiving the given frame. The control field determines the frame type, i.e., whether it is an information frame, a management frame or an unnumbered frame. The information frames contain the data to be sent between the network elements. Besides information data they can contain the data from the system network layer. The management frames contain the ARQ mechanism, whereas the unnumbered frames transmit additional information associated with the link management.

The packet field contains any number of bits which have to be sent over the link.

The CRC field is applied to check the frame parity in order to detect possible errors. In the protocol the division by the following generating polynomial specified by the ITU-T is implemented

$$g(x) = x^{16} + x^{12} + x^5 + 1$$

Using the above error detection code, the parity of the whole frame, excluding the starting and ending flags, is detected.

As we have already mentioned, the packet level of the X.25 protocol sets a virtual channel for the period of time of the information exchange between the end devices. It is started by the terminal initiating a connection by sending a special packet to the network node. Among other elements the packet contains the source and destination addresses and the request of setting up a connection. The network sends a similar packet to the end terminal with which the initiating terminal wishes to set up a connection. After sending the accepting packet by the called terminal a virtual duplex channel is established.

When one of the terminals wishes to close the information exchange, it sends the closing packet to the network node. The latter sends back an acknowledgement packet.

The X.25 protocol has been described here in a simplified manner. However, the description gives an overview of the tasks of its particular levels. The word "X.25" will appear many times in this book with reference to several mobile communication systems.

Another protocol, called LAP–D (*Link Access Protocol on the D Channel*), is closely associated with the LAP–B protocol. It is a protocol applied in the *Integrated Services Digital Network* (ISDN) for the connection management over the ISDN D channel.[11] This protocol is also used in some mobile communication networks for transmitting the control information and data sequences. The frame structure in this protocol is very similar to the frame shown in Figure 1.52. In this case, however, the address field is extended to 16 bits and the length of the information field cannot be greater than 260 bytes. The control field can have the length of a single byte or two bytes. The size of this field depends on the task of the transmitted frame and on its format. In the format applied for transmission without acknowledgement there are no information bytes (the length of the information field is zero) and the size of the control field is a single byte. In case of transmission with acknowledgement, the numbering of the subsequent frames is necessary in order to repeat the unacknowledged frame if errors are detected in it. The length of the control field is then equal to two bytes. The LAP–D protocol is applied in the data link control layer of the GSM system.

1.12 SIGNALING SYSTEM NO. 7

The performance of the digital communication networks does not depend only on the data rates offered to the users. The time of setting up, monitoring and closing a connection requires transmission of a number of control signals. The number of control signals in mobile communication networks is much higher than in fixed networks due to the mobility of the terminals and the necessity of tracking their locations. In consequence, an efficient and reliable system of control signal transmission is particularly important for the performance of a mobile network.

The best known signaling system is *Signaling System No.7* (SS7). Some mobile communication networks use this signaling system within their structure. The GSM system belongs to them.

The SS7 uses the network independently of the users' network, in which the control signals associated with many information channels are transmitted. This is known as the *common channel signalization* [18]. Thanks to the separation of the signaling network from the users' network it is possible to transmit the signaling information through the connections that are already set up. Transmission of the signalization signals is fast compared with other signaling methods. The signaling network can also be used for the transmission of maintenance and exploitation messages between the switching nodes (switching centers) and the network maintenance center. The signaling network consists of signaling links and nodes. The *Signaling Points* (SP) and *Signaling Transfer Points* (STP) are the nodes of the signaling network. The signaling points either send the signaling messages or they receive them. The signaling transfer points are the centers

[11] In the basic access to the ISDN system a user has two 64 kb/s (B) channels and a single 16 kb/s (D) channel at his disposal.

switching the signaling packets and directing them to the appropriate signaling nodes through the selected signaling links. Since the signaling system is crucial for the proper functioning of the whole communication network, the signaling points are connected with other signaling points or the signaling transfer points in a redundant manner.

The SS7 system is organized in layers, in the same way as the OSI reference model. There are four network layers which are related to the seven layers of the OSI model. The first three layers, the signaling data link, signaling bridge and signaling network, create a so-called a *Message Transfer Part* (MTP). They ensure reliable transmission and distribution of signaling messages among the signaling points. The *Signaling Data Link* corresponds to the physical layer of the OSI model, so it determines the physical link parameters between the signaling points and neighboring signaling transfer points. The second layer, the *Signaling Bridge* corresponds to the data link control layer of the OSI model. It manages the information exchange by putting messages into numbered sequences, realizes the error detection in the received packets and orders their repetition if necessary. The third layer, the *Signaling Network* takes care of the distribution of messages among the signaling points in this network. Three lower layers of the SS7 signaling system together with the *Signaling Connection Control Part* (SCCP), which belongs to the fourth SS7 layer and is responsible for the signaling connections, create the *Network Service Part.* The remaining part of the fourth layer is the so-called *User Part.*

A more detailed description of the SS7 system is beyond the scope of this introductory chapter. More interested readers are advised to study [18].

* * *

The aim of this chapter was to overview the basic elements of digital communication systems, error detection and correction codes, digital modulations and their demodulation, as well as multiple access and duplex transmission methods. We also explained the OSI reference model used later in some chapters for the description of the mobile communication systems.

The above chapter has an overview character. It mostly does not contain the derivation of the system structures or the analysis of their performance. The reader who wishes to deepen his/her knowledge in this area is advised to study the academic handbooks such as [5], [14] and [15]. The overview of the basic theoretical problems of mobile communications can be also found in [16]. General considerations on digital networks can be found, among others, in [17].

REFERENCES

1. S. R. Saunders, *Antennas and Propagation for Wireless Communication Systems*, John Wiley & Sons, Ltd., Chichester, 1999

2. C. A. Balanis, *Antenna Theory: Analysis and Design*, Second Edition, John Wiley & Sons, Ltd., Chichester, 1997

3. ITU-T Recommendation G.726, "40, 32, 24, 16 kbit/s Adaptive Differential Pulse Code Modulation (ADPCM)", Geneva, 1990

4. J. Raimer, M. McMahan, M. Arjmand, "32-kbit/s ADPCM with TMS32010", in *Digital Signal Processing Applications with the TMS320 Family: Theory, Algorithms and Implementations*, Texas Instruments, 1989

5. J. G. Proakis, *Digital Communications*, Fourth Edition, McGraw-Hill, New York, 2001

6. W. C. Wong, R. Steele, C.-E. W. Sundberg, *Source-Matched Mobile Communications*, Pentech Press, London and IEEE Press, New York, 1995

7. R. J. Sluijter, F. Wuppermann, R. Taori, E. Kathmann, "State of the Art and Trends in Speech Coding", *Philips J. Research*, Vol. 49, 1995, pp. 455-488

8. "IS-54, Cellular System Dual-Mode Mobile Station – Base Station Compatibility Standard", EIA, May 1990

9. M. Schwartz, *Information Transmission, Modulation, and Noise*, McGraw-Hill, New York, 1972

10. G. C. Clark, Jr., J. B. Cain, *Error-Correction Coding for Digital Communications*, Plenum Press, New York, 1981

11. S. Lin, D. J. Costello Jr., *Error Control Coding: Fundamentals and Applications*, Prentice-Hall, Englewood Cliffs, N.J., 1983

12. A. J. Viterbi, "Error Bounds for Convolutional Codes and an Asymptotically Optimum Decoding Algorithm", *IEEE Transactions on Information Theory*, vol. IT-13, April 1967, pp. 260-269

13. C. Shannon, "A Mathematical Theory of Communication", *Bell System Technical Journal*, Vol. 27, 1948, pp. 379-423 and 623-656

14. E. A. Lee, D. G. Messerschmitt, *Digital Communication*, Second Edition, Kluwer Academic Publishers, Boston, 1994

15. S. Haykin, *Digital Communication*, John Wiley & Sons, Inc., New York, 1994

16. J. D. Gibson [ed.], *The Mobile Communications Handbook*, IEEE Press and CRC Press, New York, 1996

17. W. Stallings, *Data and Computer Communications*, Third Edition, Maxwell Macmillan International Editions, New York, 1991

18. R. L. Freeman, *Practical Data Communications*, John Wiley & Sons, Inc., New York, 1995

19. F. Halsall, *Data Communications, Computer Networks and Open Systems*, Fourth Ed., Addison-Wesley, Harlow, 1996

20. Bluetooth Specification Ver. 1.0 B, November 1999

21. J. Haartsen, "Bluetooth – the Universal Radio Interface for ad hoc Wireless Connectivity", *Ericsson Review*, No. 3, 1998, pp. 110-117

22. G. D. Forney, *Concatenated Codes*, MIT Press, Cambridge, Massachussetts, 1966

23. C. Berrou, A. Glavieux, P. Thitimajshima, "Near Shannon Limit Error-Correcting Coding and Decoding: Turbo-Codes", *Proc. of IEEE International Conference on Communications*, ICC'93, Geneva, 1993, pp. 1064-1070

24. ETSI TR 101 683 V1.1.1, "Broadband Radio Access Networks (BRAN); HIPERLAN Type 2; System Overview", February 2000

25. T. Starr, J. M. Cioffi, P. J. Silverman, *Understanding Digital Subscriber Line Technology*, Prentice-Hall, Upper Saddle River, N.J., 1999

26. Special Issue on Very High-Speed Digital Subscriber Lines, *IEEE Communications Magazine*, May 2000

27. ETSI EN 300 744, "Digital Video Broadcasting (DVB); Framing Structure, Channel Coding and Modulation for Digital Terrestrial Television", V1.2.1, February 1999

28. J. L. Hammond and P. J. P. O'Reilly, *Performance Analysis of Local Computer Networks*, Addison–Wesley, Reading, Mass., 1986

29. Ch. Heegard, S. B. Wicker, *Turbo Coding*, Kluwer Academic Publishers, Boston, 1999

2

Overview and classification of mobile communication systems

2.1 INTRODUCTION

Mobile communication systems are characterized by a variety of features. They differ from each other in the degree of their complexity, the level of the offered services and operation costs. The attributes of all mobile comunication systems are the mobility of at least one of the connection users and the lack of wireline connection of this user's terminal with the remaining part of the system. In this chapter we will survey the basic types of mobile communication systems and their main characteristics which determine their applications.

One of the classification criteria of the mobile systems is their degree of complexity and the range of the offered services. From this point of view the mobile communication systems can be divided into the following categories:

- paging systems,
- wireless telephony,
- trunking systems,
- cellular telephony,
- satellite personal communication systems,
- systems of wireless access to local area networks.

In the following section we will shortly describe the above categories, their characteristic features and differences between them.

2.2 PAGING SYSTEMS

Paging systems are an example of broadcasting systems. In a classical paging system the communication is unidirectional. The signal is transmitted from the base station connected with the call center to a selected receiver (*pager*). All receivers listen to the broadcast signals, detecting that particular one which is addressed directly to it. In a classical system the pager is able only to receive signals. In the early phase of development the paging signal was very simple. It triggered a beeper generating an acoustic tone. At present the paging signal typically has the form of a sequence of alphanumerical symbols or is a short voice message. The pager is selected by sending the message with its unique address. Typically, messages are transmitted using a frequency modulation. Due to the fact that a pager is very simple and not able to transmit signals, its power consumption is very low and its size is small. On the other hand, the base station emits a high power signal which is able to penetrate building walls. A unidirectional connection established between the base station and the particular pager is optimized taking into account the connection asymmetry.

A typical system consists of:

- the call center to which one can submit requests to page a given user, in the form of voice or alphanumerical messages,
- the base station transmitter operating in a few hundred-MHz band,
- the set of receivers (pagers).

The paging systems range from small ones having a single antenna and short range, to regional ones with multiple antennas properly located in a determined area up to systems covering a whole country. There also exists a Paneuropean system called ERMES, which will be described in Chapter 4.

The development of paging systems has led to the introduction of a feedback channel allowing for acknowledgement of the received paging messages. In recent years a further development of paging systems has been slowed down due to the enormous success of cellular phones. It seems that the paging systems have a chance to survive in a specialized market segment and with new applications, such as remote control.

2.3 WIRELESS TELEPHONY

Wireless telephone systems appeared in the late seventies. They can be characterized as a low power wireless communication means intended for a user moving slowly and located a short distance from the base station. The basic aim of introducing a wireless telephone is to replace a wireline one. The connection quality and the cost of a wireless station should be comparable to the quality and cost of a traditional wireline phone. The base station is that part of the wireless system which is connected to the public switched telephone network (PSTN) and it is seen by the latter as a regular telephone. In most cases one base station co-works with a single mobile station. The time between successive battery rechargings should be as long as possible. As a result of the above mentioned features, the wireless telephone systems has a small number of users per unit of the assigned spectrum, a small number of users per base station, mostly a large number of base stations per area unit and a short base station range. These features properly characterize a wireless telephony implemented in analog technology. In order to avoid connections of a mobile station with the wrong base station (other than its own one), several technical means can be applied, for example, the exchange of a password (a digital sequence) known to the base station and its own mobile station. Another security means is the search for a free channel among those which are accessible for the mobile station–base station pair.

The application of the digital technology in wireless telephony introduced a substantial enhancement of the system capabilities. The coverage area was widely expanded. Previously, the range had been limited to the user's working environment or home. A second generation system known as CT-2 (*Cordless Telephony*–2), introduced in the United Kingdom, offered an additional service called *Telepoint*. The base stations were placed in highly populated areas. The user with a registered Telepoint service was able to initiate a mobile originated call within the CT-2 system. However, he/she was not able to be found except in cases when he/she was located in the range of his/her own base station. Overtaking the call of the user who initially communicated with another base station was not possible either. The CT-2 system did not become popular in the UK; however, it was quite successful in other places, such as Hong Kong and Singapore. It was also introduced in Paris under the marketing name of Bi-Bop. Some of the CT-2 mobile stations were equipped with an additional pager thus making possible to inform a CT-2 user about a desired connection with an external user.

The technical and unification works done within the European Union resulted in the EU standard of the digital wireless telephony known as DECT (*Digital Enhanced Cordless Telecommunications*), previously *Digital European Cordless Telephony*), optimized for the application in the indoor environment. The DECT base stations are connected through system controllers to private branch exchanges (PBXs). Thanks to the DECT controllers, subsequent base stations are able to handle the connection when a mobile station changes its location. A desired mobile station can also be called by any base station in whose range it is currently located. Although primarily designed for an indoor environment, the DECT systems are also installed in places featuring very intensive communication traffic, such as airports, city centers, railway stations, etc. After certain modifications the DECT technology can be successfully applied in wireless local loops. It has also been used in the wireless access to local area networks (LANs).

2.4 TRUNKING SYSTEMS

Trunking systems are mobile communication systems specializing in communications within large enterprises which manage resources dispersed in space, such as a fleet of trucks or service vehicles. Trunking systems are particularly useful in transport companies, and special services such as police, emergency services, gas and power suppliers, etc. Their characteristic feature is the existence of a dispatch and control center which manages the calls. Thus, special kinds of connections are arranged which in regular telephone networks are possible as special services only. The examples are a call from the dispatch center to all mobile stations or to a selected group of them. Another connection typical of trunking systems is a common call set-up within a group of mobile stations.

The trunking systems evolved from a system with a single base station and a common radio channel over which all mobile stations could listen to the signals sent to any of them, through the more sophisticated analog systems operating in conformity with the British standard MPT 1327 up to a fully digital system, known under the name TETRA, standardized by ETSI (*European Telecommunications Standards Institute*) which is able to transmit both voice and data signals.

The basic idea and the name of *trunking* systems stem from the rule of channel assignment applied in these systems. The system resources consist of a certain number of channels which are a common resource. Any free channel can be applied by a new connection and will be immediately returned to the common quota after ending the call. This is the basic difference between trunking systems and classical dispatch systems in which the channels were assigned to fixed groups of users. The occupancy of all the channels within one group made it impossible to get a connection with a user from that group even if some channels in other groups were free.

2.5 CELLULAR TELEPHONY

Cellular telephony is the next and perhaps the most representative example of mobile communication systems. The cellular phone system is characterized as a system ensuring bidirectional wireless communication with mobile stations moving even at high speed in a large area covered by a system of base stations. The cellular system can cover a whole country. Moreover, a family of systems of the same kind can cover the area of many countries, as in case of the GSM system.

Initially, the main task of a cellular system was to ensure the connections with vehicles moving within a city and along highways. The power used by cellular mobile stations is higher than that used by the wireless telephony and reaches the values of single Watts. In case of mobile handhelds their limited radiated power is one of the main factors deciding upon their long use between battery recharging and can be one of the main reasons for their success on the market. However, the consequence is their limited range and the necessity to locate the base stations with sufficient density.

The first generation cellular systems were implemented in analog technology. Voice was transmitted using frequency modulation with FDMA as the channel multiple access method. However, the control of the connection set-up, the change of the base stations during a connection (so-called *handover* or *hand–off*) caused by the mobile station mobility, as well as other control procedures such as mobile station power control, are implemented by transmission of digital signals. In the eighties many mutually incompatible analog cellular systems were built, such as American AMPS (*Advanced Mobile Phone System*), English TACS (*Total Access Cellular System*), Scandinavian NMT (*Nordic Mobile Telephone* system) or German C-Netz. The development of the digital technology, on one hand, and frequent cases when analog systems reached their full capacity, especially in big cities, on the other hand, led to the development of the second generation systems implemented in digital technology and using TDMA or CDMA as the multiple access method. These are the GSM system (*Global System for Mobile Communications*) operating in Europe and other continents, the IS-54/136 and IS-95 systems operating in the USA and some other countries and the Japanese PDC (*Personal Digital Cellular*) system working exclusively in Japan. The GSM and IS-95, as the most representative examples of the systems using TDMA and CDMA channel multiple access respectively, will be the subject of detailed analysis in further chapters.

The main aim in the design of the second generation systems was the maximization of the system capacity meant as the number of users per spectrum unit per area unit. On the other hand the communication with vehicles moving on highways in sparsely populated areas required high power and long-range base stations. Since these conflicting requirements had to be taken into account in the system design, the resulting systems have the following features:

- relatively large transmitter power,
- high complexity of the mobile phone, in particular its digital signal processing part,
- lower connection quality as compared with the stationary phone systems,
- high network complexity resulting from handover and mobility management procedures, necessity of finding the desired mobile station within the system and the variety of services offered.

Despite evident differences among the second generation cellular phone systems these systems are similar in some aspects, such as:

- low data rate of the digital stream representing a user's voice signal – because of complicated speech coding algorithms the data rate of the binary stream does not exceed 13 kbps, which increases the system capacity at the cost of certain loss of speech quality,
- relatively large transmission delay in both transmission directions of the order of 200 ms caused by the speech coding and decoding algorithms and complicated detection of digital signals,
- duplex transmission with the use of frequency division (FDD),
- mobile station power control which ensures equal connection quality despite the distance of the mobile station from the base station.

Further progress in the second generation cellular systems has brought a lot of novelties. In particular, the offer of data transmission services has substantially broadened. Originally, the second generation cellular phone systems were designed mainly for the transmission of voice signals. The wide application of the Internet, general development of computer networks and the popularity of portable computers (laptops, palmtops, etc.) created a demand for Internet access through a mobile station. This required much faster data transmission in the existing systems. Designers have also proposed a simplified access protocol to the Internet which with the appropriate design of web pages enables the user to surf on the Internet using a mobile phone.

In the last decade the third generation cellular systems have been developed. It was foreseen that data and multimedia transmission would constitute a larger part of the transmitted signals. Thus, much higher capacity and several types of traffic have been assumed for the new systems. It has been established that the data rate will be at least 384 kbps and can reach 2 Mbps, thus enabling video transmission. A world standard was desired for the third generation cellular systems which would allow the user to move with the mobile station all over the world. So far this task has not been successfully completed and separate standards have been established for Europe, America and some Asian countries, resulting in the whole family of ITU recommendations presented under the common name of IMT-2000 (*International Mobile Telecommunications* 2000). The basic channel access method is the code division multiple access (CDMA). A more detailed description of the third generation systems will be presented in Chapter 17.

2.6 PERSONAL SATELLITE COMMUNICATION SYSTEMS

Satellite mobile communication systems already do exist. A good example of them is the family of INMARSAT systems specialized in global maritime communications. The systems belonging to this family are also used in individual communications with a mobile station of the size of a briefcase or laptop.

The characteristic feature of currently existing satellite systems is uni- or bidirectional voice and/or data communications at a limited quality in very large areas. Their main advantage is the wide range. The access to the system is difficult inside buildings. The system capacity strongly depends on the number of satellites used. However, a large number of satellites substantially increases the system cost.

In recent years, a few new satellite systems have been proposed, among others those known under the names of Iridium, Globalstar and ICO. All of them make use of satellites which are located on low or medium orbits. In case of low orbits the number of satellites has to be large, the coverage of each of them is small but the size of the satellite and its cost are low as compared with a satellite requiring higher orbits. Additionally, the delay introduced by the way to and from the satellite is relatively small and the power of a mobile station located on the Earth surface can be low, allowing for the application of handheld phones. At the same time, a small delay ensures a higher comfort in voice signal perception. A large number of low orbit satellites having a small coverage implies a high overall system capacity due to the multiple usage of the channel frequencies. The current prospects of low orbit personal communication satellite systems seem to be still unclear due to the marketing problems of some of the systems. Despite this such systems will be the subject of our interest due to their specific concepts.

The geostationary satellite systems require the smallest number of satellites. However, each of them is very expensive. Due to a high satellite orbit (around 37,000 km above the earth) each satellite covers a high area, thus the channel reuse is much lower as compared with the low orbit satellite systems. In consequence the system capacity will be much lower as well. Additional difficulty is a substantial delay introduced due to a long signal propagation to and from a geostationary satellite (up to around 0.5 seconds). Such a delay makes the voice transmission unpleasant for the user. A compromise between low orbit and geostationary systems is the system with intermediate circular orbits (ICO). The required number of satellites is reasonable and the introduced delay is acceptable.

2.7 WIRELESS ACCESS TO THE LOCAL AREA NETWORKS

A new category of mobile systems has appeared in recent years. The wireless technology has been used to realize a wireless access to computer networks. The assumptions and tasks of such systems are different from those described so far. First of all, it is usually assumed that the range of the system and the user mobility are very limited. The power of mobile stations is very low due to assumed short distance to the nearest base station or another mobile station if direct connection between mobile stations is possible. Several frequency bands will be used in wireless LANs. Some systems use an ISM (*Industrial, Scientific and Medical*) band, whereas other systems use the spectrum in the 5 GHz range. Other frequency ranges are also investigated for wideband data transmission.

An interesting difference between the systems implementing the wireless access to LANs and other mobile systems is often the network structure. Several configurations of wireless LANs are possible starting from well-defined structures with the base stations and master controller and ending with the *ad hoc* networks, where there is no particular master station and all mobile stations can communicate with each other.

* * *

This chapter has been devoted to the general characterization and classification of the mobile communication systems. The future is seen to be the creation of a universal mobile communication system on the basis of a few subsystems applied depending on the actual location and speed of movement of the mobile user. In contrast to a traditional communication system in which a number is assigned to the localized telephone set, in the future universal radiocommunication system the number will be assigned to the user with a personal mobile phone who can be in any place of the globe. Such a type of communication will certainly have a serious impact on the future social relations.

REFERENCES

1. D. C. Cox, "Wireless Personal Communications: What Is It?", *IEEE Personal Communications*, Vol. 2, No. 2, April 1992, pp. 20-35

2. B. H. Walke, *Mobile Radio Networks: Networking and Protocols*, John Wiley & Sons, Ltd., Chichester, 1999

3. J. D. Gibson (ed.), *The Mobile Communications Handbook*, IEEE Press and CRC Press, New York, 1996

4. N. J. Boucher, *The Trunked Radio and Enhanced PMR Radio Handbook*, John Wiley & Sons, Inc., New York, 2000

5. M. J. Miller, B. Vucetic and L. Berry, *Satellite Communications. Mobile and Fixed Services*, Kluwer Academic Publishers, Boston, 1993

6. R. A. Dayem, *Mobile Data and Wireless Technologies*, Prentice-Hall, Upper Saddle River, N.J., 1997

7. B. Bing, *High-Speed Wireless ATM and LANs*, Artech House, Boston, 2000

3

Characterization of the mobile communication channel

3.1 INTRODUCTION

Specification of the transmission channel characteristics has a crucial meaning for the design of any communication system. Channel propagation properties, introduced distortions and disturbances as well as the allowable transmitted signal bandwidth determine achievable transmission rate and its quality. Thus, before the specification of any communication system, the designer has to know the properties of the channel applied in the designed system. In this respect a mobile communication channel is not an exception. Therefore, before considerations of functioning of mobile communication systems we present transmission channel properties which are specific for mobile communication systems. Characteristic features of the transmission channel strongly depend on the type of the system. Channel properties are quite different for various systems, e.g. for indoor communications, cellular or satellite communication systems. In this chapter we will concentrate on some common features of communication channels currently used in mobile communications.

We will start with a basic set of terms associated with antennae. Then we will present basic expressions dealing with signal propagation in free space and calculation of the link power budget. Subsequently we will investigate the influence of the multipath effect on the transmission properties. In turn, we will present transmission channel models allowing for evaluation of communication system quality. We will also give examples of standardized channel models characterizing signal propagation in typical types of terrain, which are used in the GSM cellular system design. We will also survey the most important channel propagation models useful for evaluation of the cellular base station coverage. At the end of this chapter we will consider the phenomena of flat and selective fading and the diversity reception as a method combatting these phenomena.

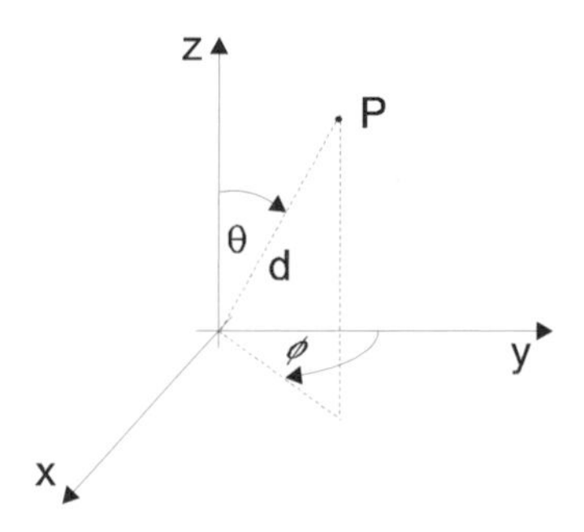

Figure 3.1 Angles φ and θ in the Cartesian system

3.2 FREE-SPACE SIGNAL PROPAGATION

Signal transmission in a radio system is based on converting the electrical signals generated by the transmitter into electromagnetic waves, propagation of waves in space and conversion back into electrical signals at the receiver side of the system. The properties of a mobile communication channel depend on many factors, in particular on the characteristics of the applied antennae, the properties of the physical medium in which the radio waves propagate, the features of the electronic circuits participating in signal transmission and reception, and the velocities of mobile stations. In order to give a simple presentation of mobile communication channel properties, we will first consider an ideal case – the signal propagation in free space. We will precede these considerations by explaining a few basic terms referring to antennae [21].

Let us consider a theoretical case of an antenna which emits the signal equally in all directions at the power level of P_T Watts. Such an antenna is called an *isotropic antenna*. It is an ideal device which cannot be built in practice; however, it serves as a reference for other antenna types. If we draw a sphere of radius r around the isotropic antenna, at each point of its surface the electromagnetic field induced by this antenna is identical. Real antennae focus the radiated energy in specific directions, therefore we usually describe a normalized radiation antenna characteristics with the expression

$$F(\theta, \varphi) = \frac{E(\theta, \varphi)}{E_{\max}} \tag{3.1}$$

where $E(\theta, \varphi)$ is the field at point P of the sphere whose coordinates are determined by the angles θ and φ, whereas $E_{\max}$ is the maximum field on the surface of the sphere. Figure 3.1 presents the coordinate system in which the angles θ and φ are denoted. The isotropic antenna is located in the origin of the coordinate system. It is easy to see that the normalized characteristic does not depend on the sphere radius r. The term *radiation density* is closely related to the normalized characteristic. It is the power radiated in the determined direction within a unit solid angle.[1] Both antenna

[1]Solid angles are measured in *steradians*. Let us recall that the full solid angle measures 4π steradians.

characteristics appear in the expression

$$U(\theta,\varphi) = U_{\max} \left| F(\theta,\varphi) \right|^2 \tag{3.2}$$

where $U_{\max}$ is the maximum radiation density. The total power P_T radiated by the antenna is the integral of radiation density with respect to the solid angle, i.e.

$$P_T = \int\limits_{4\pi} U(\theta,\varphi)\mathrm{d}\Omega = 4\pi U_{\text{mean}} \tag{3.3}$$

where $\mathrm{d}\Omega = \sin\theta\mathrm{d}\theta\mathrm{d}\varphi$. Let us note that the radiated power can be expressed as the product of the mean radiation intensity U_{mean} and the value of the full solid angle which is equal to 4π. Mean radiation intensity can be interpreted as the radiation density of the isotropic antenna radiating the same power P_T as the given antenna. The ratio of the radiation density $U(\theta,\varphi)$ to the mean radiation density is called *directivity gain* of the antenna. Its maximum value is called the *antenna directivity* D and is described by the expression

$$D = \frac{U_{\max}}{U_{\text{mean}}} = \frac{U_{\max}}{\frac{1}{4\pi}\int\limits_{4\pi} U(\theta,\varphi)\mathrm{d}\Omega} = \frac{U_{\max}}{\frac{1}{4\pi}P_T} \tag{3.4}$$

The term *directivity* means that the radiation intensity in the direction of the maximum radiation is D times larger than the radiation intensity of the isotropic antenna radiating the same power as the given antenna.

In the case of a real antenna the radiated power is only a part of the power P_{input} given to its connector. Part of the input power is dissipated and converts into heat. Thus the antenna is characterized by the *power efficiency* $\eta = P_T/P_{\text{input}}$. In order to power dissipation in the antenna characteristics, the term *antenna gain* is introduced and defined as the ratio

$$G = \frac{U_{\max}}{\frac{1}{4\pi}P_{\text{input}}} \tag{3.5}$$

Comparing (3.4) with (3.5) we get

$$G = \eta D \tag{3.6}$$

The antenna gain is often used in the determination of the *Effective Isotropic Radiated Power* (EIRP) which is described by the product $P_{\text{input}}G$. EIRP is the power which would have to be supplied to the isotropic antenna in order to achieve the same field at the reception point as that which is received at that point due to the antenna with the gain G and supplied by power P_{input}. This case is illustrated in Figure 3.2.

Another type of the reference antenna is the *half-wave dipole.* If we compare the power of the antenna with the gain G with the power of the half-wave dipole, we define the so-called *Effective Radiated Power* (ERP). The half-wave dipole has a gain equal to 1.64 equivalent to 2.15 dB as compared with the isotropic antenna. The ERP radiated by a given antenna is 2.15 dB lower than the EIRP power. Depending on the reference

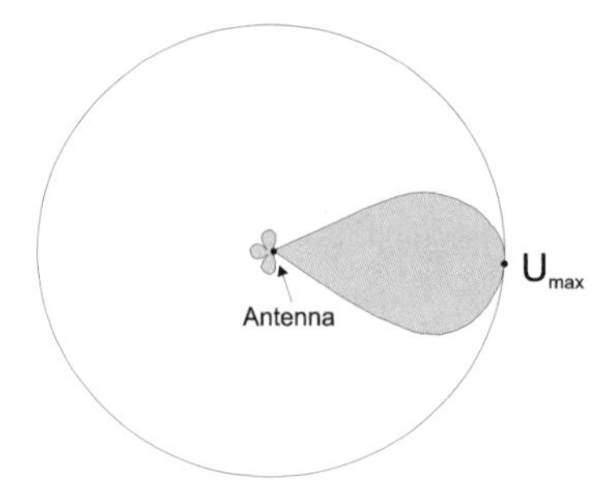

Figure 3.2 Illustration of the effective isotropic radiated power EIRP

antenna, the units of the antenna power gain are denoted as dBi for the isotropic antenna or as dBd for the half-wave dipole.

Let us consider a transmit antenna which radiates the power of P_T Watts into free space. Let the antenna gain in a given direction be G_T. The density of power sent in this direction and measured at the distance d is $P_T G_T / 4\pi d^2$ W/m^2. The receive antenna directed at the transmit antenna and located at a specified distance from the latter "collects" only a part of the radiated power. This received power depends on the *effective area of the antenna* A_R and is given by the formula

$$P_R(d) = \frac{P_T G_T A_R}{4\pi d^2} \tag{3.7}$$

Formula (3.7) shows the dependence of the received power on the distance between the transmit and receive antennae.

The electromagnetic field theory shows that the effective area of the receiving antenna can be described by the expression

$$A_R = \frac{G_R \lambda^2}{4\pi} \quad [\mathrm{m}^2] \tag{3.8}$$

where G_R is the receiving antenna gain, $\lambda = c/f$ is the wavelength of the transmitted signal, c is the light velocity and f is the frequency of the transmitted signal. Substituting (3.8) in (3.7), we obtain the formula for the power of the received signal

$$P_R(d) = \frac{P_T G_T G_R}{(4\pi d/\lambda)^2} \tag{3.9}$$

Let us define the *free-space path loss* L_s by the expression

$$L_s = \left(\frac{\lambda}{4\pi d}\right)^2 \tag{3.10}$$

which allows us to write the received power, taking into account additional loss, for example the atmospheric loss L_a, by the formula

$$P_R(d) = P_T G_T G_R L_s L_a \tag{3.11}$$

Let us note that the form (3.11) is particularly well suited for calculations. The calculation of the logarithm of (3.11) yields in the sum of the transmitted power, consecutive gains and losses expressed in the decibel scale. Thus, from (3.11) we get

$$(P_R)_{\mathrm{dB}} = (P_T)_{\mathrm{dB}} + (G_T)_{\mathrm{dB}} + (G_R)_{\mathrm{dB}} + (L_s)_{\mathrm{dB}} + (L_a)_{\mathrm{dB}} \tag{3.12}$$

The gain of the receiving antenna G_R and its effective area depend on the geometrical properties of the antenna and on the received wavelength. For example, the effective area of a parabolic antenna of diameter s is described by the expression

$$A_R = \frac{\pi s^2}{4}\alpha \tag{3.13}$$

where $\pi s^2/4$ is the real antenna area and α is the *illumination efficiency factor* contained in the range between 0.5 and 0.6. Comparing formulae (3.8) and (3.13) we obtain the expression describing the parabolic antenna gain of diameter s

$$G_R = \alpha\left(\frac{\pi s}{\lambda}\right)^2 \tag{3.14}$$

As we know from the digital communication systems theory, the principal criterion of the communication system performance is the error rate. In case of transmission over the channel introducing the additive Gaussian noise, the error rate depends on the signal-to-noise ratio and on the type of the applied receiver. Usually, the error rate (equivalently, the error probability) is given as a function of E_b/N_0 – the signal energy per bit over the power spectral density of a white Gaussian noise. The additive white Gaussian noise is the model of the thermal noise observed at the receiver input because the thermal noise power density spectrum is flat in a very wide frequency range. Let us express the ratio E_b/N_0 in terms of the received power P_R and single bit duration T_b

$$\frac{E_b}{N_0} = \frac{T_b P_R}{N_0} = \frac{1}{R}\frac{P_R}{N_0} \tag{3.15}$$

We can conclude from formula (3.15) that at the selected bit rate R and for the existing noise power density N_0, in order to achieve the E_b/N_0 required by the demanded bit error rate, we have to ensure the appropriate level of the received power P_R resulting from the formula

$$\frac{P_R}{N_0} = R\left(\frac{E_b}{N_0}\right)_{\mathrm{required}} \tag{3.16}$$

On the other hand, from (3.16) we are able to determine the maximum transmission bit rate if the remaining transmission system parameters have been set.

3.3 INFLUENCE OF THE MULTIPATH EFFECT ON SIGNAL PROPAGATION

Let us come back to the signal propagation. The power received at distance d from the transmit antenna can be expressed with respect to the power measured at a certain standard distance d_0, i.e. the reference power $P_R(d_0)$. Thus, on the basis of formula (3.9) the power received at the distance d can be expressed as

$$P_R(d) = P(d_0)\left(\frac{d_0}{d}\right)^2 \quad d \geq d_0 \tag{3.17}$$

From (3.17) we see that the received power of the signal propagating in free space is inversely proportional to the square of the distance from the transmit antenna. The reference distance d_0 has to be appropriately large in order to consider the receiver power at distance d in the antenna far-field determined by the so-called *Fraunhofer distance* d_f given by the expression

$$d_f = \frac{2L^2}{\lambda} \tag{3.18}$$

where L is the maximum physical linear size of the antenna and λ is the wavelength. In practice, in the frequency range of 1 to 2 GHz the reference distance d_0 is assumed to be 1 m for antennae used in the indoor environment (for example in the wireless telephony systems) or to be 100 m or 1 km in the outdoor environment [11].

Consider now the transmission of a sinusoidal signal of amplitude A_T and frequency f. Assume that only the direct wave reaches the receiver placed at distance d from the transmitter. The received sinusoidal signal can be described by the expression

$$\mathrm{Re}\left(A\exp(j2\pi f(t-\tau))\right) = \mathrm{Re}\left(A\exp(j\varphi_R)\exp(j2\pi ft)\right) \tag{3.19}$$

where $A = A(d)$ is the amplitude of the received signal and its phase shift is determined by formula $\varphi_R = -2\pi f\tau = -2\pi fd/c$ (c is the light velocity). Because in free space the power of received signals decreases with the square of the distance from the transmit antenna, its amplitude decreases linearly with the distance, expressed by the formula which is

$$A(d) = A_0\left(\frac{d_0}{d}\right), \quad A_0 = \sqrt{2P(d_0)} \tag{3.20}$$

In mobile communication systems the transmitted signal reaches the receiver after travelling by many paths. The signal propagating along each path is the subject of subsequent reflections and diffractions. At each reflection a part of the signal energy is absorbed by the reflecting surface. Let the j–th reflection of the i–th path be characterized by the reflection coefficient a_{ij}. The joint reflection coefficient for the i–th path is

$$a_i = \prod_{j=1}^{K_i} a_{ij} \tag{3.21}$$

where K_i is the number of reflections in the i–th path. If we assume that the number of paths along which the signal reaches the receiver is equal to L, the signal phasor can be characterized by the expression

$$A(d)\exp(j\varphi_R) = A_0 d_0 \sum_{i=1}^{L} \frac{a_i}{d_i}\exp(j\varphi_i) \tag{3.22}$$

where $\varphi_i = -2\pi f d_i / c$. Then the power of the signal reaching the receiver along L paths can be expressed by the formula

$$P_R(d) = P(d_0)d_0^2 \left| \sum_{i=1}^{L} \frac{a_i}{d_i} \exp(j\varphi_i) \right|^2 \tag{3.23}$$

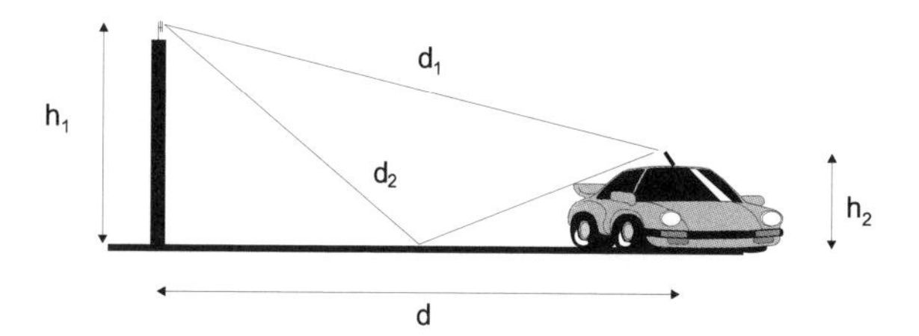

Figure 3.3 Illustration of two-path effect

Let us consider a simplified model of signal propagation observed in mobile communications. Figure 3.3 presents the model of two-path propagation, which is a simplification of the real situation. Let the transmit antenna and receive antenna be placed h_1 and h_2 meters above the ground level, respectively. The distance on the ground between both antennae is equal to d meters and is much longer than the heights of both antennae. Let us assume that the signal reaches the receiver along two paths: the direct one (line-of-sight) and one with reflection from the ground, as shown in Figure 3.3. Let the reflection coefficient $a_1 = -1$, which means that the ground behaves as a lossless reflecting surface. The power of the signal arriving at the receiver is then

$$P_R(d) = P(d_0)d_0^2 \left| \frac{1}{d_1} \exp(j\varphi_1) - \frac{1}{d_2} \exp(j\varphi_2) \right|^2 \tag{3.24}$$

Simple geometrical dependencies allow to derive the lengths of both paths. They are

$$\begin{aligned} d_1 &= \sqrt{(h_1 - h_2)^2 + d^2} \\ d_2 &= \sqrt{(h_1 + h_2)^2 + d^2} \end{aligned} \tag{3.25}$$

Figure 3.4 presents the normalized power $P_R(d)$ with respect to $P(d_0)$ as a function of distance d for a few carrier signal frequencies in case of two-path propagation. Let us note that the level of the received power depends not only on the distance but also on the signal frequency. Besides a visible tendency of the power level to decrease one can also observe fast power level fluctuations as a function of the distance. This is in fact an illustration of the fading phenomenon. For certain values of the distance between antennae the signals travelling along both paths combine at the receive antenna with the opposite phases which causes the decrease of the signal power. For some other values of d the incoming signals add to each other constructively, which results in the

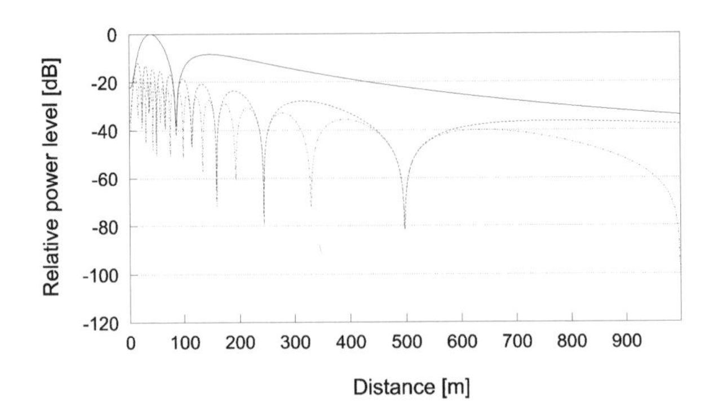

Figure 3.4 Relative power level versus distance [m] ($h_1 = 50$ m, $h_2 = 3$ m, solid line – $f_c = 100$ MHz, dashed line – $f_c = 500$ MHz, dashed line with dots – $f_c = 1$ GHz)

increase of the signal level. Of course there are many intermediate cases as well. The model of a two-path channel is a simplification of the reality; however, it shows the essence of the multipath influence on the signal reception.

Let us consider the approximate power variability as a function of the distance, which characterizes the trend of power decrease. If we assume that d is much longer than both antenna heights h_1 and h_2, the difference between distances d_1 and d_2 becomes insignificant. In that case, d_1 and d_2 in (3.24) can be replaced by the distance d and written in front of the absolute value. The phase difference between two signals arriving along both paths is

$$\Delta\varphi = \frac{2\pi f \Delta d}{c} = \frac{2\pi}{\lambda}\Delta d \tag{3.26}$$

At the above mentioned assumptions the distances d_1 and d_2 given by (3.25) can be simplified to the formulae

$$\begin{aligned} d_1 &\approx d + \frac{(h_1 - h_2)^2}{2d} \\ d_2 &\approx d + \frac{(h_1 + h_2)^2}{2d} \end{aligned} \tag{3.27}$$

These approximations result from the Taylor expansion of the square root function and from retaining the term of the first degree. We can conclude from (3.27) that the difference in distances is $\Delta d = d_1 - d_2 = 2h_1h_2/d$, so the phase difference between both path signals is

$$\Delta\varphi = \frac{2\pi}{\lambda}\frac{2h_1h_2}{d} \tag{3.28}$$

Finally, at the above assumptions the received power can be expressed by the formula

$$P_R(d) \approx P(d_0)\left(\frac{d_0}{d}\right)^2 |1 - \exp(j\Delta\varphi)|^2 \tag{3.29}$$

which for small angles $\Delta\varphi$, knowing that $|1-\exp(j\Delta\varphi)| \approx |1-(1-j\Delta\varphi)| \approx |\Delta\varphi|$, can be simplified to the form

$$P_R(d) = P(d_0)\left(\frac{d_0}{d}\right)^2 |\Delta\varphi|^2 = P(d_0)\left(\frac{d_0}{d}\right)^2 \left(\frac{2\pi}{\lambda}\right)^2 \frac{4h_1^2h_2^2}{d^2} = P_T G_T G_R \frac{h_1^2 h_2^2}{d^4} \quad (3.30)$$

Formula (3.30) indicates that the appearance of the second path besides the direct one has a serious impact on the received signal power level as a function of distance d from the transmit antenna. For two-path propagation the received power is inversely proportional to the fourth power of the distance! Thus in the logarithmic scale the power decrease is equal to 40 dB per decade of distance, as opposed to 20 dB per decade in case of a single path transmission in free space. Two-path propagation is only a theoretical case, giving the insight into the influence of the multipath propagation on the properties of the transmission channel. In reality the number of paths is much higher and depends on the specific features of the environment. Generally, the signal power received at the distance d from the transmit antenna is often described by the formula

$$P_R(d) = P(d_0)\left(\frac{d_0}{d}\right)^\gamma \quad (3.31)$$

where the value of the power γ depends on the propagation environment and is contained in the range between 2 and 5.5.

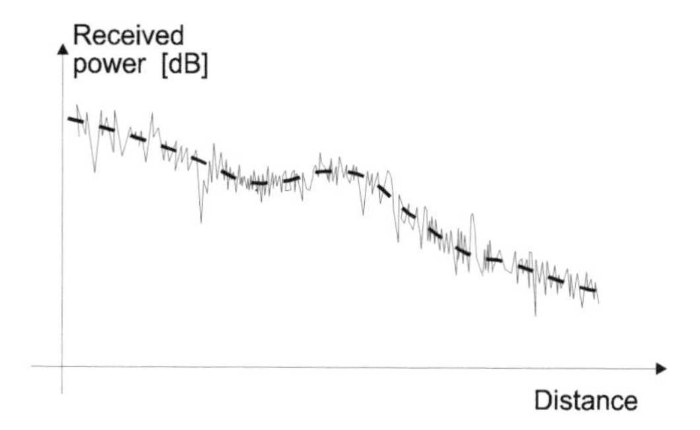

Figure 3.5 Typical path loss (dB) as a function of a distance from the transmit antenna

Figure 3.5 presents a typical plot of the power decrease against the distance between the transmitter and receiver. Similar to Figure 3.4, we notice the decreasing tendency (although not always monotonic), denoted by a bold dashed line, and fast power level changes around the mean, denoted by a thin continuous line. Variability of the mean power is called *slow fading* whereas fast changes around the mean value taking place at the distance of the order of a fraction of the carrier wavelength are called *fast fading.*

In the next paragraph we will explain the physical reasons for fast fading and show the channel model which takes it into account. In the subsequent paragraph we will survey the methods of modeling the power loss.

3.4 TRANSMISSION CHANNEL IN MOBILE COMMUNICATION SYSTEMS

Consider a cellular telephony system as an example of a mobile communication system. The area covered by the system is divided into smaller areas with base stations in their centers (see Chapter 5). Communication with mobile stations is performed through base stations. Typically, base station antennae are omnidirectional or emit the signals into three sectors of the angle width of 120 degrees each. Mobile stations, due to their movement and continuous change of location with respect to the base station, have omnidirectional antennae. These facts imply the properties of the transmission channel. Because the power is emitted in all directions (or at least within a wide angle), the signal, prior to reaching the receiver, is the subject of reflections, diffractions and dispersion caused by several obstacles. The actual environment has a crucial meaning for these propagation phenomena. Figure 3.6 presents typical signal propagation from the base station to the mobile station.

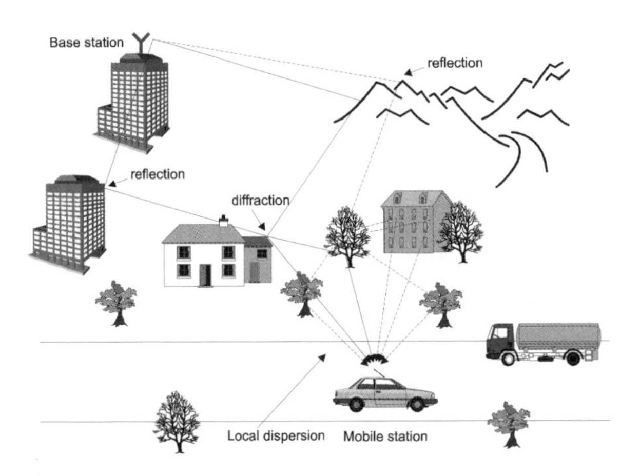

Figure 3.6 Example of signal propagation in the mobile communication system

Let us notice that the signal arrives at the receiver along a few distinguishable paths. In the direct vicinity of the mobile station each signal component is additionally dispersed due to terrain obstacles which slightly differentiate its delays and phase shifts. Very often, in particular in hilly and urban areas there is no direct visibility of the base and mobile station antennae and the signal arrives at the receiver exclusively in the form of the reflected and dispersed components. Although the channel model shown in Figure 3.6 is very simple, it reveals the basic phenomena occurring in a short time scale during digital signal transmission at the rate ranging from a few tens to a few hundred kbps in the frequency range between a few hundred MHz and 2 GHz.

Let us denote the signal emitted by the base station as $s(t)$. Its analytic[2] form is given by the expression

$$\underline{s}(t) = u(t)\exp(j2\pi f_c t) \tag{3.32}$$

where $u(t)$ is the complex baseband equivalent signal. For example, in the GSM system the carrier frequency is around 900 MHz. Let us consider a mobile station installed in the vehicle moving at a speed of 250 kmph. Due to fast speed the Doppler effect is becoming significant. The signal components arriving along different paths at the receiver undergo different frequency shifts, which depend on the value of the carrier frequency, the vehicle velocity v and the angle between the direction of arrival of the signal component and the direction of the vehicle movement. In general, the Doppler frequency f_D depends on the above mentioned factors according to the formula

$$f_D = f_c \frac{v}{c} \cos\varphi \tag{3.33}$$

where c is the light velocity. For the channel model shown in Figure 3.6 we can write the following formula describing the received signal in the analytic form

$$\underline{r}(t) = \sum_{k=1}^{M} \underline{r}_k(t) = \sum_{k=1}^{M}\sum_{i=1}^{N} \underline{r}_{ki}(t) \tag{3.34}$$

Let τ_{ki} be the delay after which the signal component $\underline{r}_{ki}(t)$ arrives at the mobile station. Let us note that for the given path index k the components $\underline{r}_{ki}(t)$ travel along almost the same path and undergo additional dispersion and reflections around the mobile station, so their delays τ_{ki} $(i = 1, \ldots, N)$ are similar. Let T_k be the mean delay of the signals originating from the same k-th path. Then the single component $\underline{r}_k(t)$ can be expressed in the form

$$\begin{aligned} \underline{r}_k(t) &= \sum_{i=1}^{N} \alpha_{ki} u(t-\tau_{ki}) e^{j2\pi(f_c+f_{Dki})(t-\tau_{ki})} \approx \\ &\approx u(t-T_k)e^{j2\pi f_c t}\sum_{i=1}^{N}\alpha_{ki}e^{-j2\pi f_c \tau_{ki}}e^{j2\pi f_{Dki} t} \end{aligned} \tag{3.35}$$

where α_{ki} is the attenuation introduced by the path denoted by indices k and i. Let us note that the simplification made in formula (3.35) is justified because the phase rotation caused by the factor $\exp(j2\pi f_{Dki}\tau_{ki})$ is negligible. For the Doppler frequency equal to 200 Hz and the maximum value of the relative delay equal to 20 μs, the angle $2\pi f_{D\max}\tau_{ki}$ is not larger than 1.6 degrees. In general, the signal received by the mobile station can be described by the expression

$$\underline{r}(t) = \left[\sum_{k=1}^{M} u(t-T_k)c_k(t)\right] e^{j2\pi f_c t} = v(t)e^{j2\pi f_c t} \tag{3.36}$$

[2]Compare formula (1.25) from Chapter 1. The analytic form of the signal $s(t)$ is denoted by $\underline{s}(t)$.

where

$$c_k(t) = \sum_{i=1}^{N} \alpha_{ki} e^{-j2\pi f_c \tau_{ki}} e^{j2\pi f_{Dki} t} \tag{3.37}$$

Equation (3.36) indicates that the baseband equivalent channel model can be considered as a tapped delay line with time-varying tap coefficients $c_k(t)$, $(k = 1, \ldots, M)$. Without loss of generality we can assume that $T_1 < T_2 < \ldots < T_M$.

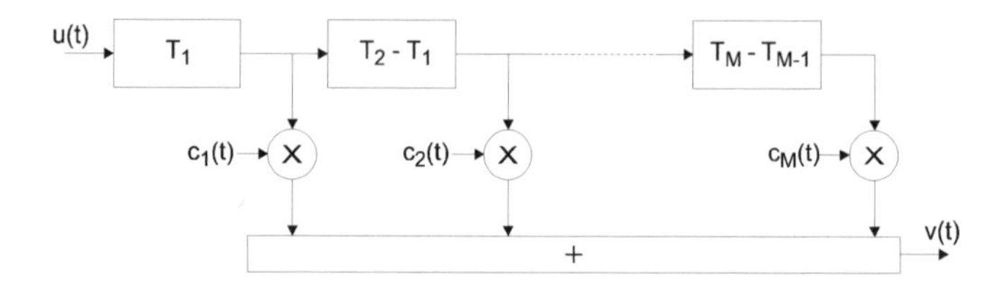

Figure 3.7 Short-term mobile communication channel model

At the given assumptions the channel model can be presented as in Figure 3.7. As we see, if the input and output signals are $u(t)$ and $v(t)$, respectively, then the channel impulse response can be described by the formula

$$h(t, \tau) = \sum_{k=1}^{M} c_k(t)\delta(\tau - T_k) \tag{3.38}$$

The variable t denotes the actual time whereas τ describes the short time run of the impulse response if the excitation pulse has appeared at the channel input at the moment t. Let us stress that such a model is adequate in short time intervals only. In a short time span the delays $T_1, T_2, \ldots, T_M$ do not change visibly. As an example confirming this observation let us consider the channel in the GSM system if the mobile station moves at maximum speed. As we will learn in Chapter 7, the GSM system transmission is organized in bursts. Each burst consists of 148 bits lasting for 3.69 μs each. At the mobile station speed of 250 km/h, during reception of a single burst the mobile changes its location by about 3.8 cm, so its place relative to the terrain topography practically does not change.

Let us concentrate on time variability of coefficients $c_k(t)$. Inspecting (3.37) we see that these coefficients depend on parameters α_{ki}, f_{Dki} and τ_{ki}, which cannot be directly measured. However, due to the mobile station movement they depend on the time variable terrain topology. For this reason these parameters are treated as random variables. At the assumption that from the point of view of a mobile station the location of dispersive and reflecting obstacles can be considered as random, the above mentioned random variables can be treated as statistically independent. Then for sufficiently large

N, taking advantage of the central limit theorem,[3] we can consider the real $c_k^R(t)$ and imaginary $c_k^I(t)$ parts of the coefficient $c_k(t)$ as random variables with approximately Gaussian probability distribution with the variance equal to $\sigma_k^2/2$. This variance can be calculated from the formula

$$\frac{\sigma_k^2}{2} = \frac{1}{2}\sum_{i=1}^{N} E\left[\alpha_{ki}^2\right] \tag{3.39}$$

Because $c_k^R(t)$ and $c_k^I(t)$ $(k = 1, 2, \ldots, M)$ are approximately Gaussian, they have a zero mean. Furthermore, assuming that for different paths the variables α_{ki}, f_{Dki} and τ_{ki} are statistically independent, then $c_k^R(t)$ and $c_k^I(t)$ are mutually uncorrelated. For a large number of dispersion components we can assume that the phase shifts $2\pi f_c \tau_{ki}$ resulting from the delays τ_{ki} are uniformly distributed in the $(0, 2\pi)$ interval. Thus we can easily find that at all above assumptions $c_k^R(t)$ and $c_k^I(t)$ $(k = 1, 2, \ldots, M)$ are the wide sense stationary processes. Because the main components $(k = 1, 2, \ldots, M)$ are also uncorrelated, such a channel model is usually called the *Wide-Sense Stationary with Uncorrelated Scattering* (WSSUS) model. Let us stress once more that these properties are fulfilled only in a short time interval.

Let us note that because $c_k^R(t)$ and $c_k^I(t)$ are independent Gaussian processes, the envelope of the process $c_k(t)$ is Rayleigh distributed, whereas the phase is uniformly distributed in the $(0, 2\pi)$ interval.

In order to find a better channel characteristics, let us determine the properties of the process $c_k(t)$ more precisely. We conclude from formula (3.37) that $c_k(t)$ is the sum of exponential signals characterized by Doppler frequencies f_{Dki}. These frequencies depend on the angle φ_{ki} between the vector of the mobile station movement and the direction of the signal arrival and are given by the formula

$$f_D(\varphi_{ki}) = f_{Dki} = f_{D\max}\cos\varphi_{ki}, \quad f_{D\max} = f_c\frac{v}{c} \tag{3.40}$$

As already mentioned, σ_k^2 is the mean power of the process $c_k(t)$. If the number N of the dispersed components arising from the k-th path tends to infinity, then the power fraction received in the angle interval $[\varphi_k, (\varphi_k + \mathrm{d}\varphi_k)]$ approaches the continuous distribution $p(\varphi_k)$. Then the power received in the given angle interval originating from the k-th component is $\sigma_k^2 p(\varphi_k)\mathrm{d}\varphi_k$. Clearly, the change of the angle is equivalent to the change of the Doppler frequency according to formula (3.40). Knowing that $f_D(\varphi_k) = f_D(-\varphi_k)$, we are able to write the following equation which involves the power density, the Doppler frequency and the angle φ_k [2]

$$G(f_D(\varphi_k))\left|\mathrm{d}f_D(\varphi_k)\right| = \sigma_k^2\left[p(\varphi_k) + p(-\varphi_k)\right]\left|\mathrm{d}\varphi_k\right| \tag{3.41}$$

It results directly from (3.40) that

$$\left|\mathrm{d}f_D(\varphi_k)\right| = f_{D\max}\left|-\sin\varphi_k\right|\left|\mathrm{d}\varphi_k\right| \tag{3.42}$$

[3]The *central limit theorem* states that the probability distribution of a random variable which is the sum of random variables approches the Gaussian distribution as the number of components tends to infinity, regardless of the probability distributions of the summed components.

In turn, from the Pythagoras theorem we can write

$$f_{D\max}|\sin\varphi_k| = \sqrt{f_{D\max}^2 - f_D^2(\varphi_k)} \tag{3.43}$$

Putting (3.43) into (3.42) and using the latter in (3.41), we receive the power spectral density as the function of the Doppler frequency and the angle φ_k

$$G(f_D(\varphi_k)) = \begin{cases} \dfrac{\sigma_k^2\,[p(\varphi_k) + p(-\varphi_k)]}{\sqrt{f_{D\max}^2 - f_D^2(\varphi_k)}} & \text{for} \quad |f_D| < f_{D\max} \\ 0 & \text{otherwise} \end{cases} \tag{3.44}$$

$G(f_D)$ is also called the *Doppler power density spectrum.* As we see, the spectrum depends on the distribution $p(\varphi_k)$ which in turn is determined by the terrain topography. If we assume the idealized case in which the signal is received at the same power from all directions, then $p(\varphi_k) = 1/2\pi$ for φ_k belonging to the interval $[0, 2\pi]$ and the power density of signal $c_k(t)$ is described by the *Jakes formula*:

$$G(f_D) = \begin{cases} \dfrac{\sigma_k^2}{\pi\sqrt{f_{D\max}^2 - f_D^2}} & \text{for} \quad |f_D| < f_{D\max} \\ 0 & \text{otherwise} \end{cases} \tag{3.45}$$

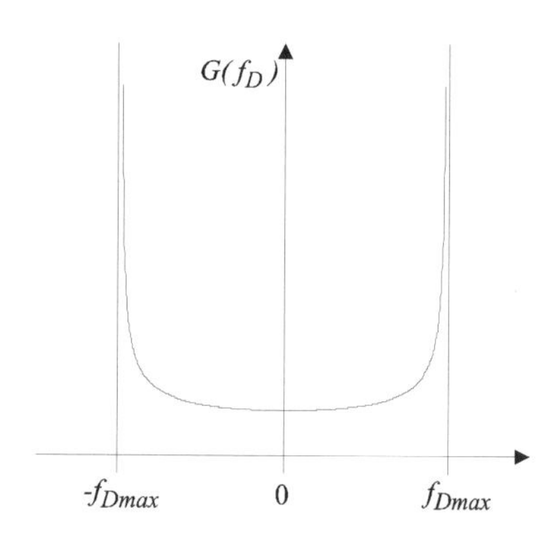

Figure 3.8 Doppler power density spectrum for the process $c_k(t)$ according to Jakes formula

Figure 3.8 presents the Doppler power density spectrum described by formula (3.45). This spectrum shows an idealized case, however, it is useful in mobile channel modeling and can be the basis for comparison of several systems and their robustness against the channel impairments. In case of the occurrence of direct path, the spectrum in Figure 3.8 has to be supplemented with the spectral line at the appropriate Doppler frequency. Figure 3.9 presents the measured Doppler power density spectra [3]. As we see, the results differ from the ideal case, however, their characteristics are still close to that shown in Figure 3.8.

The dependencies shown above do not exhaust the methods of description of mobile transmission channels. For WSSUS channels the *scatter function* is often presented in

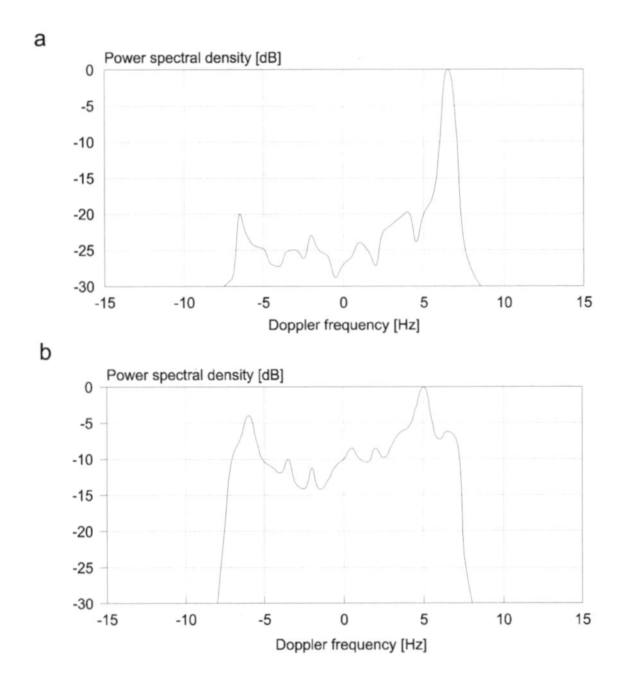

Figure 3.9 Measured Doppler power density spectra: (a) for the channel with a direct path, (b) for the channel without a direct path with a long channel impulse response [3]

the form of a three-dimensional plot. It shows the power density in function of the delay τ and the Doppler frequency f_D.

Intensive research on the GSM system performed in the eighties resulted, among others, in the unification of standard channel models which characterize the signal propagation in the 900 MHz bandwidth. ETSI GSM 05.05 standard [5] presents the tables of channel model power delay profiles recommended for laboratory experiments and simulation of transmission channels, characteristic for a few basic types of environment. Each line shown in Figure 3.10 characterizes the mean power of a discrete component which undergoes dispersion, reflections and the Doppler effect. The standard sets up the Doppler power density spectra conforming to formula (3.45) or the same spectra supplemented by a discrete line in the case of the direct path existence. The plots shown in Figure 3.10 present the standardized values of the variances σ_k^2 $(k = 1, \ldots, M)$ of each discrete component, expressed in dB scale, i.e. the power delay profiles described by the expression

$$P(\tau) = \sum_{k=1}^{M} \sigma_k^2 \delta(\tau - T_k) \tag{3.46}$$

We see from Figure 3.10 that the best transmission channel is the one characteristic for the rural area. Besides the main path the echoes concentrate in a very short time interval, which for the GSM system spans no more than a single modulation period

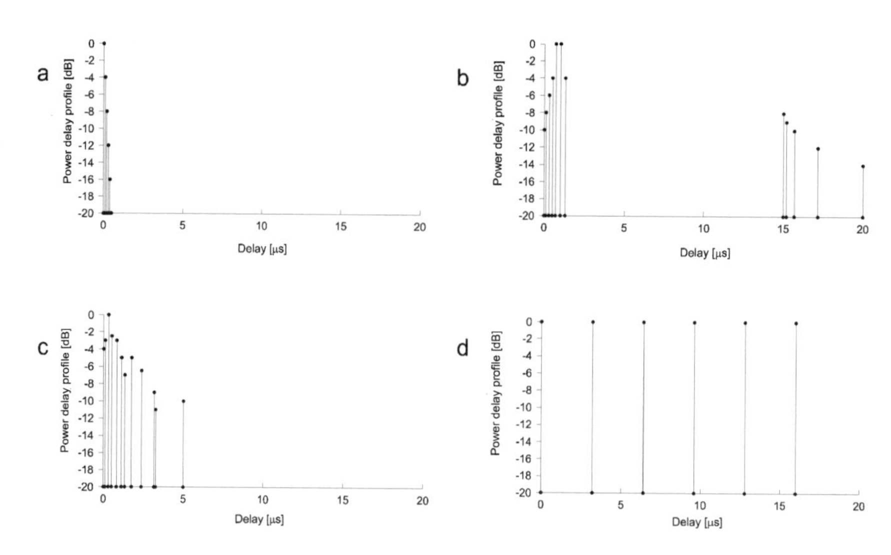

Figure 3.10 Power delay profiles standardized by ETSI GSM 05.05: (a) rural area, (b) hilly terrain, (c) urban area, (d) E-test

(3.69 μs). The model for the urban area imitates the multipath propagation caused by reflections from buildings. The channel in the hilly terrain is even worse for digital transmission. The echoes arrive at the receiver after a relatively long delay due to reflections from the obstacles located far away from the receiver. In consequence, the channel impulse response is long. Finally, the E-test model, although unrealistic, is a very difficult test channel model intended for new constructions of mobile stations.

Multipath propagation and dispersion are not the only distortions occurring in mobile communication channels. The other impairments are impulse noise, interference from other users (in case of cellular systems known as intra- or intercell interference) and thermal noise. The users operating in the same cell but using different frequency channels distort the reception of other users' signals due to imperfectly filtered sidelobes of their own signals (*interchannel interference*). Interference arising from the signal generated in the same frequency channel but in a different cell is another source of distortion. The thermal noise is the result of the self-noise of the electronic components used in the receiver and of the noise introduced in subsequent blocks of the radio link.

The basic reason of impulse noise is human activity. Car ignition systems and electrical motors are a particularly frequent source of this kind of distortions. Other causes of impulse noise are atmospheric phenomena.

Nonlinearity is another distortion encountered in radio systems. Its basic cause is the tight link power budget in mobile stations and the necessity of using a high power amplifier operating close to saturation. The influence of nonlinearity is minimized thanks to the application of constant envelope digital modulations such as FSK and GMSK. In future systems which will be characterized by higher throughput and higher spectral efficiency, multilevel QAM modulations or OFDM transmision will be applied. The influence of the amplifier nonlinearity will be minimized by the application of an amplifier predistorter or specially designed waveforms.

3.5 MODELING THE PROPAGATION LOSS

The channel model described in the previous section explains the origin of multipath propagation and time variability of the channel characteristics caused by the movement of the mobile station and by signal dispersion in its direct vicinity. The model shown in Figure 3.5 is associated with the fast power level changes around the mean value. As we already know, even small changes in the location of the mobile station can be the reason of substantial changes in the received signal level. From the point of view of mobile system design the following factor has to be taken into account as well. This factor is the mean power level as a function of the distance from the base station. Usually the measurements are averaged in the distance interval of 5λ to 40λ, where λ is the carrier wavelength [11]. In the frequency range between 1 and 2 GHz the local mean power is averaged over the distance of 1 to 10 meters. The measurement results are the function of the distance from the transmit station (compare (3.31)) and actual configuration of the main obstacles and distorting and reflecting elements along the paths to the receiver but not in close vicinity of the receiver. This type of information is necessary in the design of cellular systems. The design of cellular systems will be discussed in Chapter 5.

Let us come back to the analysis of formula (3.31). It turns out that this formula shows an agreement with the measurements if they are averaged over all possible positions of the receive antenna at distance d from the transmit antenna. After calculation of the logarithms of both sides in (3.32) we obtain the expression

$$\overline{(P(d))}_{\text{dB}} = \overline{(P(d_0))} - 10\gamma \log\left(\frac{d}{d_0}\right) \tag{3.47}$$

We conclude that the mean power decreases linearly with distance d in the decibel scale. The rate of decrease is 10γ dB per decade. As already mentioned, the parameter γ depends on the propagation environment. Table 3.1 [11] presents typical values of the path loss exponent γ for several types of environment.

Table 3.1 Values of γ for several environment types

Type of environment	Path loss exponent γ
Free space	2
Urban area cellular radio	2.7 to 3.5
Shadowed urban cellular radio	3 to 5
In building line-of-sight	1.6 to 1.8
Obstructed in building	4 to 6
Obstructed in factories	2 to 3

Formula (3.47) shows how the mean received power depends on the distance from the transmit antenna. It has been observed that the power levels measured in two different locations equally distant from the transmit antenna can be quite different due to different layout of the reflecting, attenuating and diffracting obstacles. This phenomenon is called *shadowing*. The measurements indicate that the level of the received power is random. Moreover, it has been found that the power level in dB-scale is a Gaussian random variable, i.e.

$$(P(d))_{\text{dB}} = \overline{(P(d_0))}_{\text{dB}} - 10\gamma\left(\frac{d}{d_0}\right) + X(0,\sigma) \tag{3.48}$$

where $X(0,\sigma)$ is a Gaussian random variable with zero mean and variance σ^2. Thus, in the linear scale, the received power has a log-normal distribution. In general, the log-normal distribution of the random variable X is given by the formula

$$p_X(x) = \frac{1}{x\sigma_X\sqrt{2\pi}} \exp\left[\frac{-\left(\log x - \overline{\log x}\right)^2}{2\sigma_X^2}\right] \tag{3.49}$$

Knowing the distribution in the dB-scale, in particular knowing the variance σ^2, it is possible to calculate the probability of the event that the received signal level at the given location point exceeds a selected threshold. Such calculations can be useful in the evaluation of the base station coverage area.

Radio propagation in a terrain featuring several obstacles such as buildings, terrain irregularities and trees and bushes is such a complicated process that the system designers often perform practical electromagnetic field measurements in specified terrain locations in order to determine the real base station coverage. Such measurements are very expensive, so on the basis of the measurement campaigns performed in different representative environments, several propagation models have been established which estimate the mean power loss in function of distance d from the base station, the type of environment and transmit and receive antenna heights. In the following sections we will present the most representative examples of the experimental propagation models.

3.5.1 The Lee model

W.C.Y. Lee [12] proposed a very simple signal propagation model originating from a series of measurements made in the USA at the carrier frequency f_c= 900 MHz. According to the Lee model, the mean power measured at distance d from the transmit station is determined by the expression

$$P(d) = P_0\left(\frac{d}{d_0}\right)^{-\gamma}\left(\frac{f}{f_0}\right)^{-n} F_0 \tag{3.50}$$

or equivalently in the logarithmic scale

$$(P(d))_{\text{dB}} = (P_0)_{\text{dB}} - \gamma\log\left(\frac{d}{d_0}\right) - n\log\left(\frac{f}{f_0}\right) + (F_0)_{\text{dB}} \tag{3.51}$$

The symbol P_0 is the reference median power measured at distance $d_0 = 1$ km, whereas F_0 is the correction factor selected on the basis of a series of component factors according to the formula

$$F_0 = \prod_{i=1}^{5} F_i \tag{3.52}$$

where the subsequent factors F_i are described by expressions

$$F_1 = \left(\frac{\text{Actual BS antenna height [m]}}{30.5 \text{ [m]}} \right)^2 \tag{3.53}$$

$$F_2 = \left(\frac{\text{Actual MS antenna height [m]}}{3 \text{ [m]}} \right)^\nu \tag{3.54}$$

The power $\nu = 1$ for the mobile station antenna heights lower than 3 m and $\nu = 2$ for the heights larger than 10 m. In turn

$$F_3 = \frac{\text{Actual power}}{10 \ W} \tag{3.55}$$

$$F_4 = \frac{\text{BS antenna gain with respect to half-wave dipole}}{4} \tag{3.56}$$

$$F_5 = \text{MS antenna gain with respect to half-wave dipole} \tag{3.57}$$

The parameters P_0 and γ are selected experimentally based on the performed measurements. Table 3.2 presents the values of P_0 and γ for some characteristic types of environment.

Table 3.2 Values of P_0 and γ for different environments

Environment	P_0	γ [dB/decade]
Free space	-41	20
Open (rural area)	-40	43.5
Suburban, small city	-54	38.4
Philadelphia	-62.5	36.8
Newark	-55	43.1
Tokyo	-78	30.5

The mean power loss in function of frequency is modeled by the factor $(f/f_0)^{-n}$ and the choice of power n. The value of n is contained in the range between 2 and 3 for the frequencies from 30 MHz to 2 GHz and the distances between mobile and base stations contained in the range between 2 and 30 km. The power n also depends on the terrain topography. It is recommended to select $n = 2$ for suburban and rural areas

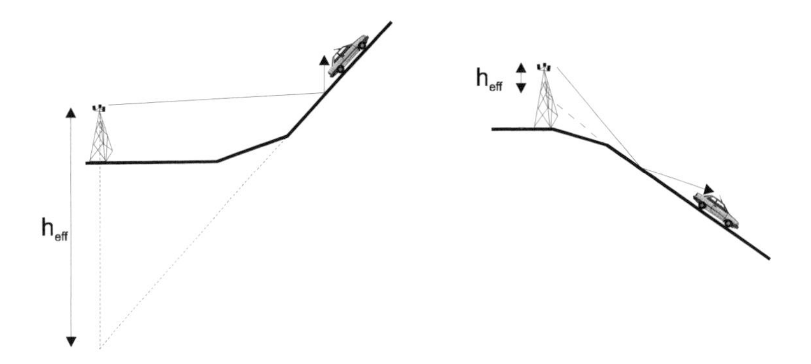

Figure 3.11 Calculation of the effective height of the base station antenna

when operating at the frequencies below 450 MHz, and $n = 3$ for urban environment and carrier frequencies over 450 MHz.

In uneven terrain the effective antenna height can be quite different from the physical height. Figure 3.11 shows how to calculate it.

The measurements which are the basis of the Lee empirical model have been performed in the system characterized by the following parameters, which appear in the formulae for the factors F_i $(i = 1, ..., 5)$:

- carrier frequency: $f_0 = 900$ MHz,
- base station antenna height: 30.48 m (100 feet),
- transmitted power: 10 W,
- base station antenna gain: 6 dB with respect to half-wave dipole,
- mobile station antenna height: 3 m,
- mobile station antenna gain: 0 dB with respect to half-wave dipole.

3.5.2 The Okumura model

The Okumura model is also a result of intensive measurements. It was presented first in [13]. The measurements in the frequency range between 150 and 1920 MHz were performed in the Tokyo area. The authors proposed the following formula for the median loss $(L_{50})_{\mathrm{dB}}$ in function of the distance d from the transmit antenna of the base station

$$(L_{50})_{\mathrm{dB}} = L_s + A(f,d) + G(h_{\mathrm{BS,eff}}) + G(h_{\mathrm{MS}}) \tag{3.58}$$

where L_s is the free space loss (compare formula (3.10)). The symbol $A(f,d)$ denotes the median attenuation relative to free space in an urban area over quasi-smooth terrain with the effective base station antenna height $h_{\mathrm{BS,eff}} = 200$ m and the mobile antenna height $h_{\text{MS}} = 3$ m. $G(h_{\text{BS,eff}})$ is the correction term in dB associated with the base station antenna and depending on its effective height if the latter is different from 200 m. $G(h_{MS})$ is the correction factor in dB associated with the mobile station antenna if its height is different from 3 m. The free space loss is calculated from (3.10) in the dB-scale. Formula (3.58) together with Figures 3.12, 3.13 and 3.14 allow evaluation of the signal attenuation in an urban area for the frequency range between 150 and 2000 MHz if the distance between the base and mobile stations is contained between 1 and 100 km and the effective base station antenna height is in the range of 30 to 1000 m.

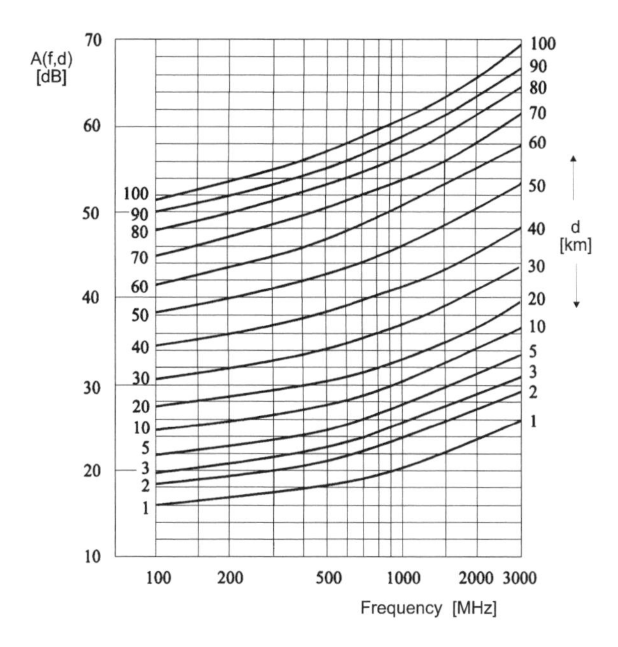

Figure 3.12 Median attenuation $A(f,d)$ relative to free space over a quasi-smooth terrain [13] (© IEEE)

In [13] additional correction terms are proposed which take into account terrain slope, type and irregularities.

In the literature one can find another equivalent to (3.58) version of the formula describing the Okumura model. It has the form

$$(L_{50})_{\text{dB}} = L_s + A(f,d) - G(h_{\text{BS,eff}}) - G(h_{\text{MS}}) - G_{\text{AREA}} \tag{3.59}$$

The parameter $A(f,d)$ is, as previously, read from the plots in Figure 3.12, whereas the correction terms $G(h_{\text{BS,eff}})$ and $G(h_{\text{MS}})$ are given by expressions

$$G(h_{\text{BS},eff}) = 20\log\left(\frac{h_{\text{BS,eff}}}{200}\right) \quad 10 \text{ m} < h_{\text{BS,eff}} < 1000 \text{ m} \tag{3.60}$$

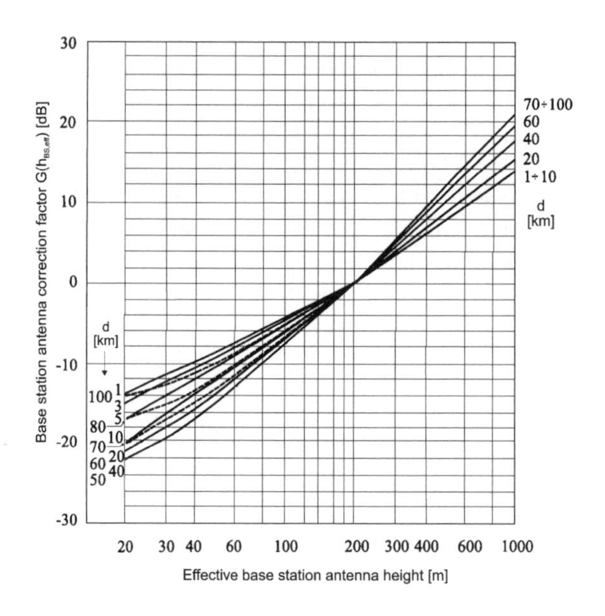

Figure 3.13 Base station antenna height gain factor in urban area as a function of range [13] (© IEEE)

$$G(h_{\mathrm{MS}}) = 10\log\left(\frac{h_{\mathrm{MS}}}{3}\right) \quad h_{\mathrm{MS}} \leq 3\ \mathrm{m} \tag{3.61}$$

$$G(h_{\mathrm{MS}}) = 20\log\left(\frac{h_{\mathrm{MS}}}{3}\right) \quad 3\ \mathrm{m} < h_{\mathrm{MS}} < 10\ \mathrm{m} \tag{3.62}$$

The correction term G_{AREA} (in dB) depends on the type of terrain and carrier frequency and is read from the plots in Figure 3.15.

The Okumura model is very simple. It is based exclusively on the measurement data collected in the Tokyo area. The characteristic of Japanese urban areas is slightly different than in Europe or the USA. Nevertheless, the Okumura model is popular and is considered one of the best models used in cellular and other land mobile systems [11]. The main drawback of the Okumura model is its slow reaction to the change of terrain type. The Okumura model is suitable for urban and suburban environment, however, it is not too practical for the rural area.

3.5.3 The Hata model

The Hata model [14] has been developed as a result of proposing empirical formulae to describe the plots created by Okumura and his collaborators. These formulae well approximate the plots for certain carrier frequency ranges and for the quasi-smooth terrain. Hata proposed the following empirical formulae for estimation of the signal attenuation.

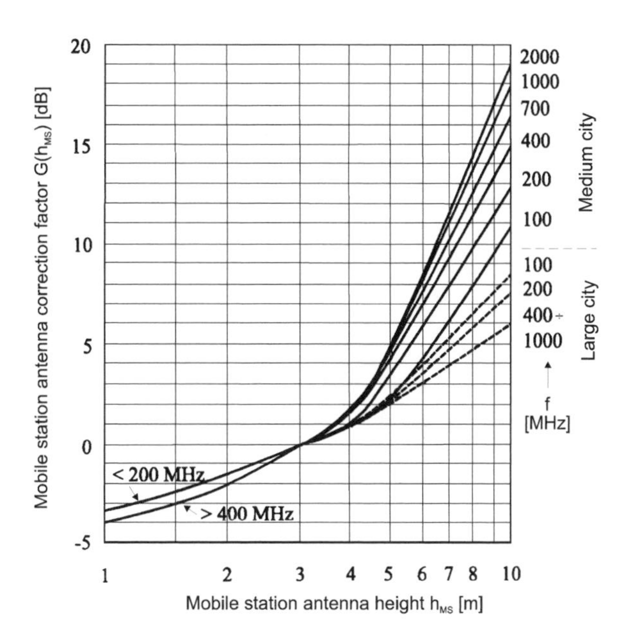

Figure 3.14 Mobile station antenna height gain factor in urban area as a function of frequency and type of a city [13] (© IEEE)

For an urban area in the frequency[4] range from 150 to 1500 MHz and for the effective base station antenna heights $30 \leq h_{\mathrm{BS,eff}} \leq 200$ m we have:

$$(L_{50})_{\mathrm{dB}}|_{\mathrm{urban}} = 69.55 + 26.16 \log f - 13.83 \log(h_{\mathrm{BS,eff}}) + \\ - a(h_{\mathrm{MS}}) + (44.9 - 6.55 \log(h_{\mathrm{BS,eff}})) \log d \tag{3.63}$$

where the correction term depends on the height of the mobile station antenna and in the range $1 \leq h_{MS} \leq 10$ m is calculated from the expression

$$a(h_{\mathrm{MS}}) = (1.1 \log f - 0.7)\, h_{\mathrm{MS}} - (1.56 \log f - 0.8) \quad [\mathrm{dB}] \tag{3.64}$$

For a large city this term is described by the equations

$$a(h_{\mathrm{MS}}) = 8.29\,(\log 1.54 h_{\mathrm{MS}})^2 - 1.1 \quad [\mathrm{dB}] \quad \text{for } f \leq 400 \text{ MHz} \\ a(h_{\mathrm{MS}}) = 3.2\,(\log 11.75 h_{\mathrm{MS}})^2 - 4.97 \quad [\mathrm{dB}] \quad \text{for } f \geq 400 \text{ MHz} \tag{3.65}$$

For a suburban area the propagation loss can be estimated according to the formula

$$(L_{50})_{\mathrm{dB}} = (L_{50})_{\mathrm{dB}}|_{\mathrm{urban}} - 2\left(\log\left(\frac{f}{28}\right)\right)^2 - 5.4 \tag{3.66}$$

[4]In the Hata formulae frequency is expressed in MHz.

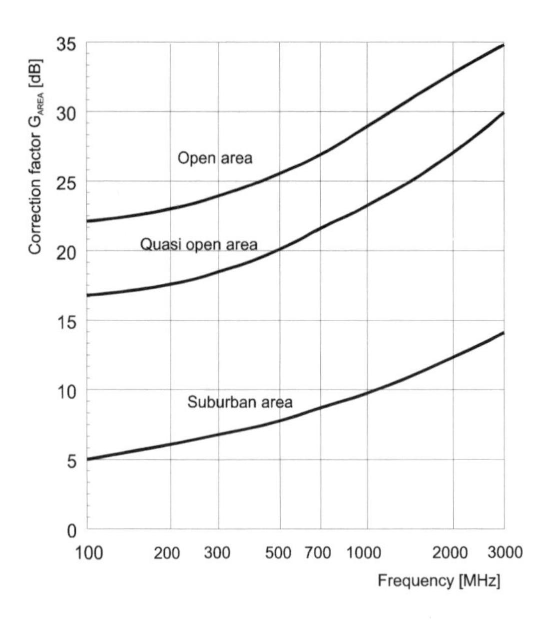

Figure 3.15 Correction term G_{AREA} in function of frequency and area type

whereas for the open area

$$(L_{50})_{\mathrm{dB}} = (L_{50})_{\mathrm{dB}}\,|_{\mathrm{urban}} - 4.78\,(\log f)^2 + 18.33 \log f - 40.94 \tag{3.67}$$

The propagation models presented so far allow for the estimation of the signal loss in function of the carrier frequency, the base and mobile station antenna heights and type of terrain. They better or worse reflect signal propagation at a large distance from the base station, exceeding 1 km. They are mostly valid for the frequency ranges reaching up to 1.5 GHz. However, *Personal Communication Systems* operate in the 1.8 to 2.0 GHz range. Two examples of such systems are DCS 1800 or PCS 1900 – two versions of the GSM system operating in Europe and the USA, respectively. Therefore, a lot of experiments and measurements have been performed in order to construct the propagation models for the band 1.8 to 2.0 GHz for the environments and at assumptions characteristic for the PCS systems. Due to larger signal attenuation in the 1.8 GHz band as compared with 900 MHz band traditionally used by cellular systems, the basic difference between PCS and classical cellular systems is the smaller size of the cells. The research on new propagation models was intensively conducted within the European Union COST#231 project.[5] There are at least two well-known propagation models reported in literature and developed within COST activity. These are the COST231-Hata model [15] and the COST231-Walfish-Ikegami model [16].

[5]COST – Cooperation for Scientific and Technical Research

3.5.4 The COST231-Hata model

Mogensen *et al.* [15] proposed an extension of the Okumura and Hata model to the frequency interval between 1.5 and 2 GHz. In this frequency range the original propagation models deliver underestimated values of the signal attenuation. The COST-Hata model is valid for the carrier frequencies between 1.5 and 2 GHz, the base station antenna heights between 30 and 300 m, the mobile station antenna height between 1 and 10 m and the distance among them between 1 and 20 km. This model allows estimation of the attenuation according to the formula

$$\begin{aligned}(L_{50})_{\mathrm{dB}} &= 46.3 + 33.9 \log f - 13.82 \log (h_{\mathrm{BS,eff}}) + \\ &\quad - a(h_{\mathrm{MS}}) + (44.9 - 6.55 \log (h_{\mathrm{BS,eff}})) \log d + C\end{aligned} \tag{3.68}$$

where the constant $C = 0$ for medium cities and suburban areas with average tree intensity and $C = 3$ for large city centers.

Formally, the Okumura, Hata and COST231-Hata models can be applied for the base station antenna heights larger than 30 m; however, they can also be applied for lower antennae under the condition that the surrounding buildings are considerably lower than the antenna. The COST231-Hata model is not suitable for estimating the attenuation for the base–mobile station distances less than 1 km. In the latter case the attenuation strongly depends on the topography of the closest terrain in which propagation takes place. This model should not be used for estimation of signal propagation along streets with high buildings, i.e. in so-called street canyons.

3.5.5 The COST231-Walfish-Ikegami model

This model can be used in cases when the base station antenna is placed either above or below the roof line in the urban area. In the set of empirical formulae defining the model several factors are taken into account, such as the base and mobile station heights, the street widths, distances between buildings, their heights and the street orientations relative to the signal propagation. In general, the formula describing the signal loss consists of three terms: the free-space path loss L_s, the *roof-top-to-street diffraction and scatter loss* L_{rts}, and *multiscreen diffraction loss* L_{ms}. The second term, L_{rts}, is caused by diffraction of the signal waves on the building roofs. As a result, the signal reaches the mobile station moving along the street. The third term, L_{ms}, is caused by multiple diffraction from the row of buildings. Finally, the median loss is determined by the formula

$$(L_{50})_{\mathrm{dB}} = L_s + L_{rts} + L_{ms} \tag{3.69}$$

The citation of the full set of formulae describing the COST231-Walfish-Ikegami model is beyond the scope of this book. The interested reader is advised to study [16]. We should add that the above model is considered by the International Telecommunication Union as a standard model for the universal third generation system IMT-2000. The model is useful for the following ranges of parameters: $800 \leq f_c \leq 2000$ MHz, $4 \leq h_{\mathrm{BS}} \leq 50$ m, $1 \leq h_{\mathrm{MS}} \leq 3$ m, $0.02 \leq d \leq 5$ km. The COST231-Walfish-Ikegami model works well for the base station antenna heights much greater than the roof heights.

3.5.6 Examples of path loss estimation using selected propagation models

In summarizing the overview of selected propagation outdoor models let us consider two examples quoted according to [17] and [18], which illustrate the application of the Hata and Lee model for the base station signal path loss estimation.

Example 1 *[17] The base station antenna installed at the height of 30 m in a metropolitan area radiates the signal at the carrier frequency $f = 1000$ MHz. This signal measured at a distance of 10 km has been attenuated by 160 dB. Compare the measured path loss with that calculated from the Hata model if the mobile station antenna height is $h_{MS} = 3$ m and the antenna is a half-wave dipole.*

According to the Hata model we apply expression (3.63) to evaluate the signal path loss. Therefore,

$$\begin{aligned}(L_{50})_{\text{dB}} &= 69.55 + 26.16 \log 1000 - 13.82 \log 30 - a(h_{\text{MS}}) + \\ &+ (44.9 - 6.55 \log 30) \log 10\end{aligned}$$

Deriving $a(h_{\text{MS}})$ for the given system parameters we obtain

$$a(h_{\text{MS}}) = 3.2 \left(\log \left(11.75 \cdot 3\right)\right)^2 - 4.97 = 2.69 \text{ dB}$$

Thus,

$$(L_{50})_{\text{dB}} = 69.55 + 78.48 - 20.41 - 2.69 + 35.22 = 160.15 \text{ dB}$$

and we observe a good agreement with the measured value of signal attenuation.

Example 2 *[18] Using the Lee model, evaluate the received signal level measured at a distance $d = 2$ km from the base station by the mobile station with the antenna height $h_{\text{MS}} = 1.5$ m and the antenna gain $G_R = 2$ dBd. The base station radiates the signal at the carrier frequency $f = 1800$ MHz at the power level $P_T = 1$ W, using an antenna installed at the effective height $h_{\text{BS,eff}} = 30$ m with the gain $G_T = 7.7$ dBi. The signal propagates in a suburban area.*

In order to evaluate the received signal level we apply formula (3.51). From Table 3.2 we select the parameters $P_0 = -54$ dBm and $\gamma = 38.4$. To calculate the correction factor F_0 from (3.52), the component factors F_i $(i = 1, \ldots, 5)$ have to be found. In order to find their values we use formulae (3.53) to (3.57). Let us note that the antenna gains used in F_4 and F_5 have to be expressed in a linear scale, so $G_T = 7.7$ dBi is equivalent to $G_T = 5.9$ in the linear scale, whereas $G_R = 2$ dBd $= (2 + 2.15)$ dBi $= 4.15$ dBi so in the linear scale $G_R = 2.6$ (let us recall that 0 dBd $= 2.15$ dBi). Finally, we get

$$F_0 = \left(\frac{30 \text{ m}}{30.48 \text{ m}}\right)^2 \cdot \left(\frac{1.5 \text{ m}}{3 \text{ m}}\right)^1 \cdot \left(\frac{1 \text{ W}}{10 \text{ W}}\right) \cdot \left(\frac{5.9}{4}\right) \cdot 2.6 = 0.1857 \tag{3.70}$$

that is $F_0 = -7.3$ dB.

Factor F_2 has been calculated in (3.70) recalling that for mobile station antenna height lower than 3 m the power $\nu = 1$. Because the carrier frequency is higher than 450 MHz, the power n in (3.50) and (3.51) is equal to $n = 3$. Finally, from (3.51) we get the received power level $(P_R)_{\text{dB}} = -73.8$ dBm.

3.5.7 Estimation of the propagation loss for indoor channels

So far we have considered propagation conditions and related predictions of the signal power loss for outdoor channels. Prediction of the propagation characteristics inside buildings is also very important, especially for the design of wireless telephony systems with base stations inside buildings and for wireless local area networks [23]. For an indoor channel, the distance between transmitter and receiver is much shorter than for an outdoor one. The reason for that is not only the geometrical parameters of a building but also the low transmitter power and high attenuation caused by internal walls and furniture. These phenomena have an impact on the channel impulse response length, so channels are characterized by much lower delay spread than the delay spread observed in typical outdoor channels. Propagation inside buildings which have fewer metal inside walls and do not have too many hard partitions (heavy walls) usually results in a small root-mean square (rms) delay spread (usually 30 to 60 ns). Larger buildings with many metal elements inside and large free space can have an rms delay spread of the order of 300 ns. Such a long channel impulse response determines the upper limit for the applied data rate or implies the application of receiver structures capable of dealing with specific channel properties. Closely related to the channel impulse response is the power delay profile. Intensive measurements performed at Lund University for the 1800 MHz band have shown that although a particular shape of the power delay profile depends on the actual objects in the environment, the average power delay profile for a cluttered environment shows strong regularities. It is well approximated by either a power function which means that the decrease is logarithmic in the dB scale, or an exponent function (the decrease is linear in the dB scale) [25]. In open area the power delay profile is well approximated by the power function, which is caused by a strong influence of the direct path. Figure 3.16 shows typical averaged power delay profiles in line-of-sight (LOS), non-line-of-sight (NLOS) and obstructed-line-of-sight[6] (OLOS) transmission.

The indoor channel can be time-varying. The reasons for time variations are the changes in the mobile terminal location, the change of the antenna orientation if the antenna is characterized by a non-isotropic radiation pattern, and the movement of scattering objects such as persons or office furniture and equipment [23].

Analysis of large measurement data has shown that the path loss for an indoor channel can be estimated by the following formula

$$L(d) = L(d_0) + 10n \log \left(\frac{d}{d_0} \right) + X_\sigma \tag{3.71}$$

where X_σ is a Gaussian random variable in the dB scale with the variance σ^2. The formula is analogous to the log-normal shadowing model. For particular carrier frequencies which are characteristic of wireless telephone or PCS systems, the parameters n and σ summarized in Table 3.3 have been measured for different types of walls and environments in different buildings [22]. The model of the indoor propagation loss described

[6]Obstructed-line-of-sight means that the direct path between the transmitter and the receiver is obstructed by a single obstacle.

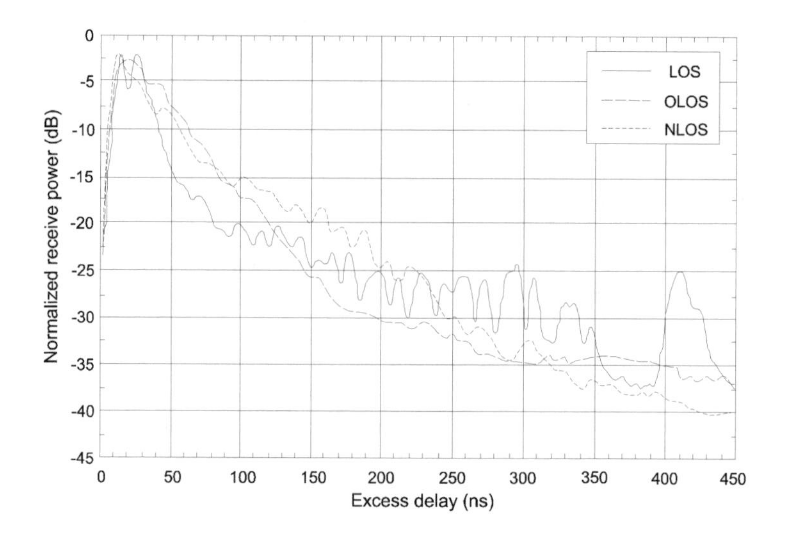

Figure 3.16 Average power delay profiles in LOS, NLOS and OLOS conditions measured in Lund (*Source*: Digital Mobile Radio Towards Future Generation Systems, COST Action 231, Final Report, EUR 18957, 1999)

by (3.71) is called a *one-slope model* (ISM) because it assumes a linear dependence between the path loss expressed in dB and the logarithmic distance.

The *multi-wall model* [24] takes into account not only the free space loss but also the losses caused by signal penetration through walls and floors along the direct path between the transmitter and the receiver. It has been found that the total floor loss is a non-linear function of the number of penetrated floors [22]. Therefore, an empirical factor b is introduced [24] and the path loss is described by the expression

$$L = L_s + L_c + \sum_{i=1}^{I} k_{w,i} L_{w,i} + k_f^{\left(\frac{k_f+2}{k_f+1}-b\right)} L_f \tag{3.72}$$

where L_s is the free space loss, L_c is the constant loss, $k_{w,i}$ denotes the number of penetrated walls of type i, k_f – the number of penetrated floors, $L_{w,i}$ – the loss in a wall of type i, L_f – the loss between adjacent floors, I – the number of wall types, usually $I = 2$ (for light and heavy walls). The values estimated on the basis of measurements of the above variables can be found in [23].

The third, very simple propagation model is called a *linear attenuation model* [23]. It is based on the assumption that the access path loss (in dB) is linearly dependent on the distance (in meters)

$$L(d) = L_{\mathrm{FS}}(d) + \alpha d \tag{3.73}$$

where the constant α in (3.73) is the attenuation coefficient.

Wireless LANs are implemented in bands over 1800 MHz. The bands around 2.5 GHz, 5 GHz, 60 GHz and recently 17 GHz are used for that purpose. Finding the channel

Table 3.3 Path loss exponent and standard deviation for different types of buildings

Building	**Frequency** [MHz]	n	σ [dB]
Retail stores	914	2.2	8.7
Grocery store	914	1.8	5.2
Office, hard partition	1500	3.0	7.0
Office, soft partition	900	2.4	9.6
Office, soft partition	1900	2.6	14.1
Factory LOS[a]			
Textile/chemical	1300	2.0	3.0
Textile/chemical	4000	2.1	7.0
Paper/cereals	1300	1.8	6.0
Metalworking	1300	1.6	5.8
Suburban home			
Indoor to street	900	3.0	7.0
Factory OBS[b]			
Textile/chemical	4000	2.1	9.7
Metalworking	1300	3.3	6.8

[a]LOS – Line-of-sight. [b]OBS – obstructed sight.

characteristics is usually one of the first activities of the system design in a new band. Intensive measurements are usually conducted (see [26] as the example) or the ray-tracing method is used to simulate signal propagation in a specific indoor environment [27]. The results of these activities are statistical properties of the particular indoor channel which allow construction of the channel models useful in the system simulation as well as in the transmitter and receiver design.

* * *

Propagation models are very wide and important from the practical point of view. There are several other propagation models used in cellular radio which have not been described in this chapter, among others the Eglie, Ibrahim and Parsons, and Longley-Rice models.The interested reader is advised to study the books [11], [16], [19] and [20].

3.6 INFLUENCE OF THE CHANNEL ON THE TRANSMITTED NARROWBAND AND WIDEBAND SIGNALS

So far we have considered physical properties of mobile communication channels only. We have shown how the multipath propagation influences the channel impulse response. In a real system the radio channel is only a single block in the transmission chain. We also have to take into account not only the properties of the channel but the properties of the transmit and receive filters and the applied signaling rate, as well. Depending on the transmitted signal bandwidth the channel can influence the signal in different ways.

In order to clarify this observation let us consider digital transmission at three different data rates through the multipath channel. Let the first data rate be 600 b/s, the second one – 270 kb/s and the third one – 1.2288 Mb/s. The first data rate is typical for paging systems (see Chapter 4), the second one – for the GSM system (Chapters 7 – 9) whereas the third one is characteristic for the pseudorandom chip data signaling in the CDMA system IS-95 (Chapter 12). For simplicity, let us consider a two-path channel model. Besides the desired signal, a single echo signal reaches the receiver. The latter is delayed relative to the former by τ_0 and is phase shifted by ρ radians. This angle can also be expressed as $\rho = 2\pi f_0 \tau_0$ using the so-called *notch frequency* f_0 and the delay τ_0. Therefore, the channel transfer function is determined by the formula

$$H(f) = a\left(1 - b\mathrm{e}^{j\rho}\mathrm{e}^{-j2\pi f\tau_0}\right) = a\left(1 - b\mathrm{e}^{-j2\pi(f-f_0)\tau_0}\right) \tag{3.74}$$

The factor a denotes the attenuation of the received signal, whereas b describes the attenuation of the echo path relative to the direct path. Figure 3.17 presents the magnitude characteristics $|H(f)|$ of the channel for $\tau_0 = 10\,\mu$s for different signal bandwidths displayed around 900 MHz. One can easily show that the maxima of attenuation occur periodically along the frequency axis with the period of $1/\tau_0 = 100$ kHz and their location depends on the frequency f_0. The depth of the notches is strictly determined by the values of b.

Let us analyze the influence of the selective fading channel on the three above mentioned types of signals. Let us note that for the first signal within the bandwidth of 600 Hz the channel characteristics are practically constant. If the values of b change, then the signal level varies approximately equally in the whole signal bandwidth. For this narrowband signal we observe that the channel is *flat fading*. The delay between both paths ($\tau_0 = 10\,\mu$s) is less than 1 percent of the modulation $T_d = 1/600$ s $= 1.66$ ms. The two-path propagation practically does not distort the data pulses; however, it causes the attenuation which is different from that which would exist if the channel were characterized by a single path propagation. The signals arriving along each path are practically unresolvable.

If the data transmission using binary signaling at the rate 270 kb/s is considered, then the signal bandwidth is much wider and in its range a few channel notches can be found. Thus, within the signal bandwidth some frequencies are substantially attenuated while others are not. Such phenomenon is called *selective fading*. For the binary transmission at the data rate 270 kb/s the modulation period is equal to 3.7 μs so it is a few times shorter than the channel second path delay $\tau_0 = 10$ μs. As a result, the channel introduces *intersymbol interference* which appears as time overlapping of the channel responses to subsequently transmitted data pulses. This phenomenon has to be compensated using complicated structures and receive algorithms. An example of such an algorithm applied in the GSM system can be found in Chapter 8.

Finally, the data transmission at the chip rate of 1.2288 Mb/s is wideband. In the signal bandwidth many maxima and minima of the channel characteristics can be found. In turn, analyzing the signal in the time domain, the second path delay of 10 μs is the reason for overlapping of the direct path signal by the echo signal delayed by 13 bit periods. Using specially selected binary sequences it is possible to extract from the

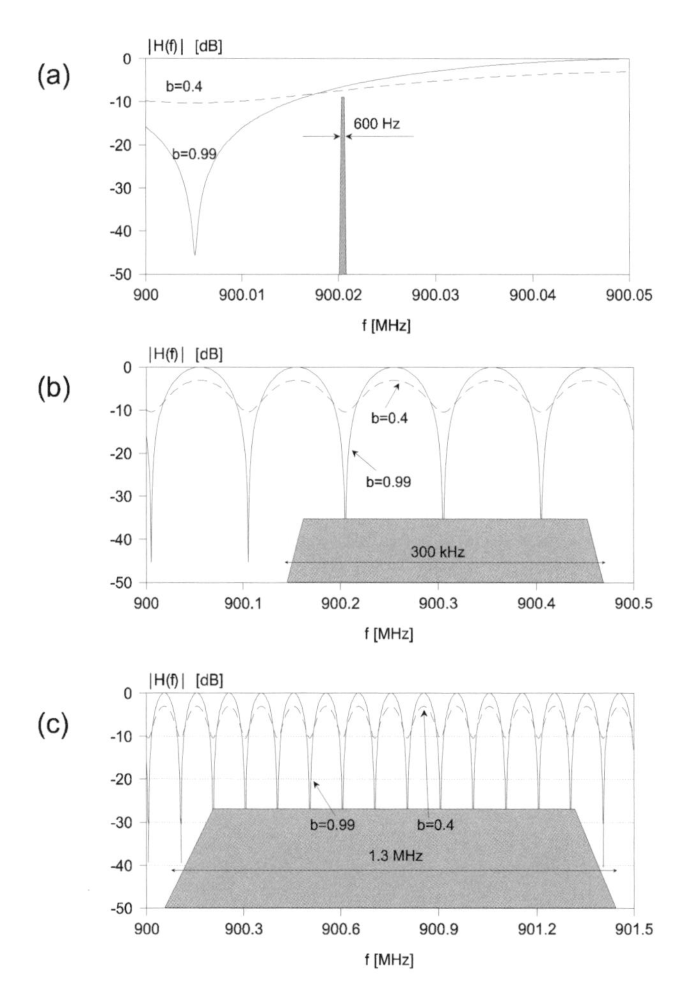

Figure 3.17 Two-path channel amplitude characteristics ($b = 0.99$ and $b = 0.4$, $\tau_0 = 10\ \mu$s, $f_0 = 10$ kHz) and the signal spectra of bandwidth: (a) 600 Hz, (b) 300 kHz, (c) 1.3 MHz

received signal the signals arriving along each path separately. In consequence, these signals can be properly combined in order to maximize the energy of the resulting signal.

The examples shown above are clearly a simplification of real signal propagation through multipath channels. The existence of more than two paths complicates the shape of the channel transfer function; however, it does not change the essence of the multipath channel behavior. Concluding, the radio channel is the reason for flat or selective fading which is the result not only of the channel physical features but also of the transmitted signal bandwidth.

3.7 DIVERSITY RECEPTION

Fading is one of the most severe distortions introduced by radio channels. Let us recall that fading can be categorized not only in the frequency domain (flat and selective fading) but also in the time domain (slow and fast fading). There are several methods to minimize the influence of fading on the system performance. Generally, if it is possible to receive a few replicas of the transmitted signal over different channels, then there is a high probability that transmission over at least one of them will result in good signal quality at the receiver. Appropriate combining of the signals received over different channels constitutes *diversity reception*. There are two categories of diversity, explicit and implicit diversity [28]. The first category is called the explicit technique. Redundant signal transmission is applied in the diversity technique belonging to the first category. An example of this is the transmission of the same signal on two appropriately separated carrier frequencies, which enables the receiver to detect separate signals and to combine them. In the second category the signal is transmitted only once but thanks to inherent features of the propagation medium and to applying special reception techniques it is possible to create multiple channels. The example of exploiting the implicit diversity is a RAKE receiver which extracts the signals arriving along each path of a multipath channel and combines them in an optimal way (see Chapter 11).

There are a few types of *diversity reception* used in several radio communication systems. They are the following [28]:

- space diversity,
- frequency diversity,
- time diversity,
- path diversity,
- polarization diversity.

In mobile communications diversity can be applied both in a transmitter and receiver. Therefore the terms *transmitter diversity* and *receiver diversity* are also introduced.

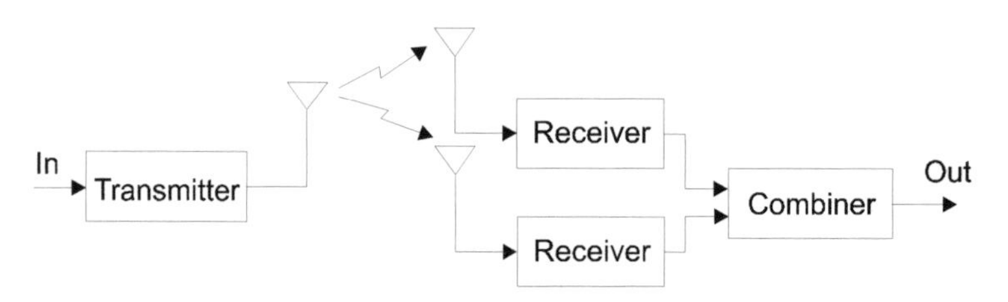

Figure 3.18 Space diversity scheme

Space diversity (Figure 3.18) relies on appropriate combining of the signals received by at least two antennae which are separated in space. The distance between antennae should be large enough to ensure the fadings occurring in the two channels between the transmitter and respective receivers to be mutually uncorrelated. Combining the diversity branch signals should be made on the vector basis. Several combining arrangements are possible from the implementation point of view. Figure 3.19 shows three solutions. The simplest one is *selection combining*. The control unit checks the carrier-to-noise ratio in each diversity branch and connects that branch to the output which currently has the highest value of C/N. In *maximal ratio combiner* the branch signals are cophased and weighted proportionally to their level. Thus, the maximum signal-to-noise ratio is received at the output of the combiner. However, the branch signal amplitudes have to be estimated. If it is difficult to perform, the branch signals can be summed with equal gain after having been cophased only. This method is called *equal gain combining*.

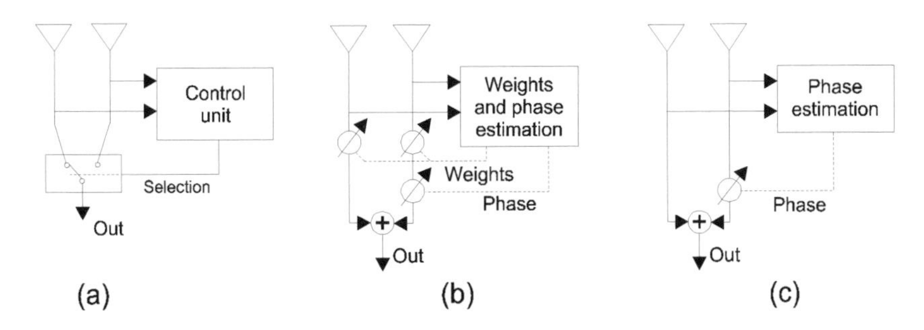

Figure 3.19 Several types of combining: (a) selection combining, (b) maximum ratio combining, (c) equal gain combining

Space diversity leads to substantial performance improvements in mobile communication systems. The price paid for that is the increased hardware complexity – at least part of the receiver branch has to be duplicated. Therefore, space diversity is applied mostly in base stations. Evaluation of the influence of space diversity on the GSM receiver performance can be found in [10].

Frequency diversity relies on transmitting the same signal on at least two sufficiently separated carrier frequencies. The frequency separation should be large enough for the fading processes on both frequencies to be uncorrelated. This diversity scheme requires additional spectral and hardware resources.

Time diversity is a method which can be used on channels with relatively fast fading. If the same signal is sent several times in the sufficiently separated time intervals, the fades in each interval are uncorrelated and repeated signals can be properly combined at the receiver, provided that not all of them are faded. Sending the same symbols a few times can be considered as a simple repetition code. As we know, much better codes exist. In general, an improved version of time diversity, as compared with a simple repetition, relies on error correction coding with sufficiently deep interleaving. However, the necessary condition of successful operation of the receiver with time diversity is that the interleaving spans a longer time interval than the fade duration. If fading were slow, the interleaver would have to be very large and it would introduce an unacceptably long transmission delay. The ARQ techniques can be also considered as a kind of time

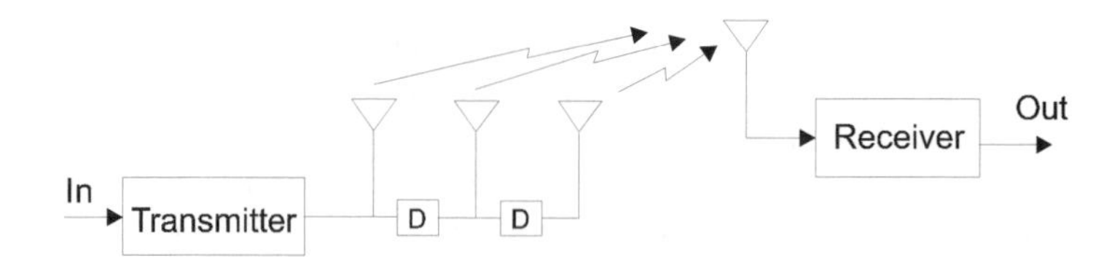

Figure 3.20 Transmitter diversity

diversity. In the ARQ techniques, diversity reception is performed in the form of the block repetition only if it is required by the receiver sending the NAK message. We can classify the reception of signals corrupted by intersymbol interference, in which the receiver decides upon a specific data symbol on the basis of signal samples collected during appropriately long time interval, as time diversity as well.

Path diversity is realized in spread spectrum systems if the receiver is able to resolve the received signal components arriving at it along different channel paths. It is possible if the relative delays among paths are not shorter than a single chip period of the spreading sequence and if the spreading sequence is white. The path signals extracted through the correlation with the properly synchronized spreading sequence are subsequently cophased, weighted and summed. The last operation is practically a maximum gain combining. Such a receiver is called RAKE and will be the subject of our analysis in Chapter 11.

Polarization diversity is usually applied in line-of-sight radio systems as the explicit technique, namely, the same signal is transmitted and received in two orthogonal polarizations. Such an arrangement is not very useful in mobile communications. Instead, implicit polarization diversity can be applied in base stations. An individual user can hold the handheld terminal at practically any orientation. Thus, the signal from the mobile station can be transmitted at varying polarization angles. In consequence, cross-polarized antennae applied in base stations substantially improve the signal reception [28].

The techniques described above relate mostly to the receiver diversity. Only the frequency diversity requires replication of some transmitter blocks and involves transmitter and receiver diversity at the same time. Special arrangement of transmitter diversity has been proposed as well. The base station transmitter emits the delayed versions of the same signal from the set of M antennae separated in space [29] (Figure 3.20). Thus, M diversity branches between each transmit antenna and the receiver are created. If the transmit antennae are sufficiently spaced, each diversity branch channel fades independently of the others. The mobile station receiver which is able to cope with intersymbol interference (ISI) collects the combined signal consisting of the attenuated and delayed version of the transmitted signal. It is highly probable that not all the branch signals are attenuated by fading so the combined signal arriving at the ISI receiver results in the improved receiver performance. Transmitter diversity has also been investigated for the application in the GSM system [30].

* * *

In this chapter we presented only selected topics associated with mobile radio channel characterization. This area continuously attracts the attention of researchers and engineers due to the fact that higher and higher spectral bands are adopted for wireless transmission. Each new band requires conducting channel measurements and building a new channel model. The literature dealing with channel modeling, propagation prediction, channel measurements and channel simulation is very rich. Special books have been devoted to this subject [31], [32]. We have to stress once more that without the knowledge of channel characteristics and its time variability the design of a high performance wireless system is not possible.

REFERENCES

1. K. Pahlavan, A.H. Levesque, *Wireless Information Networks*, John Wiley & Sons, Inc., New York, 1995

2. W. C. Jakes (Ed.), *Microwave Mobile Communications*, John Wiley & Sons, Inc., New York, 1974

3. G. Kadel, R. W. Lorenz, "Breitbandige Ausbreitungsmessungen zur Charakterisierung des Funkkanals beim GSM-System", *Frequenz*, Vol. 41, No. 7-8, 1991, pp. 158-163

4. J. G. Proakis, *Digital Communications*, 3rd Edition, McGraw-Hill, New York, 1995

5. ETSI/GSM 05.05, Radio transmission and reception, March 1991

6. J. H. Wei, "A Statistical Model for Digital Mobile Radio Channel Simulation", *The Radio and Electronic Engineer*, Vol. 1. No. 2, April 1991, pp. 12-19

7. J. D. Parsons, A.S. Bajwa, "Wideband Characterisation of Fading Mobile Radio Channels", *IEE Proceedings*, Vol. 129, Pt. F, No. 2, April 1982, pp. 95-101

8. W. Braun, U. Dersch, "A Physical Mobile Radio Channel Model", *IEEE Trans. on Vehicular Technology*, Vol. 40, No. 2, May 1991, pp. 472-482

9. H. Schulze, "Stochastische Modelle und digitale Simulation von Mobilfunkkanälen", *Kleinheubacher Berichte*, Bd. 32, 1989, pp. 473-483

10. R. Krenz, K. Wesołowski, "Simulation Study on Space Diversity Techniques for MLSE Receivers in Mobile Communications", *IEEE Trans. on Vehicular Technology*, Vol. 46, No. 2, April 1997, pp. 653-663

11. T. S. Rappaport, *Wireless Communications, Principles and Practice*, Prentice-Hall, Upper Saddle River, N.J., 1996

12. W. C. Y. Lee, *Mobile Communications Design Fundamentals*, 2nd Edition, McGraw-Hill, New York, 1993

13. Y. Okumura, E. Ohmori, T. Kawano, K. Fukuda, "Field Strength and Its Variability in VHF and UHF Land-Mobile Radio Service", *Review of the Electrical Communication Laboratory*, Vol. 16, No. 9-10, 1968, pp. 825-873

14. M. Hata, "Empirical Formula for Propagation Loss in Land Mobile Radio Services", *IEEE Trans. on Vehicular Technology*, Vol. 29, August 1980, pp. 317-325

15. P. E. Mogensen, P. Eggers, C. Jensen, J. B. Andersen, "Urban Area Radio Propagation Measurements at 955 and 1845 MHz for Small and Micro Cells", *Proc. of IEEE Global Commun. Conference* (GLOBECOM), Phoenix, 1991, pp. 1297-1302

16. G. K. Stüber, *Principles of Mobile Communication*, Kluwer Academic Publishers, Boston, 1996

17. V. K. Garg, J. E. Wilkes, *Wireless and Personal Communications Systems*, Prentice-Hall, Upper Saddle River, N.J., 1996

18. K. David, T. Benkner, *Digitale Mobilfunksysteme*, B. G. Teubner, Stuttgart, 1996

19. M. D. Yacoub, *Foundations of Mobile Radio Engineering*, CRC Press, Boca Raton, Fl., 1993

20. W. C. Y. Lee, *Mobile Cellular Telecommunications, Analog and Digital Systems*, 2nd Ed., McGraw-Hill, New York, 1995

21. S. R. Saunders, *Antennas and Propagation for Wireless Communication Systems*, John Wiley & Sons, Ltd., Chichester, 1999

22. J. B. Andersen, Th. S. Rappaport and S. Yoshida, "Propagation Measurements and Models for Wireless Communications Channels", *IEEE Communications Magazine*, January 1995, pp. 42-49

23. COST Action 231, *Digital Mobile Radio Towards Future Generation Systems*, Final Report, EUR 18957, 1999

24. C. Törnevik, J.-E. Berg, F. Lotse, "900 MHz Propagation Measurements and Path Loss Models for Different Indoor Environments", *Proc. IEEE VTC'93*, New Jersey, USA 1993

25. P. Karlsson, H. Börjeson, T. Maseng, "A Statistical Multipath Propagation Model Confirmed by Measurements and Simulations in Indoor Environments at 1800 MHz", *Proc. of IEEE PIMRC'94*, Amsterdam 1994, pp. 486-490

26. A. Bohdanowicz, G. J. M. Janssen, S. Pietrzyk, "Wideband Indoor and Outdoor Multipath Channel Measurements at 17 GHz", *Proc. IEEE VTC'99 Fall*, Amsterdam, pp. 1998-2003

27. M. Lobeira, A. Armada, R. Torres, J. L. Garcia, "Parameter Estimation and Indoor Channel Modeling at 17 GHz for OFDM-based Broadband WLAN", IST Mobile Communication Summit, Galway, October 1-4, 2000

28. A. Paulraj, "Diversity Techniques", in J. D. Gibson (Ed.) *The Mobile Communications Handbook*, CRC Press, IEEE Press, 1996

29. J. H. Winters, "The Diversity Gain of Transmit Diversity in Wireless Systems with Rayleigh Fading", *Proc. of IEEE ICC'94*, pp. 1121-1125

30. P. E. Mogensen, "GSM Base-Station Antenna Diversity Using Soft Decision Combining on Up-link and Delayed-Signal Transmission on Down-link", *Proc. of IEEE VTC'93*, pp. 611-616

31. J. D. Parsons, *The Mobile Radio Propagation Channel*, John Wiley & Sons, Inc., New York, 1992

32. J. Cavers, *The Mobile Communication Channel*, Kluwer Academic Publishers, Boston, 2000

4

Paging systems

4.1 INTRODUCTION

Basic features of paging systems have been mentioned in the chapter devoted to the survey and classification of mobile systems. This chapter contains more precise description of properties, classification and typical configurations of paging systems. Besides general characteristics we will consider a few examples of paging systems which are currently in operation.

4.2 BASIC CHARACTERISTICS OF PAGING SYSTEMS

The ITU-R Recommendations No. 539 and 584 define a paging system as a unidirectional broadcasting radio system which is used to transmit alerting signals or short numeric or alphanumeric messages, excluding voice messages [1]. In practice, some paging systems are able to transmit short voice messages as well.

Classical paging systems have the following properties:

- the use a narrow band in the range of a few hundred MHz or the VHF band in the radio broadcasting range,
- small size of mobile receivers (*pagers*),
- the lack of confirmation that the message has been received.

The last feature, which is a consequence of the unidirectional transmission, is the main drawback of classical paging systems.

The numeric message is often a telephone number which the message recipient should call. Alphanumeric messages contain a text, stock market bulletins, or other messages which can be read on the small liquid crystal display of a pager. The speech signal lasts for a few seconds and is transmitted after being converted to the digital form. In the most traditional case a text message is dictated to the paging system operator in the call center who then sends the digital message to the pager through the base station. A message can also reach the call center through a modem link or the Internet.

4.3 CLASSIFICATION OF PAGING NETWORKS

Paging networks can have various configurations, coverage area and applications. In general, paging networks are grouped into the following categories:

- private networks (e.g. based on the ETSI 300 224 standard [2]); such networks are installed and operate within an institution or a company and their coverage is restricted to the area of the company,
- public networks used by private users, which can be further divided into the following subcategories:
  - local networks whose range is limited to one city and its vicinity, and
  - national networks operating in the whole country.

In order to use the assigned spectrum efficiently and not to distort the operation of other systems, the paging system has to fulfill a set of requirements. For private networks the requirements determine the range of applied frequencies, the base station power level, selected emission parameters such as the allowable level of adjacent channel interference, the channel separation, stability of the transmitter frequency, the antenna radiation level and the exploitation parameters in extreme temperatures and extreme power supply voltages.

Since private networks are small, a single base station covers the whole network area. In case of particularly large private networks, a few base stations are installed which operate synchronously on a single carrier. Transmission methods, error detection/correction and the applied modulation are not standardized and depend only on the system supplier. The whole network equipment is usually supplied by a single vendor. The system consists of mobile stations (pagers), a base station and a paging switching center.

A typical local network uses a single carrier frequency in the whole coverage area, as private networks do. The base stations mostly work synchronously with high accuracy. Since the pager can simultaneously receive the same signal transmitted by several base stations, lack of synchronization among the base stations would cause detection problems in the pagers. In order to ensure synchronous operation of the base stations, it is necessary to control and keep transmission delays between the switching center and the base stations within allowable limits. The delay difference between the signals reaching the pager from two different base stations cannot be higher than 1/4 of the duration of one bit. For a typical data rate and the radio propagation velocity we conclude that the base stations have to be located in the distances not larger than 8 km and, in order to avoid mutual coverage from different base stations, their antennae should not be placed higher than 100 m above the ground. Sometimes the local paging system designer can avoid the synchronous operation of the base stations. In such cases, the paging signals are emitted sequentially by subsequent base stations. In consequence, the system capacity is limited.

Several protocols are applied in paging systems. The most popular protocol up to date is described in ITU-R Recommendation No. 584 and known as POCSAG (*Post Office Code Standardization Advisory Group*). In the recommendation it is denoted as CCIR Radiopaging Code No.1. It will be described in the next section.

Another protocol, which was developed in the seventies in Sweden, is called MBS (*Mobile Search*). Digital messages are transmitted to pagers, using existing FM radio broadcasting stations. The paging signal is placed around the center frequency equal to 57 kHz, above the spectrum of the stereo signal. Such a composite signal is fed to the FM modulator. At the receiver, the pager extracts the paging signal by filtering it from the FM demodulated mixture of stereo and paging signals, converts it to the baseband and detects the binary stream containing the message. Figure 4.1 presents the location of the paging signal on the frequency axis prior to FM modulation or after FM demodulation.The symbols L+R and L-R denote the sum and difference of the left and right stereo signals, respectively. The MBS protocol is not widely used nowadays.

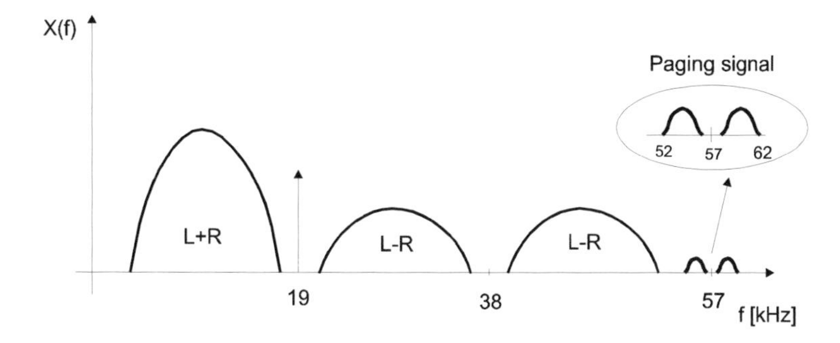

Figure 4.1 Location of the paging signal with respect to the stereo radio signal in the MBS paging format

Besides the popular POCSAG protocol, a few other more advanced protocols will be briefly described in this chapter. These protocols allow use of a paging network with at least a country coverage. They are known as

- APOC (*Advanced Paging Operators Code*), the protocol developed by Philips, which is an extension and improvement of the POCSAG protocol,

- ERMES (*European Radio Messaging System*), the protocol standardized by ETSI, in which roaming and sending acknowledgement messages are possible,

- FLEX, the protocol developed by Motorola; it is gaining popularity in the USA and many other countries; in its ReFLEX extension a pager sends back the acknowledgement messages to the message transmitter.

In the following sections we will discuss the POCSAG, ERMES and FLEX protocols.

4.4 THE POCSAG PROTOCOL

The POCSAG protocol was developed between 1975 and 1978 by an advisory group of international experts hosted by the British Post Office. This justifies the acronym of the protocol, explained in the previous section.

In the basic version of the protocol the paging signals are transmitted at the data rate of 512 bit/s. The modified versions allow operation at the data rates of 1200 and 2400 bit/s. Thus, the allowable number of messages and their length increase. However, the higher data rate implies increased complexity of the whole system structure and the necessity for synchronous operation of base stations.

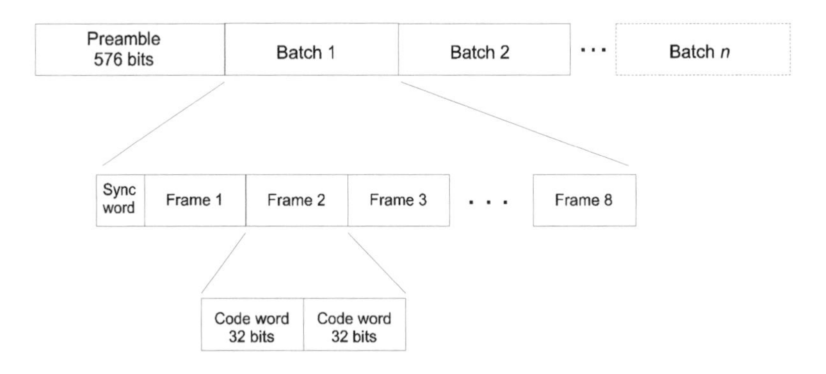

Figure 4.2 Time hierarchy in the POCSAG protocol

The transmitted data have a hierarchical structure. The basic data unit is a 32-bit code word. Two code words constitute a frame. A special 32-bit synchronization word and 8 following frames create a batch. The batches are transmitted sequentially, starting from a 576-bit preamble code that is used to wake up pagers which are in a battery-save mode. Figure 4.2 presents the time organization of the data sequences.

The preamble is a sequence of 576 alternating ones and zeros and is is used by the pager to acquire the bit synchronization. As we have mentioned, a batch starts with a synchronization code word which contains a specific binary sequence. The frames transmit the code words to the addressed groups of pagers. The pagers are divided into 8 groups. The assignment to a particular group is determined by the three least significant bits of the pager's individual address. The paging of an individual receiver takes place only in the frame assigned to its group. There are two types of words placed in the batch after the synchronization word (see Figure 4.3):

a)

1	2 - 19	20-21	22 - 31	32
F	Address	J	K	P

b)

1	2 - 21	22 - 31	32
F	Message	K	P

Figure 4.3 Word formats in the POCSAG protocol: (a) the address word (F = 0, the flag J determines the receiver function), (b) the information word (F = 1)

- the address word,
- the information word.

The flag F of the address word is set to zero. The address field contains 18 most significant address bits out of 21 address bits of the individual pager. Two bits denoted by J describe the pager mode. The address word is supplemented with 10 parity control bits K and the overall parity bit P.

The flag F of the information word is set to one. The flag is followed by 20 message bits supplemented with 10 parity control bits K and the overall parity bit P.

Although the address words occur only in the frame belonging to the appropriate group, the information words sent to the addressed pager directly follow the address word without taking into account the group structure. After finishing the message sent to the particular pager a message transmitted to another one starts from the address word placed in the next free frame belonging to group of the called pager. Transmission of messages in the form of batches implies that in case of the information sequence not fully filling the message fields of the given batch, the missing words are replaced by the stuffing bits.

As we have already mentioned, a 32-bit code word consists of the flag bit, 20 information bits, 10 parity control bits and the overall parity bit. All but the last bit constitute a code word of the (31,21) BCH code generated by the polynomial

$$g(x) = x^{10} + x^9 + x^8 + x^6 + x^5 + x^3 + 1 \tag{4.1}$$

The parity control bits are the result of dividing the following polynomial

$$m_{20}x^{30} + m_{19}x^{29} + \ldots + m_1x^{11} + m_0x^{10}$$

by the generator polynomial (4.1), where $(m_{20}, m_{19}, \ldots, m_0)$ are the flag and message bits.

The messages are transmitted in two formats. The numeric format is exclusively used to send numbers, such as the telephone number to which a user should call. The digits have a four-bit BCD format. In this case two bits J describing the receiver function have the value J = 00. One code word contains five digits. If the word is not fully filled with digits, the free place is filled with the space characters.

In the alphanumeric mode, when J = 11, the characters are coded according to the CCITT Alphabet No.5.

At the transmitter the bit stream is represented in the form of NRZ (*Non-Return to Zero*) pulses which modulate the carrier using the differential frequency shift keying (DFSK) [3].

The POCSAG protocol is used in many networks operating in the whole world and has become a popular protocol in high-volume paging applications. The POCSAG protocol can support up to 2 million individual pagers. Various carrier frequencies are used in the paging systems applying the POCSAG protocol, depending on the particular network operator. For example, the German paging system called "Cityruf" uses the following carriers: 465.970 MHz, 466.075 MHz and 466.230 MHz [4]. Several other bands are used for POCSAG paging such as 155, 148, 161, 170 MHz, etc.

4.5 ERMES - THE EUROPEAN PAGING SYSTEM

The ERMES paging system is the result of a common initiative of the European Union countries. The work on the common paging standard began in 1987. In 1990 26 operators from 16 countries signed the Memorandum of Understanding (MoU) agreeing to create an entirely new paging standard. The representatives of all countries who signed the MoU agreed to allocate the frequency range 169.4-169.8 MHz to the new paging system. In 1992 the European Standardization Institute (ETSI) approved the standard of the ERMES system, and in 1994 the International Telecommunication Union (ITU) recommended the ERMES standard to be the first world standard of the paging system with the capability of international operation and roaming.

The basic targets to be achieved by ERMES were as follows:

- ability of the pager to operate outside its own network and country,
- high information throughput and much higher number of users as compared with other existing paging systems,
- higher data rates as compared with the systems based on the POCSAG protocol,
- receiver standardization allowing use of the pagers in various ERMES networks.

The ERMES system offers a number of basic services, such as:

- tone paging – there are 8 different alerting tones,
- numeric paging – the maximum length of a numeric message is equal to 16000 digits,
- alphanumeric paging – the maximum length of an alphanumeric message is 9000 characters.

Additionally, *transparent data* transmission of the blocks not larger than 64 kbits is offered. Transmission transparency means that the system rhythmically sends the data stream out of the receiver at the same data rate as that at which the data are fed to the transmitter. Thus, despite some form of error correction coding typically used in the transmitter and receiver, the data rate is constant. However, the price paid for that is variable data quality at the receiver, depending on the current channel conditions. On the contrary, in *non-transparent transmission* special means of increasing data reliability are undertaken (such as ARQ, see Chapter 1), which results in a variable data rate at the receiver output. The reason for the variable data rate is the necessity of retransmission of the erroneous blocks. As a result, the achievable error rate is very low and practically constant.

In accordance with the set of basic services offered by ERMES, the following types of receivers are defined: tone-only, numeric, alphanumeric and transparent data receivers.

Besides the basic service set, the following supplementary services can be offered by the system operator: acknowledgements of message reception, group and collective calls, forwarding a call to another receiver, storing of incoming messages, setting the call priorities, limiting the area of paging, message encryption and others.

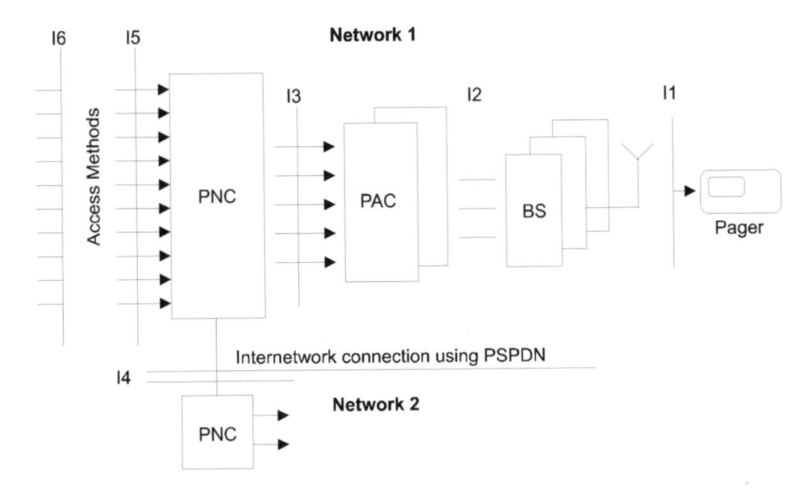

Figure 4.4 Architecture of ERMES paging network

The ERMES network structure is shown in Figure 4.4 [5]. Its basic elements are:

- a set of receivers (pagers),
- *Base Station System* (BS),
- *Paging Area Controller* (PAC),
- *Paging Network Controller* (PNC),
- *Operation and Maintenance Center* (OMC, which is a part of the telecommunications management network and is not shown in Figure 4.4).

The symbols I1 - I6 in Figure 4.4 denote the interfaces between particular network blocks and the inter-network interface. The tasks of particular elements are following.

A receiver (pager) receives the signals, demodulates these signals, decodes the receiver address, decodes the information blocks sent to it and displays them on the screen or informs the subscriber about the message by tone.

The *Paging Network Controller* (PNC) is the central unit in the network, managed by the given operator. Because networks of several operators can work at the same time, there is a possibility of interconnections among them on the PNC level. This fact is illustrated in Figure 4.4. Such interconnections are realized using a *Public Data Packet Network* (PDPN). The PNC is also connected to the access network from which it receives the paging messages and to the *Paging Area Controllers* to which it sends the received messages. The *Operation and Maintenance Center* realizes the monitoring of the PNC and network operation.

The *Paging Area Controller* manages the operation of the network in a selected paging area. This area is covered by a few base stations connected to this PAC. The PAC receives messages from the PNC which are to be sent, orders them in a queue, organizes them in groups and sets the priorities before transmitting them to a base station. The PAC also preforms some maintenance functions and cooperates with the *Mediation Device* (MD). The task of the latter is to mediate in communication between the base stations and the OMC.

The base station receives the paging messages from the paging area controller, encodes them and attaches the synchronization and identification information. The receiver address and part of the message are secured by the shortened cyclic code (30,18). The systematic code words are generated using the generator polynomial [6]

$$g(x) = x^{12} + x^{11} + x^9 + x^7 + x^6 + x^3 + x^2 + 1$$

In order to spread error bursts caused by the channel conditions the 9-word interleaving is applied. The binary signal obtained modulates the carrier in the selected channel. Then the modulated signal is emitted to the mobile receiver.

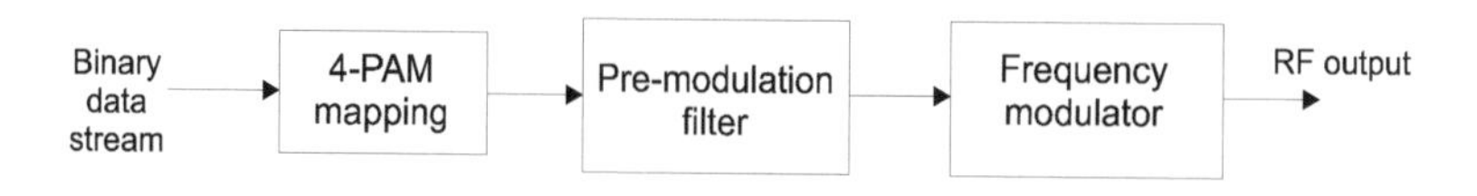

Figure 4.5 4-FSK modulator used in ERMES

In the ERMES system the 4-PAM/FM (4-FSK) modulation is applied. The binary data are grouped in dibits which determine one of four PAM amplitudes of the pulse signal which is, after shaping by the pre-modulation filter, fed to the FM modulator (see Figure 4.5).

The ERMES system can operate in up to 16 channels spaced by 25 kHz in the band between 169.4125 and 169.8125 MHz. Figure 4.6 shows a single channel along the frequency axis. Its carrier frequency is located in the middle of the 25 kHz band. The frequencies f_1, f_2, f_3 and f_4 denote the nominal frequencies for each transmitted dibit.

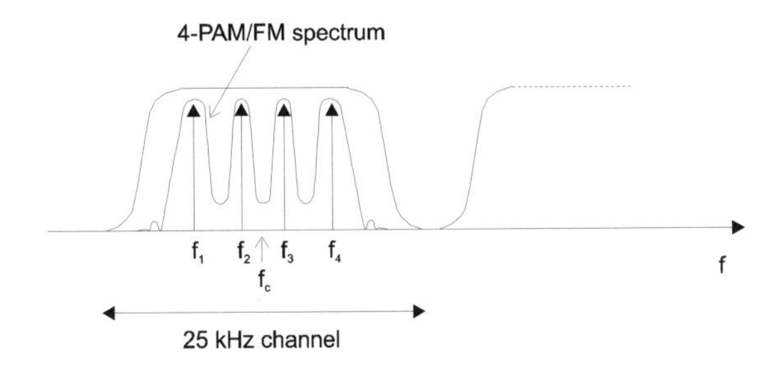

Figure 4.6 ERMES channel spectrum on the frequency axis

They differ by ±1562.5 Hz or ±4687.5 Hz from the carrier frequency. In consequence, the frequency division network operation is possible if more than one frequency channel is used by the network operator. However, time division access is an inherent ERMES access method due to the time hierarchy applied in the ERMES radio access protocol shown in Figure 4.7. If the network uses different time slots in neighboring paging areas, we can describe ERMES as a time divided network.

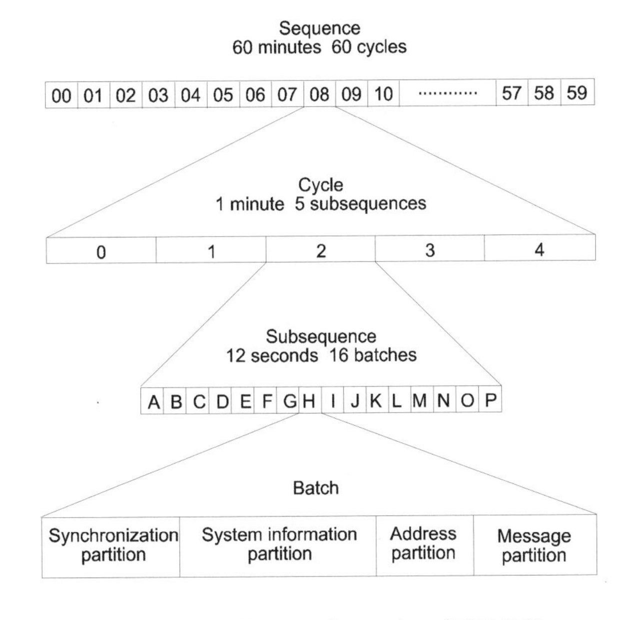

Figure 4.7 The time hierarchy of ERMES

The binary data stream is organized in 60 minute sequences. The emission of sequences is performed in synchronism with UTC (*Universal Time Coordinated*). The sequence consists of 60 one minute cycles. In turn, each cycle is divided into five subsequences. The subsequence lasts for 12 seconds. Due to the synchronism with respect to UTC, the zero subsequence is placed after the minute beacon of the UTC signal. The subsequences consist of *batches* directed to particular receivers or groups of receivers. There are 16 batches in each subsequence. They are denoted by characters A to P. The batch is a whole which has to be treated jointly in order to ensure correct reception of the message sent to a particular subscriber. The batch consists of the synchronization partition, the system information partition, the address partition and the actual message. The binary channel data rate is equal to 6.25 kbit/s. The placement of the batch in a subsequence determines its affiliation to a particular group.

Pagers also divided into 16 groups are denoted as type A to type P receivers. Their "membership" in a particular group is determined by four least significant address bits of the receiver's *Radio Identity Code* (RIC). The remaining part of the address is transmitted exclusively within the group to which the receiver belongs. In the synchronous system operation, most of the time the main receiver parts remain in a low-battery mode (see Figure 4.8). This allows them to save a substantial amount of energy. The address in a given group can be transmitted more than once, even at different carrier frequencies used in the system. After decoding its own address the receiver waits for the message directed to it. This message can be sent in the same batch in its last partition, in any following batch in the same subsequence or in the next subsequence. Each batch contains 154 code words except the last one, which has 190 code words. As we remember, the code word consists of 30 bits.

■= Active batch for a type A receiver

channel	← 12 seconds →																			
01	■	B	C	D	E	F	G	H	I	J	K	L	M	N	O	P	■	B	C	D
03	P	■	B	C	D	E	F	G	H	I	J	K	L	M	N	O	P	■	B	C
05	O	P	■	B	C	D	E	F	G	H	I	J	K	L	M	N	O	P	■	B
07	N	O	P	■	B	C	D	E	F	G	H	I	J	K	L	M	N	O	P	■
09	M	N	O	P	■	B	C	D	E	F	G	H	I	J	K	L	M	N	O	P
11	L	M	N	O	P	■	B	C	D	E	F	G	H	I	J	K	L	M	N	O
13	K	L	M	N	O	P	■	B	C	D	E	F	G	H	I	J	K	L	M	N
15	J	K	L	M	N	O	P	■	B	C	D	E	F	G	H	I	J	K	L	M
16	I	J	K	L	M	N	O	P	■	B	C	D	E	F	G	H	I	J	K	L
14	H	I	J	K	L	M	N	O	P	■	B	C	D	E	F	G	H	I	J	K
12	G	H	I	J	K	L	M	N	O	P	■	B	C	D	E	F	G	H	I	J
10	F	G	H	I	J	K	L	M	N	O	P	■	B	C	D	E	F	G	H	I
08	E	F	G	H	I	J	K	L	M	N	O	P	■	B	C	D	E	F	G	H
06	D	E	F	G	H	I	J	K	L	M	N	O	P	■	B	C	D	E	F	G
04	C	D	E	F	G	H	I	J	K	L	M	N	O	P	■	B	C	D	E	F
02	B	C	D	E	F	G	H	I	J	K	L	M	N	O	P	■	B	C	D	E

Figure 4.8 Channel synchronization and example of the scanning procedure for a type A receiver

Figure 4.9 ERMES system informatin partition

Figure 4.9 presents the system information partition contained in each batch. Its field acronyms have the following meaning:

- CC – *Country Code*,
- OC – *Operator's Code*,
- PA – *Paging Area*,
- CN – *Cycle Number*,
- SSN – *Subsequence Number*,
- BN – *Batch Number*,
- ETI – *External Traffic Indicator*,
- BAI – *Border Area Indicator*,
- FSI – *Frequency Subset Indicator*.

As mentioned, the ERMES system can operate on 16 frequency channels. A given operator offers its services in the channel subset which can consist of a single channel or a few channels. In consequence, a receiver has to be able to identify the channel on it will be paged. To perform this task, the receiver scans the channels in order to detect the frequency subset indicator (FSI) which is contained in the *Frequency Subset Number* (FSN) located in its own memory.

In the ERMES system some means have been foreseen to ensure message privacy, encryption of data transmitted in a transparent channel, subscriber's authentication, approval of the requested services and appropriate tariffing.

Each ERMES channel offers a capacity which is about five times higher than that of a traditional paging system. At the data rate of 6.25 kbit/s and assuming the mean number of calls to one subscriber equals to 0.2 per hour, each channel can support up to 500000 subscribers receiving numeric messages containing up to 10 digits or up to 160000 subscribers receiving alphanumeric messages 40 characters long. Very large addressing capabilities allow receipt of messages from various information networks concerning special information, e.g. city traffic messages, weather forecast, stock market information, sport results, etc.

ETSI considered the extension of one-way paging standard onto two-way paging. Such capabilities will considerably increase the spectrum of offered services. Two types of functionalities of two-way paging can be distinguished [7]:

- one-way messaging with either a system or a user-controlled acknowledgement of receipt,
- two-way messaging.

In the simplest case of two-way paging a return channel is created which is used to confirm the received messages. The first type of confirmation is the system acknowledgement. Its purpose is to enable the system to inform the message sender that his/her message has been successfully received. Its aim is also to improve the reliability of paging by retransmission of unacknowledged messages. The acknowledgement also creates the possibility to improve network management by transmission of a message in a selected part of the system coverage area only. The last feature can be realized by sending to the pager the message "where are you?". In response the pager will confirm the reception with the message "I am in paging area No. n".

The second type of confirmation is the user acknowledgement. This acknowledgement is initiated manually by the user of a two-way pager after receiving a message. The user can show his/her acceptance of the received message, his/her disagreement, or he/she can send back the number of one of the predefined messages, so called *canned messages*.

The pager or the pager user can also initiate a message transfer. Two types of initiated messages are possible:

- due to the system call – the pager automatically generates the message to be known in the network after switching the power on, appearing in the new non-home network (*roaming*), after re-appearing in the range of its own network, or after changing the paging area;
- due to the user call – the user-originated message contains the destination address and the actual message. The message can be a canned message, a combination of a canned message and numeric data, or can be a free format message containing numeric, alphanumeric or transparent data.

The physical aspect of the return channel has not been shown in [7].

4.6 FLEX FAMILY OF PAGING PROTOCOLS

In the early nineties Motorola developed a one-way paging protocol called FLEX. In the next years it was joined by the ReFLEX – a two-way protocol and the InFLEXion – a voice messaging protocol. Since then FLEX has become the industry standard and is a serious competitor to ERMES. Many countries such as China, Russia and Korea adopted FLEX as a national paging standard. It is difficult to find a precise description of the FLEX protocol in the open literature. Therefore, only some general information will be presented in this section.

The FLEX paging protocol is a synchronous time-slot protocol with the time clock synchronized to the *Global Positioning System* (GPS) [8]. Figure 4.10 shows the time hierarchy of the FLEX protocol. Each GPS hour is divided into 15 four-minute cycles. Cycle 0 starts at the beginning of each GPS hour. Each cycle is divided into 128 frames.

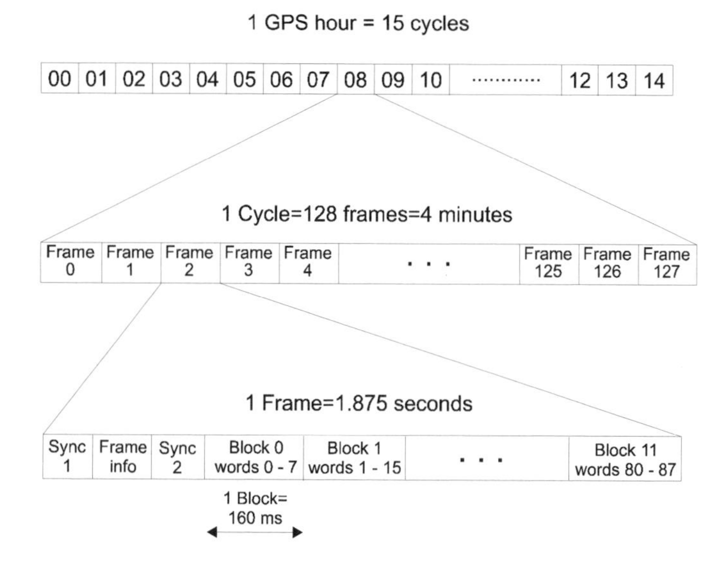

Figure 4.10 Time hierarchy of the FLEX paging system

Each FLEX frame consists of synchronization and transmitted data blocks. The frame starts with a synchronization signal followed by an 11-bit frame information word. This block allows the pager to identify the frame and cycle to which it is uniquely assigned. The second synchronization block contained in the frame indicates the rate at which the data words are transmitted. The second synchronization block is followed by 11 data blocks consisting of 8 words each. Each block lasts for 160 ms. We can easily calculate that at the data rate of 6400 bit/s each word consists of 128 bits. The words are the results of error correction coding and interleaving. Similar to the ERMES paging system, the 2- or 4-FSK modulation is applied. The data rates are 1600, 3200 and 6400 bit/s. The first two rates can be achieved using two-level FSK, whereas the highest data rate can be realized only with four-level FSK. RF transmission is realized in 25 kHz-channels.

The ReFLEX protocol provides an asymmetrical two-way data message delivery system for paging applications [8]. Its frame structure is compatible with the FLEX protocol. The data and control messages are sent to receivers over a forward channel. The reverse channel is used for transmitting message acknowledgements and short response messages from paging terminals equipped with keyboards. The reverse channel also allows monitoring of the location of the paging devices within a local geographic area.

ReFLEX systems operate in frequency bands which are multiples of 25 kHz. A single 25 kHz band supports a single FSK control and a message channel. A 50 kHz forward channel can support up to three FSK control and message channels separated by 12.5 kHz. The bands of 930-931 and 940-941 MHz have been assigned to forward channels, whereas the reverse channels work in the frequency range of 901-902 MHz. The ReFLEX protocol allows the construction of systems consisting of up to 8 forward control channels.

4.7 CONCLUSIONS: THE FUTURE OF PAGING SYSTEMS

Let us shortly discuss further possible applications of paging systems. Their simple structure, in particular of one-way paging systems without acknowledgement, has become a major limitation as compared with two-way communication systems such as cellular telephony. *Short Message Service* (SMS) messages have a similar character to paging messages. Nevertheless, it seems that paging systems are going to further improve their performance and the service offer and find some market niches. We can easily imagine persons who do not wish to be embarrassed by a mobile phone and wish to be free to answer the received message at a convenient moment. Paging systems can be easily connected to the Internet, allowing sending a copy of e-mail directly to the subscriber's pager.

Two-way paging is not paging in a strict sense any more; nevertheless, due to acknowledgement capabilities and the possibility to send short messages back to the base station, many new services are possible. A large class of services is associated with control applications. We can imagine a pager co-working with a GPS receiver installed in a car. Such a system can be an effective security system used to block the driving functions of a stolen car or to transmit its geographic position derived from GPS. Two-way paging systems with very large addressing capabilities can be applied in remote meter readings, vending machines and many other remote control and measurement applications. The non-real time nature of the communications gives very high spectrum utilization. Paging provides the highest number of subscribers per channel of any mobile system [7].

* * *

The description of basic paging protocols and systems shows us that in modern paging systems, despite simple rules of operation, complex techniques of digital transmission and error correction coding are applied. As already mentioned, paging services are offered by other mobile communication systems, including personal satellite systems. Paging systems can survive the strong competition with other systems due to specific applications and low service prices.

REFERENCES

1. A. Makiedonski, Paging Systems in Poland (in Polish), *Przegląd Telekomunikacyjny*, No. 5/6, 1995, pp. 270-280

2. ETSI ETS 300 244, Electromagnetic compatibility and Radio spectrum Matters (ERM); On-site paging service; Technical and functional characteristics for on-site paging systems including test methods, March 1998

3. R. J. Horrocks, R. W. A. Scarr, *Future Trends in Telecommunications*, John Wiley & Sons, Ltd., Chichester, 1993

4. B. Walke, *Mobile Radio Networks: Networking and Protocols*, John Wiley & Sons, Ltd., Chichester, 1999

5. ETSI ETS 300 133-1, Electromagnetic compatibility and Radio spectrum Matters (ERM); Enhanced Radio MEssage System (ERMES); Part 1: General aspects, November 1997

6. ETSI ETS 300 133-4, Electromagnetic compatibility and Radio spectrum Matters (ERM); Enhanced Radio MEssage System (ERMES); Part 4: Air interface specification, November 1997

7. ETSI TR 101 037, Radio Equipment and Systems (RES); Enhanced Radio MEssage System (ERMES); Two Way Paging system, V1.1.1, July 1997

8. D. R. Gonzales, M·CORE *Processor with On-Chip FLEX Decoder for Messaging Applications*, Motorola M·CORE Technology Center, Austin, Texas

5

The cellular system concept

In the chapter dealing with classification of the mobile systems we have already mentioned that cellular systems can be defined as mobile systems realizing two-way wireless communication between the fixed part of the system in the form of appropriately located base stations and mobile stations moving in the area covered by the base station system. The coverage area is divided into sub-areas served by the base stations usually located in their centers. The coverage area of a single base station can be symbolically represented as a regular hexagon and therefore it is often called a *cell.* Figure 5.1 presents an example of division of the system coverage area into hexagonal sub-areas. The reason for this division is insufficient system capacity which would characterize the whole system if its area was covered by a single base station having at its disposal the same spectrum interval as the system of base stations. Here, capacity is understood as the maximum number of mobile stations simultaneously served by the cellular system per Herz and per square kilometer.

Let us consider a mobile communication system which applies FDMA. Let each mobile station require B Hz for transmitting its signal and let the bandwidth assigned to the whole system be MB Hz. Each band of the width equal to B Hz can be denoted as a *channel.* Therefore, a system with a single base station emitting a high power signal which covers the whole area provides M channels and can serve only M mobile stations simultaneously. On the other hand, dividing the whole area into cells and numbering them as in Figure 5.1 allows for the multiple use of the channels if M channels are properly distributed among N different cell types ($N = 7$ in Figure 5.1). We conclude that the number of mobile stations served at the same time increases considerably, and, roughly speaking, cellular system capacity increases with respect to capacity of a single base station system in proportion to the number of a single channel use in the covered area. The area of a single cell is much smaller than the whole system area, so the

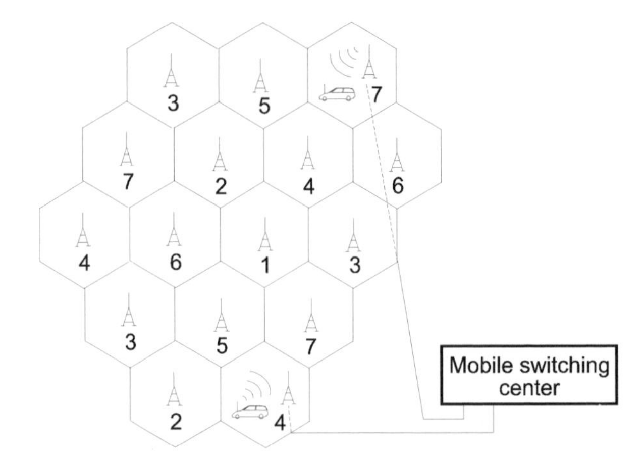

Figure 5.1 Division of the system coverage area into cells

power generated by the base station in a cell is much lower than the power emitted by a base station covering the whole system area. In consequence, the power of mobile stations communicating with much closer base stations is also much lower in a cellular system as compared with a single high power base station system. Let us note that the cells shown in Figure 5.1 have been numbered in such a way that the distance between cells denoted by the same number is maximized. This way the interference between the signals emitted in the same channel is minimized. The phenomenon of interference from other cells using the same channel as the channel in the selected cell is called *co-channel interference*. It is a characteristic feature of cellular systems and is one of the main factors which have to be taken into account in the cellular system design.

Division of the system coverage area into cells, assignment of a subset of channels to each cell and the possibility of manipulating the base station power level results in a flexible system design which takes into account the predicted traffic intensity in the given area. In the city centers, where there is a high concentration of mobile phone users, the cells are usually smaller and the channels are more frequently reused. As a result, a higher traffic per unit area can be served. In turn, in the rural areas, in which the number of the mobile phone users results, for example, from the highway traffic only, the cells are larger, they have fewer channels at their disposal and the power of the base stations is larger. In conclusion, the topography of the traffic generated by mobile users is taken into account in the mobile system deployment.

Division of the system coverage area into the sets of cells has a few important consequences. The first one, already mentioned, is the existence of the co-channel interference. The second one is the necessity to ensure automatic transfer of the connection between the base and the mobile station to the next base station if the mobile station crosses the border between the neighboring cells. Such a procedure is called *handover* or *hand-off*. In order to ensure seamless transfer to the new base station and avoid multiple transfer to the new base station and back to the old one, a rule of *hysteresis* has to be applied. This means that the connection is switched to a new base station if the level of the signal received by the mobile station from the new base station is higher by a given threshold value than the level of the signal received from the base station currently performing the link with the mobile station.

Dynamic transfer of connections of the mobile station with subsequent base stations requires complex control procedures and measuring the power of the signals generated by mobile stations and received by base stations or vice versa. If a mobile station participates in the measurements and in handover decision, which occurs in the second generation systems such as GSM and IS-136, we call such a procedure a *mobile-assisted handover*.

The important consequence of changing location of a mobile station is the necessity of finding a particular mobile station if there is an attempt to set up a connection with it. A mobile station whose power supply is on and which remains idle, should periodically update its location in order to "show its existence" in the system. In an alternative procedure, a whole group of the base stations page the selected mobile station, knowing that the searched mobile station remains within the service area of one of them.

The crucial problem to be solved in cellular radio systems is the design of the base station topography. The base stations have to be properly located and the channel subsets properly assigned to them in order to ensure a satisfactory quality of connections in the whole coverage area.

5.1 SIMPLIFIED DESIGN OF A CLASSIC CELLULAR SYSTEM

In this section we will show simplified rules of the cellular system design and channel allocation. In our considerations, the coverage area of a single base station is approximated by a hexagon. There are a few geometrical figures which ensure full coverage of a given area without either overlapping or holes. These are equilateral triangles, squares and hexagons. Hexagons best approximate the circular shape of a base station coverage in a flat terrain without obstacles and the hexagonal edges well approximate the borders between cells of the same size. We will hold the simplifying assumption about the hexagonal shape of the cells. In reality, the base station coverage does not have a regular circular shape because the coverage is a result of terrain architecture and obstacles such as houses, trees, etc. The division of the system coverage area into equal-sized cells is not possible for other than technical reasons, either. In the base stations deployment we have to take into account many constraints such as the access to the appropriate terrain locations, possibility to use natural terrain elements such as towers, high chimneys and buildings. Such elements are rarely located in the centers of the ideally planned cells. Therefore, planning of cells is a complex process, in which in modern systems the field measurements are made using specialized equipment. Cells can be planned with a certain accuracy on a basis of digital terrain map applied by specialized complex planning software tools which model the propagation of electromagnetic waves in the terrain model built upon the digital map. One of the approaches to modeling the electromagnetic waves propagation is treating the waves as light rays which reflect and diffract from various terrain obstacles with specified reflection and

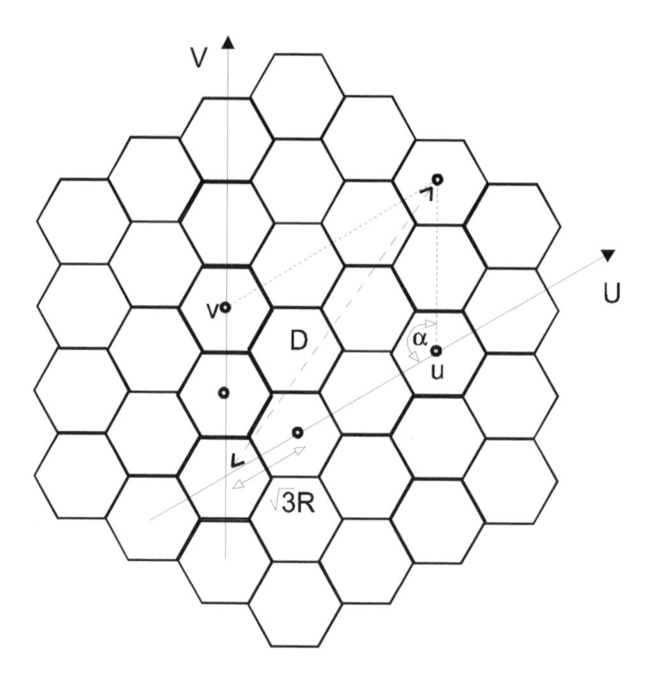

Figure 5.2 Geometry in the field of hexagons

diffraction coefficients. This method, called a *ray tracing method*, requires accurate data on the covered area and a lot of computational power. Professional software packages used for propagation modeling and cell planning often take advantage of more elaborate propagation models than those presented in Chapter 3. A limited number of measurement data allows for calibration of the propagation models used by the software package, so the results are more accurate.

Now we will analyze the rules of cell design and channel allocation "on the engineering level", discussing the factors which have to be taken into account in the design process. The results of the analysis are close to those obtained from complex evaluation by simulations and measurements. Let us stress again that in our considerations we assume the hexagonal shape of the cells and their approximately equal size.

As we know, the key principle of cellular system operation is multiple use of the same channels in different cells appropriately located in the covered area. The set of N cells using all available channels is called a *cell cluster*. In order to analyze the cluster size and the cluster features we will consider the geometrical properties of a set of hexagons [1], shown in Figure 5.2. Let the radius of the circle circumscribing each hexagon be R. Obviously, this is also the distance between the hexagon center and its node. Recalling the properties of equilateral triangles of the side length R, we can easily prove that the distance between two neighboring hexagons is equal to $\sqrt{3}R$. Let this distance be the unit of our distance measure. In a coordinate system in which the angle between the axes is 60 degrees, the distance between the center of any hexagon and the coordinate origin is[1]

$$D = \sqrt{3}R \cdot \sqrt{i^2 + ij + j^2} \tag{5.1}$$

where i and j are the coordinates of the considered hexagon center, expressed in the assumed distance unit equal to $\sqrt{3}R$. Expression (5.1) directly results from the generalized Pythagoras theorem, which states that the squared length D of the edge opposite to the angle α created by the edges of length u and v is equal to

$$D^2 = u^2 + v^2 - 2uv\cos\alpha \tag{5.2}$$

In the case shown in Figure 5.2 $u = 3\sqrt{3}R$, $v = 2\sqrt{3}R$ and $\alpha = 120°$. Therefore, $i = 3$ and $j = 2$.

Let us treat the cell which is located in the origin of the coordinate system shown in Figure 5.2 as the reference cell. Let us build a cell cluster around the reference cell. Other clusters should be placed around this cluster so that the area covered by them should have neither holes nor multiple coverage. In this context the following question arises: what is the number of cells in the cluster which ensures compact coverage? The answer to this question is the result of the following argumentation.

Let the center cells of neighboring cell clusters be located at the distance D (described by (5.1)) from the center of the reference cell. As we remember, they have the same set of channels at their disposal as the reference cell has. Each of the clusters can be represented by a single large hexagon whose area is equal to the sum of areas of all cells belonging to this cluster. Figure 5.3 illustrates this case. The area of a single hexagonal cell of the radius R is

$$(Area)_R = \frac{3}{2}\sqrt{3}R^2 \tag{5.3}$$

whereas the area of large hexagons equal to the sum of N areas of hexagons with the radius R and their respective centers located at the distance D from one another, is

$$(Area)_{D/\sqrt{3}} = \frac{3}{2}\sqrt{3}\left(\frac{D}{\sqrt{3}}\right)^2 \tag{5.4}$$

We wish the following equality to be true

$$(Area)_{D/\sqrt{3}} = N \cdot (Area)_R \tag{5.5}$$

Putting (5.1) and (5.4) into (5.5) we obtain the following expression determining the number N of cells in a cluster

$$N = i^2 + j^2 + ij \tag{5.6}$$

[1]The author hopes that the reader will not mistake the distance D for the antenna directivity denoted in Chapter 3 by the same symbol.

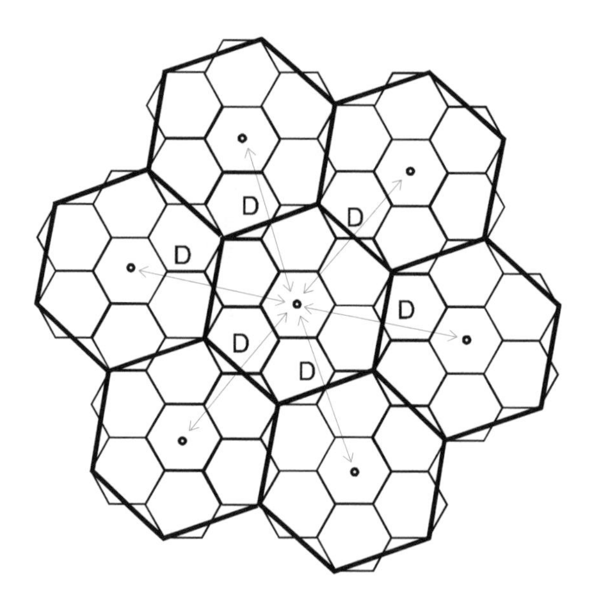

Figure 5.3 Cell cluster approximated with large hexagons

As we see, the number of cells in a cluster is not unrestricted. The cluster can consist of one, three, four, seven, twelve, etc. cells. In Figure 5.3 there is a cluster of $N = 7$ cells for which the parameters are $i = 2$ and $j = 1$.

On the basis of (5.1) and (5.6) we receive the following important relation

$$Q = \frac{D}{R} = \sqrt{3N} \tag{5.7}$$

which will be used in our future considerations.

If the system is not properly designed with respect to the number of cells in a cluster, topographic cell distribution and channel assignment, then it will experience excessive interference between the channels in different cells which use the same carrier frequencies. As we already know, such distortion is called a *co-channel interference*. It is a function of parameter Q defined in (5.7). Parameter Q is called the *co-channel interference reduction factor*. If Q increases, then co-channel interference decreases because the cells using the same channels are more distant from each other or because their size is smaller. The distance D is a function of the ratio of signal power S to interference power I. In turn, this ratio depends on the number of interfering cells K_0, according to the formula

$$\frac{S}{I} = \frac{S}{\sum_{k=1}^{K_0} I_k} \tag{5.8}$$

where I_k is the mean interference power originating from the k-th cell. Figure 5.4 presents a typical configuration of interfering cells. In case of hexagonal cells, six of them located in the first tier interfere with the central cell, which is considered as the reference one. Therefore $K_0 = 6$. We further assume that the influence of the cells situated in the second tier can be neglected due to their large distance from the central cell. Let us note that co-channel interference disturbs not only the signal reaching the base station in the central cell, but also the signals arriving at the mobile stations currently located in that cell. Assuming that base stations emit signals with equal powers, the signal-to-co-channel interference ratio calculated at the border of the central cell is

$$\frac{S}{I} = \frac{R^{-\gamma}}{\sum_{k=1}^{K_0} D_k^{-\gamma}} \tag{5.9}$$

As we remember from Chapter 3, the signal power received at the receive antenna placed at distance d from the transmit antenna is proportional to $d^{-\gamma}$. For free space propagation the power $\gamma = 2$, whereas $\gamma = 4$ in case of two-path propagation. In reality, depending on the propagation environment, γ is in the range between 2 and 5.5.

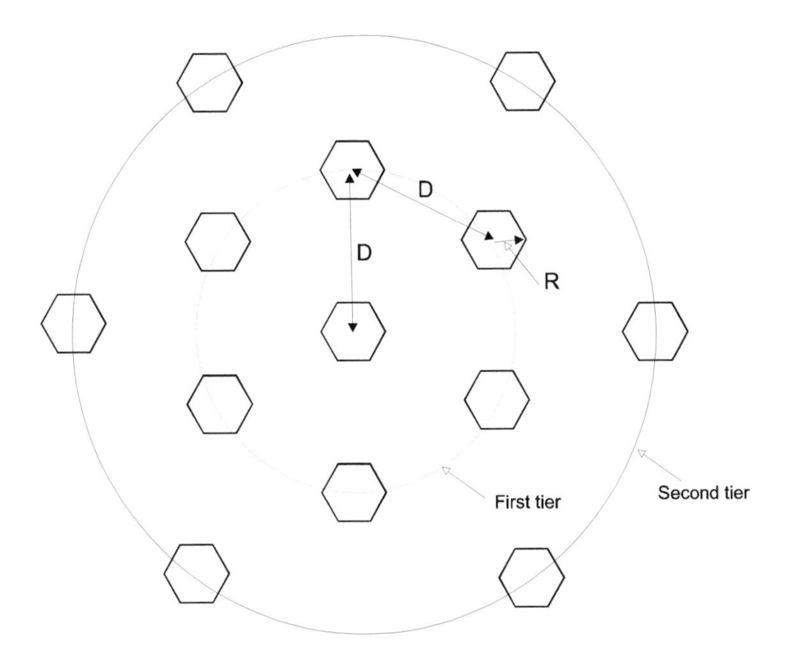

Figure 5.4 Space distribution of the interfering cells

Let us note that with the assumption of equal power generated by the base stations, the signal-to-co-channel interference ratio depends only on the geometrical properties of the cell distribution, the distances between base stations using the same channels, and the base station coverage radius.

For simplicity let us assume that for the system shown in Figure 5.4 all distances D_k are the same and equal to D. Thus, from (5.9) we conclude that

$$\frac{S}{I} = \frac{R^{-\gamma}}{6 \cdot D^{-\gamma}} = \frac{Q^{\gamma}}{6} \tag{5.10}$$

and

$$Q = \left(6 \cdot \frac{S}{I}\right)^{1/\gamma} \tag{5.11}$$

Formula (5.11) sets the relation between the ratio of the distance between the cells using the same channels and the cell radius, the signal-to-co-channel interference ratio and the type of environment. In traditional cellular systems the required ratio S/I is selected to be the value ensuring at least a good mark in a subjective speech quality test by at least 75 percent of users in 90 percent of the area covered by the system [2]. In classic analog cellular systems such as AMPS the value of S/I is selected to be 18 dB, which is equivalent to 63.1 on the linear scale. Assuming $\gamma = 4$, from (5.11) we get $Q = 4.41$. Simulation results reported in [1] yield $Q = 4.6$. Because the value of Q depends on the number N of cells in a cluster according to formula (5.7), then setting $Q = 4.6$ in this formula results in $N = 7$. As we see, if the omnidirectional antennae are applied, the rough calculations performed so far indicate that the cluster of seven cells ensures the desired value of co-channel interference reduction factor in the analog cellular radio.

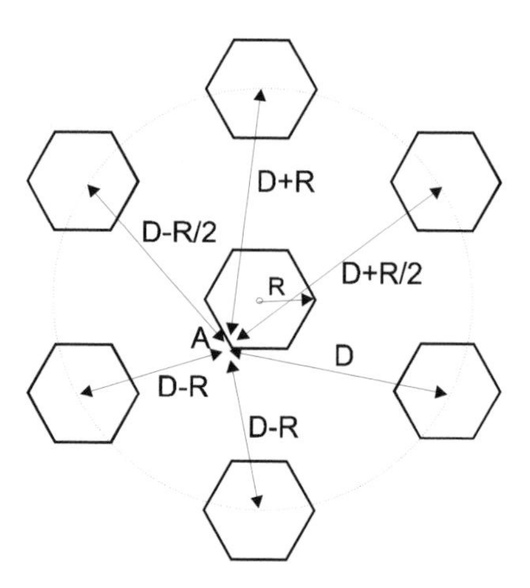

Figure 5.5 The worst case for co-channel interference if $N = 7$

Let us consider the worst case shown in Figure 5.5. Assuming $\gamma = 4$ and noting that the distances between the mobile station located at the cell border (point A) and all interfering base stations are approximately equal to $(D-R)$, $(D-R)$, $(D-R/2)$, $(D+R/2)$, D, and $(D+R)$, we get

$$\frac{S}{I}=\frac{R^{-4}}{2(D-R)^{-4}+\left(D-\frac{R}{2}\right)^{-4}+D^{-4}+\left(D+\frac{R}{2}\right)^{-4}+(D+R)^{-4}}= \tag{5.12}$$

$$=\frac{1}{\frac{2(Q+1)^4+(Q-1)^4}{(Q^2-1)^4}+\frac{\left(Q+\frac{1}{2}\right)^4+\left(Q-\frac{1}{2}\right)^4}{\left(Q^2-\frac{1}{4}\right)^4}+\frac{1}{Q^4}}$$

For $Q=4.6$ the value of $S/I=49.56$. In the dB-scale this value is roughly equivalent to 17 dB. For precise values of distances between point A and the centers of the interfering cells we get a slightly better S/I; however, it is still below the required value of 18 dB [3]. In reality, due to nonideal location of base stations, multipath propagation and distortions caused by terrain irregularities, this ratio is even worse. Therefore, the previously calculated value $Q=4.6$ is insufficient. Let us stress again that Figure 5.5 and the associated formula (5.12) illustrate the least advantageous case, because the mobile station is in the most distant place from the base station in its own cell. Therefore, the above evaluation is very pessimistic. Nevertheless, such a design approach ensures the system reliability.

There are two basic solutions to the problem of insufficient value of the co-channel interference reduction factor Q for the cluster of seven cells and omnidirectional antennae. The first is increasing the number of cells in the cluster. However, if the number N gets higher, the number of available channels in a cell decreases. Taking the next allowable value[2] of N larger than 7, e.g. $N=12$ we have to divide the available channels into 12 subsets. The second solution is decreasing the co-channel interference due to the application of sector antennae which emit signals in the 120°-angle. Each cell is divided into three sectors. Division into a different number of sectors is also possible; however, it will not be considered here.

Figure 5.6 shows the interfering cells when the sectorized cells are applied. The areas in which the same subset of channels is used are marked with a grey pattern. However, only two sectors marked by even darker pattern send the interfering signals to the central sector denoted by the darkest pattern. As we see, thanks to sectorization, the number of interfering cells has been reduced to two. Figure 5.7 shows the worst localization of the mobile station in this case.

The distance between the mobile station and the interfering base station is equal to $D+R/2$. Therefore, the signal-to-co-channel interference ratio is

$$\frac{S}{I}=\frac{R^{-4}}{2\left(D+\frac{R}{2}\right)^{-4}}=\frac{(Q+0.5)^4}{2} \tag{5.13}$$

[2]According to (5.6) taking $i=3$ and $j=0$ we get $N=9$; however, for $N=9$ we receive an irregular cell grid.

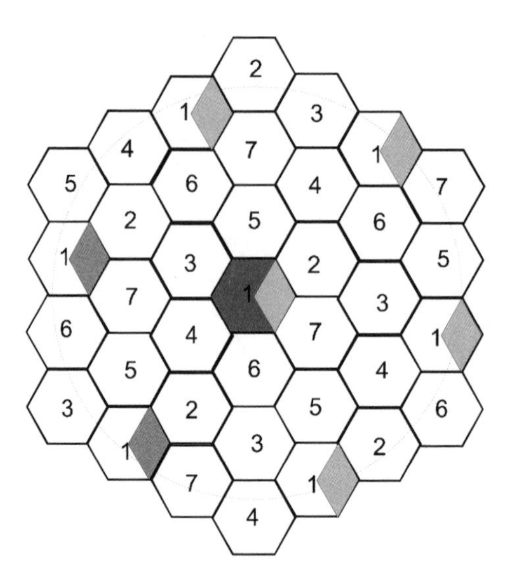

Figure 5.6 Illustration of co-channel interference in the case of 120°-sector antennae

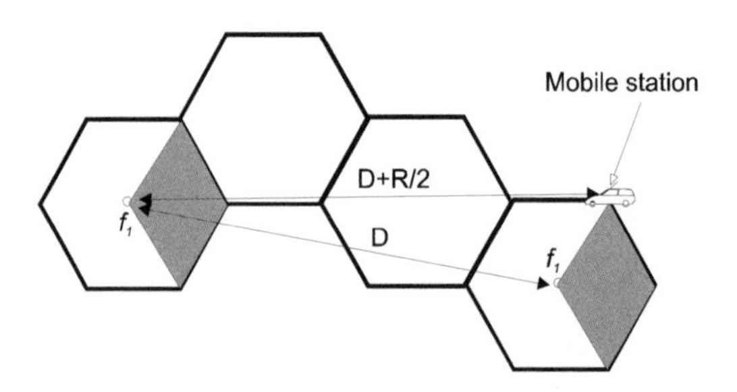

Figure 5.7 The worst case for 120° sectors

which at $Q = 4.6$ results in $S/I = 25.3$ dB. We gain about 7 dB as compared with the omnidirectional antennae. In practice, cell sectorization is applied in a classic cellular radio as well as in newer solutions such as GSM or cdmaOne (IS-95).

We have to stress that sectorization not only reduces the number of interfering base stations but also has a positive impact on the physical properties of the mobile channel. The channel impulse response delay spread is shorter as compared with that observed in the cells with omnidirectional base station antennae. Thanks to sectorization the system capacity increases; however, the achievable gain will not be fully obtained if mobile stations are not uniformly distributed in all sectors (see Section 5.4). There are some disadvantages of sectorization as well. These are:

- The sectorized cell requires more base station equipment, in particular in the RF part.
- Mobile stations moving in sectorized cells more often change channels, which results in increased control signaling.
- *Trunking efficiency* decreases. The set of channels assigned to a cell is now divided among the sectors. The number of the served users will remain the same as in non-sectorized cells only if the number of users located in each sector is proportional to the number of channels assigned to each sector.

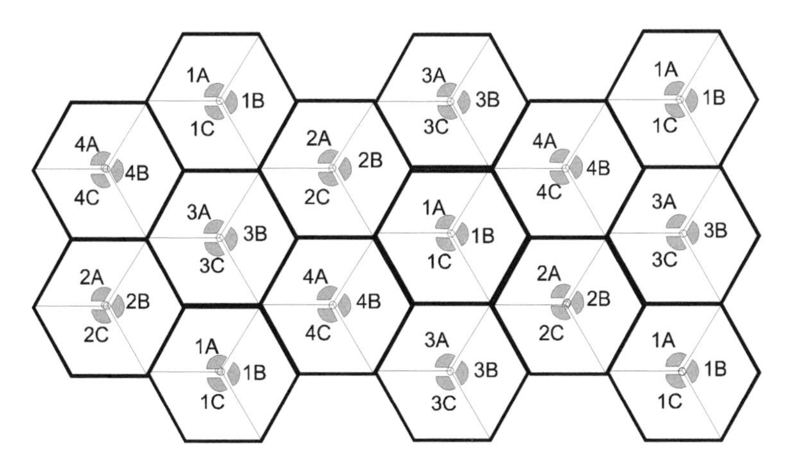

Figure 5.8 Example of the area covered by clusters with $N = 4$ and three sectors in a cell, typical for the GSM [5]

Figure 5.8 presents a typical assignment of sectors in the GSM mobile telephony [5]. Let us note that the number of cells in a cluster is $N = 4$ and the cells are divided into three 120°–sectors. The sectors are denoted by characters A, B, and C, whereas the cells in a cluster are numbered from 1 to 4. The signal in each sector is corrupted by the signals from two sectors belonging to two different clusters; therefore, the number of interfering sectors is $K_0 = 2$.

In a digital cellular system, e.g. the GSM, the criterion for selection of network topology can be the achievable error rate instead of S/I used in the design of analog cellular telephony. It has been found that, thanks to the applied digital modulation, the multiple access method, efficient digital speech coding, and error correction channel coding, the required value of the signal-to-co-channel interference ratio decreases to 9 dB [4], as compared with 18 dB for analog cellular systems. This value is also mentioned as the minimum in the GSM standard [11].

The next problem to solve in FDMA or TDMA/FDMA cellular telephony is the allocation of all carrier frequencies used by the system to the cells. Their number used in a cell or a sector, together with the multiple access method (FDMA or TDMA), decide upon the number of users who can be concurrently served in a cell. A classic solution is the assignment of a fixed subset of carrier frequencies to the cells or the sectors which are denoted by the same number in different clusters.

Before we consider this problem in detail, let us introduce some basic terms used in the traffic theory. They are necessary for evaluation of the cellular system properties.

5.2 ELEMENTS OF THE TRAFFIC THEORY APPLIED TO CELLULAR SYSTEMS

A typical telephone network is usually designed taking into account behavior of its subscribers. It has been observed that only a small fraction of them use the network simultaneously. However, the intensity of the network usage can vary in time and can differ for several subscribers. The network is usually designed with respect to the expected intensity of its use. Therefore, *trunking* is applied, which means that a given number of links remains at the disposal of a far greater number of subscribers. There is a small assumed probability that a subscriber will not be able to get a connection due to temporary lack of available links. The only way to set a connection is to repeat the connection request. The system which works this way is called a *loss system*. From the point of view of the subscriber who is not able to connect, the system is blocked. The probability of this event is called the *probability of blocking*. More sophisticated telecommunication systems, called *delay systems*, in case of the lack of available links hold the connection requests in a queue and set up the connections gradually when subsequent links are released.

Traffic considerations related to a classical telephone network and data networks can be easily extended onto cellular networks. The equivalence of a telephone link is a radio channel which is meant as a pair of carriers (in case of FDMA systems), a pair of time slots on the specified carriers (in case of TDMA/FDMA systems), or a pair of spreading sequences (in case of CDMA systems). Assignment of the radio channels should take into account the expected *traffic intensity*. Traffic intensity is a measure of channel use, expressed as average relative channel occupancy. The measure unit is Erlang. For example, the mean channel occupancy lasting half an hour in one hour time interval is equivalent to traffic intensity equal to 0.5 Erlang. The number of call requests is random and fluctuates within the 24-hour interval. Therefore, the concept of *busy hour* has been introduced. The busy hour is the one-hour time interval in which the traffic intensity is maximum.

The *grade of service* (GOS) is a measure of access to the trunked system during the busy hour. In many countries the busy hour for cellular radio systems occurs during rush hours between 4 and 6 p.m. on Thursdays and Fridays [3]. The grade of service (GOS) is a quality measure applied to determine the desired probability that a subscriber will obtain the channel access, given a specified number of channels in the cellular system. The GOS is one of the basic parameters and benchmarks in the cellular system design which has to be fulfilled in order to ensure the required system capacity and channel distribution among cells. The GOS is usually expressed in the form of probability of blocking i.e. the probability of the event that a subscriber wishing to get a connection experiences a lack of free channel or waiting for an available channel is longer than a certain prescribed time interval.

For the loss systems the probability of blocking is expressed by the following *Erlang B formula*

$$\text{GOS} = \Pr\{\text{blocking}\} = \frac{\frac{A^C}{C!}}{\sum_{k=0}^{C} \frac{A^k}{k!}} \tag{5.14}$$

The formula was derived with the assumption that channel requests have the Poisson distribution, the time of channel occupancy by a subscriber has the exponential distribution, the number of subscribers is infinite and the number of channels available in the trunking pool is finite. It was also assumed that channel requests are memoryless, which means that any subscriber can request a channel at any time. A is the *total traffic intensity* and C is the number of channels in the pool. The total traffic intensity can be calculated by multiplying the (finite) number of subscribers U by traffic intensity A_u for a single subscriber. The latter parameter can be derived as the product of the duration of a call and the average number of call requests per unit time. As already mentioned, formula (5.14) was derived at the assumption of infinite number of users. In reality this number is finite but is a few orders of magnitude higher than the number of channels, therefore the assumption is a good approximation of reality. The Erlang B formula can be shown as the family of plots (Figure 5.9) representing the blocking probability as a function of the total traffic intensity. Each plot is obtained for a particular number of channels C. The plots shown in Figure 5.9 can be used for a rough design of a cellular system. Usually we assume the specified grade of service and one of the two parameters: the total traffic intensity or the number of channels. Setting the value of one of these parameters, we read out the second one from the appropriate plot.

As we mentioned, there are more sophisticated systems than the loss systems considered so far. They are often called *delay systems*. In such systems the calls which cannot be realized due to the temporary lack of a free channel are held in a queue. They are successively served as the channels become available unless the waiting time exceeds the time limit, in which case, they are removed from the queue. Probability of blocking for delay systems is given by the Erlang C formula; however, we will not quote it here. The interested reader can find it in [3] or books devoted to telecommunication switching.

Cellular system designers often rely on computer simulations which take dynamic effects into account and allow for system verification by investigating more realistic system models [3].

Let us summarize this section with an example of an introductory design, similar to one of many interesting examples contained in [3], with data fitted to analog systems still operating in Scandinavia and Eastern Europe.

Example 1 *A certain city agglomeration has an area of 3300 km*^{2} *and is covered by a cellular system. The system applies 7-cell clusters. Each cell has a radius of 6 km. The bandwidth of* 2×4.5 *MHz is assigned to the system working in the FDMA/FDD mode. The channel bandwidth is 25 kHz. Assume that the probability of blocking in the*

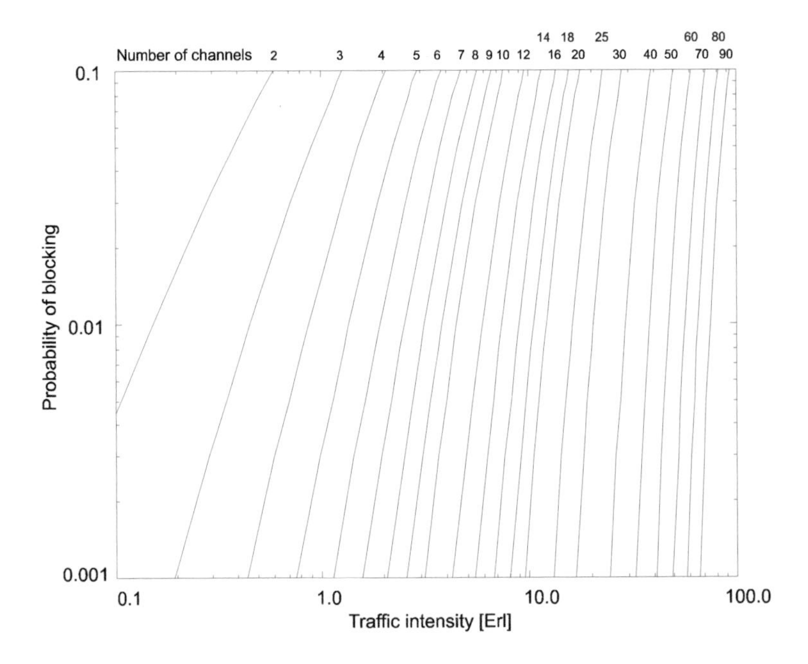

Figure 5.9 Probability of blocking as a function of the total traffic intensity a varying number of channels

cellular system is 0.02. Let the average traffic intensity offered by a user be 0.03 Erl. Recall that it means that a statistical subscriber uses a channel during 3 minutes in the time interval of 100 minutes.

*Let us calculate the number of cells covering the city area. From (5.3) we calculate the area of a single hexagonal cell. This area is equal to 2.5981*R^2. *For* $R = 6$ *km the area is* $P = 93.53$ km^2. *Therefore, the number of cells needed to cover the city area is* $N_c = 3300/93.53 = 35.28 \simeq 36$ *cells. Now let us calculate the number of channels assigned to a single cell. Because the system has* 2×4.5 *MHz at its disposal and a single connection requires two 25 kHz channels, therefore for 7-cell clusters the number of duplex channels in a cell is* $C = 9000/(50 \cdot 7) \simeq 25$ *channels.*

For $C = 25$ *channels/cell and the probability of blocking equal to 0.02, we find from the plot of the Erlang B formula (Figure 5.9) that the intensity of total traffic, which may be carried in a cell is* $A = 17.5$ *Erl/cell. Therefore, the total traffic carried in the whole system is* $A \cdot N_c = 17.5 \cdot 36 = 630$ *Erl. From this value we can deduce the number of users who can be served by the system. This number is* $N_A =$ *(total traffic)/(traffic per user)* $= 630/0.03 = 21000$ *users. The number of all channels used in the system can be found by division of the system bandwidth by the bandwidth of the channel pair. For the assumed data it is* 9 *MHz/50 kHz* $= 180$ *channels. As a result we find that the number of users per channel is* $21000/180 \simeq 116$ *users. The maximum number of users who can be simultaneously served results from the number of channels in a cell and the number of cells and is equal to* $C \cdot N_c = 25 \cdot 36 = 900$ *users. As we see, if all channels in all cells were occupied at the same time, the system could serve* $900/21000 = 4.29\%$ *of all users. We conclude that thanks to the idea of trunking the system resources can be much smaller than the number of subscribers in the whole system.*

A basic complication which has not been taken into account so far is users may move from cell to cell during a connection. If they cross the cell borders a handover is required. A new channel in a new cell has to be found and after that the channel used in the old cell can be released. In consequence, traffic calculations become more complicated. A possible solution to this problem is to construct a simulation package which takes into account the movement of mobile stations and handovers. Statistical properties of the mobile stations movement are different in cells covering urban area and in those covering the rural terrain crossed by a highway.

5.3 WAYS OF INCREASING THE SYSTEM CAPACITY

So far we have analyzed the cellular systems with uniformly distributed channels assigned to cells of equal size. In reality, the traffic intensity offered by subscribers is rarely distributed uniformly in the system coverage area. As we mentioned in the introductory section, cellular systems are able to match the traffic served by the system to the traffic offered by the subscribers. This is done by smart control of the cell sizes and channel assignment. However, experience in real systems development indicates that the demanded capacity can change in time. This is usually caused by the increasing number of users who signed up to the cellular system subscription and reside in the given area. Both problems imply the need to increase the system capacity and match its topography to the distribution of the offered traffic. This task is usually solved by applying the following solutions:

- cell sectorization,
- cell splitting, and
- using the concept of microcell zones.

In Section 5.2 we showed that introducing cell sectorization decreases the value of S/I. Therefore, it is possible that the minimum required value of S/I can be achieved for cell clusters built of a smaller number of cells. For example, we found that for $S/I = 18$ dB, required in the analog cellular telephony, the number of cells in the cluster $N = 7$ was insufficient if omnidirectional antennae were applied. Therefore, it was necessary to apply 12-cell clusters. Application of 120°-sectors in the 7-cell cluster ensured the increase of S/I by about 7 dB, as compared with omnidirectional antennae. In consequence, we were able to decrease the number of cells in a cluster to $N = 7$ or even $N = 4$, as long as the value of S/I did not fall below 18 dB. Thanks to that, the available channels are used more frequently in the same area, which results in the increase of the reuse factor $1/N$ and in the system capacity. The price paid for this is tripling the number of antennae and base stations and more frequent handovers between cells and sectors, which causes increased control traffic. As we have mentioned before, sectorization partially decreases trunking efficiency, unless the users are uniformly distributed in all sectors.

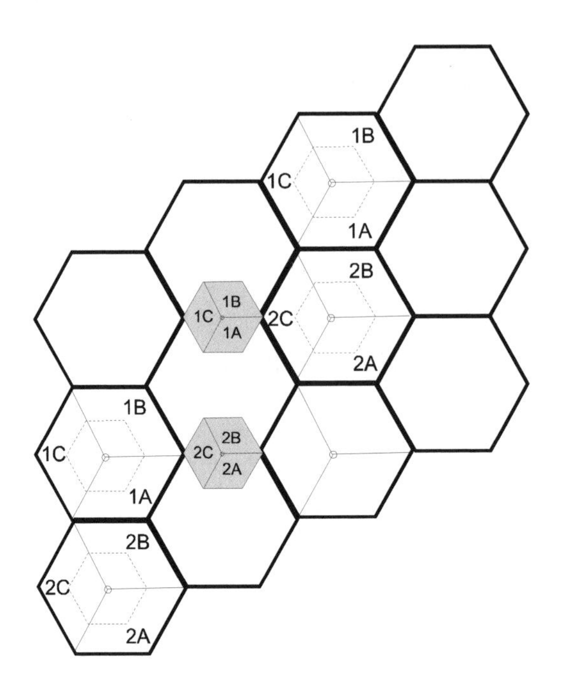

Figure 5.10 Example of cell splitting preserving sectorization

The second way of increasing capacity is splitting a cell into smaller ones [2], [3], [6]. Cell splitting is usually done by establishing smaller cells in the given locations of the covered area. The radius of these smaller cells is half the radius of the original cell. In consequence, the new cell area is four times smaller than the area of cells which have been divided. Larger cells are used in the area with low offered traffic, whereas smaller ones are used in places in which traffic is more intensive. Figure 5.10 presents an example of cell splitting for the sectorized cells. Halfway of the distance between original cells denoted by number 1, with sectors A, B, and C, a new cell is introduced with the radius equal to half the radius of the original cell. The sectorization is preserved. Let us note that locating a new small cell in the indicated place preserves the value of the co-channel interference reduction factor $Q = D/R$, because both the distance between the cells of the same type and their radii decreased twice [2]. As a result of introducing a new cell, it is possible to apply a lower power level in the base station, such that on the border of a small cell the received signal power is the same as the power level of the signal received on the border of the original cell. The same is also true for mobile stations as well. Because the power of the signal received at distance R from the base station is associated with the transmitted signal power in the formula

$$P_{R1} = k \cdot P_{T1} \cdot R^{-\gamma} \tag{5.15}$$

where k is the proportionality factor, then assuming the same received power at the border of small and large cells with the radii $R/2$ and R, respectively, we get the dependence between the transmitted power in the large cell P_{T1} and the small cell P_{T2} in the form

$$P_{T2} = P_{T1} \left(\frac{1}{2}\right)^{-\gamma} \tag{5.16}$$

For $\gamma = 4$, (5.16) indicates that the transmitted power in a small cell is 12 dB lower than that in a large cell.

Division of cells into smaller ones requires modification of channel assignment in large cells located in the neighborhood of a new small cell. Usually the process of cell splitting resulting in the increase of system capacity is performed gradually, so large and small cells coexist with one another. Then channel allocation for large and small cells of the same type is the following. The channels used in large cells (e.g. cells denoted by number 1 or 2 in Figure 5.10) are divided into two groups: the first group of channels which are also used in small cells of the same type (denoted in Figure 5.10 by number 1 or 2 as well) and the second group which consists of the remaining channels. The channels belonging to the first group are used in small cells and in the centers of large cells, denoted in Figure 5.10 by dashed lines, whereas the second group of channels is used in whole large cells including their external parts. The price paid for this modification is a higher complexity of base stations. The signals transmitted in the channels used in small and large cells have to be generated with smaller power than those generated in the channel used in the large cells only.

In 1991 W.Y.C. Lee proposed another solution to the problem of capacity increase, i.e. the concept of microcell zones.[3] In this solution, some disadvantages of cell sectorization are avoided, such as frequent handovers between sectors and the decrease of trunking efficiency.

The original cell is divided into three zones. On the edge of each zone, the zone transmitters and receivers are located. They are connected by fiber optical or microwave links with the zone selector and the common part of the base station operating in the given cell. A mobile station located in the cell is connected with the strongest zone transmitter and receiver. The pool of channels used in the cell is common for all zones. Moving the mobile station to another zone in the cell does not cause the change of the assigned channel. This channel "is taken" by the mobile station to a new zone. As a result, there is no time-consuming and resource-consuming handover. However, the main advantage of the concept of microcell zones is decreasing the co-channel interference by using the zone transmitters with a lower power than the power

[3]We should not mistake the microcells applied in the microcell zones concept with microcells used in cordless telephony and PCS systems. The idea of microcell zones is not equivalent to the placement of base stations in the tops of cells either.

generated by sector transmitters located in the middle of the cells. Therefore, without loss of trunking efficiency (all channels are in common use in all zones in the same cell) we gain the possibility of capacity increase due to decreasing the number of different cells in the cluster. Lee showed [10] that in the case of the American first generation system AMPS, the requirement of $S/I = 18$ dB while using the microcell zone concept results in decreasing the number of cells in a cluster from $N = 7$ to $N = 4$. Smaller clusters result in 2.33-times increase of the system capacity, with $S/I = 20$ dB in the worst case.

5.4 CHANNEL ASSIGNMENT TO THE CELLS

So far we have designed the system regarding the aspects of the required capacity and grade of service. The next design step is to assign particular channels (carrier frequencies) to cells or to cell sectors.

Let us consider the rule of distribution of channel frequencies among cells or sectors. *Interchannel interference* has a direct impact on our choice. This type of interference arises between the signals generated in the same cell (or sector) on different frequencies assigned to this cell (sector). This distortion should be minimized. It can be achieved by an appropriate choice of channel frequencies in each cell. Interchannel interference is also strictly associated with the movement of mobile stations in a cell and various distances of mobile stations from the common base station (*near-far effect*). Due to nonideal filtration of the channel signals, the sidelobes of the channel used by the mobile station located close to the base station may filter through the passband of the other filters separating the channel closely placed on the frequency axis and used by a far-off mobile station. A similar situation in the opposite transmission direction (from the base station to the mobile station) is also possible.

For example, if the transmitting mobile station is at the distance which is forty times longer than the distance from the distorting station using another channel in the same cell, then the ratio of the desired signal power to the power of interference measured in front of the receive filter input of the base station receiver, assuming propagation conditions for which $\gamma = 4$, is

$$\frac{S}{I} = (40)^{-\gamma} = 40^{-4} = -64 \text{ dB} \tag{5.17}$$

If other remedies against this disadvantageous situation were not applied, this ratio would have to be improved by introducing a receive filter with sharp cut-off and the large channel separation on the frequency axis.

Let the receive filter have a passband width of B Hz and the slope of 24 dB/oct. Thus, on the channel filter passband edge spaced by $B/2$ Hz from the middle of the channel, the signal is attenuated by 24 dB. Each doubling of the frequency interval introduces 24 dB attenuation. The attenuation of at least 64 dB is achieved at the frequency distance of 3.18 larger than a single channel bandwidth. In reality, four-channel bandwidth separation is required. In general, frequency separation Δf_{sep} between the channels used in the same cell can be determined from the formula:

$$\Delta f_{sep} = 2^G \cdot \frac{B}{2} \quad \text{where} \quad G = \frac{\gamma \log_{10}\left(\frac{d_0}{d_1}\right)}{L} \tag{5.18}$$

where γ depends on the propagation environment. L is the filter slope expressed in dB/oct, whereas d_0 and d_1 are the distances from the base station to the transmitting and distorting mobile stations, respectively. In the considered example $G = 64/24 = 2.67$, so $\Delta f_{sep} = 2^{2.67} \cdot (B/2) = 3.18B$. In practice the neighboring channels used in the same cell have to be separated by $4B$ Hz.

Besides the described distribution of channel frequencies among the cells, which maximizes the frequency separation between channels applied in the given cell and takes into account carrier frequencies used in the neighboring cells, there are some other means of minimizing the influence of interchannel interference. These are:

- sophisticated synthesis of transmit and receive filters which effectively attenuate the sidelobes of the transmitted and received signal spectra, which in turn ensures the increase of receiver selectivity,
- precise power control of the signal transmitted by the base stations and mobile stations on each channel.

More sophisticated filter synthesis results in increasing the value of L in (5.18). The price paid for this is the increased transceiver cost and complexity. The precise power control has a particular meaning. Thanks to it the signals arriving at the base station from various mobile stations have approximately the same and permissibly low power with respect to the S/I requirement or the error probability. All operating cellular systems apply power control in mobile stations. This is particularly important for CDMA systems (see Chapters 11 and 12) for which power control decides to a large extent the effective system capacity.

Table 5.1 Distribution of carriers for a single GSM operator

1A	2A	3A	4A	1B	2B	3B	4B	1C	2C	3C	4C
1	2	3	4	5	6	7	8	9	10	11	12
13	14	15	16	17	18	19	20	21	22	23	24
25	26	27	28	29	30	31	32	33	34	35	36
37	38	39	40	41	42	43	44	45	46	47	48
49	50	51	52	53	54	55	56	57	58	59	60
61	62	63	64	65	66	67	68	69	70	71	72
73	74	75	76	77	78	79	80	81	82	83	84
85	86	87	88	89	90	91	92	93	94	95	96
97	98	99	100	101	102	103	104	105	106	107	108
109	110	111	112	113	114	115	116	117	118	119	120
121	122	123	124								

Table 5.1 presents an example of channel distribution for the GSM system with topology shown in Figure 5.8, on the assumption that a single operator has all 124

carriers at its disposal (see Chapters 7 and 8). In reality, the GSM spectrum is divided among a few operators and a table similar to Table 5.1, containing the subset of carriers, has to be prepared for all of them. The rule applied in Table 5.1 is as follows.

Twelve different sectors are included in Table 5.1, starting from symbol $1A$ $(j = 1)$ and ending with $4C$ $(j = 12)$. The jth sector $(j = 1, \ldots, 12)$[4] uses the carriers with numbers $j+12k$, $(k = 0, \ldots, n)$, where n is the largest integer for which $(124-12n)/12 < 1$. A similar table for AMPS systems managed by two operators can be found in [3].

So far we have considered distribution of carriers (channels) from the point of view of interchannel interference minimization. If this channel allocation is done once and kept constant, this type of assignment is called a *fixed channel assignment*. Let us recall that in traditional analog cellular systems a channel is equivalent to a frequency interval with a carrier in the middle. In TDMA/FDMA systems, a channel assignment is equivalent to the assignment of a particular time slot on a specified carrier. In CDMA/FDMA systems, a channel is equivalent to a particular spreading code applied on a specified carrier.

Fixed channel assignment is the simplest strategy of system resources distribution. In reality, this strategy can be more cumbersome than in the simplest case of hexagonal cells with traffic uniformly distributed in the system area. The reason for the increased complexity is the necessity of taking into account real or expected traffic distribution and unequal size of cells or sectors. In fixed channel assignment, setting a new connection in a given cell is possible only if there are unoccupied channels in that cell. In the case of a temporary lack of available channels, the user suffers from blocking of a connection. At the same moment there can be free channels in neighboring cells. The number of call requests may often vary considerably depending on the day, hour, or particular events. Thus, fixed channel assignment can be an inefficient solution resulting in high probability of blocking during busy hours. There are, however, more sophisticated channel assignment methods [6], [8], which take into account a dynamically changing demand for channels. Figure 5.11 presents the classification of the channel assignment strategies.

The *simple borrowing* strategy is an improvement of the basic fixed strategy, which adds some dynamics to the latter. If all channels permanently assigned to the cell are busy, a channel can be borrowed from a neighboring cell, provided that this channel does not interfere with current calls. If the channel is borrowed by the given cell, some other cells are not allowed to use the same channel because of inter- and co-channel interference. The mobile switching center supervises the borrowing process, locking the borrowed channels in the cells located one or two cells away from the borrowing cell. The switching center keeps a record of free, borrowed and locked channels and informs appropriate base stations about them. In consequence of the applied strategy, the probability of blocking decreases to a certain threshold value of traffic intensity. If the actual traffic exceeds this threshold, the channel utilization degrades because borrowing a single channel causes locking that channel in about five other cells.

[4]In case of the GSM system, if the number of cells in a cluster is $N = 4$ and each cell is divided into three sectors, there are 12 different sectors.

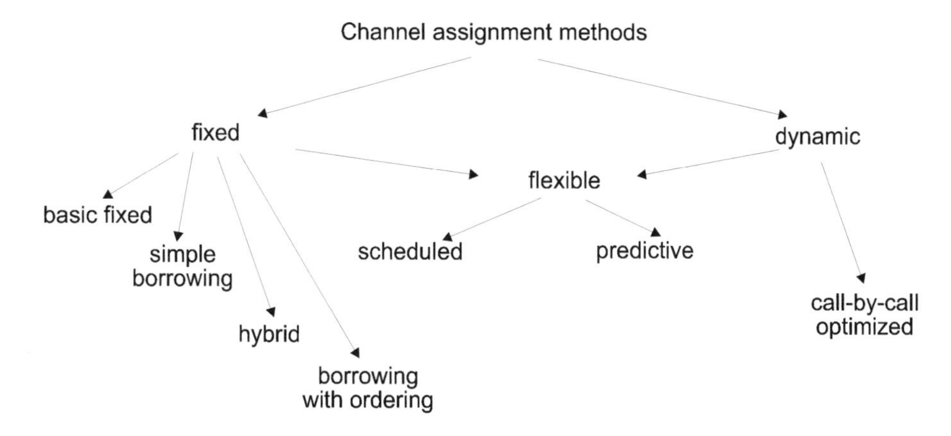

Figure 5.11 Classification of channel assignment strategies

The *hybrid channel assignment* strategy improves the properties of the simple borrowing strategy. In this strategy, channels used in a cell are divided into two categories: the channels belonging to the first category cannot be borrowed and are used only within the cell, whereas the channels belonging to the second category can be borrowed. The ratio of channel numbers from both categories is set on the basis of the expected traffic. Each channel is labeled with respect to the category to which it belongs.

Further improvement of the channel utilization is achieved due to the *borrowing with ordering* strategy. In this strategy the number of channels belonging to both categories changes dynamically with respect to the traffic conditions. The probability of borrowing is assigned to each borrowable channel and these channels are ordered with respect to the decreasing value of this probability. The probabilities of borrowing are updated on the basis of the number of the channel borrowings.

In the *dynamic channel assignment* strategy there are no channels permanently assigned to the cells. The channel is assigned to a particular call on a call-by-call basis. Decision upon the assigned channel can be made either by the mobile switching center or by the mobile station. In the first case we talk about centralized control, whereas the second case is described as the distributed control of the channel assignment process. The mobile switching center selects the particular channel on the basis of a criterion minimizing the selected cost function. The cost function depends on the probability of blocking, the frequency of utilization of the potentially assigned channel, the distance to the cell currently using the same channel, etc. In the process of channel assignment, the mobile switching center also takes into account the measurements of the signal received from a mobile station.

The *flexible channel assignment* strategies combine the advantages of fixed and dynamic channel assignment strategies. Each cell has at its continuous disposal a set of channels which is sufficient to serve the traffic of moderate intensity. The mobile switching center manages the remaining set of channels which can be assigned to those cells in which the current traffic intensity is too high to be served by the permanently assigned channels.

In the *flexible channel assignment with scheduling* the assignment of additional channels is planned in advance, taking into account the time of the day and location of the cell. The changes in the channel assignment are made at the predetermined moments at which the traffic achieves extreme values.

In the *predictive flexible channel assignment* strategy the traffic intensity is measured in real time and channel reassignment can be made by the mobile switching center at any moment.

The channel assignment strategies presented so far are only a representation of a rich set of strategies which can be found in literature. A wide survey of these strategies can be found in [9] which a particularly interested reader is advised to study. Comparing the two main groups of channel assignment strategies, we can state that there is a compromise between the achievable service quality (meant here mostly as the probability of call blocking), implementation complexity, and efficient use of the assigned spectrum. We conclude from [9] that the simulation and analysis results indicate that for low traffic intensity dynamic channel assignment strategies are better than fixed channel assignment strategies. However, the latter show their superiority at high traffic intensity and the uniform distribution of mobile stations in the covered area. In the fixed channel assignment, the channels are assigned to ensure the maximum number of their multiple use. This is not possible for dynamic channel assignment strategies which have a statistical character resulting from the statistical nature of the call and channel requests.

REFERENCES

1. V. H. Mac Donald, "The Cellular Concept", *Bell System Technical Journal*, Vol. 58, No. 1, January 1979, pp. 15-41

2. W. C. Y. Lee, "Elements of Cellular Mobile Radio Systems", *IEEE Trans. Vehicular Technology*, Vol. VT-35, No. 2, May 1986, pp. 48-56

3. T. S. Rappaport, *Wireless Communications*, Prentice-Hall PTR, Upper Saddle River, N.J., 1996

4. K. David, T. Benkner, *Digitale Mobilfunksysteme*, B. G. Teubner, Stuttgart, 1996

5. A. Mehrotra, *GSM System Engineering*, Artech House Publishers, Boston, 1997

6. G. L. Stüber, *Principles of Mobile Communication*, Kluwer Academic Publishers, Boston, 1996

7. D. Rutkowski, "The Rules of Cellular Network Construction and their Progress Perspectives" (in Polish), *Przegląd Telekomunikacyjny*, No. 4, 1996, pp. 241-249

8. S. Tekinay, B. Jabbari, "Handover and Channel Assignment in Mobile Cellular Networks", *IEEE Communications Magazine*, November 1991, pp. 42-46

9. I. Katzela, M. Naghshineh, "Channel Assignment Schemes for Cellular Mobile Telecommunication Systems: A Comprehensive Survey", *IEEE Personal Communications*, June 1996, pp. 10-31

10. W. C. Y. Lee, "Smaller Cells for Greater Performance", *IEEE Communications Magazine*, November 1991, pp. 19-23

11. ETSI GSM 05.05, *Radio Transmission and Reception*, March 1991

12. A. Wojnar, *Land Mobile Communication Systems* (in Polish), WKŁ, Warszawa, 1991

6

First generation cellular telephony - NMT and AMPS examples

6.1 FIRST GENERATION CELLULAR SYSTEMS

First generation cellular systems using analog voice transmission came into operation in the early eighties. The first cellular system, which is still in use, is the American *Advanced Mobile Phone System* (AMPS) [4]. Although there are technologically much more modern systems than AMPS, it is still popular in the USA, Canada, Central and South America and some other countries, due to its wide coverage. The first mobile cellular system implemented in Europe was the *Nordic Mobile Telephone* (NMT) system. This was the first-generation cellular system which operated in a unified way in more than one country and allowed mobile communications in the whole of Scandinavia. It is still used in Scandinavia and some East European countries.

NMT and AMPS are only two examples of first-generation cellular systems. Since those systems were mutually incompatible, a common digital cellular telephony standard was proposed in Europe. The other first-generation cellular systems are known as *Total Access Communication System* (TACS)[1] which was used in the United Kingdom, Ireland, Spain, China, New Zealand, Hong Kong and other countries, and *C-Netz* used in Germany, South Africa and Portugal. A similar system was also established in France. All cellular analog systems were similarly advanced and worked according to approximately the same rules. Due to their generally declining tendency they will not be the subject of a detailed description. However, we will shortly concentrate on NMT and AMPS systems, as they still retain some popularity. Table 6.1 presents the basic technical parameters of both systems.

[1]TACS is a European modification of AMPS.

Table 6.1 Basic technical parameters of AMPS and NMT cellular systems

System parameters	**AMPS**	**NMT 450**	**NMT 900**
Transmission frequency [MHz]			
– Base station	869-894	463-467.5	935-960
– Mobile station	824-849	453-457.5	890-915
Frequency separation between transmitter and receiver [MHz]	45	10	45
Spacing between channels [kHz]	30	25	25
Number of channels	832	180	1000
Base station coverage radius [km]	2–25	1.8–40	2–20
Modulation of audio signal	FM	FM	FM
– Frequency deviation [kHz]	±12	±5	±5
Control signals			
– Modulation	FSK	MSK	MSK
– Frequency deviation [kHz]	±8	±3.5	±3.5
Data transmission rate of control signals [kbit/s]	10	1.2	1.2
Transmitter output power [W]			
– Maximum for base station	100	50	25
– Medium for mobile station	3	1.5	1

6.2 NMT ARCHITECTURE

The NMT system was developed jointly by the Scandinavian countries, i.e. Denmark, Norway, Sweden and Finland. In the first phase of development the system operated in the 450 MHz band. After some time the system gained so much in popularity that it was close to saturation. Therefore, it was necessary to assign a new band in the range of 900 MHz and a new, modified system version denoted as NMT 900 was created.

Originally, the designers assumed existence of the following mobile stations:

- mobile stations installed in vehicles,
- portable mobile stations,
- coin-operated wireless terminals.

These assumptions are specific for Scandinavian countries where one of the reasons for introducing cellular telephony, and the reason for its enormous success, was low population density. Technological progress made since the eighties resulted in the replacement of the portable mobile stations by *handhelds*. The system was aimed at land communications; however, it is possible to use it in short distance maritime communications along the sea coast as well. It is also used as one of the alternative means of wireless subscriber access to PSTN (see Chapter 15). The following basic requirements were defined for the NMT system [3]:

- automatic setting up and charging of a connection to and from a mobile station,
- the possibility of setting up a connection with any subscriber of a PSTN network in any country or with any subscriber of the mobile network,
- the cost of a call is charged to the calling subscriber, the cost being calculated on the basis of the duration of the call and the dialled number,
- from the point of view of a mobile subscriber the system should be similar to a standard PSTN network in the aspects of the network use, reliability of signaling, charging and call secrecy,
- the introduction of the NMT system should not require any significant changes in the PSTN network.

The NMT system consists of three basic group of elements: *Mobile Telephone Exchanges* (MTX), *Base Stations* (BS) and *Mobile Stations* (MS).

The MTX is the main control element of the system. It provides the interface with the PSTN. The interface is possible on the local, transit or international exchange levels. The preferred level of interface is a transit exchange. The MTX is a specialized digital switching exchange. Therefore, besides standard connections it allows additional services such as short dialing, transfer of a connection, etc.

The base stations realize the interface between the fixed part of the system and mobile stations. They are connected with the MTX using four-wire transmission systems based on wireline or line-of-sight radio links. The areas covered by base stations are grouped into *traffic areas*. Each traffic area is connected through the MTX with the fixed network. It is also possible for the MTX to control a few traffic areas. Such a configuration is shown in Figure 6.1. The area covered by base stations controlled by a single MTX is called a *service area*. Such an area is divided into traffic areas.

Each base station manages a subset of channels assigned to the cell according to the channel distribution plan. From the system point of view, the following types of channels exist:

- paging (calling) channel,
- traffic channels,
- access channel (in NMT 900),
- combined paging and traffic channel,
- data channel.

In a typical system arrangement, each base station has a single paging channel, a single access channel (in NMT 900), a single data channel and a limited number of traffic channels.

The *paging channel* is used by the base station for transmission of a continuous identification signal. Mobile stations located in a given traffic area and remaining in

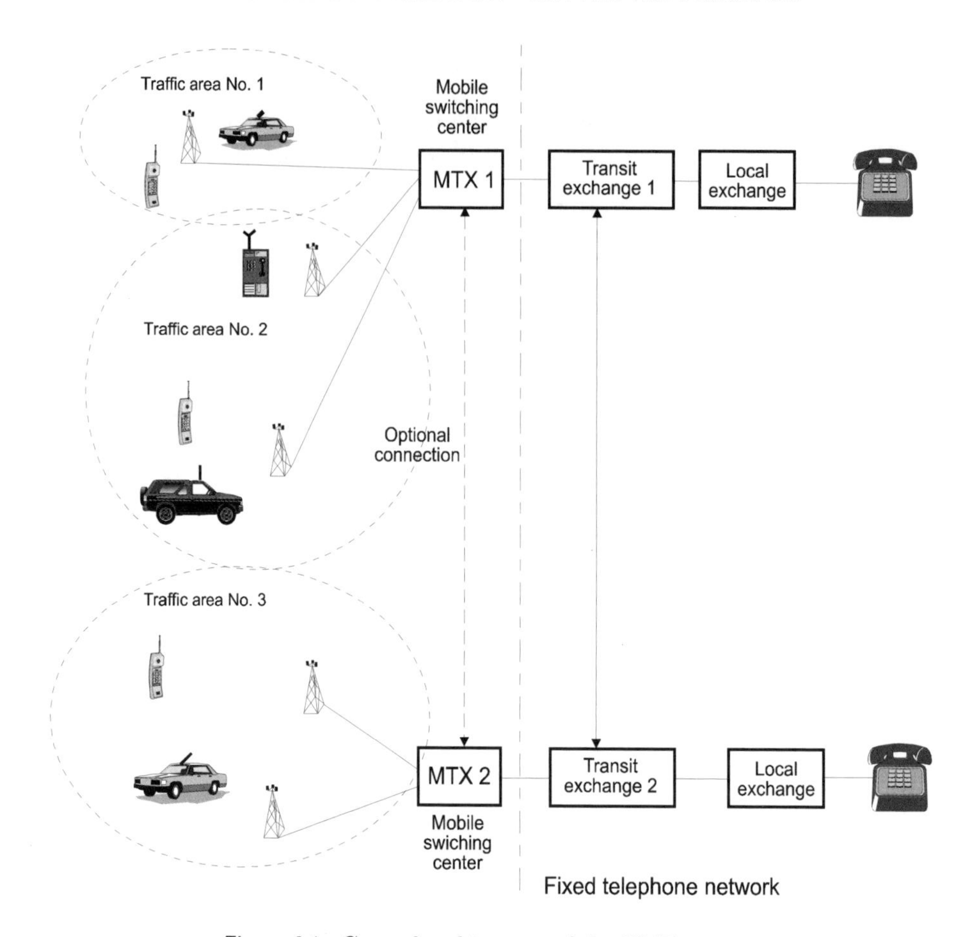

Figure 6.1 General architecture of the NMT system

the *idle state*[2] are locked to the paging channel and wait for call requests directed to them. After the identification of the call request, a traffic channel is assigned to the mobile station to perform the remaining part of the connection set up and to realize the call.

Traffic channels are used to perform calls and to manage a part of the call request. A traffic channel can remain in three different states, marked with special flag signals:

- in a free state – the mobile station can use it to initiate a call request to the base station,
- in a busy state – the call is currently performed,

[2]In the idle state the power supply is switched on but only the parts of the mobile stations necessary for call request detection are powered.

- in an idle state – the channel is neither in a free state nor in a busy state.

Let us recall that a link consists of a pair of traffic channels: one channel from base station to mobile station (*downlink*) and another channel shifted by 10 MHz and used for communications in the opposite direction (*uplink*). A mobile station wishing to set a connection looks for a downlink channel marked as a free channel. Thus, an uplink channel shifted by 10 MHz can be used for the call request procedure.

The *combined paging and traffic channel* has the features of both channels. In a regular mode it is used as a paging channel. However, if all traffic channels are occupied, it can be temporarily used by selected, high priority subscribers as a traffic channel.

The *data channel* allows measurement of the power level of the signal of the mobile station being in the active connection state. The measurement results are used by the MTX in the handover process.

The *access channel* is a special channel in the NMT 900 version of the system used to perform a call request instead of a traffic channel marked "free".

The NMT 450 operates in the FDMA/FDD mode. The frequency ranges for the downlink are 463-467.5 MHz for NMT 450 and 935-960 for NMT 900. The uplink is realized in the frequency range of 453-457.5 MHz for NMT 450 and 890-915 MHz for NMT 900 [5]. Channel carrier separation is 25 kHz, which in the assigned spectrum allows for allocation of 180 channels in NMT 450 and 1000 channels in NMT 900. Voice signals are transmitted using FM modulation with 5 kHz deviation [5].

Because the number of channels in NMT 450 was equal to 180, the system designers foresaw that the system capacity would be too small in densely populated urban areas. Thus, the idea of cell splitting was applied. One of the possible solutions was a construction of concentric sector cells with the sector angles equal to 60° and the center located at the point close to the highest traffic density. Around the center the cells are very small. The cells in subsequent rings have a larger area and theoretically have a shape of a ring sector. Diversified cell sizes and the base station powers require a dynamic power level control at the mobile stations in order to minimize interchannel interference and the *near-far effect*.

All control signals are transmitted directly in the traffic channel using the MSK modulation (see Chapter 1), at the data rate of 1200 bit/s. When control signals are transmitted, regular speech transmission is interrupted, which can be heard by the subscriber and deteriorates the overall transmission comfort. The control signals are blocks consisting of 16 hexadecimal characters, i.e. they are 64-bit long. In order to ensure reliable reception, the Hagelberger error correcting code is applied. The code is designed to correct burst errors which are not longer than 6 bits. The subsequent error bursts have to be interlaced with error free sequences which are at least 20-bit long. The applied code ensures error correction of most of the errors due to fading occurring at a typical mobile station velocity.

The control frame consists of 166 bits and is divided into the following blocks:

- 15 bit synchronization bits (101010101010101),
- 11 frame synchronization bits (11100010010),
- 140 bits containing control data bits and Hagelberger parity bits.

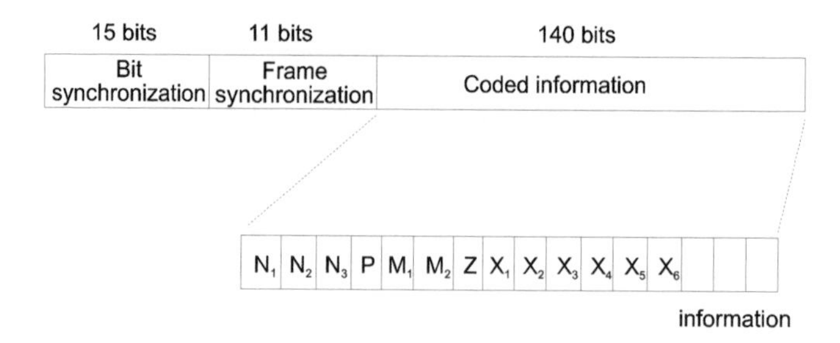

Figure 6.2 Signalling frame structure in the NMT system

Figure 6.2 presents the control frame structure. The coded information contains a three digit channel number, a prefix character, a two-digit traffic area number, a seven-digit subscriber number and three-character user's information.

6.3 INFORMATION FLOW CONTROL IN THE NMT SYSTEM

The flow control functions often have a big influence on the system performance. They are associated with connection set-up, handover, paging of mobile stations and their location update.

Let us consider the system operation at the moment of a network originated connection set-up. The PSTN subscriber wishing to call a mobile subscriber makes a connection with the MTX. There are two possibilities of setting the path to the mobile station. The first possibility is a connection with the mobile's own MTX, i.e. that MTX in which the mobile station is permanently registered. The mobile station usually resides in the service area of that MTX. The data on the current location of the paged mobile station can be found in the MTX subscriber register. On the basis of these data the MTX directs the connection to the proper MTX in whose service area the mobile station is currently located. The current location of the mobile station in the service area of a particular MTX is marked in its subscriber register. The data of the called mobile station are finally analyzed in the destination MTX and the latter manages the final steps of the connection set-up.

The second possibility of path routing is a connection to the closest MTX being the nearest gateway to the NMT system. This exchange analyzes the form of the called subscriber number and sends the request to the mobile subscriber's own MTX to find out the current mobile subscriber location. After reception of this information, the shortest route to that particular MTX is set up which manages the area in which the called mobile station is located. Both possibilities of a connection are shown in Figure 6.3 [5]. The standard connection is shown using a solid line, whereas the connection in which the closest MTX is used is shown by dotted lines. MTX H (*Home* MTX) denotes the home location exchange containing the permanent data and data on the current location of the called mobile station and MTX V (*Visited* MTX) is the exchange serving the traffic area in which the mobile station is currently located. MTX G (*Gateway* MTX)

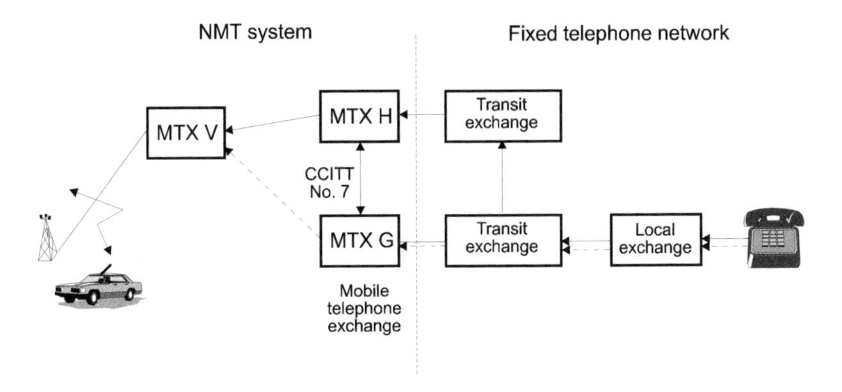

Figure 6.3 Path routing during connection set-up: standard procedure (solid line) and procedure involving the use of the nearest MTX (dotted line)

denotes the NMT exchange which is closest to the calling subscriber. As we see, the second method of path routing requires transmission of additional signaling information between MTXs. In turn, the costs of signal transmission are minimized. This method can be applied if in the mobile network a numbering plan is applied which allows for identification of a mobile subscriber at the early stage of the call set-up.

If the mobile station receives the call request containing its own identification number in the paging channel, it responds in the uplink paging channel. Based on this response, the MTX identifies the base station which is closest to the mobile station and belongs to the traffic area in which the mobile station has recently been registered. In turn, the MTX assigns the number of a free traffic channel to which the called mobile station subsequently tunes. From that moment the paging channel is available for other call set-ups. If the called mobile station does not acknowledge the call, the MTX generates a tone or a voice announcement indicating that the called mobile station cannot be reached. A similar signal is also sent if there is no free channel for the call set-up.

If the mobile station has been identified and a free traffic channel has been assigned to it, the subsequent call set-up steps follow. The selected base station begins to send a test tone to supervise the connection quality. In case of NMT 900, the base station transmits the request for mobile station identification number and its password. The mobile station uses the assigned traffic channel to send these parameters. If mobile station identification is positive, the base station sends a command to generate a ringing signal. After the response of the mobile subscriber (he/she symbolically takes the phone off the hook) the call is finally realized. The disconnection of one of the subscribers triggers a signal for release of the assigned traffic channel and the mobile station turns to the idle mode. In this mode it is tuned to the paging channel of the cell where it is currently located.

Now let us consider the mobile-originated call request. The necessary condition for making the call request is the current registration of the calling mobile station in the cellular network and being in the power-on mode in which the mobile station is tuned to the paging channel for possible detection of the paging signal. The initiation of the mobile-originated call request implies the search of the mobile station for a free traffic channel. Such a channel is found after monitoring the downlink traffic channels transmitting a "free channel" marker. In the uplink traffic channel associated with the first found free downlink channel, the mobile station initiates the exchange of control signals. It transmits the identification data and the number of the called mobile or fixed network subscriber. The MTX serving the traffic area in which the mobile station has issued the call request checks the category of the calling subscriber.

In NMT 900 the mobile station performs the same procedure in the access channel. Additionally, the MTX checks the subscriber's password. Then a free traffic channel is assigned to the connection. In case of system overloading, high priority subscribers (emergency service, police, medical doctors) are granted temporary usage of the combined paging and traffic channel.

During the call the base station sends to the mobile station a continuous tone signal of the frequency around 4000 Hz (3955, 3985, 4015 or 4045 Hz). The mobile station reverses this signal back to the base station. The base station evaluates the quality of the received signal in order to realize the handover procedure or disconnect the call. The base station informs the MTX about the quality of the received tone signal. If the tone level is too low, the MTX initiates measurements in the neighboring cells. The measurement results are the basis for the MTX's decision upon possible transfer of the connection to a new cell (handover). If the measurement results are no better in any surrounding cell than those in the current cell, after 20–30 seconds another attempt of handover is made. If the level of the measured signal falls below a critical value, the link is released. However, if handover is initiated and a new cell is selected, a new traffic channel has to be found. After finding it, the MTX transmits the order to the mobile station to transfer the connection to the new traffic channel in the new cell.

Let us note that during handover the mobile station remains passive. Only the MTX and co-operating base stations are involved in the measurements, subsequent analysis of these measurements and the handover decision. This is a serious load for the control part of the system. In newer systems the measurement and decision processes are more distributed. Mobile stations take part in them, too.

If the MTX decides to transfer the connection to one of the neighboring cells, it sends the order to the surrounding base stations to perform the power signal measurement on that channel which is currently used by the mobile station. For this reason all base stations are equipped with a monitoring receiver, which is able to measure the power level of all the system channels. The information obtained from this receiver helps the MTX to decide to which base station the connection with the specified mobile station should be transferred. Let us note that handover can occur not only between two cells in the same traffic area but between two traffic areas or two service areas. In the last case the change of MTX takes place as well.

The measurement procedure is also ordered to the base station at the beginning of the call initialization in order to check if this base station is appropriate for a connection with the given mobile station. This measurement result is also useful in checking if the mobile station located closely to the base station is not sending the signal whose power level is too high. If this is detected, the MTX commands the mobile station to decrease the transmitter power in order to balance the signals received from different mobile stations. The measurement procedure is also useful if the connection is transferred to another traffic channel in the same cell when the currently assigned channel is out of order or does not ensure adequate quality due to the propagation conditions.

Location update is one of the procedures closely related to the system control functions. As we have already mentioned, if the mobile station is in the idle mode it is tuned to the paging channel in its cell. This channel transmits the code of the traffic area to which the cell belongs. If the mobile station changes the location, it tunes to a new paging channel. In consequence, if the mobile station detects a new traffic area code, it initiates the call request in order to update its location in the subscriber register of the home MTX.

If due to mobility the mobile station moves to a new service area which is managed by another MTX, the registration procedure has to be initiated in the new "visited" MTX. This is so called *roaming*. The newly visited MTX sends an information sequence containing the new location of the mobile station to its home MTX. In response, the home exchange sends to the new MTX the sequence with the current subscriber status.

In Scandinavian countries a mobile station is equipped with circuits preventing tuning to base stations from a foreign country.

The system should complete call set-up and handover procedures within the following time intervals [5]:

- mobile originated call set-up time – 4 seconds,
- mobile terminated call set-up procedure:
  - occupation of a paging channel – 1 second,
  - occupation of a traffic channel – 1 second,
- call clearing time – 0.75 second
- handover time: 1 second (NMT 450) or 0.3 second (NMT 900).

6.4 SERVICES OFFERED BY NMT

Services offered by NMT partially follow from the call management performed by electronic telephone exchanges applied in the NMT system, which have been modified for operation in this mobile cellular system. In general, the system offers similar services to those offered by a typical fixed telephone network. Besides a standard voice connection, there is a list of additional services such as abbreviated dialing, immediate call diversion or call diversion on no reply from the subscriber, malicious calls tracing, etc. These services are common for the NMT and fixed networks. However, there are services specific for the mobile system as well.

The first one called *Data Mobile Station* is low rate (600 bit/s) data and fax transmission. As we remember, in a regular connection, the signaling sequences are transmitted in the traffic channel during interruptions of a voice signal. Such an operation would be disastrous for data transmission. Thus, data transmission is possible only after switching off the signaling in the traffic channel, necessary for handover and power control.

The data transmission service is activated and deactivated by a subscriber and requires special equipment interfacing between a data or fax terminal and a mobile station providing the modem interface.

The second service is called *Coin Box Mobile Station*. This service provides the mobile station with charging information during mobile-originated calls. It can be used, for example, by coin-operated mobile terminals.

Other services offered to the subscriber are priority calls, DTMF[3] signaling and a list of additional improvements in the mobile station (push-button dialing, on-hook dialing, telephone book in the mobile station, last number redial, electronic lock and others). A very important service offered to the subscribers traveling in vehicles is hands-free operation.

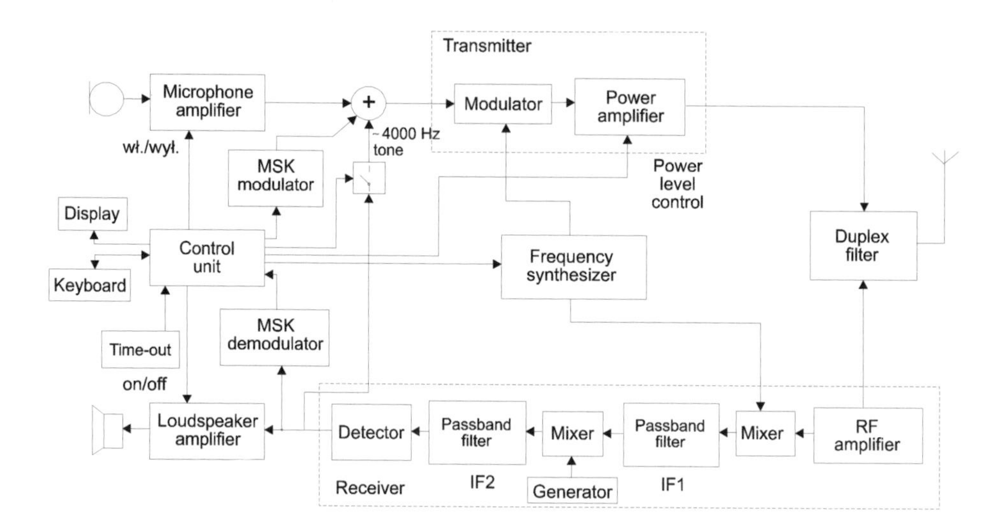

Figure 6.4 Block diagram of a typical NMT mobile station

6.5 TYPICAL MOBILE STATION AND BASE STATION DESIGN

A block scheme of a typical mobile station is shown in Figure 6.4. It consists of three basic parts: a transceiver part, a logical block, and a user interface.

The voice signal received from the microphone is amplified and fed to the FM modulator whose carrier frequency is generated in the frequency synthesizer. This synthesizer is controlled by the control unit which determines the number of the applied channel. The modulated signal is amplified by the power amplifier with adjustable gain. The output signal, after passing through the duplex filter is fed to the antenna. Instead

[3]Dual-Tone Multiple Frequency

of a voice signal, the MSK modulated signal carrying the control commands can be transmitted. As we have already mentioned, besides the voice or data signals, the mobile station retransmits the test tone received from the base station at the frequency of around 4000 Hz.

The signal received by the antenna is directed through the duplex filter to the mobile station receiver. First it is amplified and then moved down in two steps to the intermediate frequency at which the frequency discriminator operates. The detector output voice signal is fed to the loudspeaker amplifier. If a digitally modulated control signal is received, it is fed to the MSK demodulator. The test tone is extracted from the received signal and fed back to the transmitter.

On the basis of the messages received from the base station, the control unit decides upon the traffic channel number. Depending on the type of transmitted or received signal, it switches on and off the microphone and loudspeaker amplifiers, and supervises the keyboard and display. This unit controls the process of call set-up in the mobile station, in particular it searches for a free traffic channel. The unit also controls the process of channel release after completing the call and manages the procedure of automatic registration of the mobile station in the system.

The base station consists of a certain number of units which are called *channel equipment* and the circuits which are common for all the units such as transmitter combiner, antenna multicoupler, supervisory unit, measurement receiver, RF test loop and power supply.

The channel equipment comprises a transmitter, receiver and control unit. The transmitter contains a pre-emphasis filter, FM modulator, test tone generator and amplifier. The maximum transmitted signal power level is equal to 50 W for NMT 450 and 25 W for NMT 900. The receiver consists of an input amplifier, FM detector, de-emphasis filter, a stopband filter cutting off the test tone and a measurement circuit supervising the level of the received signal. The control unit consists of the controlling microprocessor, a modem communicating with the MTX and a test tone generator. The control unit participates in the signal transmission between the MTX and the RF part of the base station. It also controls the self-test loop and performs alert functions and evaluates the received test tone quality.

A transmitter combiner allows connection of the transmitters of all channel equipments to a common antenna without mutual loading. This block also contains the filters cutting off the spectrum outside the band of a given channel.

A multicoupler distributes the received signals to the respective receivers. It is performed using passband channel filtration and active amplification of the channel signals.

A supervising block cooperates with the measurement unit, evaluating the level of the received signals. The measurement results are quantized on 64 levels and this information is transmitted to the MTX. It can be used to make a decision about possible handover.

* * *

The NMT system which is already outdated played a significant role in introducing cellular telephony in Scandinavian, East European and many other countries. Now it has mostly been replaced by GSM. However, it can still be useful in rural areas with low population density, in implementing wireless local loops. Thus, huge investments made in its deployment in the eighties and early nineties are not completely wasted.

6.6 THE OVERVIEW OF AMPS

The *Advanced Mobile Phone System* was developed in the seventies [4]; however, the first deployment took place in the Chicago area in 1983. The system gained wide popularity in the USA, Canada, Mexico, South America, Australia and Israel. It was also installed in some areas of the former Soviet Union. Although outdated, it is still useful in the large territory of the North American continent. AMPS was allocated 40 MHz spectrum in the 800 MHz band by the Federal Communications Commission, later extended by the next 10 MHz (see Table 6.1 for spectrum intervals). The spectrum allocation is also shown in Figure 6.5. Typically, due to promoted competition, the spectrum is shared by two service providers: the nonwireline service provider (denoted by character A) and the wireline service provider (denoted by B). The extended parts of the spectrum assigned to each service provider are denoted by A' and B', respectively. Each service provider has 416 channels at its disposal, including 21 control channels. We have to note that channel allocations can be different in other countries. In some countries a single service provider exploits the system, so there is no division of the available channels between two providers.

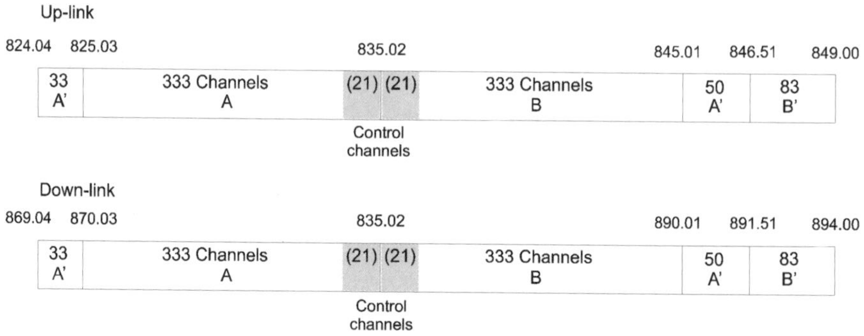

Figure 6.5 AMPS spectrum allocation

The system uses 7-cell clusters, mostly with 120°-sector antennae, to ensure at least 18 dB of the signal-to-interference ratio. The 30-kHz channels are divided into the following categories:

- *Forward Voice Channels* (FVC) – used to transmit voice signals originated in the fixed network from a base station to mobile stations and to send some control signals during the call,

- *Reverse Voice Channels* (RVC) – applied to transmit voice signals and control signals in the opposite direction,

- *Forward Control Channels* (FCC) – used to broadcast system information and paging of the called mobile stations,
- *Reverse Control Channels* (RCC) – used to carry signaling information from the mobile station to the base station in the set-up phase of a connection.

A typical base station has eight or more duplex voice channels, a forward control channel and a reverse control channel; however, the number of voice channels can be as large as 57 and the number of control channels can be also higher than one. The actual channel configuration depends on the expected traffic and on specific environment. Each base station has a control channel transmitter using FCC. This transmitter continuously sends a digital FSK modulated signal at the data rate of 10 kbit/s, containing system information such as the *System Identification Number* (SID) specific for each service provider and the system status containing power control handling, roaming capabilities, system ability to process other related standards such as DAMPS[4] (IS-54) and NAMPS[5] (narrowband AMPS). The system and paging information has the form of overhead messages, mobile station control messages, or control file messages. In the USA 21 control channels for each service provider have been standardized, so mobile stations wishing to "enter" the AMPS network and tune to the strongest FCC need to scan the preset channels only. As we have mentioned, the reverse control channel (RCC) is mainly used by mobile stations to initiate calls and to acknowledge the received page messages.

6.7 AMPS RADIO INTERFACE

The details of AMPS radio interface depend on the type of the channel. They are different for voice channels and control channels. Let us consider signal transmission on a voice channel first.

The basic function of the voice channel is to perform transmission of a user's voice signal. This is realized applying FM modulation. Therefore, the transmitter and receiver blocks in the voice transmitting channel are typical of FM systems (see Figure 6.6).

In the transmitter part, the signal representing voice is received from a microphone or a PSTN source. Such a signal is filtered and limited in amplitude. Then, it is fed to a compressor. The compressor is a variable gain circuit which controls the effect of speech level variability and decreases the dynamic range of the speech signal. It improves the subjective quality of transmission over distorted channels. The compression is performed in such a way that a 2 dB increase in input power level produces a 1 dB increase in output power level. The compressed signal is the subject of pre-emphasis and amplitude limiting in order not to exceed the specified frequency deviation of 12 kHz in

[4]DAMPS (*Digital* AMPS) is a TDMA system in which three users share each 30 kHz channel.
[5]N-AMPS (*Narrowband* AMPS) is the modification made by Motorola allowing for the processing of three calls in a single 30 kHz channel by applying FDMA and 10 kHz band for each user. It is mainly installed in highly populated areas.

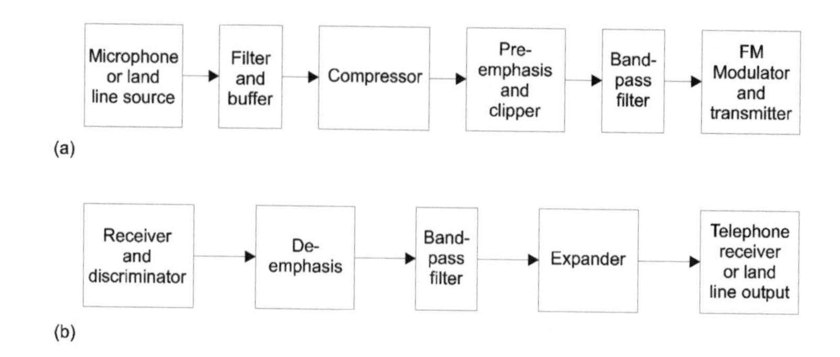

Figure 6.6 Transmitter (a) and receiver (b) blocks in the voice processing chain

the FM modulator. In turn, the signal is fed to the bandpass filter which ensures that the signal bandwidth is limited to the specified spectral interval. Finally, the signal is supplemented with the special *Supervisory Signal* (SAT) performing a similar function to that of the test tone in the NMT system, and is fed to the FM modulator. The SAT tone has one of three possible frequencies: 5970, 6000 or 6030 Hz. Only one of them is applied in a given cell.

In the receiver, specific blocks perform operations parallel to those in the transmitter. After down-conversion to the intermediate frequency, the received FM signal is discriminated, processed by the de-emphasis filter, bandpass filtered and expanded. The characteristics of the expander is reciprocal to that of the compressor, so both operations cancel each other.

Now let us consider data transmission. It is performed at the data rate of 10 kbit/s. There are two categories of data transmission in AMPS: continuous transmission and transmission in the form of discontinuous bursts. The first one is applied on forward control channels. The discontinuous data transmission is performed, similar to the NMT system, on forward voice channels and reverse control channels. On FVC the base station sends information concerning call release, handover orders, and mobile station power control. The method of transmission is called *blank-and-burst signaling*. A mobile station realizes a discontinuous data transmission on reverse control channels (RCC) if it sends a request to set up a connection.

In order to transmit data at 10 kbit/s, the FSK modulation with frequency discrimination at the receiver is applied. The data are line coded using the Manchester (bi-phase) code. In such a code data bits are represented by a pair of pulses. A binary one is assigned a bi-pulse 0,1 and a binary zero is represented by a bi-pulse 1,0. Thanks to that, the maximum of signal energy is concentrated around 10 kHz and the spectra of voice and data signals, where the latter is transmitted in the burst mode, do not overlap too much. Thus, the data bursts are barely heard by a user, although during data burst transmission the voice signal is blanked for a period of 50 ms when a data burst is sent. However, during data burst transmission on the voice channel, the SAT signal must be suspended because it would interfere with the data signal.

Digital information is secured by application of forward error correction in the form of BCH codes. A (40,28) BCH code is applied on forward channels whereas (48,36) BCH code is used in reverse channels.

Continuous transmission on a forward control channel is organized in blocks of 463 bits. The block starts with a 10-bit bit synchronization sequence followed by an 11-bit synchronization word. Then BCH encoded words A and B are transmitted in the interleaved manner in order to avoid correlation of potential errors. Each word has 40 bits and is formed by a BCH code word. Words A and B are directed to mobile stations having even or odd identification numbers, respectively. 42 bits remaining in the 463-bit block are so-called busy-idle status bits. They are inserted between the synchronization sequences and after each block of 10 message bits, as shown in Figure 6.7. They are intended to show the status of the reverse control channel in order to minimize the probability of collisions which can occur if more than one mobile station attempts to send a connection set-up request at the same time.

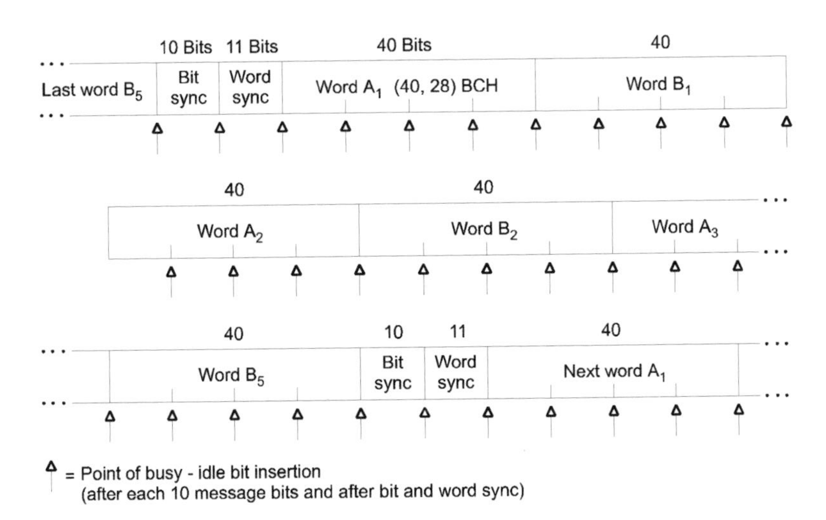

Figure 6.7 Data format for the forward control channel

Data sent on the reverse control channel take the form of packets. On this channel the mobile station competes to initiate a call. Therefore, the moment of the start of packet transmission is more or less random. The randomness is limited by monitoring the busy-idle status bits on the FCC channel. Due to discontinuous packet transmission, a bit synchronization word is longer than that on the FCC channel. It has 30 bits and is followed by an 11-bit word synchronization. The next 7 bits make up the so-called coded Digital Color Code (DCC) word. After the message precursor described above, the coded words constituting the message follow. The message consists of 1 to 5 words of 48 bits each, repeated 5 times. The base station performs a bit-by-bit, 3-out-of-5 voting to determine the form of the received 48-bit sequences. These sequences are finally decoded by the decoding algorithm of the BCH (48,36) code, which corrects one error or rejects the message if it remains uncorrectable. The format of the data packet used on a reverse control channel is shown in Figure 6.8.

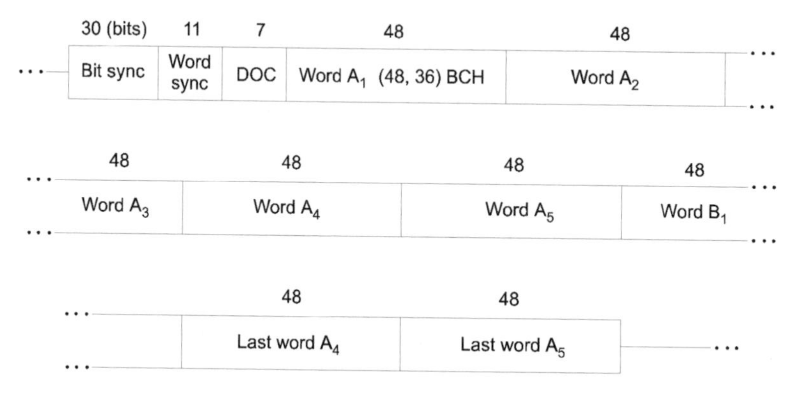

Figure 6.8 Data format for reverse control channel

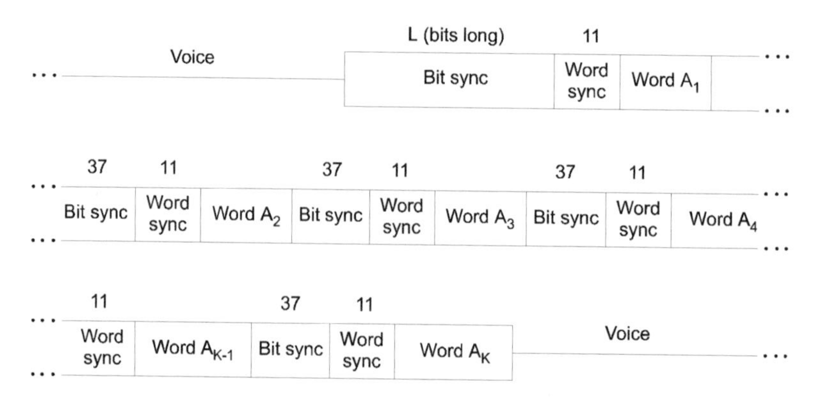

Figure 6.9 Data format used by packets transmitted on voice channels

As we have already mentioned, data transmission in a blank-and-burst mode also takes place on voice channels. The format of such a packet is shown in Figure 6.9. The packet starts with an L-bit long bit synchronization sequence ($L = 100$ for forward transmission and $L = 101$ for reverse transmission) followed by 11-bit word synchronization sequence. Then the message word is sent (40-bit long in forward direction and 48-bit long in reverse direction) interleaved with the bit and word synchronization sequences and repeated K times, where $K = 11$ for forward direction and $K = 5$ for reverse direction. Such a large number of repetitions performed in the forward direction is necessary to achieve high transmission reliability needed for handover handling.

6.8 CALL PROCESSING IN AMPS

Knowing some details of the AMPS radio interface, we are able to consider call processing performed in the system. As in the NMT system, two cases are possible. The first one is a mobile terminated call, where the source of the call request is a PSTN subscriber. The PSTN network issues a call request sending the number of a mobile subscriber to the AMPS mobile switching center (MSC). The MSC sends the paging message, which includes the called subscriber's *Mobile Identification Number* (MIN), to all base stations. They emit the page on the forward control channels. If the called mobile station is in the idle mode, it is listening to one of these channels. If it receives the paging message successfully, it acknowledges its reception on the reverse control channel by sending back to the base station its MIN, its serial number or the *Equipment Serial Number* (ESN). This way the MSC learns where the mobile station is currently located. In turn, the MSC instructs the selected base station to assign the unused voice channel pair (in forward and reverse directions) to the connection with the mobile station. The base station also assigns the mobile station a SAT tone and a *Voice Mobile Attenuation Code* (VMAC). The VMAC determines the power which will be used by the mobile station during transmission on the reverse voice channel. As a result, the mobile station changes its carrier frequencies to the assigned channel frequencies. In turn, it starts to retransmit the SAT tone sent by the base station. After receiving it back, the base station issues the *alert signal* on the forward voice channel and the mobile station starts to ring. The reception of the alert message is acknowledged by sending a 10 kHz *Signaling Tone* (ST) through the reverse voice channel until the mobile station answers the call. Finally, the call takes place over the assigned voice channels.

Let us now consider the second case, i.e. a mobile-originated call. If a mobile subscriber wishes to make a call, he/she enters the called user's number and the mobile station (which so far has remained in the idle mode) starts transmitting data on the reverse control channel. The message contains the MIN, *Equipment Serial Number* (ESN), *Station Class Mark* (SCM) and the called subscriber's number. If this message is received correctly, the data are sent to the MSC in order to verify the registration of the calling subscriber and the MSC connects the mobile subscriber with the PSTN subscriber, assigning a pair of voice channels to the base station – mobile station pair.

Handover is the crucial feature of cellular systems. It is also performed in the AMPS. As in the NMT system, the mobile station behaves to a large extent passively and the whole procedure is performed by the network. During a connection with the mobile station, the signal received on the reverse voice channel is periodically measured by the serving base station. If the signal level drops below a preset threshold or the reception quality of the SAT tone is insufficient, the MSC uses measurement receivers in the neighboring base stations to determine which base station receives the signal from the mobile station at the highest level. If such a base station is found, the connection with the mobile station is re-directed to it. It is done by assigning a new channel pair to the connection by a new base station. The mobile station acknowledges receipt of this message by sending the short ST tone on the reverse voice channel and the mobile station transmitter turns off. In the meantime the new base station starts to transmit the SAT tone. The mobile station tunes to the new traffic channels and begins to retransmit the new SAT tone. If the new base station detects this tone, the base station acknowledges correct handover procedure to the MSC and the MSC releases the old voice channel pair. The full procedure takes approximately 0.2 seconds.

* * *

The AMPS has been described without going into details. The reader interested in a more detailed description is advised to study [7] and [8].

The analog AMPS will be fully replaced by the second generation digital systems such as DAMPS (IS-54) IS-136 or cdmaOne (IS-95). Due to the fact that in the USA the second generation systems have not been allocated a new spectrum interval, a method of harmonized evolution from the AMPS to them has been developed. The DAMPS system uses the same channels as analog AMPS, assigning them to three users in a TDMA manner. The IS-95 system is designed to operate in a part of the AMPS spectrum, applying CDMA mode and using dual mode mobile stations.

REFERENCES

1. R. Steele (Ed.), *Mobile Radio Communications*, Pentech Press Publishers, London, 1992

2. A. Mehrotra, *Cellular Radio Analog and Digital Systems*, Artech House, Boston, 1994

3. *Nordic Mobile Telephone, Brief System Description*, Swedish Post Administration, Stockholm, 1981

4. *Bell System Technical Journal*, special issue on AMPS, Vol. 58, No. 1, January 1979

5. D. Westin, NMT: "The Nordic Solution" in D. M. Balston, R. C. V. Macario, *Cellular Radio Systems*, Artech House, Boston, 1993, pp. 73-111

6. Z. C. Fluhr, P. T. Porter, "Advanced Mobile Service: Control Architecture", *Bell System Technical Journal*, Vol. 58, No. 1, January 1979, pp. 43-69

7. R. V. C. Macario, *Cellular Radio. Principles and Design*, Macmillan New Electronics, Houndmills, Basingstoke, Hampshire, 1993

8. J. M. Hernando, F. Pérez-Fontán, *Introduction to Mobile Communications Engineering*, Artech House, Boston, 1999

7

GSM cellular telephony - architecture and system aspects

7.1 INTRODUCTION

In the eighties many different mutually incompatible analog cellular telephony systems were in use in Europe, such as TACS (*Total Access Communication System*), NMT (*Nordic Mobile Telephony*), C-Netz, Radiocom-2000 and their variants. In consequence, the users were assigned exclusively to their own operator and their mobile phones stopped operating when leaving the coverage area of their own mobile communication system.[1] As a result of cooperation within the European Union, a special working group for mobile telephony (*Groupe Special Mobile* GSM) was established, the task of which was to specify the objectives and the standards of a common mobile cellular telephony system. Industrial, scientific and R&D organizations from 17 countries, which signed the Memorandum of Understanding (MoU), participated in the development of a new system. As a result of common research and discussions a series of standards has been established covering all the required layers of the OSI (*Open Systems Interconnection*) reference model. Currently, the GSM acronym stands for *Global System for Mobile Communications* which stresses the fact that the application of GSM spans far beyond the European continent.

We will start the chapter devoted to the GSM system with a description of its general architecture and we will present its basic radio parameters. Then we will concentrate our attention on time and logical organization of the radio transmission. We will briefly present a few examples of the main procedures performed within the system, such as

[1]NMT is the only exception. It has coverage in the whole of Scandinavia and the users are able to get a connection through the other NMT operator when leaving their own country to another Scandinavian country.

registration of a mobile station in the network and setting up a connection. The aim of this chapter is to give an overview of the GSM system operation rather than to present details that can be found in the ETSI/GSM standards or books devoted exclusively to GSM. Two other chapters devoted to GSM will describe the system on the signal processing level and will concentrate on data transmission and system enhancements and modifications meant for high speed data and multimedia transmission.

7.2 BASIC GSM ARCHITECTURE

A basic scheme of the GSM architecture is shown in Figure 7.1.

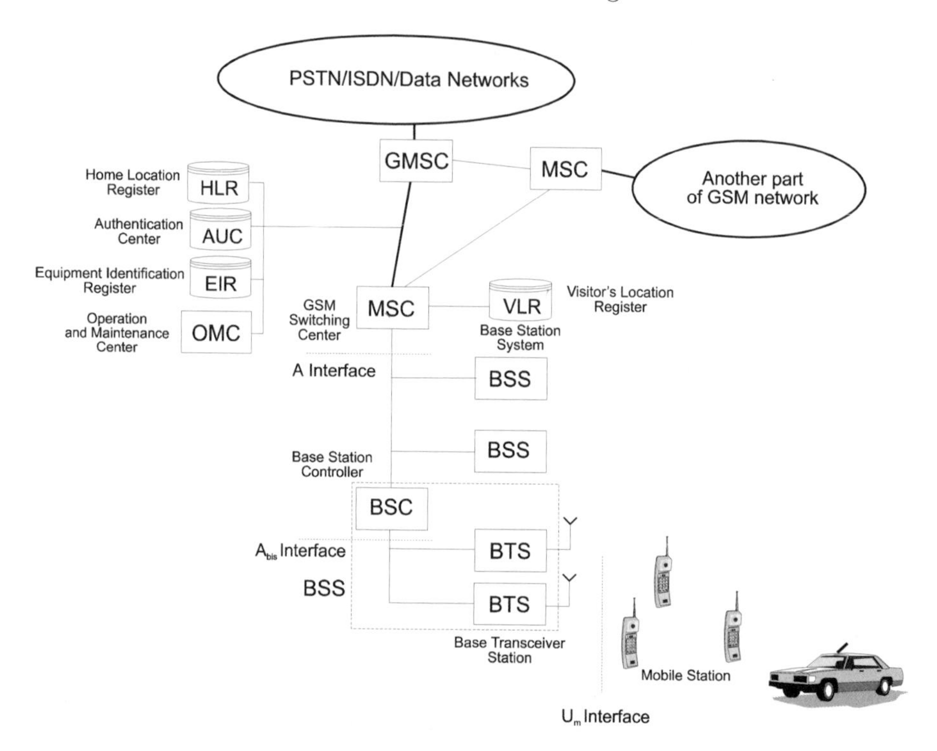

Figure 7.1 General GSM system architecture

The area of GSM operation is divided into subareas managed by particular *Mobile Switching Centers* (MSC). The MSCs are specialized electronic switching offices supplemented by functional blocks performing tasks that are specific for a mobile cellular system. Each MSC is connected with a respective VLR *(Visitor's Location Register)* database. This is a register containing necessary information regarding mobile stations that are temporarily located in the area serviced by the local MSC. Apart from the VLRs, the GSM system that is administered by a particular operator is equipped with three other databases. They are:

HLR (*Home Location Register*), the database of mobile stations permanently registered in the system administered by a particular operator,

AUC (*Authentication Center*), the database that allows checking if the user with the assigned SIM (*Subscriber Identity Module*) card (see further paragraphs in this chapter) is allowed to perform a call,

EIR (*Equipment Identification Register*), the database of serial numbers of the mobile phones used in the system. The phones that have been lost or stolen are placed on a black list, which prevents them from being used in the system.

The HLR register is the central database which stores the permanent parameters of the users and information on their current location. In large systems there can be more than one HLR. However, the individual user's data is stored only in one of them. The HLR register contains all the data on the users permanently registered in the GSM network, which allows the system to establish a connection path to them, even if at the moment of connection they are temporarily registered in a different GSM network operated in another country. Thus, the user's record in the HLR contains his/her status, the *Temporary Mobile Identification Number* (TMSI) and the address of the VLR register which is associated with the user's current location area. Among the data stored for each user, there is a list of additionally booked services and the encryption keys for digital signal transmission and user's authentication.

The VLR register is, as already mentioned, the database consisting of records describing mobile stations currently registered in the service area of a particular MSC. The VLR and HLR registers exchange data concerning the users currently located in the area served by that VLR. Such an arrangement and information exchange allows for identification of a current location of a called user by reading information on his/her current location area in the HLR and routing the connection to that MSC which co-works with the VLR currently containing the data on the called user. The VLR register also stores data necessary for initiation of a mobile-originated call.

The mobile switching centers (MSCs) are connected with each other. One or more MCSs, which are then called *Gateway Mobile Switching Center* (GMSC), play the role of a gate to the external networks such as PSTN, ISDN and packet data networks. Each MSC controls at least one *Base Station System* (BSS) which consists of a *Base Station Controller* (BSC) and a certain number of *Base Transceiver Stations* (BTS) or simply *Base Stations* (BS). The base station consists of a subsystem performing the basic signal transmission and signal reception functions and of a unit performing simple control functions. GSM-specific speech coding/decoding as well as the data rate adaptation are performed here as well. The base stations are usually located in the centers of the cells that cover a whole system operation area. Within such cells, a certain number of *Mobile Stations* (MS) operate with a possibility of changing their location dynamically. They perform the information exchange with the closest (or the strongest at their point of location) base station.

The main task of the MSC is to coordinate the call set-up between two mobile GSM users or between a GSM user and a user of an external network such as PSTN, ISDN or PSDN (*Public Switched Data Network*). This task is realized, among others, by performing the following functions:

- calling the user, setting up and maintaining a connection,

- dynamic network resource management in the system area managed by that MSC,
- rerouting of the connection to a new cell served by a base system controller (BSC) different from the controller which supervised the cell within which the MS has been operating so far (so-called handover),
- performing the interface with other networks (in case of GMSC),
- encryption of a user's binary stream,
- re-assignment of the BTS carrier frequencies in order to redistribute the network resources with regard to a particular load of the local part of the network.

Information exchange between MSC and BSS is normalized by defining the so-called *A interface*, whereas A_{bis} *interface* standardizes the data exchange between the base station controller and its transceivers (BTSs).

The A interface mostly deals with network and switching aspects, for example with the functions performed by the MSC, HLR and VLR, with management of fixed links, with network management, controlling and encryption of user data and signaling information, with management resulting from the necessity of MS authentication and location updating caused by MS geographical movement and with management due to calling the users.

The A_{bis} interface is associated with the information exchange related to the radio transmission, such as the distribution of radio channels, connection supervising, the queuing of messages before their transmission, carrier frequency hopping control (FH - *Frequency Hopping*), if applied, channel coding and decoding, the coding and decoding of speech signals, message encryption and radiated power control.

The rules of information exchange between BTS and MS are normalized by the definition of the radio interface denoted by U_m. It will be the subject of detailed considerations in further parts of this chapter.

The *Operation and Maintenance Center* (OMC) supervises the operation of particular GSM system blocks. It is connected with all the switching blocks of the GSM system and performs management functions such as tariff accounting, traffic monitoring, management in case of failures of particular network blocks. One of the most important tasks of the OMC is HLR management. In case of large networks there are more than one OMC and the whole network is managed by *Network Management Center* (NMC). The communication between the OMC and the network blocks is realized by a special communication management network implemented by leased telephone links or other fixed networks. Message transfer is performed using the SS7 signaling protocols and X.25 protocol for A and A_{bis} interfaces (see Figure 7.2). Figure 7.2 features some other interfaces denoted by capitals C, D, E. They are normalized by ETSI/GSM standards and set the message exchange rules between particular blocks of the signaling network. These interfaces, although important at the moments of the connection set-up and connection rerouting between different MCSs, are outside our interest.

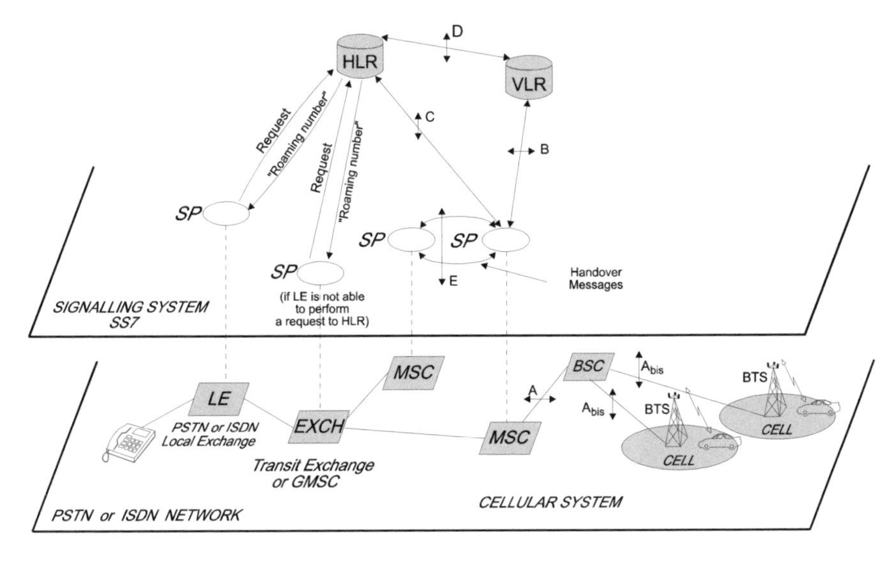

Figure 7.2 Transmission and signaling layers of the GSM network with interfaces among network elements

7.3 BASIC RADIO TRANSMISSION PARAMETERS OF THE GSM SYSTEM

Two bands, each 25 MHz wide, have been assigned to the GSM system. The frequency band between 890 and 915 MHz is used for transmission by mobile stations to base stations (*Uplink*) whereas the band between 935 and 960 MHz is used for transmission from the base stations to mobile stations (*Downlink*). As we can see, a duplex transmission is realized in a *Frequency Division Duplex* (FDD) mode. Both bands are divided into 124 frequency intervals of 200 kHz each with carrier frequencies in their centers. Time along each carrier is divided into 8 slots. Thus, multiple access is realized through assigning the connection a particular carrier frequency (or a sequence of them if frequency hopping is performed) and a selected time slot. As a result, the GSM can be treated as a system with TDMA/FDMA multiple access scheme. According to the latter, a *physical channel* is a sequence of time slots (denoted by the assigned slot number) which are placed on the selected carrier. Physical channels are arranged in pairs. Each pair consists of one physical channel in each direction (uplink and downlink). They are marked with the same time slot number and their carriers differ by 45 MHz.

Due to frequency planning, a subset of carrier frequencies is assigned to each cell or its sector in accordance with the general rules presented in Chapter 5. Figure 7.3 presents the frequencies and time slot assignment in the GSM system. Let us note that the numbering of the time slots is delayed by three in downlink as compared with uplink transmission. Due to that, a mobile station realizing a connection to a base station in the assigned time slot never transmits and receives the signals at the same time (as

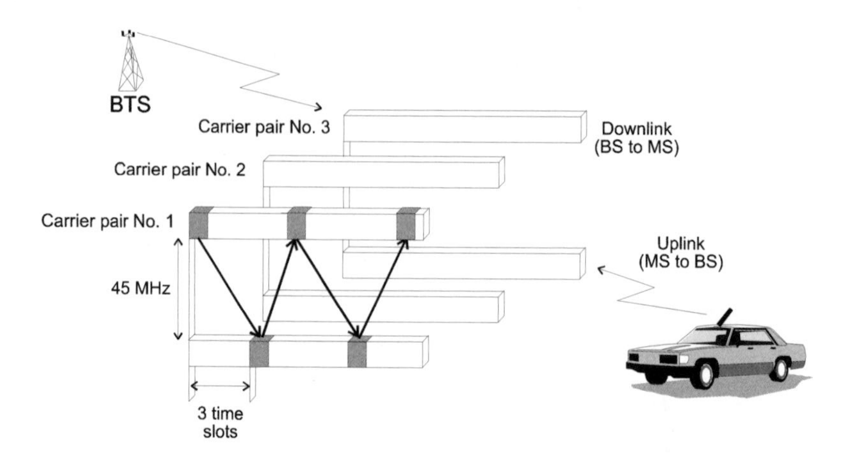

Figure 7.3 Frequency and time structure of GSM channels [5]

happens in the first generation analog systems). This way, an electromagnetic feedback between a mobile station transmitter and receiver is avoided and design requirements for RF and DSP blocks are not so stringent. Thus, the computational power can be shared between the transmitter and the receiver.

None of the carriers and time slots are devoted to a particular exclusive use. It means that the carriers and time slots can be used to perform different functions depending on the carrier and the current state of the system. For that reason we introduce a *logical channel* – a structure performing a particular task in our system.

7.4 LOGICAL CHANNEL DESCRIPTION

The physical channel assigned to a connection is used for transmission of data packets. They have a different time structure and meaning. They implement particular logical channels arranged to organize the information exchange.

In each cell a mobile station finds a particular carrier on which in its zero time slot the system information important for all the mobile stations located within that cell is transmitted. In turn, on the associated carrier always shifted down by 45 MHz, in the zero time slot the mobile stations can declare their need for information exchange with the system or even their need to set up a call. This carrier pair in their zero slots is permanently utilized for that purpose so it does not belong to the carrier set for which slow frequency hopping is realized.

System information is transmitted using a whole series of logical channels in the zero time slots of the downlink carrier of the above mentioned carrier pair. We will call it simply a *broadcast carrier*. In larger cells more than only the zero slots on that selected carrier are needed for the system information exchange. Thus, the first slots and sometimes even the second ones of that carrier are used for signaling purposes. In general, the following logical channels can be listed:

- *Broadcast Control Channel* (BCCH) – used for transmission of information that controls the network, a particular cell and the neighboring cells,
- *Frequency Correction Channel* (FCCH) – applied by the mobile station for tuning their reference carrier frequency,
- *Synchronization Channel* (SCH) – used for frame synchronization and identification of the base transceiver station, in whose service area the mobile station is located,
- *Common Control Channel* (CCCH) – serves for call set-up and other information exchange procedures. It consists of the following logical channels:
  - *Random Access Channel* (RACH) – used by mobile stations on the uplink carrier containing the control time slots to declare their need to be connected with the base station,
  - *Access Granted Channel* (AGCH) – serves to inform a particular mobile station on the downlink carrier that access to the system is allowed and will be performed,
  - *Paging Channel* (PACH) – used to call a particular mobile station in the set of neighboring cells called *Location Area.*

Transmission of user information and control information associated with it is performed using the following logical channels both in downlinks and uplinks:

- *Traffic Channel* (TCH) – applied to transmit the user data characterizing speech or data; there are *Full Rate Traffic Channels* (TCH/F) and *Half-Rate Traffic Channels* (TCH/H),
- *Slow Associated Control Channel* (SACCH) – used for transmission of control data and commands associated with a particular traffic channel. Examples of such commands are: change the power emitted by the mobile station, prepare for handover and send reports on measurements done by the mobile station.
- *Fast Associated Control Channel* (FACCH) – used to send urgent messages associated with a particular link realized on the traffic channel.

Before traffic channels in downlink and uplink are assigned to the call to be realized, a *Dedicated Control Channel* (DCCH) is implemented. It consists of the following logical channels:

- *Stand-Alone Dedicated Control Channel* (SDCCH) – used for information exchange preceding a connection, such as user's authentication, localization update and traffic channel assignment,
- Non-associated version of FACCH used for transmission of short messages.

Application and meaning of all logical channels will be much clearer after the presentation of the GSM time hierarchy.

7.5 GSM TIME HIERARCHY

Time division into eight time slots does not fully describe the time structure of the GSM system. The smallest time element is a single binary pulse (bit) lasting 3.69 μs, so the GSM data rate is equal to 270.833 kbps. In each time slot a burst of 148 bits is transmitted. The only exception is an 88-bit burst used for implementation of a random access channel (RACH). The length of a typical time slot is equal to 0.577 ms, which is equivalent to the duration of 156.25 bits. The difference between the effective burst length and the time slot length is called a guard time. Its existence originates from the necessary time reservation for switching the transmitter amplifier on and off at the beginning and the end of the transmitted burst. This time is also needed to ensure the necessary accuracy of the time alignment of the burst within the slot. The burst propagation time is non-negligible within the cell area and is proportional to the distance of the mobile station from the base station.

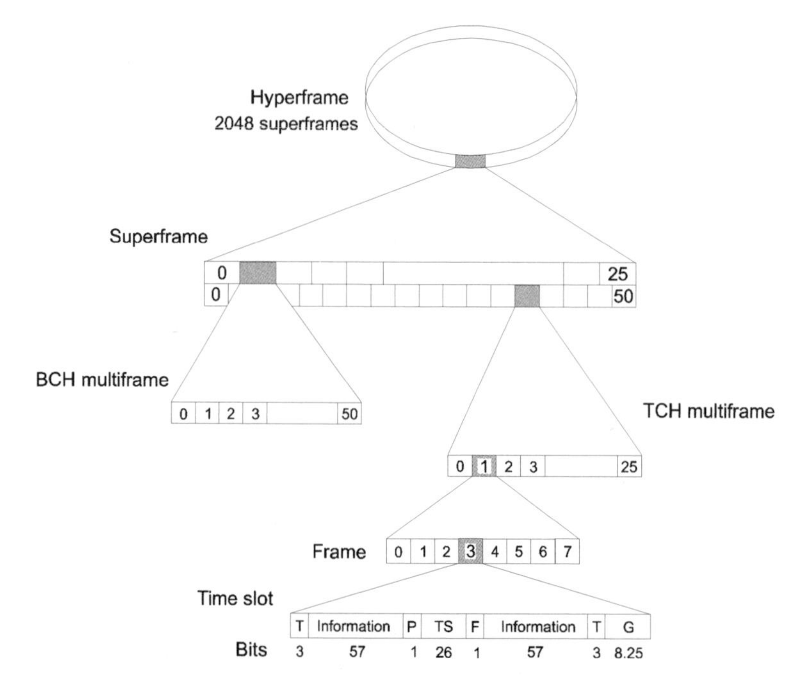

Figure 7.4 Time slot hierarchy of GSM system [5]

Eight time slots constitute a *frame*. In case of traffic channels and the control channels associated with them, 26 frames create a *multiframe*. In turn, 51 such multiframes constitute a *superframe* which lasts for 6.12 seconds. In case of broadcast channels and control channels located in the same slots a multiframe consists of 51 consecutive frames and a superframe is built of 26 multiframes. In both cases 2048 superframes create the highest level unit of the GSM time hierarchy - a *hyperframe*. It lasts for 3 hours 28 minutes 53 seconds and 760 milliseconds. The system clock returns to its initial state after this time. Such a long period of the GSM system clock results mostly from the application of a data encryption algorithm using the current frame number to generate the encryption sequence. Application of a long system clock period prevents illegal deciphering of the user data and results in the increase of safety and call privacy. In Figure 7.4 the GSM time hierarchy is summarized.

7.6 GSM BURST STRUCTURES

Each logical channel is realized by transmission of a specific type of data packet (burst) in the assigned time slots. As it has already been mentioned, frequency correction, synchronization and broadcast logical channels are sent in the zero time slot of the broadcast carrier together with some other specific control channels. In order to realize all these channels, a normal burst, frequency correction burst and synchronization burst are emitted. Their structure is presented in Figure 7.5.

Frequency Correction Burst consists of three zero tail bits at the beginning and at the end of the burst and of 142 zero bits between them. Due to GMSK modulation[2] applied in the GSM system, supplying the modulator with a long sequence of zeros results in the output of the non-modulated sinusoid shifted by 1625/24 kHz with respect to the carrier frequency. This sinusoid allows for identification of the broadcast carrier with the frequency correction channel realized on it. As a result, the mobile station carrier frequency is tuned allowing for reception of the other broadcast channels.

Synchronization Burst consists of three zero tail bits at the beginning and at the end of the burst, 64-bit training sequence in the middle of the burst and two portions of 39 encoded system information bits on both sides of the training sequence. The training sequence is applied to determine the samples of the channel impulse response which are used in the data detection process. A shorter training sequence of 26 bits is also applied in the *Normal Burst*. A longer training sequence of the synchronization burst allows for more precise channel estimation and more reliable detection of the system information. Two 39-bit information sequences contained in the burst are used to identify the base station color code (0 - 7) and the color code of the public land mobile communication network. It allows identification of the network operator to which a user is assigned. The system information bits also serve for synchronization within the time hierarchy of the GSM system.

Let us recall that the system clock period is almost 3.5 hours. The total number of frames is $26 \times 51 \times 2048 = 2715648$. The frames are numbered from 0 to 2715647. The current frame number is transmitted in the synchronization burst in a reduced form. The so-called *Reduced Frame Number* (RFN) consists of 19 bits describing the state of three counters, denoted by T_1, T_2 and T_3. They are described by the following equations in which the *Frame Number* (FN) is involved:

[2]See the next chapter

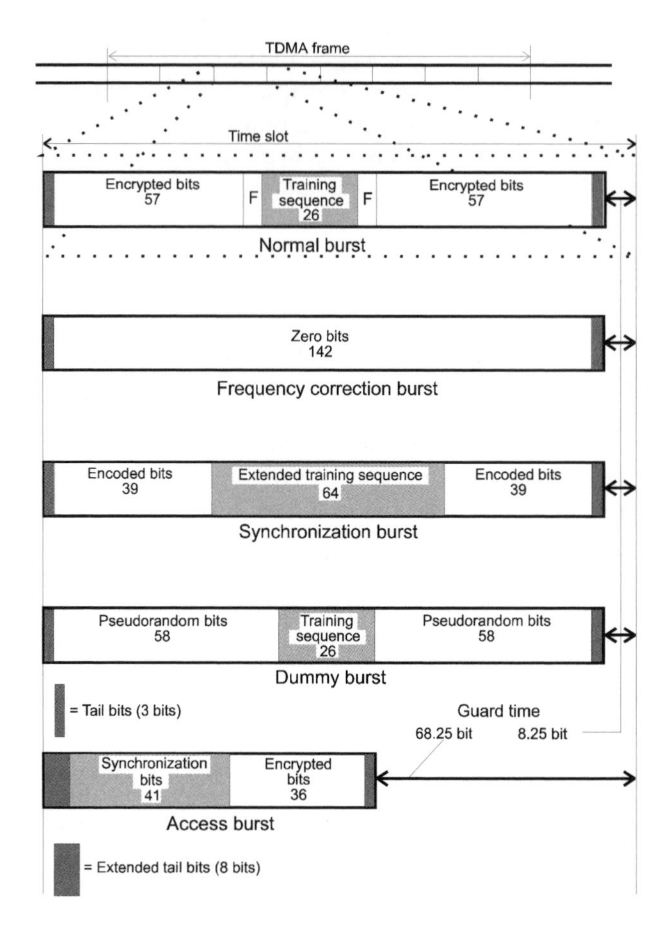

Figure 7.5 GSM burst structures

T_1 (11 bits)	$T_1 = FN \mathrm{div}(26 \times 51)$	range from 0 to 2047
T_2 (5 bits)	$T_2 = FN \bmod 26$	range from 0 to 25
T_3 (3 bits)	$T_3 = (FN \bmod 51 - 1)\mathrm{div}(10)$	range from 0 to 4

The frame number FN is derived on the basis of the counters T_1, T_2 and T_3. FN is needed in the process of data encryption and decryption.

Normal Burst is used to bear several logical channels of the control and traffic type. As in previous cases, it starts and ends with three zero tail bits. In the middle of the burst a 26-bit training sequence is placed. On its both sides of this sequence there are 57-bit long information blocks with one-bit flags. The value of the logic flag F indicates if the information block carries user data or if it contains the sequence of a control logical channel (SACCH or FACCH).

A system access request issued by the mobile station is realized by sending an *Access Burst*. It is shorter than any other burst used in the system. The burst starts with 8 zero tail bits followed by 41-bit synchronization sequence necessary for identification of the channel properties and synchronization of the base station receiver. The encoded message characterizing the connection request is 36-bit long. The burst ends with 3 tail bits. An exceptionally long guard time is justified by the fact that the mobile station, when sending its access burst for the first time within a new cell, does not know the necessary timing advance. When a normal burst is sent, it is necessary to start it properly early in order to compensate for the propagation time from the mobile to the base station and place the burst inside the time slot in the receiver. The access bursts can be transmitted by several mobile stations in the same time slot. In such a case collisions occur and none of the bursts can be accepted. A mobile station attempts to send another access burst after a pseudorandom number of frames.

In GSM transmission a *Dummy Burst* is also applied. Its structure is the same as that of a normal burst. The only difference is that two 57-bit information blocks do not carry any useful information, however, they have good statistical properties. Dummy bursts are placed in those time slots of the broadcast carrier which are temporarily unassigned to any user. This carrier should have higher mean power than any other carrier applied within the same cell. Thanks to this fact, this carrier is recognized by the mobile stations which attempt to activate themselves in the system after switching on their power. Thus, the dummy bursts serve as a kind of signal stuffing to keep the mean power of the broadcast channel high enough.

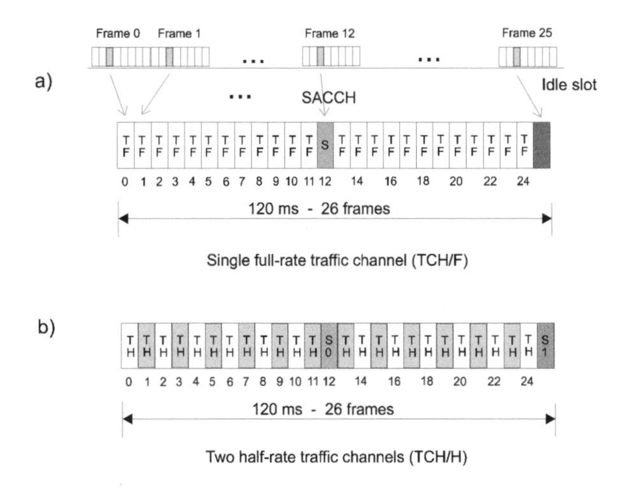

Figure 7.6 Multiframe for traffic channels with: full rate traffic channels (TF) (a), half-rate traffic channels (TH) (b), S0 - SACCH for the first TH, S1 - SACCH for the second TH

Consider now a structure of a multiframe that consists of traffic channels. As we remember, 26 subsequent frames constitute a multiframe. A pair of users has an access to the time slot assigned to them in each frame of the multiframe. The logical meaning of the transmitted bursts in subsequent frames is explained in Figure 7.6.

As we see in Figure 7.6a, a full rate traffic channel is sent in the assigned time slot during twelve consecutive frames (numbered from 0 to 11). In the twelfth frame a slow

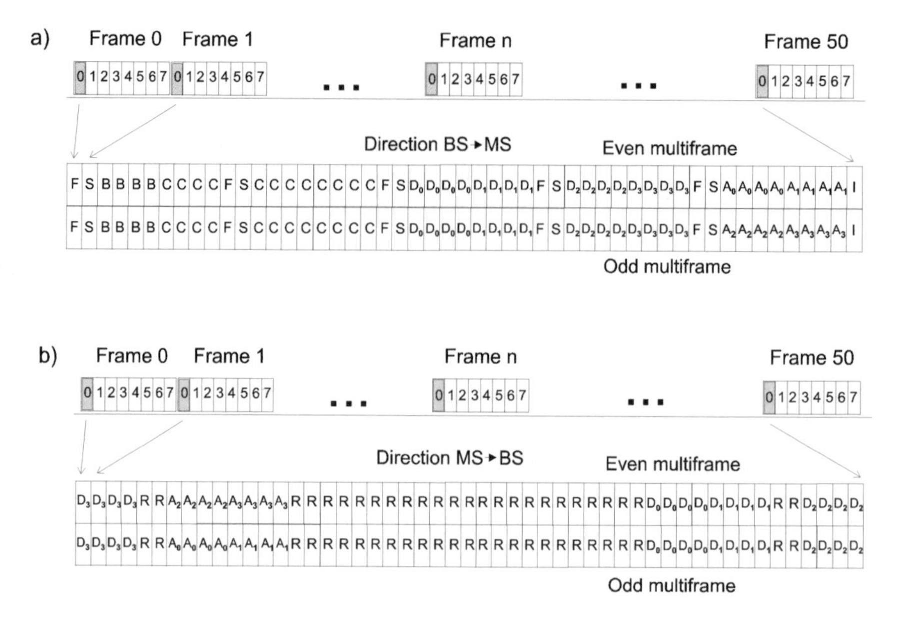

Figure 7.7 Multiframe structure of broadcast and control channels for cells with a single carrier

associated control channel (SACCH) is placed. In frames 13 to 24 a full rate traffic channel is performed again. In the last frame no signal is transmitted in the assigned slot.

In Figure 7.6b a multiframe organization for half-rate traffic channels is presented. Due to double effectiveness of speech encoding algorithm as compared with the full rate channel, two traffic channels are interwoven in subsequent frames. In the twelfth frame the control channel SACCH associated with the first traffic channel is located. The SACCH associated with the second channel is placed in the 25th frame.

The frame structure just described is valid for traffic channels transmission both in the downlink and uplink. Let us stress that transmission of two traffic channels assigned to neighboring time slots is in fact shifted in time with respect to each other by $97=12\times 8+1$ slots. In consequence, if a SACCH is transmitted in a particular slot in the twelfth frame, then the next slot related to the neighboring channel remains empty because its frame number equals 25.

As we remember, in case of broadcast channels and control channels associated with them, the multiframe consists of 51 frames. The organization of the multiframe can take different forms dependent on expected cell capacity expressed in the number of used carrier frequencies. In the cells where low traffic is expected and a single carrier is applied, all broadcast and control channels are transmitted in the zero time slot and the remaining slots are used for users' traffic channels organized as in Figure 7.6. Figure 7.7 presents the control channels' organization in the zero slot. In the downlink the multiframe consists of five ten-frame blocks and the frame for which in the zero slot no burst is transmitted. Each ten-frames block starts with a frequency correction channel

(FCCH) followed by a synchronization channel (SCH). In the first block four broadcast control channels (BCCH) and four common control channels (CCCH) are subsequently transmitted. Within the CCCH, the base station realizes a paging channel (PCH) or access granted channels (AGCH). In the following two ten-frame blocks, FCCH and SCH channels are followed by four stand-alone dedicated control channels (SDCCH) used during the call set-up, user authentication and mobile station registration procedures. The last ten-frame block contains slow control channels (SACCH) associated with the dedicated channel. Let us note that their data rate is so slow that four SACCH channels use the last ten-frame block in two subsequent multiframes.

In the uplink the multiframe of the control channels consists mostly of random access channels (RACH) realized by sending a random access burst from the mobile station to the base station. In the zero slots of selected frames, stand-alone dedicated control channels and slow associated control channels are performed. Let us stress once more that Figure 7.7 shows the organization of control channels for zero time slots. In other slots regular traffic is carried (see Figure 7.6).

The multiframe structure becomes more complicated for those cells where more carrier frequencies are applied due to expected intensive traffic. The time resources provided by the zero time slot are insufficient to carry the control signals of all mobile stations that can be served within such a cell. Thus, on the carrier on which the broadcast channels are realized in the zero slot, the first time slot is also used for control purposes. Such a situation is shown in Figure 7.8.

In the downlink in the zero time slots of the broadcast carrier only the following channels are realized: the frequency correction channel (FCCH), synchronization channel (SCH), broadcast control channel (BCCH) and common control channel (CCCH) (Figure 7.8a). In the first slots of the same carrier the stand-alone control channels (SDCCH) and slow associated control channels (SACCH) are implemented taking into account organization of even and odd multiframes. In the uplink carrier matched to the downlink broadcast carrier, the zero time slots are used exclusively to realize random access channels (RACH) whereas in the first time slots SACCH and SDCCH are placed.

Figures 7.6, 7.7 and 7.8 do not exhaust all the possible arrangements of the control channels. In case of the cells serving very intensive traffic, even more control channels are needed. The examples of such control channels configuration can be found, among others, in [2].

7.7 DESCRIPTION OF THE CALL SET-UP PROCEDURE

The necessary initial condition to make a mobile station active is to switch its power supply on. After "power on" the mobile station (MS) realizes a sequence of operations which are needed to "find itself" in the GSM network. In the first step, the MS looks for the carrier on which in the local cell the broadcast channel is transmitted. In order to find this carrier the MS measures the received power of all 124 carriers applied in the GSM system. The carrier containing the broadcast channel is emitted at a higher power than any other carrier in the same cell. The MS lists the measured carriers according to their decreasing power. In the next step, the MS listens to the subsequent carriers

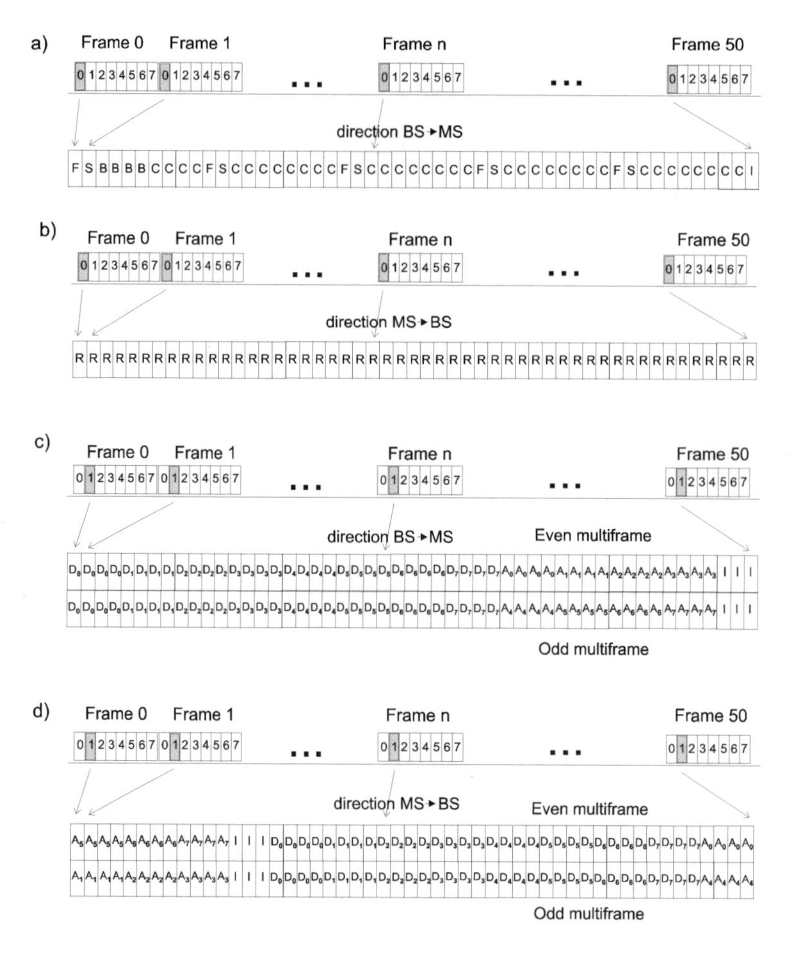

Figure 7.8 Multiframe structure for broadcast and control channels using the zero-th and first time slots

from the list, searching for the frequency correction channel (FCCH). Its identification is equivalent to finding the zero time slot of the broadcast carrier. The MS carrier frequency generator is adjusted to that frequency. In the zero time slots of the next frame, the MS finds other important control channels, as shown in Figures 7.7b or 7.8a. In the zero time slot of the next frame the MS finds the synchronization channel (SCH) and decodes the information contained in it, in particular the *Base Station Identity Code* (BSIC) and the *Reduced Frame Number* (RFN). Next, the MS decodes the information carried by the broadcast control channel (BCCH). That information contains:

- up to 16 carrier frequencies containing BCCH in the neighboring cells that enables the MS to monitor them for possible selection of the best one when crossing the cell boundary,

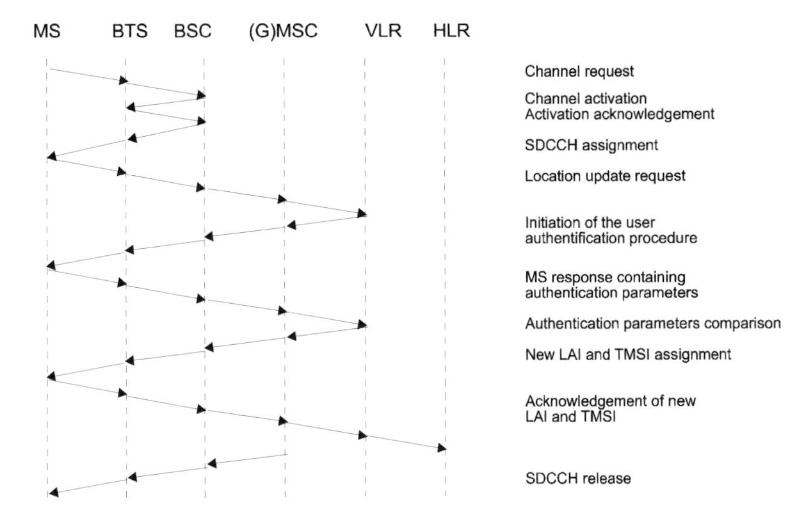

Figure 7.9 Simplified typical MS registration in the GSM network

- *Cell Global Identity* (CGI) - a sequence consisting of the country code, mobile network code, location area code and cell identity,
- the number of common control channels applied in the cell,
- other parameters, e.g. maximum power that can be used by the MS during connection set-up, minimum power level required for signal reception in the cell (information for mobile stations in the idle state), the value of hysteresis that prevents switching back and forth between cells at cell boundaries in the handover procedure.

At this moment the passive part of the MS activation in the network is completed. In order for the MS to be paged or to initiate the call set-up it has to be registered in the network. In Figure 7.9 a simplified registration procedure is shown. Registration is one of the reasons why the MS updates its location in the system. Registration takes place if the number received by the MS from the base station, called *Location Area Identity* (LAI), differs from the latest one that has been stored in the MS. The MS location update also occurs periodically, which allows for monitoring of the MS mobility and its current status.

When a mobile station starts the location update procedure resulting, for example, from the initiation of its registration procedure following switching on the power, it transmits a RACH channel to the base station in which it asks for a channel to be used for the registration procedure. The BS transfers this request to the base station controller. The latter responds by sending a command to the base station with the request to assign a free stand-alone dedicated channel. The BS sends the confirmation of the reception of this command to the BSC and informs the MS about the SDCCH assignment using the AGCH channel. In turn, the mobile station requests the location update by sending this request by the newly assigned SDCCH. This request contains the type of location update to be performed, the *Temporary Mobile Subscriber Identity* (TMSI) - a number temporarily assigned to the user in the network, and the *Location Area Identity* (LAI). The last two parameters are currently stored in the mobile station. The MS request reaches the Mobile Switching Center (MSC) and the Visitor's Location Register (VLR) through the base station and base station controller. If the VLR already contains the user's TMSI, it updates its data about him/her by noting the start of his/her activity. If there is no TMSI of the user in the current resources of the VLR, the *Location Area Identity* sent by the MS is decoded. The LAI indirectly describes the VLR that previously served the mobile station. After communication of the current VLR with the previous one the latter submits the user's parameters to the current VLR, including his/her *International Mobile Subscriber Identity* (IMSI) and data necessary for his/her authentication. If the previous VLR does not contain the required parameters, which happens at the first entry to the network, the MSC requires the mobile station to send its IMSI number, using the SDCCH. The MS sends its IMSI through the air only once. This message reaches the MSC and VLR. The IMSI number determines the address of the user data in the HLR. It ensures the possibility of retrieval of necessary information from the HLR user's record needed for user authentication. This information is loaded to the current VLR. Except for the first registration in the network, the necessary data about the user are already contained in the current or previous VLR.

Now the authentication procedure can be initiated. It is described in detail in Section 7.9. After acknowledgement of the MS and user authenticity new parameters of the MS are set. They are: *Temporary Mobile Subscriber Identity* (TMSI, not in all cases) and *Location Area Identity* (LAI). Information about the user is updated in HLR and VLR. Strictly speaking, the HLR initiates the erasure of the information about the user from the previous VLR and its transfer to the new VLR. The acceptance of the new parameters is acknowledged by the MS. The new TMSI has to be sent over the radio channel to the MS in an encrypted form, so the encryption mode is initiated. As a result, the BS and MS have to exchange information, using SDCCH assigned to them once more. This phase of the registration procedure has not been shown in Figure 7.9. The current TMSI number is valid only during the presence of the mobile station in the area served by the current VLR. This solution is in fact one of the steps towards call privacy. The SDCCH channel used for the registration and location update is released after performing these procedures.

After full synchronization of the MS with the network and its registration, the initiation of the call set-up is finally possible. The mobile station listens to the common control channel (CCCH) in order to find a paging channel (PCH) directed to it.

In case of the MS-originated call, the following steps are performed:

- The mobile station sends a random access burst to the base station realizing the RACH logical channel.

- Three bursts performing the Access Granted Channel (AGCH) realized within the CCCH channel have to be received by the MS. The AGCH contains the number of the SDCCH assigned to the mobile station for the call set-up procedure.

- In case of collision with other mobile stations during transmission of the RACH channel the RACH has to be repeated up to a fixed number of times after a pseudo-randomly selected time period. The repeated sending of RACH is done taking into account two parameters: randomization period TX-INTEGER and allowable maximum number of retransmissions MAX-RETRANS that are retrieved from the broadcast control channel (BCCH).

After the reception of the AGCH, the SDCCH is assigned to the MS for the time of the call set-up. Using that channel, the MS sends a message through the BS to the base station controller (BSC) informing it about the call set-up request. It also contains the user's TMSI number. The BSC transfers that message to the MSC switching center, supplementing it with the cell identification code of the cell where the MS is currently located. Subsequently, the MSC informs the VLR associated with it about the call set-up request issued by the MS. In turn, the VLR initiates the MS authentication procedure (see Section 7.9) and the SDCCH is used once more between MS and BS to realize it. Positive authentication is followed by starting the encrypted transmission mode. Subsequently, a new TMSI number assigned to the mobile station is transmitted to the MS. The mobile station acknowledges its reception. The next message sent by the MS contains the number of the called user and the type of the requested service. This message reaches the MSC and VLR. If the requested service is on the list of the user's subscribed services stored in the VLR, the network sends the message to the calling mobile station about the start of the call set-up procedure. The MSC assigns a fixed link to the call and a traffic channel to the MS–BS transmission. As a result of this procedure, the mobile station receives the following information:

- the number of the assigned time slot,
- the code of one of the eight possible training sequences used in the normal burst for channel identification,
- the *Absolute Radio Frequency Channel Number* – the number determining the particular carrier that will be used during the transmission,
- the *Timing Advance* parameter (TA) – the number determining the necessary time advancement of the start of the transmitted burst with respect to the system clock, ensuring an appropriate time alignment of the burst in the time slot at the base station,
- the parameters of the *Frequency Hopping* (FH) procedure if it is implemented.

Subsequently, the mobile station acknowledges the channel assignment. This information reaches the MSC. Additional steps of the call set-up are performed if the called user belongs to the same GSM network, however, these details are beyond the scope of this chapter.

After setting up the link through the MSC, the GSM gateway MSC (GMSC) and the switching offices of the public telephone network, the requested PSTN user is called. After his/her response, the mobile station receives the message about the acceptance of the call by the called user. From this moment a digital voice or data transmission

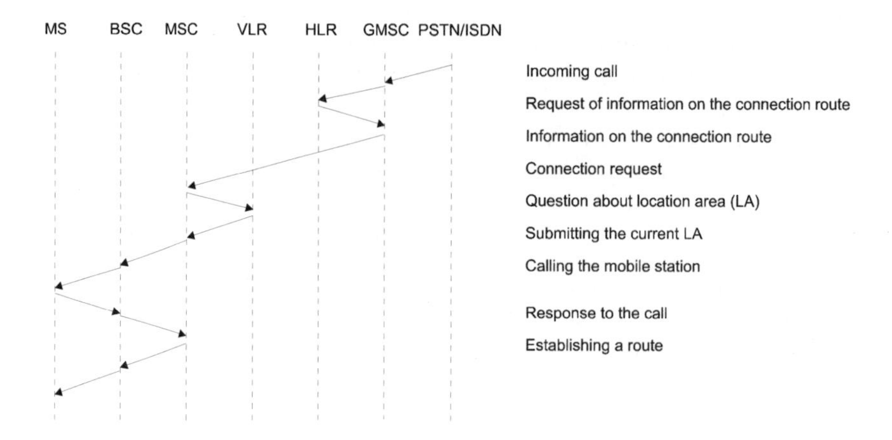

Figure 7.10 Simplified mobile terminated call set-up

is possible between both users applying the assigned time slot at the assigned carrier frequency according to the multiframe structure shown in Figure 7.6. We recall that besides the traffic channel, a slow associated control channel (SACCH) is assigned to this call. In this control channel the mobile station transmits the following data to the base station:

- the power of the received signal and quality evaluation of the serving cell,
- the power of the six strongest carriers transmitting the BCCH channel in the neighboring cells,
- the power of the received signal in the idle slot during the measurement period,
- the time advancement applied by the MS during the measurement period.

In turn, the base station uses the SACCH to transmit the following information to the mobile station: the description of the neighboring cells, the local cell identity, the location area identity and control signals for the power level to be applied by the MS.

Let us consider a simplified call set-up of a mobile terminated call. The subsequent phases of establishing a connection have been shown in Figure 7.10.

In the first step a fixed network user dials the number of a GSM user. The PSTN, ISDN or public data network switching office (depending on the type of a network which serves the calling user) issues a call request to the GSM gateway MSC (GMSC). In order to set up a connection, it is necessary to find the current location of the called mobile station. For this reason the GSMC sends a request to the Home Location Register (HLR) that in turn informs the GMSC on the MS location area and current Visitor's Location Register (VLR) in which the called MS is registered. As a result of the HLR request, the VLR submits the *Mobile Subscriber Roaming Number* (MSRN) - a temporary MS number that is transferred to the GMSC. The GMSC starts to establish a path to the appropriate MSC. The latter checks in the VLR if the requested service is on the list of allowable services of the called user. The MSC also retrieves the current MS location area from the VLR. In fact, the MSC obtains from the VLR the parameters of the called MS, such as *Location Area Code* (LAC) and its TMSI.

Let us note that the location area is served by a whole group of base stations. After determining this area and checking in the VLR that the mobile station remains at least in the idle mode, the MSC sends a paging request to all base station controllers (BSC) operating in the determined location area. It is realized by sending a logical paging channel on the broadcast carriers of all cells served by the BSCs. As we see, the MS is searched for concurrently in a few cells. If the mobile station detects the paging directed to it, it responds by sending the RACH channel to its nearest base station. Thus, the MS has been localized in a particular cell of the location area. The BSC assigns the SDCCH channel to the called mobile station and informs the MSC about that. From that moment the MS will respond to the MSC using the assigned SDCCH. In turn, the base station controller sends the service request to the MSC. The request contains the cell identity and the temporary mobile subscriber identity. As a result, the MSC informs its VLR about the response of the MS to the call directed to it. The VLR initiates the MS authentication procedure that involves the MSC, BSC and MS. After completing this procedure, the VLR issues a command the MSC to start data encryption. The latter transfers this command to the base and mobile stations. After the initiation of data encryption in the MS, a new TMSI number is assigned to the mobile station for the time of connection. The MS acknowledges reception of this number and this information reaches the VLR. At this moment the MSC starts the connection set-up with the selected MS, sending a *Set-Up Message* to it. The MS acknowledges the receipt of this message. Then the MSC assigns a fixed link between itself and the base station controller and requires the BSC to assign a free traffic channel on the radio link. Thus, a carrier number, a time slot number and a training sequence are assigned to the connection. The MS sends the acknowledgement of these parameters that is further transferred to the MSC. At the same time the MS starts to ring and the MSC is informed about that. The latter starts sending the ringing signal to the calling user. After accepting the call by the mobile user the actual connection starts.

One of the most interesting features of the GSM system is the possibility of making a connection with a mobile station located in a foreign country in the area served by another GSM operator. If there is a formal agreement between the local operator and the foreign one, it is possible to set up a connection and to receive a call in the foreign network. This feature is called *roaming* and is also implemented in other Paneuropean mobile networks such as TETRA and ERMES.

If the connection request arrives from a fixed network, the GMSC checks the current location of the called user in the HLR. Let us consider the case when the called user's mobile station is located in the area covered by another network and is temporarily registered in it. The address of the mobile station is retrieved from the HLR and sent to the switching center appropriate for the calling user. That switching center sets an international connection through the fixed network to the called user operating in another GSM network.

The call set-up has been described in short, leaving out some details. They can be found in books exclusively devoted to the GSM, such as [2], [6], [10] and ETSI/GSM standards. In the next section we will describe the handover - a very important procedure realized in the cellular networks.

7.8 HANDOVER

Handover is one of the most important control procedures in the cellular network. It allows for continuing the call when the mobile user crosses the cell boundaries and moves to a neighboring cell. It is this feature that makes a user really mobile, however, the handover control signals heavily load the network. The main task of the efficient transfer of the connection between the cells is to perform it in a very short time. An efficiently performed handover allows the user not only to continue the call but it is done so quickly that he/she cannot even notice that it has just been performed. There are a few types of handover that will be described below. In Figures 7.11, 7.12, 7.13 and 7.14 typical situations where handover takes place are illustrated.

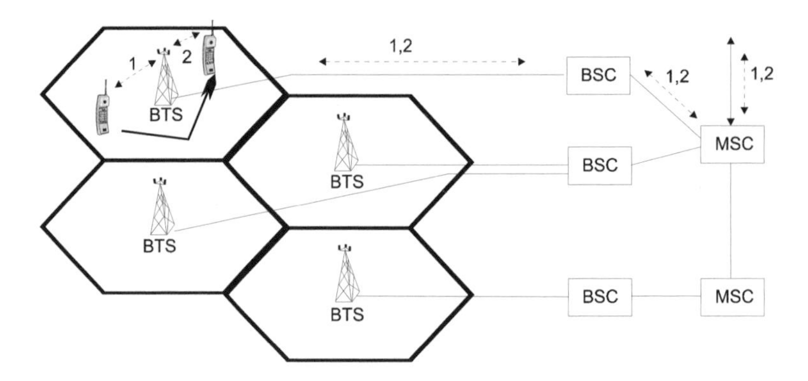

Figure 7.11 Intra-cell handover

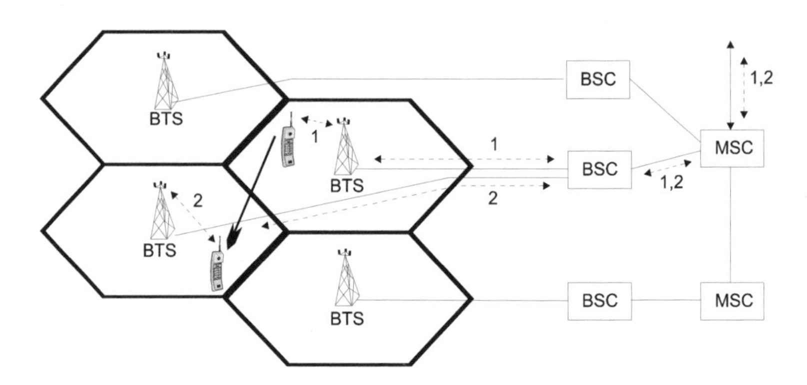

Figure 7.12 Intra-BSC handover

The first figure presents the *intra-cell handover*. This handover procedure is performed in order to optimize the traffic load in the cell or to improve the quality of a connection by changing the carrier frequency. The remaining figures deal with the situations when the user moves from one cell to another. In the simplest case (Figure 7.12 – *intra-BSC handover*) only one base station controller (BSC) is engaged and only the base stations are changed. In the second case (Figure 7.13) the handover is performed between two BSCs (*intra-MSC handover*) and in the third one the procedure implies rerouting of the whole path changing also the MSC. The latter type of handover is called *inter-MSC handover*. This type of handover sets particularly high requirements on the cellular network and digital transmission is performed in the fixed network that does not often belong to the cellular system operator. Let us stress that the MSCs can be located more than 100 km from each other.

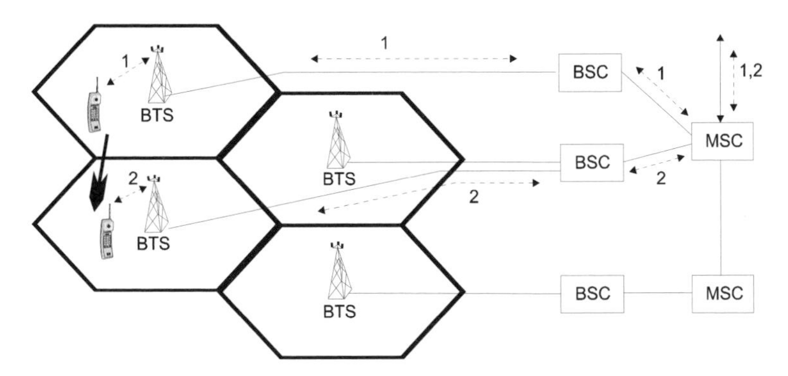

Figure 7.13 Inter-BSC handover

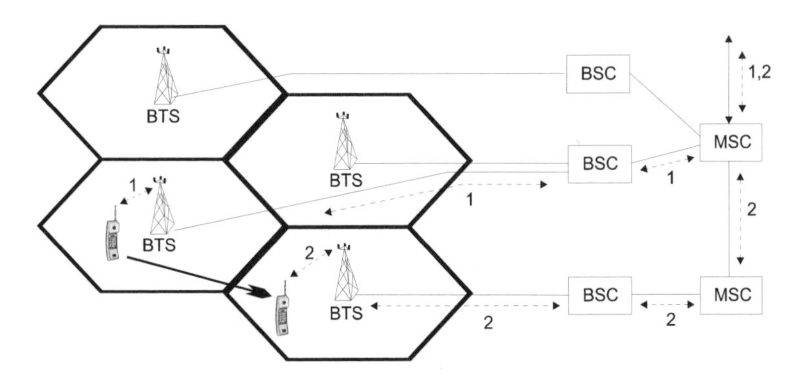

Figure 7.14 Inter-MSC handover

Handover in the GSM network differs from similar procedures performed in the first generation analog networks. In the latter ones the handover procedure is initiated on the basis of the quality or field strength measurements in the uplink made by the network.

This leads to a substantial network load. Additionally, in case of the application of the analog testing and of distorted control signals, the procedure becomes unreliable.

In the GSM system the measurements are performed both in downlink and uplink. This implies that the mobile station actively participates in the handover. The quality and level of the received signals are measured in both transmission directions. The measurement results are denoted by symbols[3] RXLEV-U, RXQUAL-U, RXLEV-D, RXQUAL-D and DISTANCE. The last value describes the distance between the mobile station and the base station. Let us recall that the measure of distance is the timing advance of the burst transmission that enables the burst to arrive within the assigned time slot at the receiver.

The mobile station performs regular measurements of the 16 strongest carriers transmitting the BCCH. The measurement results of the six strongest carriers are transmitted to the currently assigned base station every 0.48 seconds. The base stations also perform interference measurements in those time slots that are currently free. The *Operation and Maintenance Center* monitors current levels of the cell traffic. Their values can also be used in the handover procedure.

As already mentioned, the procedure of handover is distributed between mobile and base stations. The mobile station can be the initiator of this procedure. Handover is not a subject of normalization. In general, it has to be finished within 1 second. When it lasts less than 100 ms, it is almost unnoticeable to the user. If a handover error occurs, the connection has to be overtaken by the previous base station. In case of a transmission breakdown, the radio connection of the MS is re-activated within 4 to 8 seconds, with the base station sending the strongest signal to the mobile station.

7.9 ENSURING PRIVACY AND AUTHENTICATION OF A USER

In the first generation cellular system, ensuring privacy of the call was very difficult due to the analog nature of the transmitted speech signal. A relatively simple device can be used for tapping of the calls that are not secured by any type of coding. In digital cellular systems the task of ensuring call privacy is much easier to achieve due to digital speech coding transforming the samples of analog speech into the binary stream, data encryption, scrambling and other known methods of prevention against unauthorized access to the transmitted information.

The meaning of the protection against tapping and unauthorized access to the digital network is stressed by the digital cellular network construction. In the public cellular network there is no path to the called user known in advance due to the nature of radio interface and user's mobility. It is necessary to verify the user's authenticity in order to set up a connection with the terminal authorized to be used in the network and to ensure that the billing will be directed to a proper address.

The following means have been undertaken in the GSM system to ensure transmission security:

[3] RXLEV-U denotes *Received Signal Level in the Upper Link*, RXQUAL-U - *Received Signal Quality in the Upper Link*, D at the end of the remaining abbreviations stands for *Down Link*.

- access to the network based on the verification of the user's authenticity,
- secrecy of the transmitted data (speech, data, fax and control signals) due to encryption,
- anonymity of the users due to the temporary mobile user identity numbers utilized inside the network.

Another important element ensuring security in the GSM system is the *Subscriber Identity Module* (SIM) card. It is an intelligent plastic card with a microcontroller that constitutes an inherent part of a mobile terminal. The user receives the SIM card from the network operator. The card contains a list of individual user data, encryption programs and keys. Due to the possibility of separating the SIM and the mobile phone, the card can be treated as an additional security means against an unauthorized usage of the stolen or lost phone. It also allows for usage of the replacement phone during the repair of the user's own mobile phone. One has to stress that a mobile station consists of two strictly related parts: a mobile phone and a SIM card.

As already mentioned, the SIM card contains a microcontroller with ROM, RAM and NVM (*Non-Volatile Memory*). The ROM stores the programs implementing A3 and A8 encryption algorithms. The first of them is used in the user's authentication process, whereas the second one is applied for derivation of the key for encryption of transmitted data. The ROM has the capacity of 4 to 6 kB and cannot be copied. The RAM is relatively small and its capacity does not exceed 256 bytes. The NVM of the size 2–3 kB contains individual user's parameters and data, such as:

- K_i – user's authentication key,
- IMSI (*International Mobile Subscriber Identity*) – 15-bit long user's individual identification number consisting of the country code, network code and number of the user,
- TMSI (*Temporary Mobile Subscriber Identity*) – temporary identification number assigned to the user after each registration in a new VLR,
- LAI (*Location Area Identifier*),
- PIN (*Personal Identification Number*) – 4 or 8-digit code identifying the user with respect to the SIM card,
- the personal telephone book – a list of telephone numbers entered by the user,
- the list of foreign cellular networks where roaming is allowed,
- the received short messages (SMS).

The SIM card communicates with the phone in the asynchronous serial mode in half-duplex transmission at the data rate equal to 3.2 kbps in each direction.

The network begins the verification of the user's authenticity during his/her registration (compare Figure 7.9 and Figure 7.15). The network sends a 128-bit pseudorandom

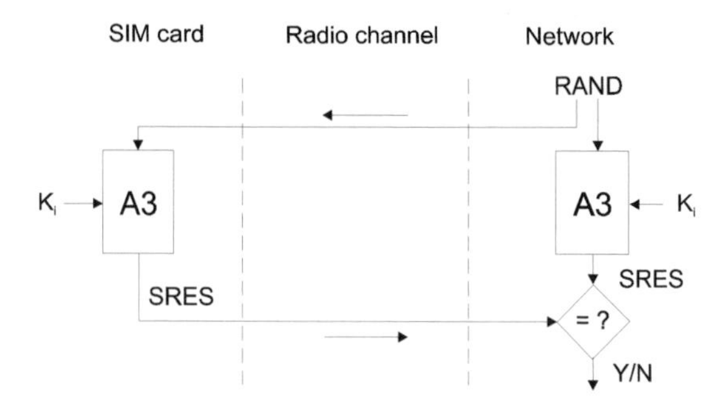

Figure 7.15 Verification of the user authenticity

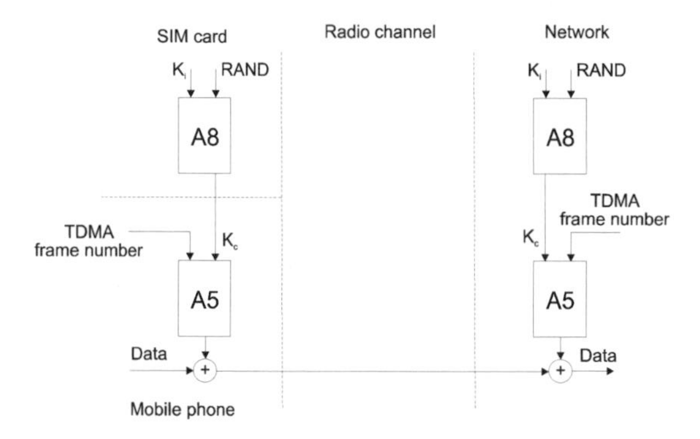

Figure 7.16 Data encryption process

number RAND to the mobile station. In the network and inside the SIM card of the mobile station the 32-bit electronic signature SRES is derived, using the A3 encryption algorithm and the individual user's key K_i. The SRES calculated by the mobile station is transmitted to the fixed part of the network, where both signatures are compared. If they are identical, the authentication process is completed. In the next step, the VLR assigns the TMSI and LAI to the user (compare Figure 7.9). Both numbers are transmitted in the encrypted form to the mobile station and stored in its SIM card. The individual IMSI number is transmitted only "once in the user's lifetime" in the system. It occurs when the mobile station registers for the first time in the network and there is no data regarding it in any VLR register.

As already mentioned, starting from a certain phase of the connection set-up the user data and control signals are transmitted in the encrypted form. The encryption process applies a public encryption algorithm called A5, which is a European standard. The access to this algorithm is limited to the manufacturers of cellular devices. Ensuring secrecy of the information transfer does not rely only on the very limited access to the encryption algorithm but mainly on the application of the encryption key that is never transmitted over the air.

The A5 encryption algorithm is realized inside the mobile phone. The activation of the algorithm is initiated by the network. The SIM card derives the key for the data encryption algorithm, on the basis of the individual key K_i and the RAND number sent by the network during the authentication process. This is done using the A8 algorithm stored in the SIM card. On the basis of the derived key K_c and the current 22-bit TDMA frame number, the A5 algorithm generates 114 bits that are modulo-2 added to the information bits of the normal burst. Let us recall that the two information fields of the normal burst contain 57 bits each. In the fixed part of the network the same operation is performed for the received block of 114 encrypted data bits. When no errors occur during transmission, the modulo-2 addition of the received block and the generated encryption sequence results in the recovery of the original data sequence generated in the mobile station. This way a decryption process is realized. Graphical representation of the encryption and decryption processes is shown in Figure 7.16.

7.10 MODIFICATIONS AND DERIVATIVES OF GSM

The basic version of the GSM system was implemented in the early nineties. The first version of the GSM standard was set after a few years of work of the committee called *Groupe Spécial Mobile* operating within CEPT (*Conférence Européene des Postes et Télécommunications*) in 1990. In the same year, due to British request, the work on adaptation of the GSM standard for the 1800 MHz band was initiated. Simultaneously, perspectives of system modifications leading to its improved reliability and enlarged capacity were acknowledged. Attention was also paid to the seamless introduction of new services. In fact the system evolution and the process of system specification can be divided into three phases [7]:

- Phase 1 (1991-1994) – the system was operating in its basic version,
- Phase 2 (1994-1995) – the system specification was verified in order to allow future gradual modifications,
- Phase 2+ (from 1995) – the modifications are being introduced.

One of the GSM modifications that can potentially influence its capacity is introducing the half-rate speech coding standard. This standard results in a binary representation of a speech signal at the rate of 6.5 kbps. As we see, this is half of the original data rate of the output signal generated by the standard 13 kbps speech encoder. The speech quality is unfortunately diminished, particularly if the connection is set between two GSM users. In the cascade of speech coders and decoders applied between two users, the process of double transcoding from GSM to PCM standards and backwards turns out to be the reason of deterioration of speech quality. On the other hand, a new high quality speech encoder (*Enhanced Full Rate Encoder*) has been introduced, based on the CELP principle (see Chapter 1). The new speech encoder will be shortly described in Chapter 8. Very often modern handhelds are equipped with speech encoders fulfilling all three coding standards.

The main modifications of the GSM system lead to substantial enhancement of data transmission capabilities and open wide possibilities of providing new services. They will be considered in detail in Chapter 9.

One of the most important enhancements of the original GSM system is its 1800 MHz version called *Digital Cellular System* (DCS 1800) placed in the PCN (*Personal Communications Network*) category [8]. The idea of PCN is related to the concept of wireless network in which a user can receive and initiate a call using a small and light personal terminal independently of his/her location. DCS 1800 does not fulfill all the PCN assumptions but it is closer to it than regular cellular networks. DCS 1800 is primarily devoted to the operation in urban and suburban environments in which the traffic density is very high. This system targets the mass market. It has turned out that the GSM 900 technology is so universal that after minor modifications it can be applied in the 1800 MHz band (or 1900 MHz band in the USA).

Table 7.1 Basic differences between DCS 1800 and GSM 900

Feature	GSM 900	DCS 1800
Frequency range Uplink (MS→BS) Downlink (BS→MS)	 890-915 MHz 935-960 MHz	 1710-1785 MHz 1805-1880 MHz
Number of duplex channels	124	374
Frequency interval between uplink and downlink frequencies	45 MHz	95 MHz
Maximum BS power	320 W (55 dBm)	20 W (43 dBm)
Maximum MS power	8 W (39 dBm)	1 W (30 dBm)
Minimum MS power	0.02 W (13 dBm)	0.0025 W (4 dBm)
MS classes	20 W (not implemented) 8 W (car/transportable phone) 5 W (car/transportable phone) 2 W (handheld) 0.8 W (handheld)	1 W (handheld) 0.25 W (handheld)
Maximum vehicle speed	250 km/h	130 km/h

The main difference between GSM 900 and DCS 1800 lies in the lower power of the base and mobile stations, which implies the lower signal range and the smaller size of cells. The bandwidth assigned to the DCS 1800 is larger than that of the original GSM 900. Up to 374 carrier frequency channels can be assigned to the DSC 1800. Recalling also the smaller cell size, we can conclude that the DCS 1800 features much higher traffic capacity than the GSM 900 system. On the other hand, applying twice as high carrier frequencies in the DCS 1800 as compared with GSM 900 results in twice as high sensitivity to the Doppler effect. Thus, the maximum vehicle speed is limited to 130 km/h.

Another essential enhancement of the DCS 1800 is the possibility of roaming inside a given country. This means that if more than one DCS network operate in a country, a user can use the resources of a network that is different from his/her mother network if the latter does not ensure the coverage in the user location. The return to the mother network occurs automatically if the user enters his/her own network coverage area. This is impossible in GSM 900 networks due to organization reasons. Mobile operators exploiting both the GSM and the DCS system offer dual band mobile phones that select the band automatically without the user's knowledge.

Table 7.1 summarizes the basic differences between DSC 1800 and GSM 900.

There is another promising modification and enhancement of the original GSM 900 system. It has been observed that the first generation systems working in the 450 MHz band gradually lose their customers who move to the second generation system. After closing the old analog systems the frequency range could be utilized by the GSM system as well. For this reason ETSI standardizes the GSM system operating in the bands around 450 and 480 MHz band and called GSM 400. The 850 MHz band can also be allocated to GSM (GSM 850). The whole network infrastructure will remain unchanged, however, new software will be required in some network elements. The basic technical features of GSM 400 are following:

- frequency allocation for GSM 400:
  - 450.4–457.6 MHz (uplink) and 460.4–467.6 MHz (downlink)
  - 478.8–486.0 MHz and 488.8–496.0 MHz
- frequency spectrum: 7.1 or 7.2 MHz
- duplex separation: 10 MHz
- carrier spacing: 200 kHz
- support of fragmented usage of the frequency band (resulting from NMT 450 frequency variants).

In turn, the frequency allocation to GSM 850 is 824–849 MHz (uplink) and 869–894 MHz (downlink).

As we remember from Chapter 3 the attenuation of the electromagnetic waves depends on their frequency. For GSM 400 the carrier frequencies are twice as low as for GSM 900. As a result the cell coverage is extended to approximately 67 km. Work is being conducted on further range extention between 70 to 140 km. GSM 400 will support all the ETSI standardized capacity enhancing features presently being supported by GSM 900 and DCS 1800 such as frequency hopping, discontinuous transmission, MS and BTS power control and adaptive multirate.

* * *

The structure of the GSM system and its communication procedures show the level of complexity of a digital mobile cellular system. As compared with the digital wireline system, the higher level of the GSM complexity results from the necessity of locating a called mobile station and the possibility of its movement within the area served by different base stations during a connection. As we will see in the following chapters, the GSM system is not the last word in the enormous development of mobile communications. However, the GSM is a subject of intensive modifications that are the element of seamless transition to the Universal Mobile Telecommunication System (UMTS) that will also be a subject of our considerations.

REFERENCES

1. R. Steele (Ed.), *Mobile Radio Communications*, Pentech Press Publishers, London, 1992
2. M. Mouly, M.-B. Pautet, *The GSM System for Mobile Communications*, 1992
3. K. Kakaes, Global System for Mobile Communications (GSM), Tutorial No. 5, International Conference on Communications, Geneva, May 23-26, 1993
4. K. David, T. Benckner, *Digitale Mobilfunksysteme*, B.G. Teubner, Stuttgart, 1996
5. D. Picken, "The GSM mobile-telephone network: technical features and measurement requirements", *News from Rohde & Schwarz*, No. 136, 1992, pp. 28-31
6. A. Mehrotra, *GSM System Engineering*, Artech House Publishers, Boston, 1997
7. M. Mouly, M.-B. Pautet, "Current Evolution of the GSM Systems", *IEEE Personal Communications*, October 1995, pp. 9-19
8. A. R. Potter, "Implementation of PCNs Using DCS1800", *IEEE Communications Magazine*, December 1992, pp. 32-36
9. A. Hadden, "DCS1800: The Standard for Personal Communication Networks", *Telecommunications*, June 1991, pp. 61-63
10. G. Heine, *GSM Networks: Protocols, Terminology, and Implementation*, Artech House, Boston, London, 1999
11. B. H. Walke, *Mobile Radio Networks: Networking and Protocols*, John Wiley & Sons, Ltd., Chichester, 1999

8

GSM cellular telephony – physical layer

8.1 INTRODUCTION

Speech and data transmission in the GSM and associated control data signaling system require advanced digital signal processing and modern methods of coding, transmission and reception applied previously in specialized professional equipment. These requirements arise mostly from the available channel bandwidth and the applied signaling rate. The design of mobile stations is a particularly demanding task. Competition on the market of mobile phones pushes the producers towards continuous improvement of those mobile station parameters which are crucial for market success. They are phone size and weight, time between battery recharging and standby time. Improved parameters are achieved thanks to the progress in the IC technology and smart division of mobile station functions between particular blocks, which allows a large part of the transceiver to stay in the low power mode.

The ETSI standards dealing with transmission aspects precisely define speech coding and decoding as well as channel coding and applied modulation; however, they leave freedom in the realization of many algorithms, in particular those applied in the receiver.

In this chapter we will describe the algorithms and blocks realizing subsequent transmit and receive functions.

8.2 CONSTRUCTION OF A TYPICAL MOBILE STATION

Figure 8.1 presents the block scheme of a typical mobile station. Similar blocks can also be found in base stations, so we will limit ourselves to the description of mobile phone blocks.

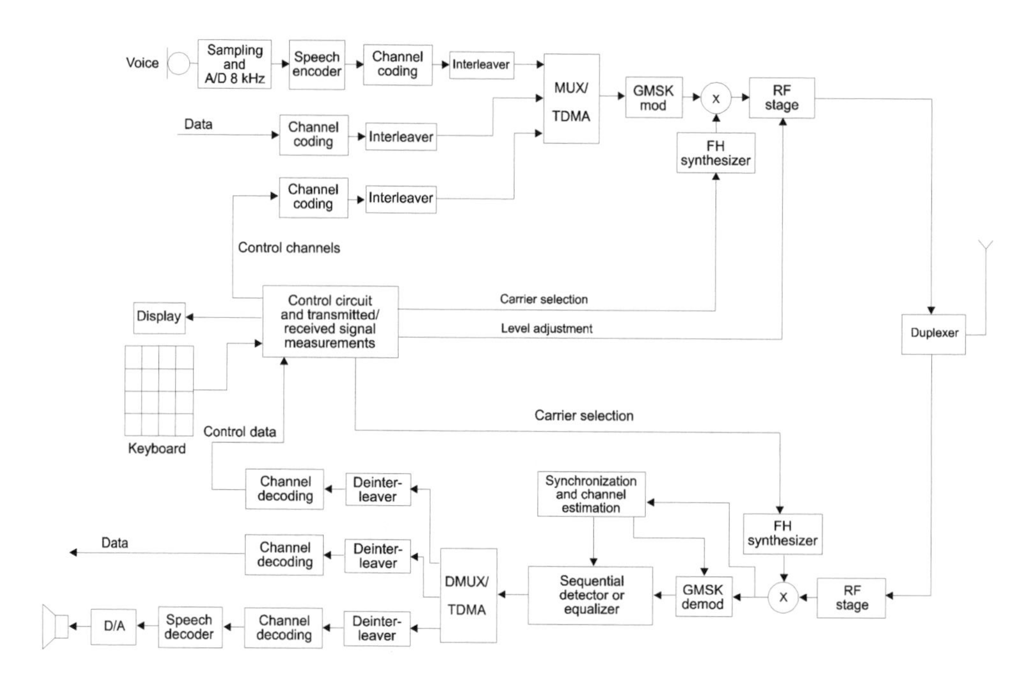

Figure 8.1 Block scheme of a typical mobile station

A voice signal is converted to an electrical signal in the microphone and its bandwidth is limited to 4 kHz by the anti-aliasing filter. It is then sampled with the frequency of 8 kHz and represented by a sequence of binary symbols, using a 13-bit analog-to-digital converter. The A/D converter must be linear due to the type of operations performed by the speech encoder. Let us note that the binary stream at the output of the A/D converter has the data rate of 104 kbit/s, so it is almost twice as fast as the binary stream at the output of the 8-bit nonlinear PCM converter. The sequence of 13-bit samples is processed in 20-ms blocks by the speech encoder applying the RPE-LTP (*Regular Pulse Excitation Long Time Prediction*) method. The speech encoding algorithm will be presented later in this chapter. As a result of application of this algorithm, we receive 260 bits representing each 20-ms block, so the data rate has been reduced from 104 kbit/s to 13 kbit/s at the encoder output. In the 260 bits of each block we can determine a binary block of crucial importance for the decoding of the whole block (*class* 1a bits), a block of moderate validity (*class* 1b bits), and those bits (*class* 2 bits) whose erroneous detection does not significantly influence the quality of voice generated by the speech decoder. Such a division of the data block allows application of non-equal error protection.

In turn, the 13-kbit/s binary stream is the subject of channel coding [2]. Errors in 50 bits of class 1a are detected thanks to the block polynomial code which supplements the data block with 3 parity bits. If the parity check performed in the receiver detects errors in the block, the whole 260-bit block is not taken into account in the speech decoding process and interpolation of speech samples is performed instead.

The binary block of moderate validity consists of 132 bits. The block built of class 1a bits, 3 parity bits and class 1b 132 bits appended with 4 zero bits is convolutionally encoded using $R = 1/2$ code of the constraint length equal to 5. As a result, 378 bits are received, which together with the remaining 78 bits of class 2 constitute a 456-bit block representing a 20 ms long speech signal. Thus, due to the channel coding the data rate increases to 22.8 kbit/s. Figure 8.2 presents the scheme of channel coding for a block of speech signal.

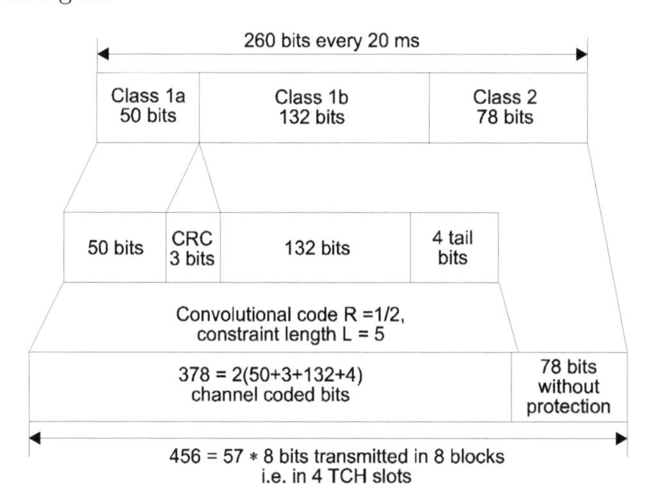

Figure 8.2 Channel coding of a binary block representing 20 ms of encoded speech signal

In order to realize effective decoding and to spread errors grouped in bursts due to fading, interleaving is applied. As a result, the bits generated in the coding process are reordered. In the receiver the reverse process takes place, which recovers original ordering of the binary stream. In GSM block interleaving is applied. The block of 456 bits resulting from encoding a single 20-ms block is divided into 8 sub-blocks of 57 bits each. As we remember from the previous chapter, a normal burst used to transmit traffic channels (TCH) contains two 57-bit blocks transmitting user data. Each of these two 57-bit blocks transmits a fragment of a different 20-ms data block. The interleaving process can be summarized as follows.

456 bits representing a 20 ms speech interval are read into an 8-column by 57-row matrix RAM, filling its subsequent rows. Then the bits are read out in columns. This way 8 sub-blocks of 57 bits each are formed. Next, the sub-blocks are distributed over 8 successive bursts placed in the assigned time slots. In each burst two 57-bit sub-blocks are filled with bits from adjacent 456-bit blocks. The bits from sub-blocks 0 to 3 occupy even bit positions, whereas the bits from sub-blocks 4 to 7 are placed in odd bit positions. This way, the first four bursts contain bits from the current and the preceding 456-bit data blocks and the last four bursts carry bits from the current and the following blocks.

Taking into account the relatively time-consuming procedure of speech coding and decoding (20 ms), channel coding and decoding using the Viterbi algorithm (10-20 ms)

and block interleaving/deinterleaving (40 ms), we end up with the non-negligible delay of 70-80 ms.

In the GSM system the *Voice Activity Detection* (VAD) has been applied. This means that besides regular speech coding, the speech encoder checks if the coded signal is really a speech signal or a noise signal generated by the speaker's environment. During the speaker's activity both his/her speech and noise are encoded. If the speaker is not active the binary signal representing the noise is not transmitted continuously. The pause in the speech signal causing a pause in the signal encoding would imply large fluctuations in the received acoustic signal level. This effect is very unpleasant for a listener. To reduce it, the frames describing the surrounding noise are sent in regular time intervals. They allow for generation of an artificial noise signal at the receiver. This noise, called a *comfort noise*, is similar to that arising in the speaker's vicinity. Because the frames describing noise are sent much more rarely than the bursts containing binary speech representation, generally the signal-to-interference ratio is improved.

In its standard version, the GSM system supports data transmission at data rates of 2400, 4800 and 9600 bit/s. For each data rate an appropriate convolutional code with the coding gain between 1/6 and 1/2 is applied. Interleaving is also used; however, in these cases it is much deeper than in the case of a speech signal. The delay introduced by the interleaver is not as critical as for speech transmission. On the other hand, a much lower bit error rate is required, which can be achieved thanks to the application of strong error correction channel coding and deep interleaving. More detailed information on data transmission in GSM can be found in Chapter 9.

In the previous chapter we learnt that besides the user data stream, the control data are sent between base and mobile stations and vice versa. Their errorless detection is crucial for reliable system operation. In consequence, stronger error correction coding is applied than in the case of speech transmission. Figure 8.3 presents the scheme of two-level coding applied for control signal transmission.

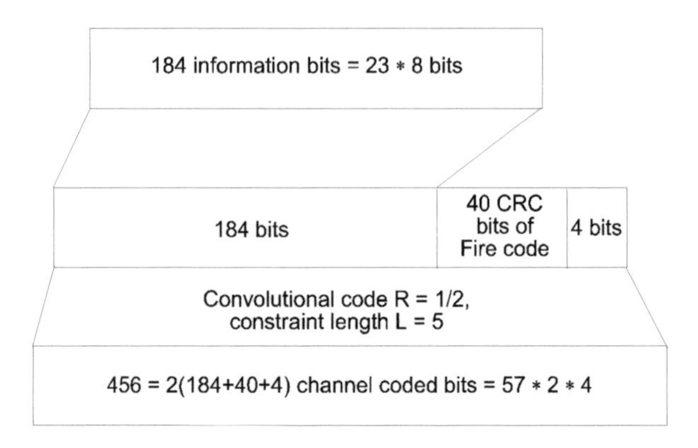

Figure 8.3 Coding procedure for control data block

The control messages are 23 bytes long. Thus, the data block has 184 bits. The outer error correcting code is the shortened Fire block code. The Fire code was invented to correct error bursts of a specified length occurring after appropriately long error-free intervals. The applied code is able to correct a single burst of length equal to 12 bits. Due to the coding process the data block is appended by 40 CRC bits. The use of this burst error correcting code is motivated by the application of the inner code being a $R = 1/2$ convolutional code of the constraint length equal to 5. This inner code is decoded using the Viterbi algorithm explained in Chapter 1. Possible errors contained in the output bits of the Viterbi decoder are often grouped in bursts. They are the result of a wrong selection of the path on the code trellis diagram. The four zero bits appended to the Fire code word allow setting of the encoder state at the end of the data block. This is very useful in decoding the end part of the 456-bit block.

Encoded data blocks are subsequently interleaved and finally formed into the TDMA bursts. These bursts are fed to the GMSK modulator which is triggered at the accuracy of a 1/4 bit length to generate the GMSK signal in the assigned time slot. The moment of generating the GMSK signal is determined by the synchronization subsystem. The GMSK signal is subsequently shifted to the frequency range of the assigned channel and amplified in the RF amplifier. After transferring it through the duplexer it is emitted by the antenna.

Many processes in the receiver are parallel to those taking place in the transmitter. In the mobile station receiver, the same antenna which was used for signal transmission receives the incoming bursts which are shifted by three time slots with respect to the transmitted bursts. After low noise filtration the burst signal is down-converted to the baseband. Then, the in-phase and quadrature components are derived from the received signal. Usually synchronous detection is applied. It means that the GMSK signal is treated as a linearly modulated signal. Subsequently, the stream of the baseband in-phase and quadrature samples is processed by the data signal detector. Usually the Viterbi sequential detector is applied, which uses channel impulse response samples derived by the channel estimator. This impulse response is calculated from the signal being the response of the mobile transmission channel to the known midamble contained in the transmitted burst. The detection algorithm will be described in one of the later paragraphs. The Viterbi algorithm delivers a stream of binary decisions with the associated likelihood information. After deconvolution these data are applied in the soft-decision Viterbi decoder used both in decoding the binary stream representing the speech signal and in decoding the control data stream. The received data are then decoded in the outer code decoder and, if this process is successfully completed, the resulting 260 bits or 184 control message bits are obtained. In case of the speech signal the data block is delivered to the speech decoder operating as the speech synthesizer. Its output, after conversion to the analog form, is transferred to the loudspeaker of the mobile phone. In case of the control data, the message is accepted by the control unit and appropriate activity is initiated by it.

Similar transmitter and receiver blocks can be found in a base station as well. The main difference is the source and sink of the speech signal processed by the base station. In the transmitter, the speech signal is delivered by the fixed network in the form of nonlinearly quantized 8-bit PCM samples. Therefore, these samples are first converted into 13-bit linearly quantized binary representation. A similar procedure in the opposite direction is performed in the receiver, which in the last step generates 8-bit nonlinearly quantized PCM samples and sends them to the fixed network. It has been noticed that this transcoding process has a negative influence on the overall speech detection, in particular if the half-rate speech coding is applied.

8.3 CODING AND DECODING OF A SPEECH SIGNAL

As we know from Chapter 1, speech coding is in fact a process of waveform compression. Instead of generating quantized and processed samples of the input signal, the encoder determines the quantized parameters of the speech source model, which allows the receiver to generate the speech signal very similar to that which was the subject of processing in the transmitter. In the GSM three speech coding standards have been established:

- *Full Rate* speech coding (FR) [3],
- *Half Rate* speech coding (HR) [4], and
- *Enhanced Full Rate* speech coding (EFR) [5].

Modern GSM mobile phones contain speech encoders operating in accordance with all three standards. We will now discuss these standards.

8.3.1 Full rate speech coding

This type of speech coding applies a modified linear prediction RPE-LTP method. The encoder scheme is shown in Figure 8.4. The input signal sampled in the 13-bit linear analog-to-digital converter or transcoded from a nonlinear 8-bit PCM to a linear 13-bit representation is first pre-processed to eliminate possible offset. Then it is processed by a first order pre-emphasis filter. As we remember, 160 samples represent a 20 ms speech signal. The samples from the output of the pre-emphasis filter are analyzed to determine the coefficients of the *Short-Term Prediction* (STP) filter. This is in fact the process of LPC analysis. The determined filter parameters are subsequently applied in the short-term analysis filter to filter the same 160 signal samples. As a result 160 signal samples of the short-term residual signal are obtained. The determined filter parameters, which are eight reflection coefficients of the STP lattice filter (see Figure 1.8), are transformed to logarithmic area ratios, LARs, because the logarithmically compounded LARs have better quantization properties than the reflection coefficients.[1] The LARs in the quantized form are represented by 36 bits which constitute a part of the encoder output block.

[1] If we denote the ith reflection coefficient as a_{ii} then

$$LAR(i) = \log_{10}\left(\frac{1 + a_{ii}}{1 - a_{ii}}\right)$$

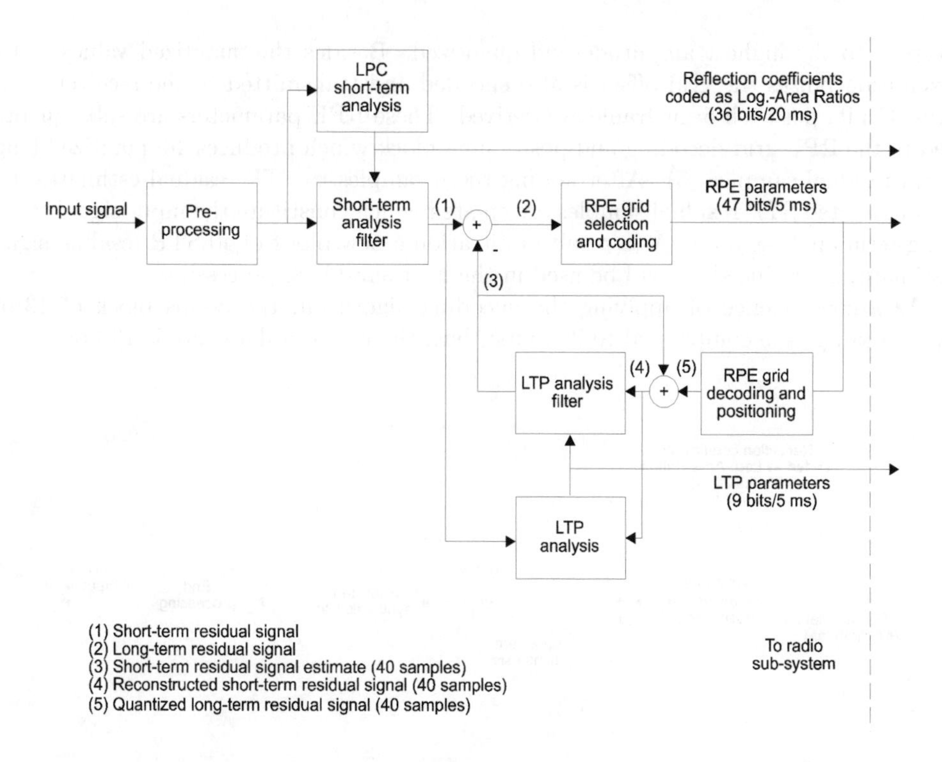

Figure 8.4 Block scheme of the GSM full rate speech encoder

For further processing, the 20-ms frame of STP residual samples is divided into 4 sub-frames containing 40 samples each. Each sub-frame is processed blockwise by the subsequent encoder blocks.

Before processing each 40-sample sub-frame, the speech encoder determines the parameters of the *Long-Term Prediction* filter, the LTP lag and LTP gain.[2] It is done on the basis of the current sub-frame of the STP residual samples (see signal (1) in Figure 8.4) and a stored sequence of previous 120 reconstructed short-term residual samples (signal (4) in Figure 8.4). A block of 40 samples of the long-term residual signal (2) is the difference between 40 estimates of the STP residual signal (3) and the 40 STP residual signal (1). As a result, a block of 40 long-term residual samples is obtained. After discarding the last sample, the block is subsequently fed to the *Regular Pulse Excitation* (RPE) analysis block. The RPE block decomposes the processed block into three 13-sample excitation sequences. It is performed by sample decimation and selection of the signal grid. Subsequently, the energy of the three decimated sequences is computed and that sequence which has the highest energy is selected as a representative of the whole block of the LTP residuals. The selected excitation pulses are normalized with

[2]See, for example, the lag L and the gain G_3 in the LTP filter shown in Figure 1.11.

respect to the highest amplitude and quantized. Besides the quantized values of the excitation pulses the grid offset is also encoded and transmitted to the receiver. This way 47 bits per each 5-ms frame are derived. These RPE parameters are subsequently fed to the RPE grid decoding and positioning block which produces 40 quantized long-term residual samples (5). After adding these samples to STP residual estimates, the reconstructed STP residual samples are created which constitute the input signal to the long-term analysis filter. As a result of filtration a new block of 40 STP residual signal estimates is produced. It will be used in the next sub-block processing.

As a consequence of applying the encoding algorithm, the 20-ms block of 13-bit speech samples is compressed to 260 bits; thus, the received data rate is 13 kbit/s.

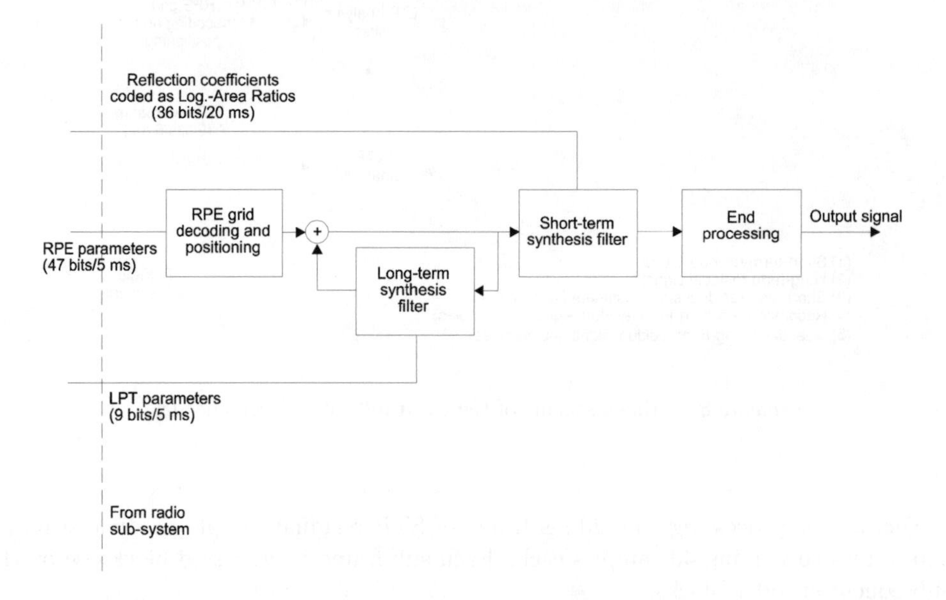

Figure 8.5 Block scheme of the RPE-LTP speech decoder

Figure 8.5 shows the simplified RPE-LTP decoder. It contains the same structure as the structure of the feedback loop in the encoder. In case of errorless transmission the output signal of this decoder part reconstructs the sequence of the short-term residual signal samples. These samples are subsequently fed to the input of the short-term synthesis filter. The output samples are processed by the de-emphasis filter generating the reconstructed samples of the speech signal.

8.3.2 Half rate speech coding

Now let us briefly describe the GSM half rate speech encoder [4]. It applies an analysis-by-synthesis approach in the version called VSELP which was briefly considered in Chapter 1. The analysis-by-synthesis procedure is used to find the best code word characterizing the excitation signal for each 20 ms signal frame. This code word is found by applying each code word as an excitation for the CELP synthesizer. The synthesized speech signal is subsequently compared with the input speech signal and a difference signal is calculated. This difference signal is weighted by a spectral weighting filter with the characteristics $W(z)$ and a second weighting filter $C(z)$. As a result, the error signal $e(n)$ is obtained. That code word which ensures the lowest power of the error signal $e(n)$ is selected as the best code word characterizing the frame. The characteristics of the weighting filter were chosen to ensure the best subjective human perception of the synthesized speech signal. The second weighting filter $C(z)$ controls the amount of error in the harmonics of the speech signal. In Figure 8.6 showing the general block scheme of the half rate encoder, both weighting filters have been placed in the branches of the input signal and synthesized signal before the subtractor comparing the signals from both branches. The filter denoted by $A(z)$ is a so-called short-term spectral filter, whereas $B(z)$ denotes the long-term filter with the lag L and filter coefficient β. In the process of speech analysis-by-synthesis, the encoder calculates 18 parameters characterizing each 20 ms frame. The parameters of a single frame are represented by 112 bits, which is equivalent to the 5.6 kbit/s data rate on the output of the half rate encoder.

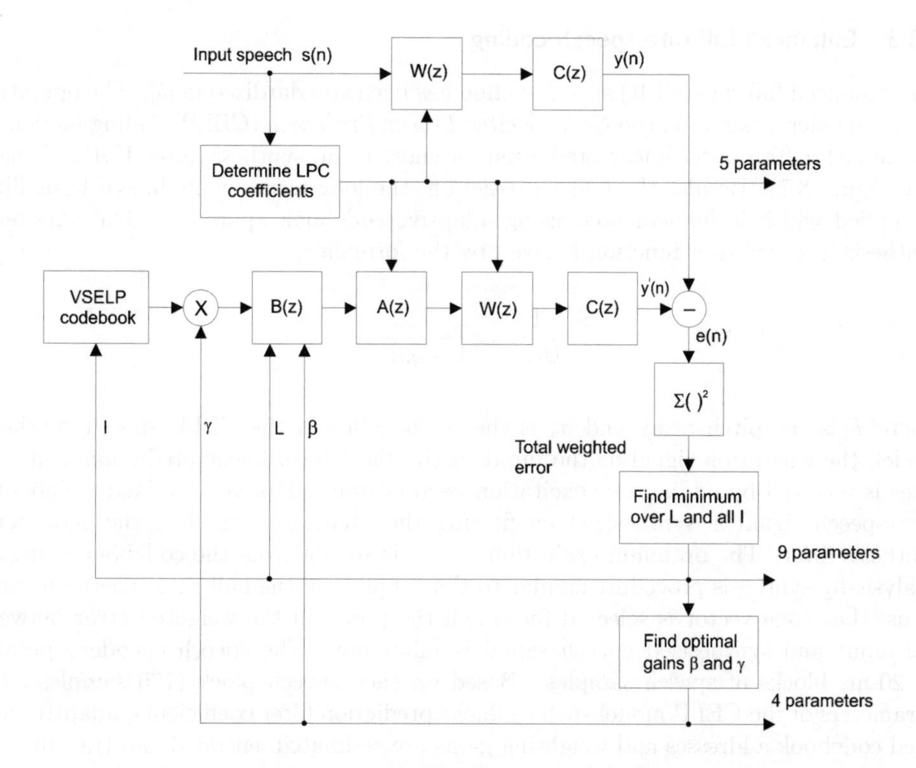

Figure 8.6 Simplified block diagram of the GSM half rate speech encoder

The half rate speech decoder is a subset of the encoder. On the basis of the received parameters the same speech synthesizer as that contained in the encoder generates the synthesized voice.

Because the number of bits representing a 20-ms frame is much lower as compared with the full rate coding, errorless decoding of these bits is more important for the speech quality at the receiver. Therefore, a stronger convolutional code is applied than that used for encoding of the full rate encoder output block presented in the previous section. Due to convolutional coding and introduction of the block parity check and some tail bits, the number of bits representing the 20-ms frame increases to 228. This number is equivalent to the 11.4 kbit/s data rate at the output of the channel encoder, which is exactly the half data rate at the output of the channel encoder co-operating with the full rate speech encoder.

The main advantage of the half rate speech encoder is doubling the physical channel capacity. The same slot can be shared by two half rate traffic channels realized in the interlaced manner. The reason for applying half rate speech coding was an attempt to avoid capacity problems in highly populated areas. The consequence of its introduction is the necessity of using double-standard encoders in mobile phones. The main disadvantage of the half rate speech coding is lower speech quality as compared with the full rate coding.

8.3.3 Enhanced full rate speech coding

The enhanced full rate (EFR) speech coding has been standardized in [5]. The operation of the encoder is based on the *Code-Excited Linear Predictive* (CELP) coding model. In this model a 10th order linear prediction, or short-term, synthesis filter $1/A(z)$ is used (see Figure 8.7). Besides the linear prediction, the long-term, or pitch, synthesis filter is applied which is implemented using adaptive codebook approach. The long-term synthesis filter transfer function is given by the formula

$$\frac{1}{B(z)} = \frac{1}{1 - g_p z^{-T}}$$

where T is the pitch delay and g_p is the pitch gain. In the CELP speech synthesis model, the excitation signal at the input of the short-term linear prediction synthesis filter is received by adding two excitation vectors from adaptive and fixed codebooks. The speech signal is synthesized by fitering the excitation signal in the short-term synthesis filter. The optimum excitation vector is sought from the codebook using an analysis-by-synthesis procedure similar to that applied in the half rate speech encoder. Thus, that code vector is selected for which the power of the weighted error between the input and synthesized speech signal is minimum. The speech encoder operates on 20-ms blocks of speech samples. Based on each speech block (160 samples), the parameters of the CELP model such as linear prediction filter coefficients, adaptive and fixed codebook addresses and weighting gains are estimated, encoded and transmitted. The decoder uses the received parameters to synthesize the speech signal in the same CELP synthesizer as that applied in the speech analysis in the transmitter.

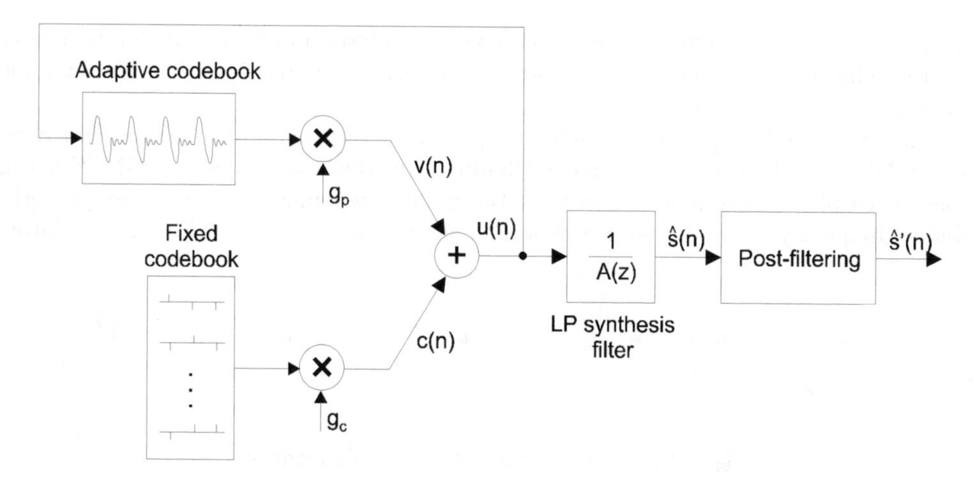

Figure 8.7 Simplified block diagram of the EFR encoder

The EFR encoder generates the data stream at the rate of 13 kbit/s. The channel coding and interleaving are identical to those applied for the RPE-LTP standard encoder. The tests have shown that the voice quality achieved through the EFR coding is much higher than in the RPE-LTP coding. This type of encoder has been mainly adopted in newly established networks, in particular in PCS 1900 networks in North America. In newer mobile phones the EFR encoder is one of the possible choices of voice encoders.

Table 8.1 [6] shows the comparison of GSM speech encoders in the computational complexity aspect. As we see, the standard encoder is substantially less complicated than the other two. The required program and data memory are also a few times smaller.

Table 8.1 Comparison of three GSM speech encoders

Speech Codec	Maximum Mips	Program and data ROM	RAM
GSM FR	2.5–4.5	4–6 kWords	1–2 kWords
GSM HR	17.5–22	16–20 kWords	∼ 5 kWords
GSM EFR	17–22	15–20 kWords	∼ 5 kWords

8.4 GMSK MODULATION

As we remember from the previous chapter, the data within a single TDMA burst are transmitted at the rate of 270.833 kbit/s, whereas the channel spacing is 200 kHz only. This sets stringent requirements for the applied modulation, with respect not only to the width of the mainlobe but also to the attenuation of the sidelobes. Another requirement related to efficient use of the nonlinear power amplifier at a mobile station is preserving the constant signal envelope. This feature is particularly important for the energy budget of a mobile station.

The GSM standard [7] recommends the application of *Gaussian Minimum Shift Keying* (GMSK) modulation. Let us recall Chapter 1 in which we stated that GMSK is the continuous phase FSK modulation with the modulation index $h = 1/2$ and properly shaped frequency pulse. Figure 8.8 presents a generic scheme of the GMSK modulator.

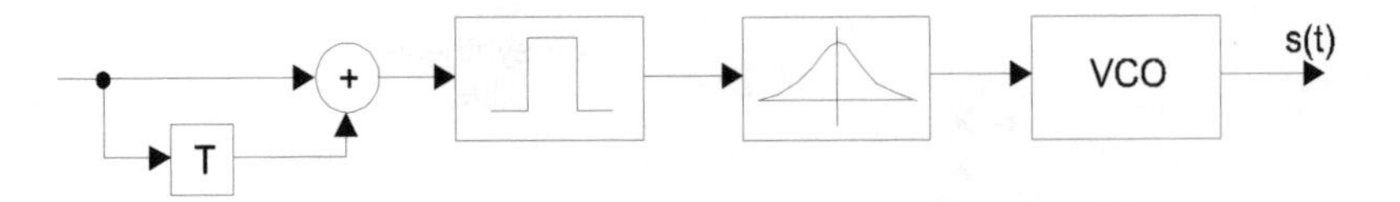

Figure 8.8 Block diagram of the GMSK modulator

The binary symbol at the nth timing instant is modulo-2 added to the symbol from the previous time instant. This is a kind of differential encoding applied in the GSM modulator; however, it is not an inherent part of the GMSK modulator in a general case. The resulting symbol, after conversion to the NRZ (*Non-Return to Zero*) form, is fed to the input of the filter with the Gaussian impulse response, i.e.

$$h(t) = \frac{1}{\sqrt{2\pi}\sigma T} \exp\left(\frac{-t^2}{2\sigma^2 T^2}\right) \tag{8.1}$$

where $\sigma = \sqrt{\ln 2}/(2\pi BT)$, $BT = 0.3$. B is a 3 dB bandwidth of the pulse $h(t)$, and T is the time duration of a single bit. The effective duration of the filter impulse response is $4T$ to $5T$. The output signal of the filter excites the voltage-controlled oscillator. Figure 8.9 shows the power spectral density at the output of the ideal GMSK modulator ($BT = 0.3$) normalized with respect to the signaling period T. In a real system the modulator works only during the burst transmission in the assigned time slot. A certain technical problem is to ensure sufficiently fast switching of the power amplifier on and off so that the length of the transient processes meets the GSM standard [7]. The allowable limits of the transient process during switching on and switching off the power amplifier are shown in Figure 8.10.

8.5 SEQUENTIAL DATA DETECTION

We have already mentioned that the Viterbi sequence estimator is usually applied in GSM mobile and base stations. This is a sequential detector, i.e. the data decisions worked out by the agorithm are made on the basis of the whole data sequence, as opposed to symbol-by-symbol detectors which generate subsequent decisions based on each individual data symbol.

At this point we have to recall a few features of a GSM transmission channel which determine the character of the detection algorithm. As we know, the signal arrives at

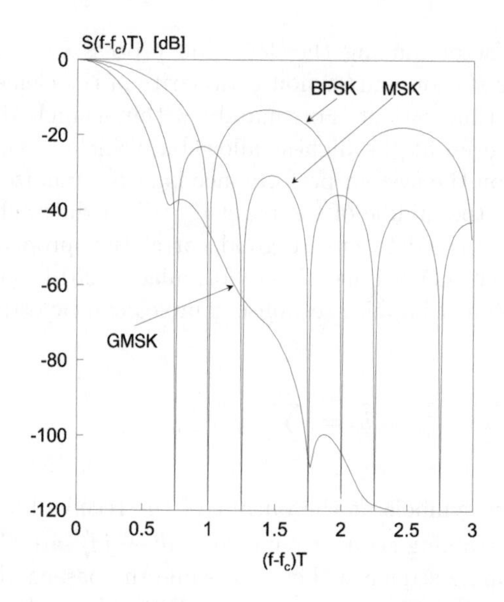

Figure 8.9 Power spectral density of GMSK, MSK and BPSK modulated signals

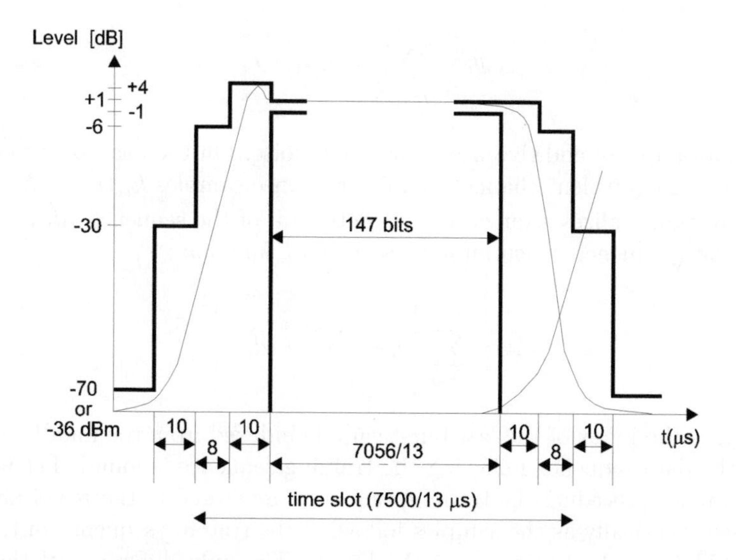

Figure 8.10 Transient processes during switching the GSM power amplifier on and off

the receiver in the form of reflected and dispersed rays. Particular signal paths have different lengths so the signal echoes have various delays. In particularly disadvantageous cases the delay between the earliest and latest echo reaches 20 μs (compare Figure 3.10 in Chapter 3) at the symbol period equal to 3.69 μs; thus, the phenomenon of intersymbol interference is evident. The Doppler effect, being the result of the mobile station movement, is another factor making the detection process even more difficult. The movement of the mobile station and/or nonstationarity of the elements of environment cause the channel to be time variant. Fortunately, within a single time slot the channel variability is moderate even at the highest allowable vehicle speed. The influence of the channel variability on the system performance is also minimized thanks to placing the training sequence in the middle of the burst [9]. Eight different training sequences listed in [9] have been selected to ensure good correlation properties of the samples of the baseband equivalent GMSK signal (see formulae (1.25) to (1.28) in Chapter 1). Thanks to them the channel impulse response samples can be estimated according to the formula

$$\widehat{h}_i = \sum_{j=-N}^{N} y_j d_{i-j}^{*} \tag{8.2}$$

where y_j are subsequent symbols at the output of the receive filter, indexed starting from the middle of the training sequence and $d_i = d_i^I + jd_i^Q$ are GMSK data symbols associated with the training sequence. Let us assume the baseband equivalent channel model in the form of a finite impulse response (FIR) filter. It is described by the equation

$$y_i = \sum_{j=-N}^{N} h_j d_{i-j} + \nu_i \tag{8.3}$$

where ν_i is a sample of additive white Gaussian noise. On the basis of the estimates $\widehat{h}_i$ of the baseband equivalent channel impulse response samples h_i $(i = -N, \dots, N)$ the Viterbi algorithm realizes sequential data detection of the sequence $\{\widehat{d}_n\}$ by searching for such a data sequence which minimizes the cost function

$$D = \sum_{i=0}^{N_{\max}} \left| y_i - \sum_{j=-N}^{N} \widehat{h}_j \widehat{d}_{i-j} \right|^2 \tag{8.4}$$

where $N_{\max}$ is the index of the last burst sample indexed from the middle of the burst. This way the data sequence following the training sequence is found. Let us note that the data samples preceding the training sequence are stored in the RAM and they can be processed identically as the samples following the training sequence in the direction from the middle to the beginning of the burst. The only difference of the backward processing is the reversing of the channel impulse response samples.

Figure 8.11 presents a general scheme of the receiver which applies the Viterbi algorithm and the channel estimator. In the Viterbi detector the minimization of the cost function D is performed sequentially, taking advantage of the channel description in the form of the automaton characterized by a trellis diagram similar to that featured by the convolutional encoder (compare Figure 1.18). Because GMSK is a binary modulation, this automaton has 2^{2N} states, where $2N + 1$ is the length of the channel impulse response. Actual GSM detectors have $2^4 = 16$ states. The trellis diagram presents

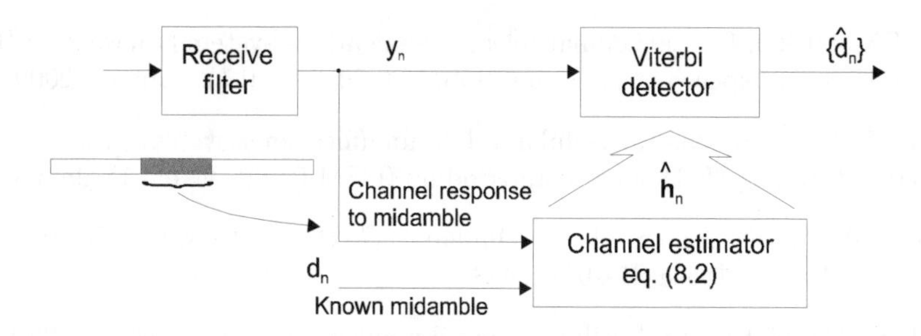

Figure 8.11 General scheme of the sequential detector applying the Viterbi algorithm

all possible paths from the states at the nth moment to the states at the $(n+1)$st moment. Let us recall from Chapter 1 that the shortest path to the considered state on the trellis diagram at the nth moment characterized by the smallest cost function D consists of the shortest route to one of the states at the $(n-1)$st moment and the path from that state to the considered state at the nth moment. This rule allows for iterative operation in which the current results (in the form of the shortest paths to each state at the previous step) can be used in the calculation of the shortest paths in the next step.

Minimization of the cost function D is the practical realization of the *Maximum Likelihood* reception. A GSM mobile station is probably the first product of massive use in which the Viterbi algorithm used for convolutional code decoding and data sequence detection found its application.

The detection method described above is a standard solution. Several simplifications of this method are possible, for example, those described in [10], in which the number of the trellis states is reduced. The reader interested in GSM terminals is also advised to study [11]. The reduction of the number of states is particularly important if other than binary modulation is applied. As we will see in Chapter 9, the enhanced form of the GSM system updated for high speed data transmission, called EDGE, applies 8-PSK modulation. Thus, the number of trellis states would be 8^{2N}. Even for a short channel impulse response the number of states woud be excessive. Therefore, simplified detection algorithms have to be applied.

REFERENCES

1. R. Krenz, K. Wesołowski, "Pan-European System of GSM Cellular Telephony" (in Polish), *Przegląd Telekomunikacyjny*, No. 6, 1993, pp. 279-284

2. ETSI EN 300 909, Digital cellular telecommunications system (Phase 2+); Channel coding (GSM 05.03 V. 8.3.0.) April 2000

3. ETSI EN 300 961, Digital cellular telecommunications system (Phase 2+); Full rate speech; Transcoding (GSM 06.10, V. 7.0.2), December 1999

4. ETSI EN 300 969, Digital cellular telecommunications system (Phase 2+); Half rate speech; Half rate speech transcoding (GSM 06.20, V. 7.0.1), January 2000

5. ETSI EN 300 726, Digital cellular telecommunications system (Phase 2+); Enhanced Full Rate (EFR) speech transcoding (GSM 06.60, V. 6.0.1), January 2000

6. S. M. Redl, M. K. Weber, M. W. Oliphant, *GSM and Personal Communications Handbook*, Artech House, Boston, 1998

7. ETSI EN 300 959, Digital cellular telecommunications system (Phase 2+); Modulation (GSM 05.04 , V. 7.1.0), January 2000

8. ETSI EN 300 910, Digital cellular telecommunications system (Phase 2+); Radio transmission and reception (GSM 05.05, V. 8.3.0), April 2000

9. ETSI EN 300 908, Digital cellular telecommunications system (Phase 2+), Multiplexing and multiple access on the radio path (GSM 05.02, V. 8.3.0), April 2000

10. K. Wesołowski, R. Krenz, K. Das, "Efficient Receiver Structure for GSM Mobile Radio", *International Journal of Wireless Information Networks*, Vo. 3, No. 2, 1996, pp. 117-123

11. Z. Zvonar, R. Baines, "Integrated Solutions for GSM Terminals", *International Journal of Wireless Information Networks*, Vol. 3, No. 3, 1996, pp. 147-161

9

Data transmission in GSM

9.1 INTRODUCTION

Cellular systems were first deployed in the early eighties and their basic task was to transmit voice signals. The increasing demand for data exchange, caused by the progress in computer networks, created the necessity to offer data transmission as a new service in cellular systems. First generation systems which transmit voice using analog FM modulation are not well fitted for data transmission. They ensure relatively reliable data transmission at the rate of up to 1200 bit/s [1] or do not offer this service in their standard version at all. Second generation cellular systems are better prepared for data transmission due to the applied digital technology.

In this chapter we will concentrate on data transmission realized in the GSM system. Let us recall that the binary data rate in a GSM burst is equal to 270.833 kbit/s. As we know, this high data rate is the result of the applied TDMA mode in which time is divided among eight users. We have to recall that the binary stream generated at the output of the channel code encoder has the data rate of 22.8 kbit/s regardless of whether voice or data is transmitted. Voice and data should not be differentiated on the level of burst transmission. Therefore, the organization of data transmission is more or less associated with the transmission procedures of digitally represented voice.

The GSM standard [2] sets the rate of data transmission at most at 9600 bit/s and in some special conditions at 14.4 kbit/s. In order to match the data rate with the rate of 22.8 kbit/s used in data bursts, special error correction codes depending on the input data transmission rate have to be used. In fact, data transmission blocks create a parallel chain to the blocks operating in voice transmission. Let us mention that connecting a modem to the acoustic input of a mobile station would not make any sense because the voice signal encoder is optimized with respect to human speech and the statistical and spectral properties of the data modulated signal at the output of the acoustic modem are quite different from the speech signal.

9.2 ORGANIZATION OF DATA TRANSMISSION IN THE GSM SYSTEM

The GSM network can be considered as the access network to other communication networks such as the *Public Switched Telephone Network* (PSTN), *Integrated Services Digital Network* (ISDN), *Circuit-Switched Public Data Network* (CSPDN) or *Packet-Switched Public Data Networks* (PSPDN). Peculiarity of digital transmission in the GSM network requires a kind of interface which modifies the data stream generated in the GSM network to the properties of other networks. Therefore, the mobile switching center (MSC) is equipped with an *Interworking Function* (IWF) block. The task of this block is to implement the exchange protocol between the GSM and other networks. Data rate adaptation, ARQ procedures and acoustic modem functions are the most important tasks realized by the IWF. Since data exchange between a GSM terminal and the modem connected to the PSTN can take place in the synchronous or asynchronous mode at the data rate of 300, 1200, 1200/75, 2400, 4800 and 9600 bit/s, the acoustic modems located in the IWF operate in accordance with ITU-T V.21, V.22, V.22bis, V.23, V.26ter, V.32 and V.29 (in the case of the facsimile transmission) recommendations.

On the opposite end of the link, a computer or a measurement device connected to the GSM mobile station is both a source and destination of the data stream. From the point of view of data exchange, mobile stations are divided into three types which are differentiated by the location of the *Terminal Adaptation Function* (TAF) block, which matches the mobile station to the external data terminal. The types of mobile stations are the following:

- MT0 (*Mobile Termination type* 0) – a mobile station which is fully integrated and which does not have any interface to external devices. It can be a simple mobile station which does not offer data transmission at all or can be a very sophisticated mobile station integrating the data terminal, fax and regular voice transmission in a single unit.

- MT2 – a mobile station which features the V.24/RS-232 serial interface, so the TAF block is integrated with the MS.

- MT1 – a mobile station equipped with the ISDN 64 kbit/s S-interface. In order to connect a typical terminal with a modem interface (V.24) a special terminal adapter is necessary.

Figure 9.1 presents the location of TAF and IWF blocks in the GSM transmission chain.

9.3 DATA SERVICES IN GSM

In general, the services offered by the GSM system can be divided into two classes:

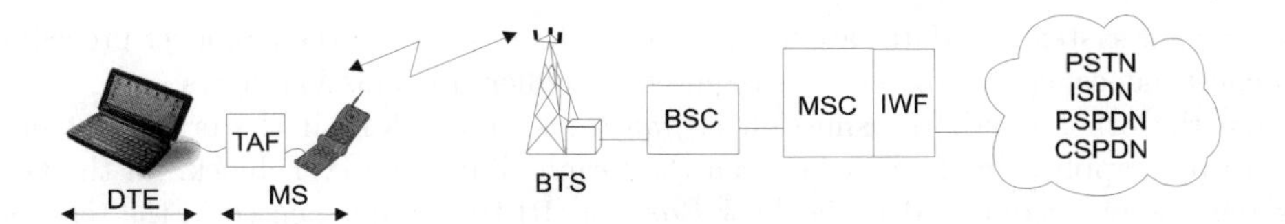

Figure 9.1 Location of Terminal Adaptation Function and Interworking Function blocks in the data transmission chain in GSM

- *Teleservices* – the GSM system provides full services within the network; there is no need for external devices to realize this type of services. The examples of teleservices are: speech transmission, SMS (*Short Messages Service*) and facsimile transmission.

- *Bearer services* – the GSM implements lower layers of services or realizes the transport mechanism between access points. *Terminal Equipment* (TE) such as a laptop, palmtop or a computer server is necessary on both ends of the link to implement this kind of services. The bearer services have their own upper layer protocols. The GSM system only transports the data, without differentiating which particular service is realized.

From the point of view of this division, data transmission is a bearer service. The bearer services set different requirements on the network depending on the kind of service. As we have already mentioned, the data service can have different rates. Both synchronous and asynchronous data transmission can be applied. The information transfer can be realized in the form described as *Unrestricted Digital Information* (UDI) transfer or it can apply a 3.1-kHz wide acoustic modem signal for information exchange. In the first case, transmission between the GSM user and ISDN or the packet network subscriber takes place. The acoustic representation of the transmitted data occurs if the GSM subscriber is linked to a PSTN subscriber or a PSTN is the intermediate network to which the GSM network is connected. In general, the data transmission link can connect two individual terminals (*point-to-point transmission*) or the link can be established between a particular terminal and many receiving terminals (*point-to-multipoint transmission*). For the data exchange between two terminals a physical link can be realized to transmit data packets exclusively between these two terminals (*circuit-switched transmission*) or the packets supplied with the source and destination addresses can travel through the network, being sent between subsequent network nodes from the source to the destination terminal (*packet-switched transmission*). In the standard GSM system, point-to-point and circuit-switched data transmission is always realized.

The data transmission can be transparent or non-transparent. We say that transmission is *transparent* (T) if the link between the terminal equipment, the mobile station and the IWF block on the other end of the GSM system does not use any procedure of automatic repetition of erroneous or lost frames or packets. As a result, the data rate and transmission delay are constant; however, the error rate is variable. Additionally, in the cellular system the data rate increases considerably during the handover procedure. In consequence, error protection is required in higher transmission layers.

On the other hand, transmission is *non-transparent* (NT) if the error protection protocol is applied in the link between the terminal and the IWF block. In the GSM system the protocol called *Radio Link Protocol* (RLP) is implemented in the terminal, mobile station or the TAF block. This protocol is based on the ARQ technique. As a result, constant high transmission quality is assured; however, the data rate and transmission delay are variable. The data rate substantially decreases during handover due to the repetition of the frames which are lost during the change of the serving base station.

We have to stress that both transparent and non-transparent types of transmission applied in the GSM system use error detection and correction coding in the physical layer, as described in the previous chapter.

Table 9.1 lists all the bearer services offered by the standard GSM system. PAD stands for *Packet Assembler/Disassembler* block. Using the PAD block is one of the possible ways to communicate with packet networks through PSTN or ISDN networks. As we see in Table 9.1, there are 21 different services associated with data transmission, starting from very slow asynchronous transmission at 300 bit/s and ending with direct packet access in synchronous mode at the data rate of 9.6 kbit/s.

9.3.1 Rate adaptation

In all data services the data rate has to be adjusted to the rate of 22.8 kbit/s used in GSM bursts on the one hand and to the data rate applied in the fixed part of the GSM network linking BTS, BSCs and MSCs on the other hand. In the latter case it is advantageous to multiplex the data streams of different users in the 16 or 64 kbit/s link applied between BSS and MSC. For this purpose the ITU-T V.110 Recommendation [4] has been adapted to GSM. Different adaptation functions are applied depending on the configuration of the actual data transmission. We will shortly describe them below [5].

- **RA0** – the adaptation function used in an asynchronous interface. This function realizes the transformation of the asynchronous stream generated by the user to a synchronous stream at the first allowable higher data rate. This operation is performed by adding or subtracting some stop bits contained in the asynchronous stream.

- **RA1** – the adaptation function used to adapt the data stream to intermediate rates. Its input is the output of a RA0 block or the synchronous data. The block performs the transformation of the data stream to an 8 or 16 kbit/s stream, using bit repetitions. The resulting frames contain synchronization bits, control bits and repeated data bits. Table 9.2 shows transformation of data rates for RA1.

- **RA2** – the adaptation function used to convert the data stream at the output of RA1 block to a 64 kbit/s stream (A interface) according to ITU-T V.110 Recommendation. 8 kbit/s data bits are placed in the 64 kbit/s data stream as the first

Table 9.1 Bearer services offered in GSM [3]

Bearer service		Access		Information	
No.	Name	Structure	Rate	Transfer	Attribute
21	Asynch 300 bit/s	Asynch	300 bit/s	UDI or 3.1-kHz	T or NT
22	Asynch 1.2 kbit/s	Asynch	1.2 kbit/s	UDI or 3.1-kHz	T or NT
23	Asynch 1200/75 bit/s	Asynch	1200/75 bit/s	UDI or 3.1-kHz	T or NT
24	Asynch 2.4 kbit/s	Asynch	2.4 kbit/s	UDI or 3.1-kHz	T or NT
25	Asynch 4.8 kbit/s	Asynch	4.8 kbit/s	UDI or 3.1-kHz	T or NT
26	Asynch 9.6 kbit/s	Asynch	9.6 kbit/s	UDI or 3.1-kHz	T or NT
31	Synch 1.2 kbit/s	Synch	1.2 kbit/s	UDI or 3.1-kHz	T
32	Synch 2.4 kbit/s	Synch	2.4 kbit/s	UDI or 3.1-kHz	T or NT
33	Synch 4.8 kbit/s	Synch	4.8 kbit/s	UDI or 3.1-kHz	T or NT
34	Synch 9.6 kbit/s	Synch	9.6 kbit/s	UDI or 3.1-kHz	T or NT
41	PAD access 300 bit/s	Asynch	300 bit/s	UDI	T or NT
42	PAD access 1.2 kbit/s	Asynch	1.2 kbit/s	UDI	T or NT
43	PAD access 1200/75 bit/s	Asynch	1200/75 bit/s	UDI	T or NT
44	PAD access 2.4 kbit/s	Asynch	2.4 kbit/s	UDI	T or NT
45	PAD access 4.8 kbit/s	Asynch	4.8 kbit/s	UDI	T or NT
46	PAD access 9.6 kbit/s	Asynch	9.6 kbit/s	UDI	T or NT
51	Packet access 2.4 kbit/s	Synch	2.4 kbit/s	UDI	NT
52	Packet access 4.8 kbit/s	Synch	4.8 kbit/s	UDI	NT
53	Packet access 9.6 kbit/s	Synch	9.6 kbit/s	UDI	NT
61	Alternate speech/data				
81	Speech followed by data				

Table 9.2 Input and intermediate data rates used in RA1 rate adaptation function

Synchronous data rate bit/s	Intermediate rate kbit/s	Duplication (times)
600	8	8
1200	8	4
2400	8	2
4800	8	1
9600	16	1

Table 9.3 Transformation of synchronous data rates to data rates before coding on the radio interface

Synchronous data rate [kbit/s]	Data rate before coding on the radio interface [kbit/s]
$\leq$ 2.4	3.6
4.8	6
9.6	12
14.4	14.5

bits in each octet, whereas 16 kbit/s data bits are placed in the 64 kbit/s data stream in the first two octet positions. The remaining bits in each octet are filled with ones. This inefficient representation of the data stream from the output of RA1 block in the form of 64 kbit/s data stream can be avoided if multiplexing of a few 8 or 16 kbit/s data streams is applied. Thus, the multiplexed data streams are placed in subsequent bit positions in the octets.

- **RA1'** – the adaptation function which transforms the rates of data received from the terminal (DTE) to the rates used in a radio interface. The data stream is supplemented with control bits. The source of data for RA1' is either the output of RA0 or a synchronous stream. Table 9.3 presents data rate transformation between DTE and radio interface. Let us note that the values of 3.6, 6, 12 and 14.5 kbit/s are related to the data stream before channel coding.

The process of channel coding applied in data transmission services will be described in the next subsection. Then the motivation for the selection of data rates on the radio interface will be clear.

Let us illustrate the rate adaptation for asynchronous transparent data transmission. This is only one example of many possible configurations of data exchange. Figure 9.2 presents the rate adaptation for three input data rates. In case of the synchronous input data the first step of RA0 is omitted.

The asynchronous data stream is turned into a synchronous stream of the first higher data rate (performed by RA0 function). As a result, a synchronous data stream of 2.4, 4.8 or 9.6 kbit/s data rate is received. RA1' function transforms this stream to the data stream of the rate equal to 3.6, 6.0 or 12.0 kbit/s. Next, channel coding with puncturing and interleaving matched to a particular data rate is applied, giving the 22.8 kbit/s data stream. This data is packed into the GSM bursts and sent over the air. On the fixed side of the GSM network the data is decoded to the 3.6, 6.0 or 12.0 kbit/s data stream, respectively. This stream is subsequently transformed into 8 or 16 kbit/s data stream (RA1'/RA1) and it is finally placed in the ISDN octets (RA2).

The non-transparent data transmission is organized similarly. In non-transparent transmission the highest rates of the radio interface are applied. Let us discuss the data rate adaptation for synchronous data exchange. In the case of non-transparent transmission data blocks are protected by the *Radio Link Protocol* (RLP), which will

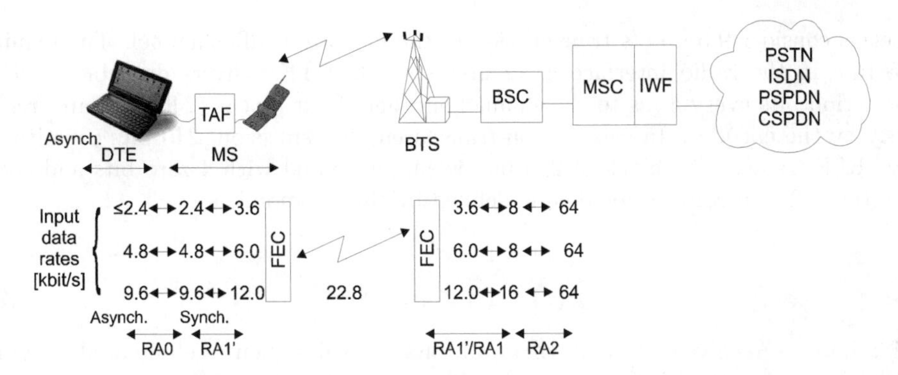

Figure 9.2 Example of rate adaptation functions applied in asynchronous data transmission in GSM

be explained later in this chapter. Figure 9.3 presents the subsequent steps of rate adaptation if a full-rate channel is used in the radio interface. A half-rate channel can be applied as well, resulting in a lower data throughput. Let us note that the RLP protocol is applied externally with respect to the channel coding.

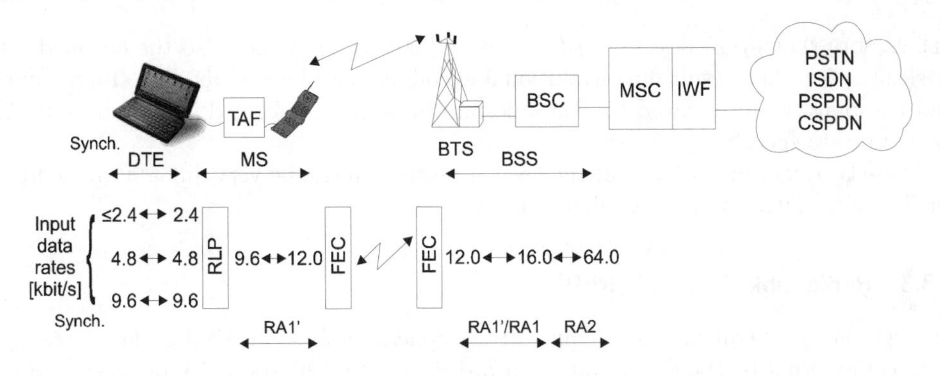

Figure 9.3 Rate adaptation functions for non-transparent synchronous transmission using full rate channel

9.3.2 Channel coding

Channel coding applied in data transmission depends on the data rate of the radio interface. As we have mentioned above, as a result of RA1' adaptation function we receive a data stream of the rate 3.6, 6.0 or 12 kbit/s. This data stream can be transmitted using the full- or half-rate traffic channel. All possible channel coding configurations can be found in [2]. Below we show only two examples of channel coding procedures applied in data transmission over GSM.

Let us consider 9.6 kbit/s transmission in the full-rate traffic channel. The required data rate in the radio interface is 12 kbit/s. The DTE delivers data blocks of 60 information bits every 5 ms to the channel encoder. Four blocks (240 bits) are treated jointly by the encoder. In case of non-transparent transmission, 240 bits constitute a single RLP frame. The block of 240 bits is supplemented with 4 zero bits and coded using the 1/2-rate convolutional code defined by the polynomials:

$$\begin{aligned} g_1(x) &= 1 + x^3 + x^4 \\ g_2(x) &= 1 + x + x^3 + x^4, \end{aligned} \tag{9.1}$$

so 488 bits are received. Among them, 32 bits are subsequently eliminated by puncturing, decreasing the block size to 456 bits. Next, the coded bits are reordered and interleaved. Finally, the 456-bit block is transmitted using 22 consecutive data bursts.

As the second example, let us consider channel coding applied in data services at the rate of 2.4 kbit/s and less. The data rate of radio interface is now 3.6 kbit/s. The DTE delivers 36 information bits every 10 ms. Two blocks (72 bits) are treated jointly by the coding process. Such 72-bit block is increased by four zero bits and is a subject of rate 1/6 convolutional coding defined by the polynomials:

$$\begin{aligned} &g_1(x) = 1 + x + x^3 + x^4, \quad g_2(x) = 1 + x^2 + x^4, \quad g_3(x) = 1 + x + x^2 + x^3 + x^4 \\ &g_4(x) = 1 + x + x^3 + x^4, \quad g_5(x) = 1 + x^2 + x^4, \quad g_6(x) = 1 + x + x^2 + x^3 + x^4 \end{aligned} \tag{9.2}$$

Let us note that $g_1(x) = g_4(x)$, $g_2(x) = g_5(x)$ and $g_3(x) = g_6(x)$, so the encoded bits are sent twice. The result of convolutional encoding is a block of 456 bits. Interleaving and mapping the interleaved bits on a normal burst are performed in the same way as for a full-rate speech signal.

These two examples show that the level of protection can be very different, depending on the input data rate and required quality.

9.3.3 Radio Link Protocol (RLP)

Transparent data transmission requires the application of some kind of data exchange protocol external to the GSM system in order to ensure high quality of service. Such a solution may not be feasible due to high time consumption and a large amount of system resources used because the block chain for transmitting the signaling information can be very long. The block chain consists of a data terminal (DTE), mobile station (MS), base station system (BSS), mobile switching center (MSC), interworking function (IWF) block, the external fixed network and the DTE of the second user. If the second user is also a subscriber of a mobile system the blocks listed above are again used in the transmission chain in the reversed order. Therefore, non-transparent transmission using *Radio Link Protocol* (RLP) is more advantageous in many cases. If the RLP is applied, error protection is performed between TAF and IWF blocks using a kind of ARQ procedure.

The 240-bit RLP frame has a very simple structure. It is shown in Figure 9.4. The frame starts with a 16-bit header which performs control functions such as establishing or releasing an RLP link, counting the frames and signaling the presence of erroneous

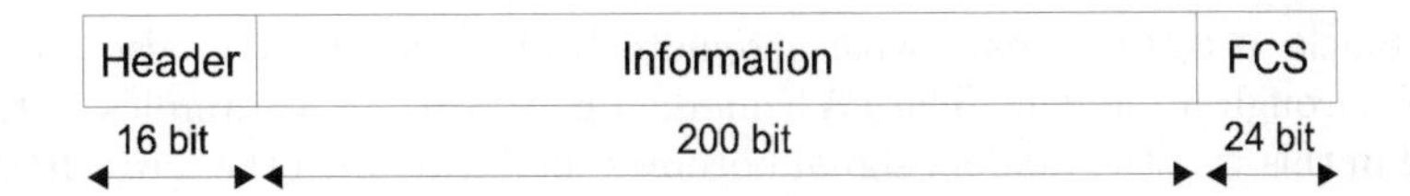

Figure 9.4 RLP frame structure

frames which have to be retransmitted. The next field has the length of 200 bits and contains user data. The frame is ended with a 24-bit *Frame Check Sequence* (FCS) calculated by dividing the frame contents by the cyclic code polynomial (see Chapter 1).

The RLP protocol works in two modes. The first one, called *Asynchronous Disconnected Mode* (ADM), indicates that no RLP link has been established yet. A necessary condition which has to be fulfilled before establishing the RLP link is the allocation of the traffic channel to a given connection. If the RLP is in the *Asynchronous Balanced Mode* (ABM), the exchange of numbered frames and their acknowledgement takes place.

Three types of frames appear in the RLP protocol. The first type is called unnumbered frame (U) and its header contains, among others, the commands for establishing and releasing the RLP link. The supervisory frames (S) are used to send supervisory commands during frame exchange, which inform about the status of the received frame, the receiver status and the number of the received frame. The third type of a frame is a supervisory frame combined with information. Its header contains the commands indicating the numbers of the received and sent frames as well as the status of the received frame and the receiver.

9.3.4 Data transmission in the aspect of access to different networks

As we know, the GSM subscriber transmitting data can communicate with the subscribers of several types of networks. The PSTN network is so far the most popular among them. The ISDN or packet data network can participate in a digital link as well. We will shortly describe the specific features of the links arranged using the networks of different types. Figure 9.5 presents the data link arranged with a PSTN subscriber or a host connected via PSTN to the outside world. Although nowadays the PSTN is

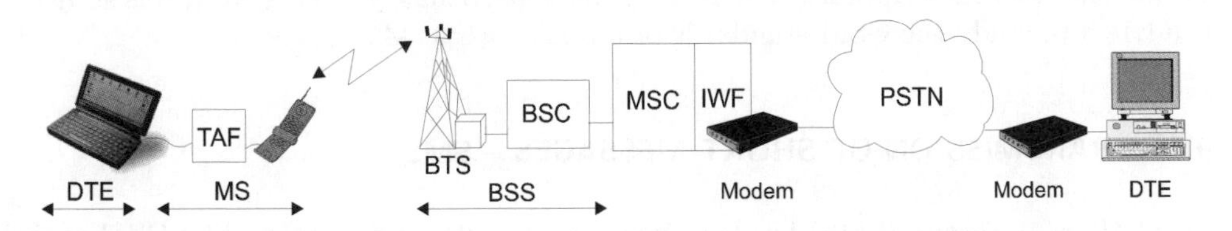

Figure 9.5 Data transmission over GSM and PSTN

very often digital, during the connection set-up it is not known what the properties of the PSTN link are. Therefore, the GSM network uses the analog modem contained in the IWF block to communicate with a remote DTE. The latter is also connected to the PSTN through a modem. The IWF modem generates PCM samples of the analog signal and in this way the modem signal becomes similar to any other speech or acoustic data signal represented digitally in the PSTN. The types of modems supported by GSM and used in the IWF block have already been mentioned in Section 9.2.

The ISDN is the second type of network to which the GSM system can be connected. In this case the link is fully digital so analog modems are not needed any more. The rate adaptation blocks have to be applied instead, in order to fit the data rate of the terminal connected with the GSM network to the data rate in the ISDN network. Figure 9.6 illustrates this case.

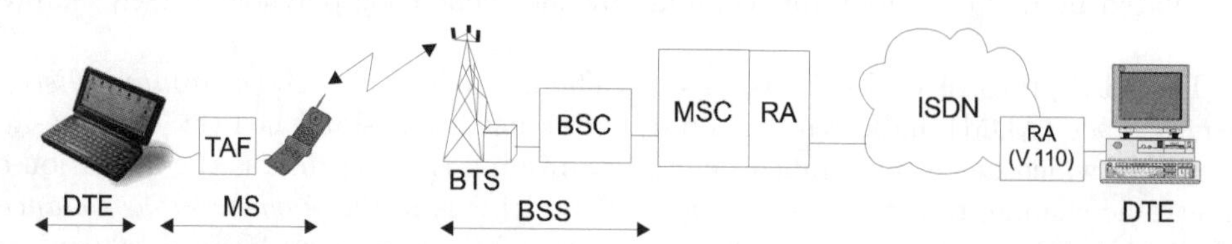

Figure 9.6 Data transmission over GSM and ISDN networks

Cooperation of a GSM network with a PSPDN is more complicated. A GSM user can set up a connection with a packet data network user through the intermediate PSTN network or can do it directly with the desired PSPDN. In the case of PSTN functioning as an intermediate network, analog modems have to be used in the IWF and at the PSTN/PSPDN interface point. Figure 9.7 illustrates the case when the GSM network and ISDN are interfaced directly. This time modems are not needed. The PSPDN can be accessed by the *Packet Assembler/Disassembler* (PAD) block or through a *Packet Handler* (PH) which is located inside the PSPDN. The type of the applied block depends on the protocol which is supported by a mobile terminal (DTE+MS). If the transmission from the mobile terminal is possible in accordance with the X.32 packet protocol, then the PH block is applied. If a mobile terminal does not support a packet protocol, the PAD block has to be used.

The GSM standards normalize the rules of operation of a data link with *Circuit-Switched Public Data Networks* (CSPDN) and of the fax transmission as well; however, we will skip their descriptions. The reader who is particularly interested in this subject is advised to study the GSM standards or a book such as [3].

9.4 TRANSMISSION OF SHORT MESSAGES – SMS

Short Message Service (SMS) has become a very popular service offered by GSM operators. It is particularly attractive to young subscribers due to its low price. From the point of view of the service type it is classified as a teleservice. The SMS is a kind of paging with acknowledgement of message delivery. If the SMS recipient is outside the network range or its mobile station is switched off, the system will store the message for him/her and will deliver it as soon as the MS reappears in the network. This is the

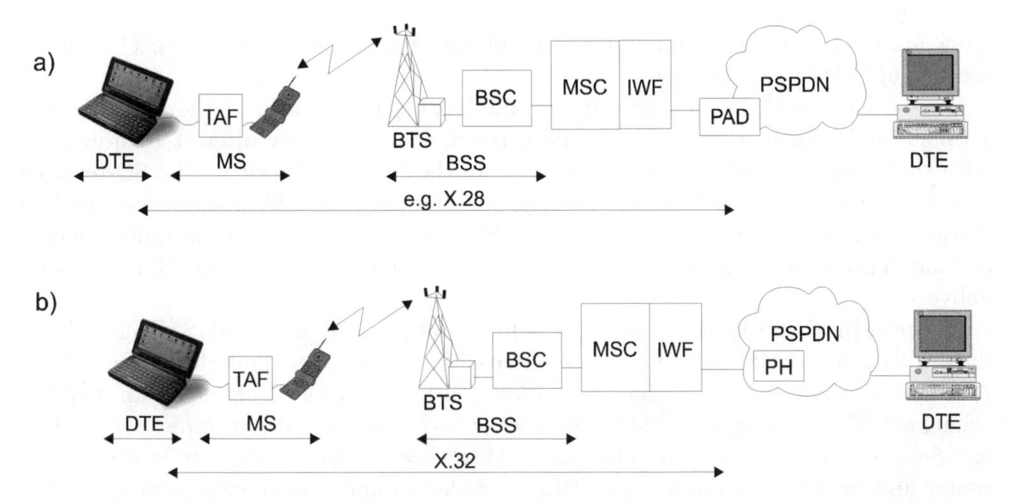

Figure 9.7 Data transmission using GSM and PSPDN: through PAD (a), applying packet handler (PH) (b)

so called *store-and-forward* mechanism. The SMS can be delivered in parallel to the ongoing speech transmission. An individual message is limited to 160 characters which are mapped into 140 bytes thanks to applying the special alphabet.

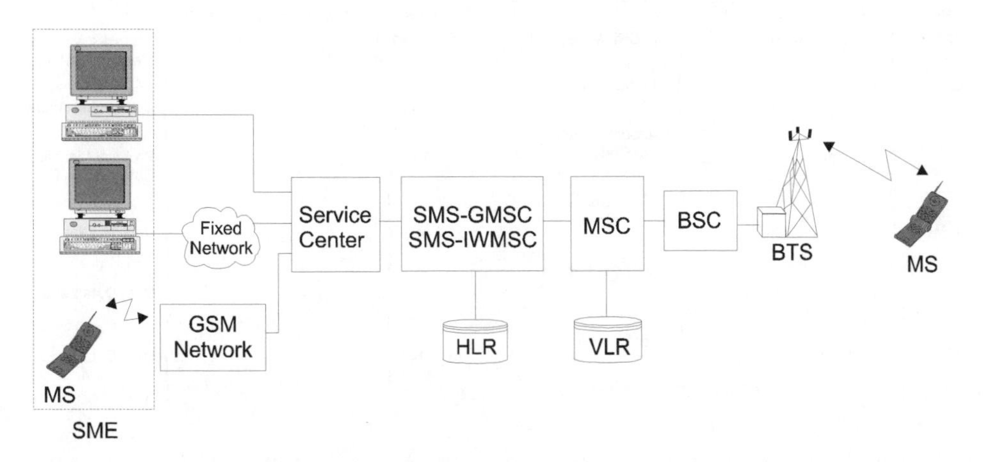

Figure 9.8 Network blocks taking part in the SMS service

Figure 9.8 presents the network architecture for the SMS service. The network element which is the source of the SMS message is called *Short Message Entity* (SME). It can be a mobile station registered in the same or different network, a computer directly connected to the SMS *Service Center* (SC) (owned by the network operator) serving as the access point to the SMS service, a computer delivering updated news (e.g. current stock market prices) or a computer located in the fixed network (e.g. Internet) allowing sending of SMS messages as well.

The service center is the most important block participating in the SMS service. Formally it is not a part of the mobile network; however, very often it is integrated with the mobile switching center. The tasks of the SC are following. The SC receives messages from SME and forwards them to mobile recipients. If the addressed mobile station is not available at the moment, the SC stores the message for a limited period of time. The SC informs the sender about the successful or unsuccessful SMS message delivery.

The next block in Figure 9.8 denotes the functions associated with SMS which are performed within the MSC. For mobile-terminated SMS messages the SMS-GMSC (*Short Message Service-Gateway Mobile Switching Center*) is applied, whereas for mobile-originated SMS messages the SMS-IWMSC (*Short Message Service-Interworking Mobile Switching Center*) is used. The SMS-GMSC receives the messages from the service center and investigates the network HLR in order to find the mobile recipient of the SMS message. After finding in the HLR the appropriate MSC serving the location area in which the called mobile station is currently registered, it directs the message to this MSC. The MSC delivers the message to the addressed mobile station. The SMS-IWMSC is a function which allows its own service center (see Figure 9.8) to receive SMS messages from the MSC which then can be transferred to that service center to which the receiving mobile station is assigned.

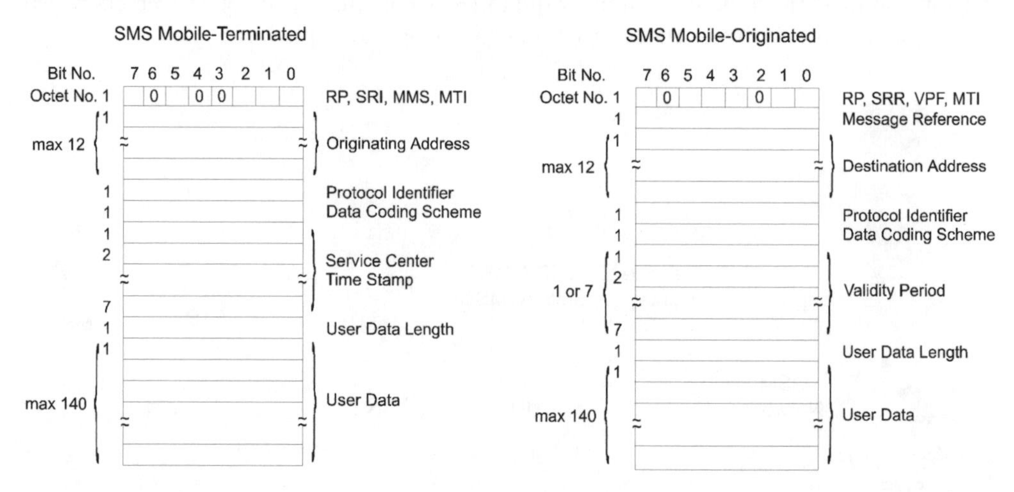

Figure 9.9 Frame structures for mobile-terminated and originated SMS messages (RP - Reply Path, SRI - Status Report Indication, MMS - More-Message-to-Send, MTI - Message Type Indicator, SRR - Status Report Request, VPF - Validity Period Format)

Figure 9.9 presents frame structures for mobile-terminated and mobile-originated SMS messages.

Let us briefly describe the frame used in mobile terminated SMS messages. The first byte contains some useful flags and a three-bit zero pattern indicating that the message is mobile-terminated. The *Reply Path* (RP) flag indicates that the answer to the received message will be paid by the message sender. The *Status Report Indication* (SRI) indicates that the message sender wishes to receive acknowledgment of his/her message delivery. The *More-Message-to-Send* (MMS) flag indicates that a certain number of SMS messages is waiting for delivery to the receiving MS. Finally, the *Message Type Indicator* (MTI) flag describes the type of message contained in the frame. Other fields of the frame are more or less self-explanatory. The *Originating Address* field normally contains the telephone number of the calling station. The *Protocol Identifier* shows the type of protocol which has to be used in order to deliver the SMS message. It is particularly important when the message has to be sent to the recipient located in another type of network. The *Data Coding Scheme* informs what kind of data representation has been applied in the user data. A standard alphabet needs 7 bits for a character representation. The *Service Center Time Stamp* indicates the time at which the SMS message arrived at the service center. The *User Data Length* indicates the length of the user data, counted in the number of characters. The *User Data* contains up to 160 characters (no more than 140 bytes if the 7-bit character representation is applied).

Similar parameters and flags are contained in the frame used by SMS mobile-originated messages. The first byte again contains some useful flags and a two-bit zero pattern indicating that the message is mobile-originated. The *Status Report Request* flag states whether the SMS originating station wishes to receive acknowledgement of the message delivery. The *Validity Period* flag indicates if the validity period is contained in the transmitted frame. The *Destination Address* is the telephone number of the receiving station. The *Validity Period* indicates the validity of the sent message. It determines how long the service center should try to send the SMS message to the recipient.

Let us trace now the SMS message on its way from the originator to the receiving mobile station. Let the SMS message be generated in one of the possible message sources (SME), e.g. a mobile station or a specialized computer. The message reaches the service center where the validity period, priority and the number of messages for the same recipient are checked. The time stamp is also added. The service center informs the SMS-GMSC functional block that there is a message for a particular mobile station. The SMS-GMSC finds in the HLR the MSC serving the area in which the desired mobile station is currently located. Is subsequently sends the message to this MSC. The appropriate MSC checks the location area of the mobile station in its VLR and the base stations belonging to this area page the desired mobile station. The base station in whose area the mobile station has been found establishes a channel and sends the SMS message to the mobile station.

The kind of channel established for SMS message delivery depends on the current status of the mobile station. If the mobile station is outside the network area or if it is switched off, the message cannot be delivered and attempts to send it till the validity period expires. If the mobile station is in the idle state, it monitors the common control channels. Therefore, the SMS is sent using a stand-alone dedicated control channel (SDCCH) which allows to transmit 184 bit during 240 ms. If the mobile station is in the active mode (the speech or data transmission is taking place) the slow associated control channel (SACCH) can be applied and the SMS data rate is about two times lower than in the previous case.

This short description of the SMS does not present all details of this service. More detailed considerations on the SMS can be found in [3]. The highly motivated reader is asked to study ETSI standards [6] and [7].

9.5 HIGH-SPEED CIRCUIT-SWITCHED DATA SERVICE - HSCSD

Although the number of different data transmission services in the GSM is quite high, their main drawback is a slow data rate. The data rate is mostly insufficient for Internet applications and the connections can be very expensive due to their duration. Therefore, a lot of effort has been made to improve data transmission capabilities of the GSM system. Generally, two solutions have been found:

- increasing the data rate with the minimum system modifications – this is realized by the assignment of several time slots to a single connection, preserving the circuit-switched communications so, as the result, the link capacity is multiplied,
- better usage of the system resources by application of the packet-switched communications with the possibility of assigning several time slots in a frame for transmission of a data packet. As a result, substantial system modifications are required.

In this section we will consider the first, more traditional approach called *High-Speed Circuit-Switched Data Service* (HSCSD). Its detailed description can be found in [8] and [9].

As we have mentioned, the main idea exploited by HSCSD is the co-allocation of multiple full-rate channels (TCH/F) in a single HSCSD link. Theoretically, if all 8 time slots were assigned to a single connection, then using a full-rate traffic channel TCH/F9.6 with 9.6 kbit/s data rate we would achieve the user data rate of 8×9.6 kbit/s $= 76.8$ kbit/s. In practice, the maximum data rate of 57.6 kbit/s is received if 4 slots using TCH/F14.4 channels are assigned to the connection. A typical data rate is $38.4 = 4 \times 9.6$ kbit/s, if 4 TCH/F9.6 channels are used. All applied channels follow the common frequency hopping procedure, the same training sequence in the midamble, the same channel coding, interleaving and rate adaptation; however, they have their own enciphering keys derived from the same K_c key contained in the SIM. A slow associated control channel (SACCH) is assigned to each composite channel. Each HSCSD link uses a single fast associated control channel (FACCH) denoted as *Main HSCSD Subchannel* (MHCH). Logically, all the traffic channels applied in the HSCSD radio interface belong to the same HSCSD configuration and they are treated by the network as one radio link. This is important for all performed cellular operations, e.g. handover.

Figure 9.10 presents the architecture of the HSCSD. The main modification of the system is made in the TAF block on the mobile system side and in the IWF block in the MSC. Both blocks perform the combining and splitting of n composite data streams ($n = 1, 2, \ldots, 8$) transmitted in different traffic channels. Some modifications in the

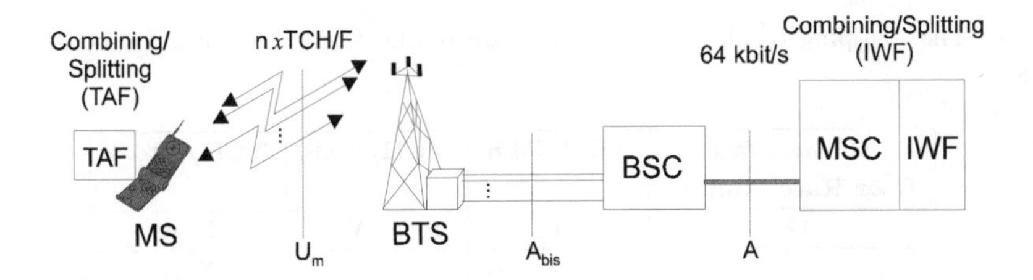

Figure 9.10 Architecture of the HSCSD

base station system (BSS) are also necessary due to joint handling of traffic channels involved in the single HSCSD link. In the A interface the composite data streams should be multiplexed in one 64 kbit/s circuit. Figure 9.11 illustrates this process.

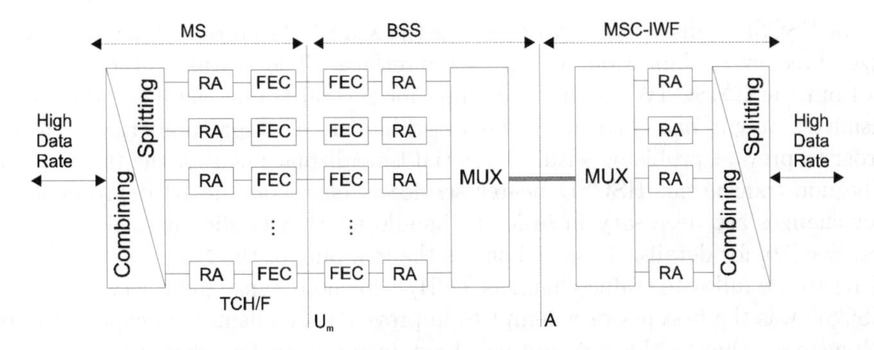

Figure 9.11 The process of splitting, combining and multiplexing in the HSCSD transmission

When the HSCSD connection is set up, a user's mobile station indicates the maximum number of traffic channels (*Desired Number of Channels* - DNC) and the minimum number of channels (*Required Number of Channels* - RNC) needed for the realization of the selected service, the acceptable channel coding, the allowable type of the remote modem and the values of the fixed network user data rate. If the connection is non-transparent, the desired radio interface user rate is indicated. The user can modify his/her needs during the connection if such an option has been negotiated during the connection set-up. All selected parameters constitute a HSCSD characteristic which is applied by the network to allocate the network resources to the HSCSD connection.

In the HSCSD service both transparent and non-transparent connections can be established. A connection can have a symmetric or an asymmetric configuration. For a symmetric configuration, the same number of traffic channels is assigned to the downlink and the uplink, whereas in an asymmetric configuration, the downlink uses higher number of the traffic channels than the uplink. The asymmetric configuration is selected if the requirements of the user cannot be met in the symmetric configuration. Above all, the network attempts to fulfill the data rate requirements in the downlink.

Table 9.4 The mapping of air interface user rate to the traffic channels for non-transparent services

Air Interface User Rate [kbit/s]	TCH/F4.8	TCH/F9.6	TCH/F14.4
4.8	1	N/A	N/A
9.6	2	1	N/A
14.4	3	N/A	1
19.2	4	2	N/A
28.8	N/A	3	2
38.4	N/A	4	N/A
43.2	N/A	N/A	3
57.6	N/A	N/A	4

In the HSCSD transparent bearer services the X.30/V.110 protocols are applied which realize three-level adaptation to the user interface. The status of the modem V.24 lines from the GMSC-IWF is now common for several traffic channels, therefore it is transmitted only in one of them. The released bits are used for numbering the channels in order to prevent problems with a potential time displacement of the traffic channels.

The non-transparent HSCSD bearer service is based on the RLP protocol. Some minor changes are necessary in order to handle up to 8 traffic channels by the RLP block. See [10] for details. Table 9.4 shows the mapping of the *Air Interface User Rate* (AIUR) to the full-rate traffic channels TCH/F for non-transparent services [8].

HSCSD was the first major attempt to improve data transmission capabilities of the GSM system. Due to the assumption about introducing few changes in the system, the HSCSD remains a circuit-switched system, as the standard GSM is. Therefore, it is not well suited to the packet traffic which is characteristic for data transmission. The reservation of more than one time slot during the whole data exchange session (unless the dynamic system resources have been negotiated) increases the probability of blocking in a cell. Charging for the HSCSD service is determined by the number of the used channels and the duration of the session.

9.6 GENERAL PACKET RADIO SERVICE - GPRS

General Packet Radio Service (GPRS) is a major improvement and extension of the standard GSM system. The reasons for its development were manifold. Data transmission rates in the existing mobile networks were insufficient and the connection set-up time was long. The circuit-switched transmission was not well fitted to the bursty and asymmetric nature of the data traffic, therefore the existing system resources were used inefficiently. As a result, it was decided that packet-switched transmission should be applied. Thus, the system users can share the same physical channels and the system resources can be used much more efficiently due to *statistical multiplexing*. The consequence of packet-switching is tariffing based on the number of transmitted packets.

The session can last for a long time but the user pays for the transmitted data volume only.

9.6.1 GPRS system architecture

In a sense, introducing GPRS results in a network parallel to the GSM network. The network uses a lot of GSM resources; however, the main network elements are connected with each other by a separate backbone network, which is based on the IP protocol. Figure 9.12 presents the basic GPRS architecture. User data and signaling data flowing through the main network are denoted by solid lines whereas signaling data transmitted through the signaling network are shown as dashed lines. The main novel elements in the GPRS network are *GPRS Support Nodes* (GSN). They deliver data packets and determine their route between a mobile station and external packet data networks.

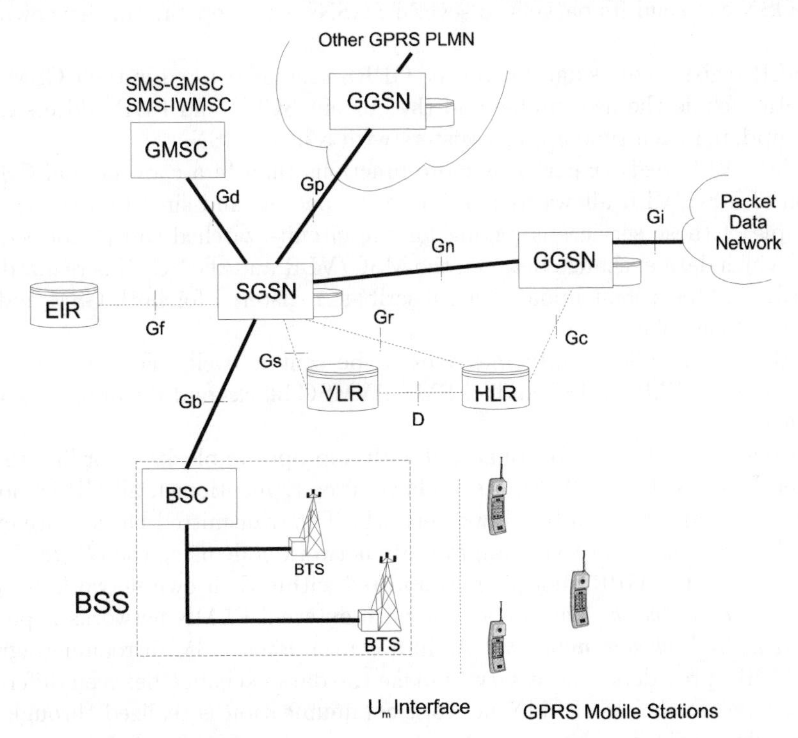

Figure 9.12 GPRS system architecture

The *Serving GPRS Support Node* (SGSN) is responsible for delivery and reception of packets to and from mobile stations in the SGSN service area. It works in the similar way as a mobile switching center in a regular GSM system. It determines the route of transmitted packets and transfers them to appropriate nodes. It also manages MS mobility and is responsible for logical link management. The SGSN performs MS authentication functions and stores all information about all the GPRS subscribers registered in this SGSN in the location register. Such information contains the index of the current cell, the current VLR and a subscriber profile consisting of the international mobile subscriber identity (IMSI) and the address used in the packet network.

The *Gateway GPRS Support Node* (GGSN) constitutes an interface between the backbone GPRS network and external packet data networks. This node performs the conversion of the GPRS packets arriving from SGSN to the appropriate format of the *Packet Data Protocol* (PDP), depending on the type of the interfacing network. As a result, the converted packets are sent out to the destination network. The processing of packets which arrive from external networks relies on the conversion of the address from the PDP format to the GSM format and sending the processed packet to the appropriate SGSN. In order to perform this task, GGSN recalls the stored subscriber profile and the currently serving SGSN from its location register. GPRS ensures interface with several external packet networks, therefore different GGSNs are possible. On the other hand, a single GGSN can send its packets to several SGSNs which operate in their own service areas.

The HLR register stores data on all the GPRS users registered in their GSM system. These data include the user profile and the current SGSN and PDP addresses. These data are updated each time a user registers with a new SGSN.

The MSC/VLR register performs more functions than in a conventional GSM. The operation of MSC/VLR allows to coordinate the packet- and circuit-switched services. An example of these services is paging for the circuit-switched call performed by the SGSN in which data exchange between the MSC/VLR and the SGSN is realized. In the MSC/VLR register a joint update for subscribers registered for both GSM and GPRS services also takes place.

A GPRS system allows SMS messages to be sent as well. For that reason data exchange between SMS-GMSC and/or SMS-IWMSC blocks and the appropriate SGSN takes place.

All GPRS system blocks are connected with appropriate blocks according to the defined interfaces (see Figure 9.12). As we have already mentioned, all GPRS nodes are connected using an IP-based backbone network. The transmitted packets are encapsulated by the GSNs and sent to the appropriate network node using the *GPRS Tunneling Protocol* (GTP). The GPRS nodes are connected within their own network using *intra-PLMN backbone network.* The connection with external PLMN networks is performed via *inter-PLMN backbone networks.* As in a conventional GSM, a roaming agreement between GPRS providers is necessary to make the data exchange between different networks possible. Such inter-PLMN network communication is realized through *Border Gateways* which protect their own networks against the unauthorized use.

9.6.2 GPRS physical layer

In a GPRS system the physical layer is very similar to that in a standard GSM; however, packet transmission and asymmetric traffic require some changes and extensions in this layer. First of all, the GPRS system allows for *multislot operation,* in which up to 8 slots in the frame can be used by a single mobile station. The consequence of the asymmetric packet traffic is a separate resource assignment to uplink and downlink. The channel allocation is performed only for the duration of packet transmission or reception. Other mobile stations can use the same channel between subsequent packet transmissions. Thus, the rule of *capacity on demand* is realized. It means that the number of assigned channels is a function of the traffic load, service priorities and the class of the multislot operation. The channels can be shared by the standard GSM and GPRS.

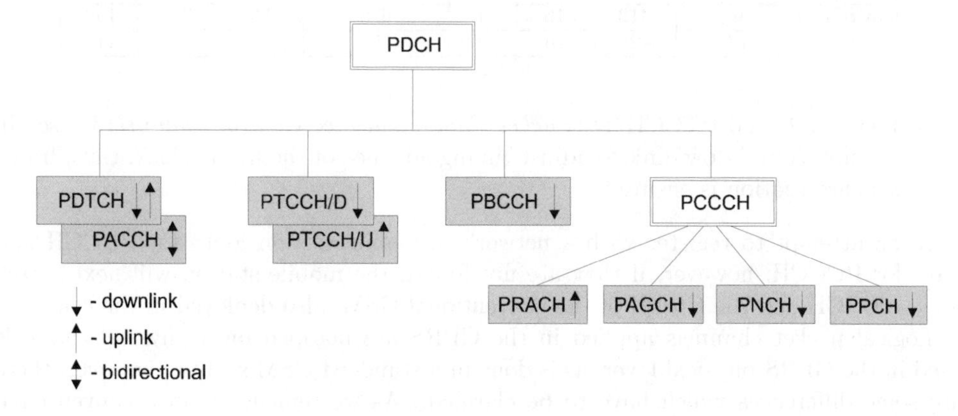

Figure 9.13 Logical channels in GPRS

A basic physical GPRS channel is called a *Packet Data Channel* (PDCH). Several logical channels are used in the GPRS and their organization is shown in Figure 9.13. We describe their functions below.

- PDTCH (*Packet Data Traffic Channel*) is applied to transport user data; one or more PDTCHs can be allocated to a GPRS terminal,
- PBCCH (*Packet Broadcast Control Channel*) is used by a base station system to inform all mobile stations located in the cells about the organization of GPRS and GSM,
- PCCCH (*Packet Common Control Channel*) is a common control channel which is further divided into the following channels:
  - PRACH (*Packet Random Access Channel*) used by a mobile station to request one or more PDTCHs,
  - PAGCH (*Packet Access Grant Channel*) applied to confirm the allocation of one or more PDTCHs to the mobile station,
  - PPCH (*Packet Paging Channel*) used by the base station to page the desired mobile station (MS) and find the cell in which the MS is currently located,
  - PNCH (*Packet Notification Channel*) used to inform the mobile station about messages in multicast or group calls,
- PACCH (*Packet Associated Control Channel*) is used to send signaling information associated with one or more PDTCHs used by a mobile station; it is bidirectional,

Table 9.5 Basic parameters of GPRS coding schemes

Coding scheme	USF precoding	RLC block	CRC	Tail bits	Conv. code output	Punc. bits	Code rate	Data rate
CS-1	3	181	40	4	456	0	1/2	9.05
CS-2	6	268	16	4	588	132	2/3	13.4
CS-3	6	312	16	4	676	220	3/4	15.6
CS-4	12	428	16	0	456	0	1	21.4

- PTCCH/U and PTCCH/D (*Packet Timing advance Control Channel*) is used in the uplink and downlink to adjust timing advance of the frame clock, thus, frame synchronization is ensured.

In an attempt to register with a network, a mobile station looks for PBCCH and later for PCCCH; however, if they are not found, the mobile station will next search for the BCCH and the CCCH of the conventional GSM, also deployed in the cell.

Logical packet channels applied in the GPRS are mapped onto physical channels used in the GPRS physical layer, as is done in a standard GSM system; however, there are some differences which have to be clarified. As we remember, in a conventional GSM system, the mutiframe consists of 26 frames for traffic channels or 51 frames for signaling channels. In the GPRS the multiframe is built of 52 frames. The structure of a GPRS multiframe is shown in Figure 9.14. Four subsequent frames constitute a block. A block is a unit from the point of view of the applied channel coding. There are 12 blocks in the multiframe (B0–B11). Two frames are idle, two other frames are used to update the frame timing advance. As we remember, a single GSM normal burst carries 114 bits of user data. Therefore, one block consisting of 4 frames contains $4 \times 114 = 456$ bits and each coding scheme applied in GPRS results in a 456-bit block.

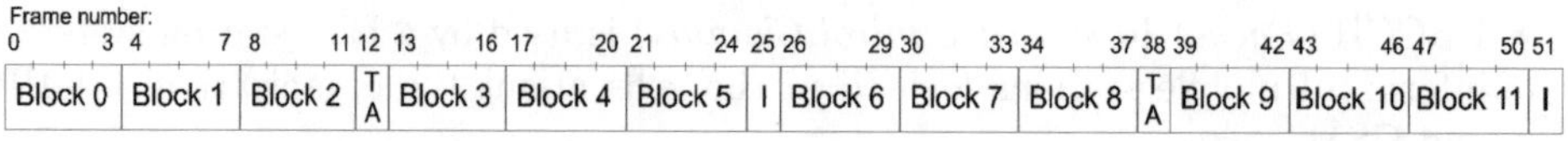

Frame number:
0 3 4 7 8 11 12 13 16 17 20 21 24 25 26 29 30 33 34 37 38 39 42 43 46 47 50 51

Block 0	Block 1	Block 2	TA	Block 3	Block 4	Block 5	I	Block 6	Block 7	Block 8	TA	Block 9	Block 10	Block 11	I

Figure 9.14 GPRS multiframe structure (TA - *Timing Advance* frame, I - *Idle* frame)

Two-level channel coding is generally applied in the GPRS. The outer code is a block code. Some tail bits are usually added at the end of the code word. In turn, this code word is convolutionally coded and some bits can be punctured in order to the code length of 456 bits. Table 9.5 presents the basic data concerning four channel coding schemes applied in the GPRS. The USF (*Uplink State Flag*) is applied in the downlink to inform the MS if the uplink channel is free.

The coding schemes CS-1–CS-4 are fitted to different channel conditions. CS-1 to CS-4 are applied in traffic packet data channels, whereas CS-1 is also used to protect signaling channels (except PRACH). The choice of the coding scheme depends on the channel conditions and service requirements. Therefore, CS-1, a strong coding scheme, is used in poor channel conditions, resulting in 9.05 kbit/s per slot. On the other extreme, CS-4 can be used in very good channel conditions. No convolutional coding is applied in this case. Since the resulting data rate is 21.4 kbit/s, therefore, at the assignment of 8 slots to a user, a 171.2 kbit/s data rate is achievable. Figure 9.15 presents graphically all four coding schemes.

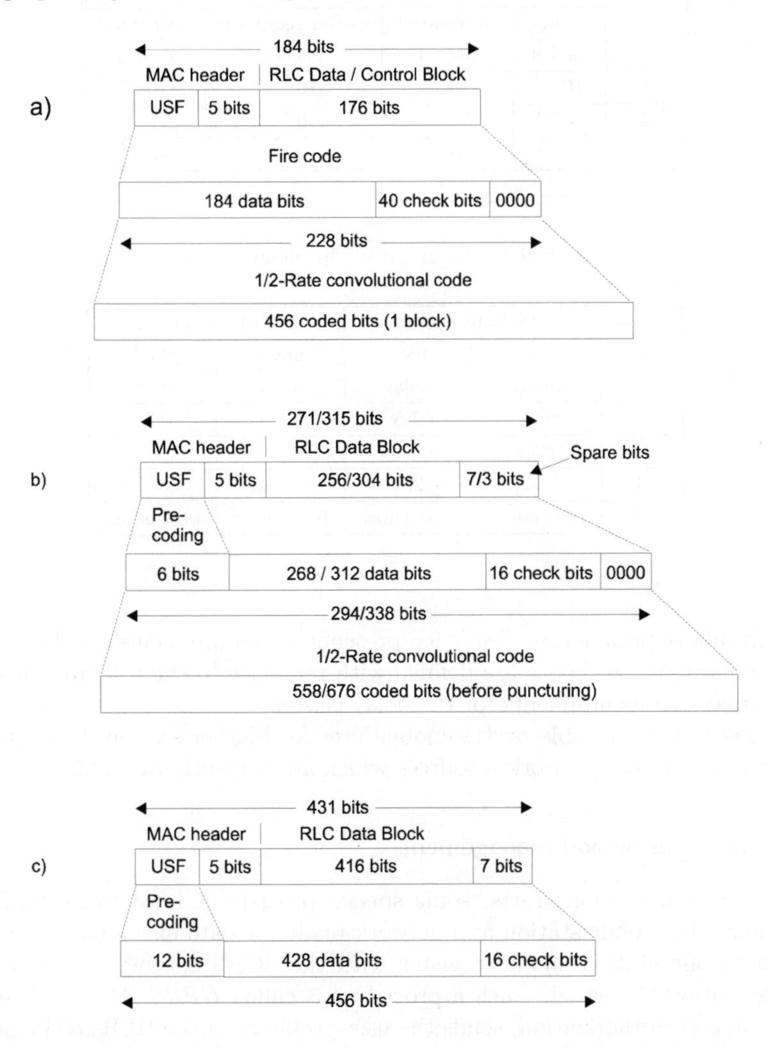

Figure 9.15 GPRS coding schemes: CS-1 (a), CS-2/CS-3 (b), CS-4 (c)

Different coding schemes and data rates ensure a whole set of different services characterized by appropriate *Quality of Service* (QoS). Quality of service requirements strongly depend on the applications. In the GPRS system QoS profiles have been defined using such parameters as priority of the service, reliability, delay and throughput. The throughput specifies the maximum and mean bit rates. Depending on the kind of service, various probability of packet loss, packet duplication and packet corruption can be tolerated. Three reliability classes have been defined, characterized by specific

Table 9.6 Requirements for reliable classes in GPRS packet services

Class	Probability of			
	lost packet	duplicated packet	out of sequence packet	corrupted packet
1	10^{-9}	10^{-9}	10^{-9}	10^{-9}
2	10^{-4}	10^{-5}	10^{-5}	10^{-6}
3	10^{-2}	10^{-5}	10^{-5}	10^{-2}

Table 9.7 Requirements for delay classes

Class	128-byte packet		1024-byte packet	
	mean delay	95% delay	mean delay	95% delay
1	<0.5 s	<1.5 s	<2 s	<7 s
2	<5 s	<25 s	<15 s	<75 s
3	<50 s	<250 s	<75 s	<375 s
4	best effort	best effort	best effort	best effort

values of the above parameters. Table 9.6 presents the requirements for these reliability classes. Similarly, three classes are defined with reference to the tolerable delay. Table 9.7 summarizes the requirements for the delay classes.

QoS profiles are negotiable by the mobile user for his/her session depending on the desired services and the network resources which are currently available.

9.6.3 GPRS transmission management

Before GPRS transmission starts, some special procedures have to be performed in order to make the mobile station and network ready for data exchange.

First, a mobile station should register with the SGSN serving the area in which the mobile station is located. Such a procedure is called *GPRS Attach.* The network performs the user authorization, sends the user profile from the HLR to the SGSN and assigns a *Packet Temporary Mobile Subscriber Identity* to the user. For some MS classes the combined GSM/GPRS registration can be performed.

Before performing GPRS Attach procedure, the MS is not visible in the GPRS network and is in the *idle* state. Due to the network attachment, the MS switches to the *ready* state in which it sends information to the SGSN after every movement to a new cell. As a result, the location of the MS in the ready state is known with the accuracy of a single cell. In the ready state the MS is able to send and receive packets if an appropriate initiation of the data exchange is done (see the next paragraph). If the mobile station does not send or receive packets for some time it goes to the *standby* state. The location of the MS in the standby state is traced with accuracy of a cell group, called *Routing Area* (RA). In order to send a packet to the MS in this state, paging must be performed in order to find the cell in which the MS is located.

In order to start data packet exchange with a packet data network, such as an IP or X.25 network, the MS applies for the address used in this data network called *Packet Data Protocol (PDP) address*. Next, a PDP context is created for each session. This context consists of the PDP type, PDP address of the mobile station, the requested QoS and the address of the Gateway SGN (GGSN) connecting the GPRS system with the appropriate data network. The context is stored in several units such as the mobile station, the SGSN and the appropriate GGSN. This way the mobile station is visible to the external data network.

Let us consider the communication of the MS with the IP network. The MS sends packets to the destination host through the GPRS and external networks. The MS transmits packets through the base station system to the appropriate SGSN. The SGSN checks the PDP context of the MS, encapsulates the IP packets received from the MS and sends them through the IP backbone network to the gateway GSN (GGSN) connecting the GPRS system with the desired data network. The GGSN decapsulates the packets and sends them further to the data network which overtakes transmission of the packets to the destination host.

9.6.4 GPRS services

Data transfer in a GPRS system is performed within the selected bearer or supplementary services. The bearer services can be divided into two categories:

- the *Point-to-Point* (PTP) service – a connection between two individual users, which can be realized in a connectionless mode (using the IP network) or a connection-oriented mode (using the X.25 network), and
- the *Point-to-Multipoint* (PTM) service – a connection between one user and a specified number of other users. The users can be selected by their location in a specified area (*multicast service*) or can be addressed according to a specified list (*group service*).

SMS is another service offered by the GPRS system. Some other complementary and non-standard services are also planned.

9.6.5 GPRS protocol architecture

Figure 9.16 presents the GPRS protocol stack in the transmission plane for the mobile station, the base station system, the serving GPRS support node (SGSN) and the gateway GPRS supporting node (GGSN), through which the data is transferred to and from the external packet data network. The interfaces between the network blocks are also shown in Figure 9.16.

Let us briefly describe these layers starting from the MS side. Signal transmission, including modulation and demodulation, is performed within the *RF Physical Layer* (RFL). Channel coding, interleaving and detection of physical link congestion are re-

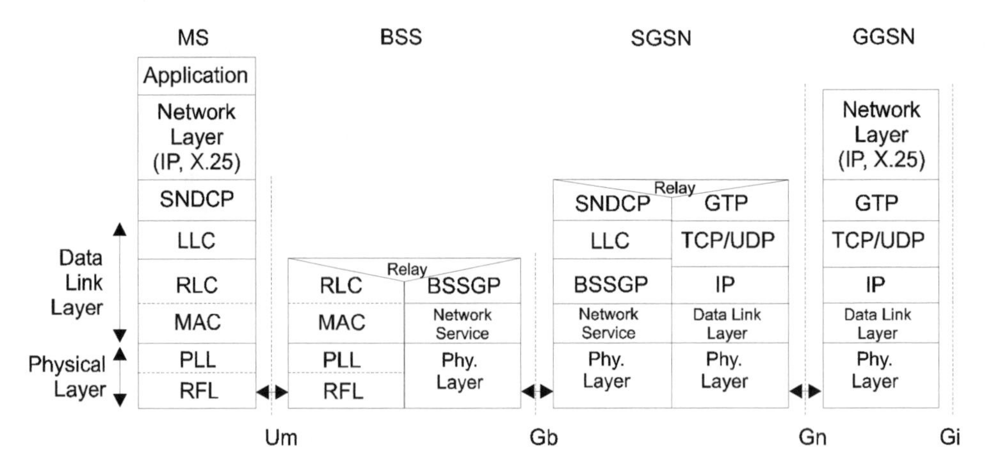

Figure 9.16 GPRS protocol stack in the transmission plane; Acronyms: SNDCP - Subnetwork Dependent Convergence Protocol, LLC - Logical Link Control, RLC - Radio Link Control, MAC - Medium Access Control, PLL - Physical Link Layer, RFL - Physical RF Layer, BSSGP - BSS GPRS Application Protocol, GTP GPRS Tunneling Protocol, TCP - Transmission Control Protocol, UDP - User Datagram Protocol, IP - Internet Protocol

alized within the *Physical Link Layer* (PLL). The access of the MS to the system resources is controlled within the *Medium Access Control* (MAC) layer. As in the GSM, the GPRS MAC algorithm is based on the *slotted Aloha* principle. A reliable link between the MS and the base station system is established within the *Radio Link Control* (RLC) layer. The most important tasks performed in this layer are the segmentation of the frames created in the *Logical Link Control* (LLC) layer into RLC data blocks and performing the ARQ for uncorrectable codewords. The operation of the LLC layer is based on the *High-Level Data-Link Control* (HDLC) protocol. The sequence control, appropriate ordering of packets, flow control, error detection, retransmission and enciphering are performed within this layer. The *Subnetwork Dependent Convergence Protocol* (SNDCP) manages the data transfer between the mobile station and the appropriate GGSN. The protocol multiplexes several possible connections of the network layer into a single logical connection of the LLC layer and compresses and decompresses user data and headers. The procedures of the network layer result from the connection realized by the MS with the appropriate external data network. Finally, the user tasks (e-mail, web browsing, ftp, etc.) are realized in the application layer.

As seen in Figure 9.16, many layer protocols described so far also occur in the protocol stacks of the other network elements such as BSS, SGSN and GGSN. *BSS GPRS Application Protocol* (BSSGP) is a new layer of the BSS protocol stack, which manages routing and information at the required QoS between BSS and SGSN. The *Network Service Protocol* performs the transfer of data or signaling units, network congestion indication and status indication. The Network Service protocol is based on the *Frame Relay* protocol.

The data transfer between the SGSN and the GGSN is done through the backbone IP-network. For this purpose appropriate layers in the SGSN-GGSN interface have been designed. On the bottom of the protocol stack, the physical and *Data Link* layers realize Ethernet, ISDN or ATM transmission. The *Internet Protocol* (IP) is applied in the IP layer. Above this layer, *Transmission Control Protocol* (TCP) or *User Datagram Protocol* (UDP) are applied depending on the kind of network applied (X.25 or IP, respectively). Finally, the *GPRS Tunneling Protocol* (GTP) is applied to transfer the user data packet between SGSN and GGSN.

Protocols can also be considered in the signaling plane. The interested reader is advised to study the tutorial paper [11].

* * *

The GPRS can be considered a major improvement of the GSM system, which takes into account the increasing demand for data transmission over mobile networks. The system has been optimized for packet data transmission occurring e.g. in data transmission between individual users and the Internet. As a result of using the GSM radio infrastructure and applying new specialized blocks such as SGSNs and GGSNs, a new characteristic has been achieved. It is well fitted to the bursty nature of data transmission. It is worth noting that tariffing based on the number of transmitted packets has replaced tariffing based on the duration of the connection used in the regular channel-switched GSM system.

9.7 EDGE - ENHANCED DATA RATE FOR GLOBAL EVOLUTION

GPRS allows transmission of bursty data in the packet switched mode using more than one time slot in a frame, if it is desired and possible from the system resources point of view. However, data rates are still moderate as compared with the wireline Internet connections. EDGE is the response to the demand for higher data rates. Originally, EDGE was treated as an extension of the GPRS system. Now this acronym has a manifold meaning. For GSM, EDGE is the evolution of GPRS and HSCSD to EGPRS (*Enhanced GPRS*) and ECSD (*Enhanced Circuit Switched Data*), respectively. In the USA, EDGE is the basis for extension of the TDMA personal mobile communication system IS-136, resulting in the IS-136 HS (High Speed) outdoor component.

EDGE for TDMA is currently realized in two modes, *EDGE Compact* and *EDGE Classic* [15], [16]. EDGE Compact employs a new 200 kHz control-channel structure. The base stations are synchronized and the system can be implemented in a minimum spectrum bandwidth of 1 MHz. The packet traffic channels are realized in the 1/3 frequency-reuse pattern. EDGE Classic employs the traditional 200 kHz control-channel structure typical for the GSM. The carriers containing the broadcast control channel are used with a typical 4/12 frequency-reuse pattern (see Chapter 5). The minimum deployment of EDGE Classic requires 12 carriers or 2.4 MHz with guard bands.

Below we concentrate on the European version of EDGE.

9.7.1 Main improvements in the physical layer

EDGE for GSM features some improvements which allow transmission of data packets at much higher rates as compared with a regular GSM or GPRS. The first improvement is the application of 8-PSK modulation for higher data rate modes. At lower data rates GMSK is still used. By retaining the symbol rate of 270.833 kbit/s and the 200 kHz frequency raster of a regular GSM, three times higher data rate is ensured if 8-PSK is used, because three bits are transmitted per single data symbol. The 8-PSK signal applied in EDGE can be described by the equation

$$y(t) = \mathrm{Re}\left\{\sum_k b(k)\exp\left[j\frac{3\pi}{8}k\right]p(t-kT)\exp\left[j2\pi f_c t\right]\right\} \tag{9.3}$$

where the data symbols belong to the set

$$b(k) \in \left\{\exp\left[j\frac{2\pi}{8}l\right]\right\}_{l=0}^{7} \tag{9.4}$$

The index l in the data symbol $b(k)$ is determined by the current three-bit block. Bit block-to-symbol mapping is designed according to the Gray rule. The variable k is the time index of a data symbol. Let us note that, besides the phase shift controlled by the information bits, the phase is additionally shifted at each symbol period by $3\pi/8$. Thanks to this, the low levels of signal envelope are avoided and the peak-to-average power ratio equals only 3.2 dB. In order to fit to the channel spectral width and retain the shape of the GMSK spectrum, the baseband pulse $p(t)$ is applied. It is a linearized version of the GMSK pulse shape calculated according to the Laurent method [12]. The pulse shape shown in Figure 9.17 is similar to the Gaussian curve and is calculated numerically. As we remember from Chapter 8, for GMSK modulation the Viterbi maximum-likelihood sequence estimation is usually applied at the receiver. In case of 8-PSK the number of states in the Viterbi algorithm would be excessive, therefore, a selected suboptimum sequential algorithm has to be implemented.

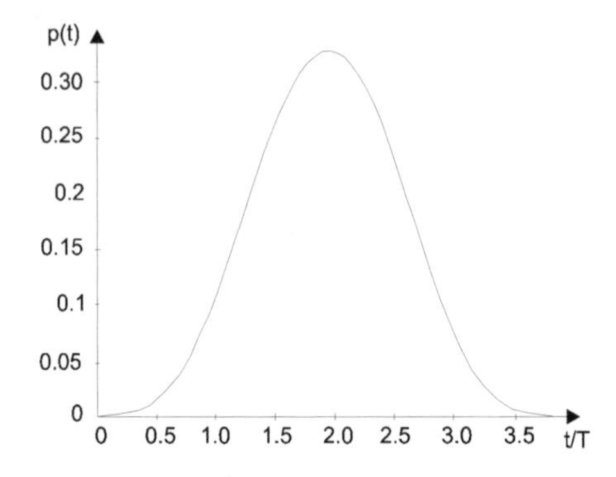

Figure 9.17 Impulse response of the EDGE pulse shaping filter used in 8-PSK modulator

Another feature of EDGE is slow frequency hopping. This feature is optional in a standard GSM system. Frequency hopping can be interpreted as a form of frequency diversity which fights against channel fading. It also has a substantial influence on the level of co-channel interference, and consequently on the overall system capacity. Due to frequency hopping each encoded block is transmitted on four different carriers.

The next important feature of EDGE is link quality control. Mobile stations inform the base station about the channel quality. On the basis on this information a decision is drawn as to which combination of modulation and channel coding should be used. In the EDGE system two modulations (GMSK and 8-PSK) and nine coding rates can be applied. Each applicable combination has a characteristics shown as a throughput (per slot) versus SNR. If switching between the combinations of coding and modulation is possible, the throughput is maximized. Figure 9.18 illustrates this principle.

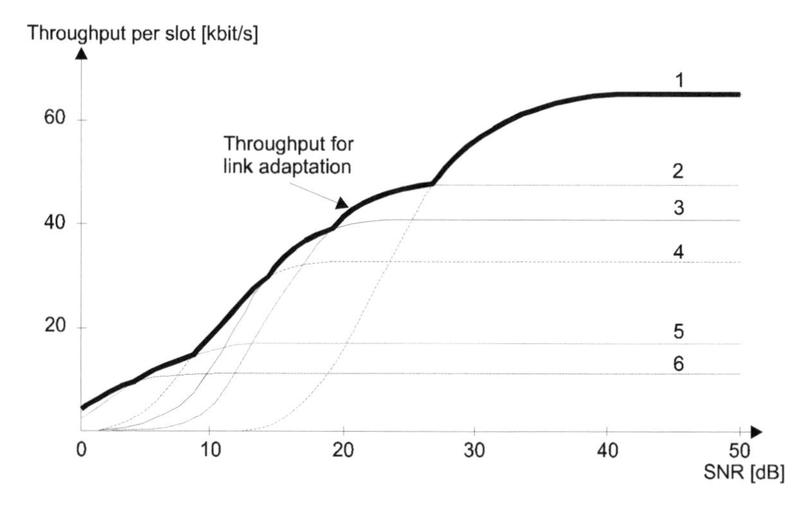

Figure 9.18 Throughput per time slot for different combinations of coding and modulation - illustration of link adaptation principle [13]

The data burst applied in EDGE has the same format as in a standard GSM system; however, if 8-PSK is used, 8-level data symbols replace binary symbols. The burst contains a 26-symbol midamble, 3 tail symbols on both ends of the burst and two portions of 58 symbols of user data. For 8-PSK it results in 348 information bits per burst as compared with 116 information bits per burst if GMSK is used. The multiframe structure used in EDGE is the same as that applied in GPRS. It consists of 52 frames (see Figure 9.14) in which 12 data blocks are placed. Every 13th frame does not carry user data and is used for timing advance correction or measurements. Thus, taking into account all the above mentioned factors, the maximum data rate per carrier (if all time slots are applied) is 556.8 kbit/s for 8-PSK and 185.6 kbit/s for GMSK. Obviously, maximum user data rates are lower due to applied channel coding. The used channel coding rates are ranged between $R = 0.38$ and $R = 1$.

Table 9.8 Modulation and coding schemes in EDGE (EGPRS)

Scheme	Modulation	Max.rate [kbit/s]	Data code rate	Header coding	PDUs/20ms	PDU size [bytes]	Family
MCS-9	8-PSK	473	1.0	0.36	2	74	A
MCS-8	8-PSK	435	0.92	0.36	2	68	A
MCS-7	8-PSK	358	0.76	0.36	2	56	B
MCS-6	8-PSK	234	0.49	1/3	1	74	A
MCS-5	8-PSK	179.2	0.37	1/3	1	56	B
MCS-4	GMSK	141	1.0	1/2	1	44	C
MCS-3	GMSK	119	0.80	1/2	1	37	A
MCS-2	GMSK	90	0.66	1/2	1	28	B
MCS-1	GMSK	70.4	0.53	1/2	1	22	C

Link Quality Control applied in EDGE and realized through adaptive selection of modulation and coding is fully justified by the information theory results. They show that in order to achieve maximum throughput, the data rate should be high when good channel quality (high SNR) is observed and low when the channel is temporarily of a low quality. We have to stress that such an adaptation is possible if feedback between receiver and transmitter exists. Table 9.8 summarizes applicable combinations of coding and modulation schemes. They are divided into three families (A, B and C). The change of the coding rate can occur exclusively within the same family.

Radio blocks are the smallest transmitted data units within the EDGE system. Each radio block contains one or two packet data units (PDU). The number of PDUs in a radio block depends on the selected modulation and coding scheme. The radio blocks are transmitted after interleaving over four frames, each of which is transmitted on a different carrier due to frequency hopping. Encoded radio blocks contain 1392 bits if 8-PSK is used or 464 bits if GMSK is applied. For modulation and coding schemes applying 8-PSK, the structure of a radio block follows a general scheme shown in Figure 9.19. The structure of uncoded and coded data blocks differs depending on the selected modulation and coding scheme; however, the general rule shown in Figure 9.19 remains valid.

In case of non-transparent data transmission, an ARQ method is typically applied. As we have explained in Chapter 1, a standard ARQ relies on appending CRC parity bits calculated in the transmitter to a data block. The CRC bits are once more calculated in the receiver on the basis of the received block. If the calculated bits are the same as the received ones, the receiver sends positive acknowledgement (ACK) and the transmitter starts to transmit the next block. In case the CRC bits calculated in the receiver differ from the received ones, the whole block is eliminated, so the information contained in the block is wasted and the whole block has to be sent again. Instead of this simple ARQ rule, the idea of *Incremental Redundancy* (IR) ARQ can be applied.

IR ARQ relies on reusing the erroneous blocks in repeated error detection. In the coding process puncturing using two or three different schemes (P1, P2 or P3) is ap-

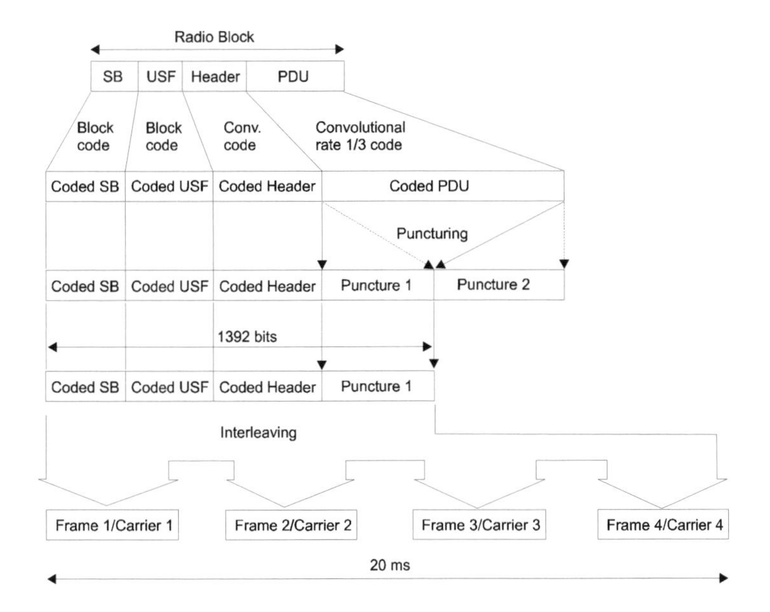

Figure 9.19 Example of a format and coding of a radio block for EDGE using 8-PSK modulation [16]

plied at the output of the convolutional encoder. Initially, the encoded block after P1 puncturing is transmitted. If errors are detected, the bits obtained in the P2 puncturing process are transmitted and the block supplemented with the previously received bits is decoded again. As we see, the number of parity (redundancy) bits has been incremented and decoding of a full block results in much better error correction. If all bits calculated according to all puncturing schemes have been received and errors have been detected again, the whole process of block transmission is repeated.

* * *

Above we briefly described selected aspects of GSM/GPRS extension known as EDGE. As we have mentioned, the EDGE concept has been intensively investigated for American TDMA cellular telephony IS-136, resulting in a substantial convergence of these two widely used technologies. In fact, due to high data rates achievable in EDGE, it is considered one of the third generation (3G) system proposals. We will see in the near future how the implementation of EDGE in the GSM network will influence the deployment of UMTS.

REFERENCES

1. J. Jayapalan, M. Burke, "Cellular Data Services Architecture and Signalling", *IEEE Personal Communications*, Second Quarter 1994, 44-55

2. ETSI EN 300 909 V. 8.3.0, Digital Cellular Telecommunications System (Phase 2+); Channel Coding (GSM 05.03, Release 1999), April 2000
3. S. M. Redl, M. K. Weber, M. W. Oliphant, *GSM and Personal Communications Handbook*, Artech House, Boston, 1998
4. ITU-T Recommendation V.110, Support by an ISDN of Data Terminal Equipments with V-Series Type Interfaces, Geneva, October 1996
5. ETSI EN 300 945 V.7.0.3, Digital Cellular Telecommunications System (Phase 2+); Rate Adaptation on the Mobile Station – Base Station System (MS-BSS) Interface (GSM 04.21), December 1999
6. ETSI TS 100 901, Digital Cellular Telecommunications System (Phase 2+); Technical Realization of the Short Message Service (SMS) (GSM 03.40, version 7.4.0, Release 1998)
7. ETSI TS 100 942, Digital Cellular Telecommunications System (Phase 2+); Point-to-Point (PP) Short Message Service (SMS) Support on Mobile Radio Interface (GSM 04.11, version 7.0.0, Release 1998)
8. ETSI TS 101 625, Digital Cellular Telecommunications System (Phase 2+); High Speed Circuit Switched Data (HSCSD); Stage 1 (GSM 02.34, Version 7.0.0, Release 1998)
9. ETSI TS 101 038, Digital Cellular Telecommunications System (Phase 2+); High Speed Circuit Switched Data (HSCSD) - Stage 2 (GSM 03.34, Version 7.0.0, Release 1998)
10. B. H. Walke, *Mobile Radio Networks: Networking and Protocols*, John Wiley & Sons, Ltd., Chichester, 1999
11. Ch. Bettstetter, H.-J. Vögel, J. Eberspächer, "GSM Phase 2+ General Packet Radio Service GPRS – Architecture, Protocols, and Air Interface", *IEEE Communication Surveys and Tutorials*, http://www.comsoc.org/ tutorials, 3rd Quarter 1999
12. P. A. Laurent, "Exact and Approximate Construction of Digital Phase Modulations by Superposition of Amplitude Modulated Pulses (AMP)", *IEEE Trans. Commun.*, Vol. COM-34, 1986, pp. 150-160
13. A. Furuskär, J. Näslund, H. Olofsson, "EDGE - Enhanced Data Rates for GSM and TDMA/136 Evolution", *Ericsson Review*, No. 1, 1999, pp. 28-37
14. A. Furuskär, S. Mazur, F. Müller, H. Olofsson, "EDGE: Enhanced Data Rates for GSM and TDMA/136 Evolution", *IEEE Personal Communications*, June 1999, pp. 56-66
15. Ch. Lindheimer, S. Mazur, J. Molnö, M. Waleij, "Third-Generation TDMA", *Ericsson Review*, No. 2, 2000, pp. 68-79
16. R. Ramesh, K. C. Zangi, "Enhanced Data Rates for Global Evolution: A Tutorial", *Proceedings of a Tutorial Course*, IEEE VTC-Fall, 2000

10

CDMA in mobile communication systems

10.1 INTRODUCTION

In the previous chapters we presented mobile communication systems which use FDMA and TDMA access schemes. In recent years *Code Division Multiple Access* (CDMA) was the focus of attention of industrial and academic research centers, resulting in great development in multiple access communications using the spread spectrum technique with individual spreading sequences applied by the users. Nowadays, CDMA is the dominating method of multiple access in third-generation (3G) mobile communication systems. Most of the proposals for IMT-2000 (*International Mobile Telecommunications*)[1] family of standards of International Telecommunication Union (ITU) rely on CDMA as a multiple access method. Therefore CDMA and its advantages and disadvantages have to be explained.

10.2 MOTIVATION FOR CONSIDERING CDMA AS A POTENTIAL MULTIPLE ACCESS METHOD

As we remember, analog FDMA systems and digital FDMA/TDMA systems require frequency planning in order to make possible the multiple use of the same channel frequencies in sufficiently spaced cells. The distance between the cells which use the same carrier frequency is finite and multiple frequency use leads to interference among the mobile stations applying the same carriers in different cells. Multiple use of the

[1]More information on IMT-2000 and 3G systems is included in Chapter 17.

same frequencies is expressed by a *frequency reuse factor*, which is the reciprocal of the cell cluster size N. The division of cells into sectors (usually three) does not lead to improvement of the frequency reuse factor (each sector has a different subset of the carrier frequencies assigned to the cell); however, as we remember, the application of directional antennas decreases co-channel interference. The application of disjoint subsets of carrier frequencies used in neighboring cells/sectors in TDMA and FDMA systems necessitates rapid switching of the connection from the current base station to the neighboring one at the moment of crossing the border between the cells. This process is associated with the necessary change of carrier frequency and is usually called *hard handover*. With certain system inaccuracies and imperfections, handover may result in distortions or even in the loss of a connection.

One of the dominating distortions occurring in mobile communication systems is the multipath. As we remember, the result of multipath is fading. Flat fading occurs in narrowband systems, whereas selective fading is observed in the systems using wider spectrum; however, the spectrum in the systems considered so far is so wide that the signal components arriving at the receiver have relative delays which are lower than the reciprocal of the signal bandwidth. Amplitude distortions resulting in intersymbol interference have to be reduced using complex adaptive equalizers (see Chapter 8).

All disadvantages of FDMA and TDMA systems, i.e.

- limited system capacity,
- hard handover, and
- sensitivity to flat and selective fading

are reasons for investigation of spread spectrum systems with CDMA as the multiple access method.

10.3 SPREADING SEQUENCES

Before we analyze the basic scheme of the CDMA transmitter and receiver, let us review the main properties and types of pseudorandom sequences used in spread spectrum systems.

The spread spectrum systems were originally built for the military. The aim of spread spectrum systems was to hide the fact that digital transmission was taking place, to make the transmission difficult to intercept and to make it robust against intentional jamming. Pseudorandom (pseudonoise) sequences play a crucial role in performing these tasks. Pseudorandom sequences are fully deterministic periodic digital sequences with such a long period as compared with the duration of a single element of the sequence that to an external observer they look random. In fact, they cannot be fully random, because they have to be replicated in the receiver. In order to imitate noise well, their power spectrum should be white. As a result, their correlation function should be a peak, i.e. the symbols of an ideal pseudorandom sequence should be uncorrelated. Since the mean of a white sequence is equal to zero, the autocorrelation function should be equal to zero for non-zero arguments. The last property also ensures successful reception, when the signal arrives at the receiver in the form of echoes mutually shifted in time. Particular echoes can be effectively extracted by performing correlation of the composite signal with the pseudorandom signal which is synchronized with the signal contained in the desired echo. Summarizing, the white spectrum and the zero-autocorrelation function are the desired features of an ideal spreading pseudorandom sequence.

Next, the following question arises – what are the required features of spreading sequences if more users transmit in the same band, as happens in a CDMA system? In order to differentiate the users, each of them should apply a different sequence which would allow extraction of the signal of a particular user from the mixture of all signals arriving at the receiver. Thus, a zero cross-correlation function of different sequences is a required property. It can be obtained in two ways. In non-military applications all users apply the same pseurorandom periodic sequences which are uniquely shifted in time. Efficient generation of time-shifted versions of a given pseudorandom sequence is a task to solve here. In military applications such an approach is insufficient and different mutually uncorrelated sequences have to be applied.

Let us review the basic types of pseudorandom sequences used in spread spectrum CDMA systems. The interested reader is advised to study the excellent tutorial chapter on theory and application of pseudonoise sequences in [2] and chapters devoted to pseudonoise sequences in [3] and [4]. Those who know German are advised to study [5]. Let us also recall the fundamental book on pseudorandom sequences by Golomb [6].

10.3.1 m-sequences

So called *m-sequences* have gained a lot of interest and have found applications due to their simplicity of generation. One of the easiest methods of generation of pseudorandom binary sequences, including m-sequences, is the application of a *Linear Feedback Shift Register* (LFSR).

Figure 10.1 shows the example of an LFSR. If the LFSR generator is initialized with a non-zero content of its N memory cells, out of 2^N possible states of the N-stage register at most $2^N - 1$ can occur. The zero state has to be excluded because the LFSR would remain in it permanently. Thus, the period of the generated sequence cannot exceed $2^N - 1$. If the period reaches its maximum, the sequence is called a *maximal length sequence* or in short m-sequence.

The taps used to calculate the feedback are selected according to a specified polynomial. It can be proved that in order for the LFRS to generate an m-sequence, the polynomial determining the LFSR structure should be irreducible and primitive. The polynomial is irreducible if it cannot be factorized using polynomials of a lower order with the same set of coefficients (binary in our case). The polynomial is primitive if its root is a primitive element of the extension field GF(2^N) (see Section 1.4.4). The potential polynomials indicating the taps of LFSRs which allow generation of m-sequences are tabulated. The number of different irreducible primitive polynomials quickly rises with the increasing order N of the polynomial. The m-sequences feature the following properties:

- *The balance property.* In a full period of the m-sequence of length $2^N - 1$ the number of binary "1s" is 2^{N-1} and the number of "0s" is $2^{N-1} - 1$.

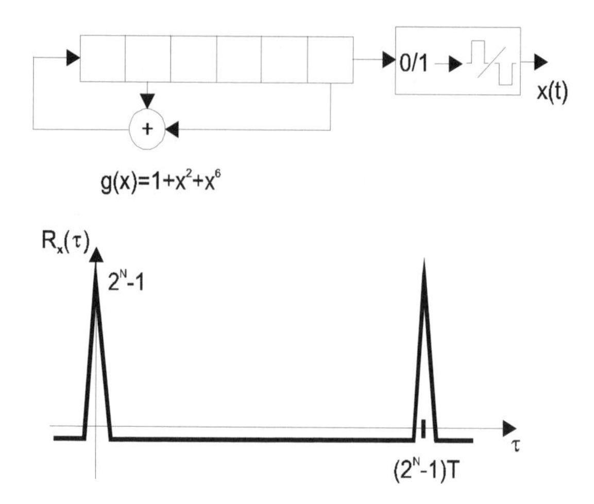

Figure 10.1 Example of an LFSR and the autocorrelation function of the bipolar LFSR output signal

- *The run property.* There are 2^{N-1} consecutive "1s" or "0s"; half of the runs are of length 1, $1/2^2$ runs are of length 2 and generally $1/2^k$ runs are of length k. Finally, there is one run of zeros of length $N-1$ and one run of ones of length N.
- *The correlation property.* The autocorrelation function of the m-sequence shaped in the form of a bipolar sequence is equal to -1 for arguments different from zero, whereas the peak value of the autocorrelation function is $2^N - 1$.

The above properties allow us to apply m-sequences in the spread spectrum and CDMA systems. However, let us note that the autocorrelation function of an m-sequence is not exactly equal to zero. Although for large values of N its value $R(n) = -1$ is very small as compared to $R(0) = 2^N - 1$, the m-sequence is not fully orthogonal with its shifted replicas. It can create a problem if many mutually shifted sequences are used in the same area. Their non-perfect orthogonality results in the increase of the noise level and limits the number of simultaneous users.

From the coding theory point of view an m-sequence can be considered as a single code word of a linear cyclic code[2] $(n,k) = (2^N - 1, N)$ known as a *maximum-length code*, which is dual to the Hamming code. As we know, in a linear code each code word can be synthesized as a linear combination of k properly selected code words. In case of the maximum-length code these $k = N$ code words are observed at subsequent moments on consecutive positions of the LFSR. Because the code is cyclic, by adding the outputs of selected LFSR memory cells we are able to synthesize a replica of the reference code word (m-sequence) shifted by any desired number of cycles. The logical circuit performing a linear combination of the LFSR outputs giving a time-shifted m-

[2]Let us recall that in a cyclic code each code word is a cyclic permutation of another code word.

sequence is often called a *mask*. As we will see in the next chapter, this type of circuit is used in the transmitter and receiver of the IS-95 CDMA system. Figure 10.2 illustrates the generation of a time-shifted replica of the reference m-sequence using a mask. The appropriate combination of the LFSR positions is selected by setting the entries of the vector $(m_1, m_2, \ldots, m_N)$ to logical "1" or "0". The symbol $\oplus$ denotes modulo-2 addition.

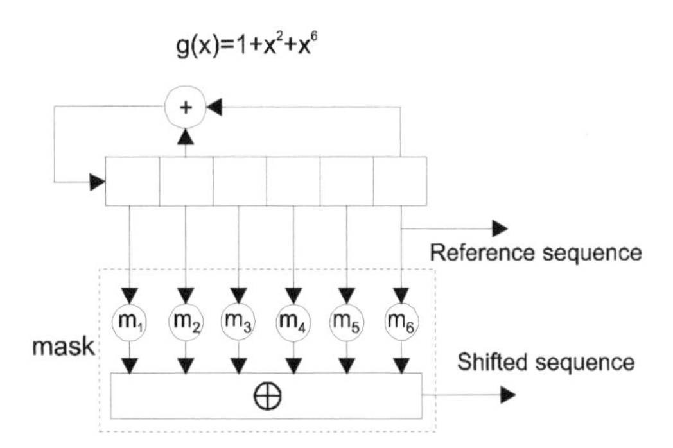

Figure 10.2 Using a mask for generation of the shifted replica of the m-sequence

An implementation problem occurs if the period of the applied m-sequence is much longer than the number of chips contained in a single information data period. This is exactly the case in most spread spectrum systems. Therefore, a partial correlation is performed in the receiver (see the next section) and the values of the autocorrelation function can differ from those obtained when the autocorrelation function is calculated over a whole sequence period. If the autocorrelation is calculated, starting at the $m-$th moment, on M chips a_j of the sequence with the period of length $P = 2^N - 1$, then one can show [2] that the mean value of the partial autocorrelation function $R_M(k,m)$ is

$$E\left[R_M(k,m)\right] = \sum_{i=0}^{M-1} \overline{a_{i+m}a_{i+m-k}} = \begin{cases} M & \text{for } k = 0 \\ \dfrac{-M}{P} & \text{for } k \neq 0 \end{cases} \tag{10.1}$$

whereas the variance of the partial autocorrelation function is

$$Var\left[R_M(k,m)\right] = \begin{cases} 0 & \text{for } k = 0 \\ M\left(1 + \frac{1}{P}\right)\left(1 - \frac{M}{P}\right) & \text{for } k \neq 0 \end{cases} \tag{10.2}$$

We conclude from (10.1) and (10.2) that for synchronized sequences the partial autocorrelation gives the exact result; however, for $k \neq 0$ the partial autocorrelation may substantially differ from the full-period autocorrelation value. This fact has to be kept in mind when we consider distortions existing in real CDMA systems.

10.3.2 Gold and Kasami sequences

The m-sequences are easy to generate and they possess good autocorrelation properties; however, the cross-correlation of two different[3] sequences of the same length can achieve relatively high values as compared with the value $2^N - 1$ equal to the maximum of the autocorrelation function of an m-sequence. This property is a serious drawback in CDMA systems in which users apply different sequences. *Gold sequences* and *Kasami sequences* are the examples of the solutions to this problem.

Gold [7] found that certain pairs x, y of m-sequences of length $2^N - 1$ exhibit a three-valued cross-correlation function $R_{xy}(k)$. The values of are $R_{xy}(k)$ as follows,

$$R_{xy}(k) \in \{-1, -t(N), t(N) - 2\} \tag{10.3}$$

where

$$t(N) = \begin{cases} 2^{(N+1)/2} + 1 & \text{for } N \text{ odd} \\ 2^{(N+2)/2} + 1 & \text{for } N \text{ even} \end{cases} \tag{10.4}$$

Such pairs x,y are called *preferred sequences.* The values given by (10.3) are much smaller than maximum values of the cross-correlation function of any pair of m-sequences of the same length. The Gold sequences are generated on the basis of a pair of preferred sequences by taking the modulo-2 sum of the first m-sequence with any cyclic shift of the second m-sequence. As a result, a new periodic sequence with the period $2^N - 1$ is achieved. The number of sequences obtained in this way and forming a family of Gold sequences is $2^N + 1$ because the number of possible shifts of the second sequence is $2^N - 1$ and both preferred sequences without a shift are also included in the family. Figure 10.3 presents an example of the generator of Gold sequences of length 63. It is worth noting that Gold sequences are applied in the UMTS system (see Chapter 17).

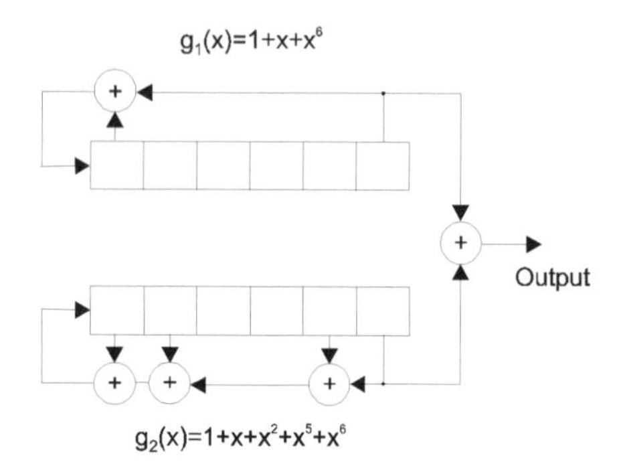

Figure 10.3 Generator of Gold sequences based on the polynomials $g_1(x) = 1 + x + x^6$ and $g_2(x) = 1 + x + x^2 + x^5 + x^6$

Kasami sequences are obtained in the same way as Gold sequences. Let us consider the m-sequence x of length $2^N - 1$, where N is an even number. Let us form the

[3] i.e. generated by the LFSRs determined by different polynomials of the same order

sequence y by the decimation of the bits of the sequence x by taking every $2^{N/2}+1$ bit of the sequence x. One can prove that the sequence y is periodic, with the period equal to $2^{N/2}-1$. Kasami sequences are obtained by modulo-2 adding of the sequence x and shifted and repeated sequences y. The number of possible shifts of sequence y is $2^{N/2}-1$; therefore, if the sequence x is also included in the family of Kasami sequences, their number is $2^{N/2}$. One can also prove that the autocorrelation and cross-correlation functions of the sequences belonging to the family take the values from the set $\{-1, -(2^{N/2}+1), 2^{N/2}-1\}$ and the Kasami sequences achieve a lower bound on the cross-correlation of any pair of binary sequences of period $n = 2^N - 1$ from a set of M sequences ($M = 2^{N/2}$ for Kasami sequences).

10.3.3 Walsh sequences

Walsh sequences were given a lot of attention in the seventies when they were treated as a serious alternative to sinusoidal signals forming the basis for the Fourier analysis. The reason of the interest was their exact mutual orthogonality. This property is also crucial for their application in CDMA systems. Unlike m-sequences, the cross-correlation of two different Walsh sequences of the same length is perfectly equal to zero. Therefore, Walsh sequences are applied in the CDMA IS-95 system (see the next chapter). The Walsh sequences can be created by the following recursion of the Hadamard matrices

$$H_1 = \begin{bmatrix} 1 & 1 \\ 1 & -1 \end{bmatrix}, \quad H_k = \begin{bmatrix} H_{k-1} & H_{k-1} \\ H_{k-1} & -H_{k-1} \end{bmatrix} \tag{10.5}$$

Each sequence is formed by a row of the matrix H_k. Figure 10.4 graphically illustrates the Walsh sequences of length 64 for $k = 5$.

Although exactly mutually orthogonal, the Walsh sequences have some drawbacks, too. The main one is a non-zero value of the cross-correlation function of the Walsh sequence with its own cyclic shift or a cyclic shift of another Walsh sequence of the same length. This disadvantage comes out at the CDMA receiver if differently delayed signal replicas in the form of echoes caused by the multipath arrive at the receiver.

10.4 BASIC TRANSMITTER AND RECEIVER SCHEMES IN THE CDMA SYSTEM

As we remember from Chapter 1, in the CDMA access scheme all users share the same band. Their signals are spectrally much wider than the information data rate makes it necessary. The reason for that is the use of a spreading sequence of much higher data (*chip*) rate than the information data rate. Ideally, the sequences of different users do not interfere with each other because they are mutually orthogonal. Thus, taking advantage of this property, through the correlation of the received signal generated by many users with the spreading sequence uniquely assigned to a particular user, we extract the interesting signal, zeroing the remaining signals due to the orthogonality of applied spreading sequences. Mutual orthogonality of spreading sequences is that

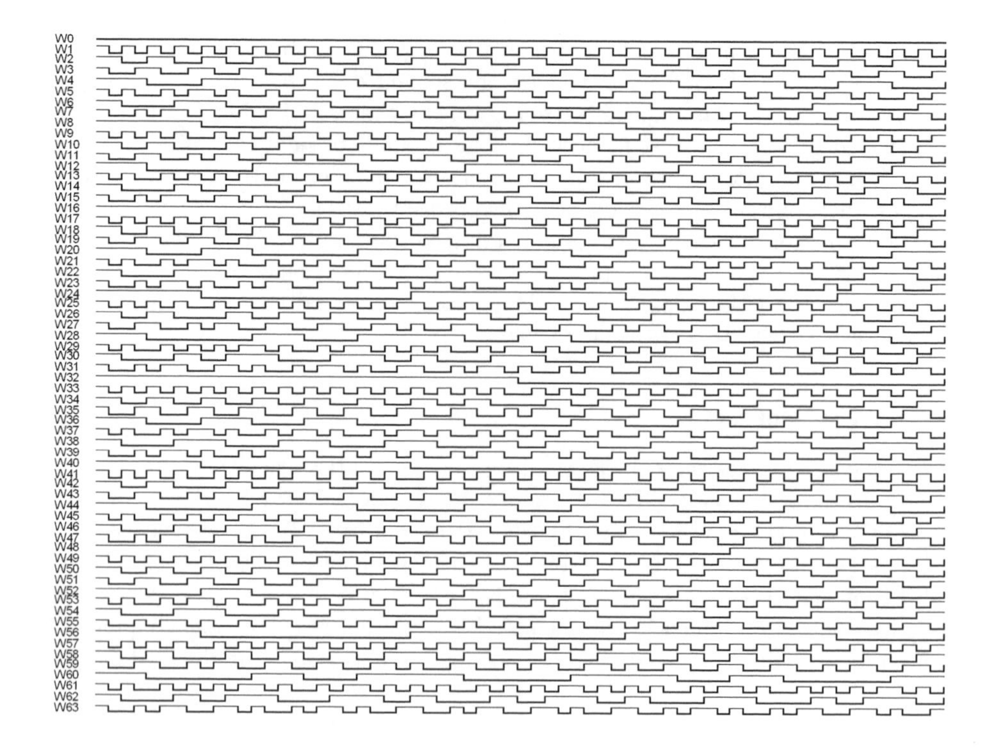

Figure 10.4 Set of the Walsh functions of length 64

feature which is crucial for the successful operation of the whole system based on the CDMA scheme.

Figure 10.5 presents generic structures of a CDMA transmitter and receiver [1]. A binary signal representing the user speech or data is the subject of error correction coding and interleaving. The resulting sequence is multiplied by a pair of pseudorandom sequences with the chip period equal to T_c. Multiplication by pseudorandom sequences performs spreading and considerably widens the data signal spectrum. The chip rate of these sequences is much higher than the data rate of the sequence at the encoder output. After transforming the spread data sequences into a bipolar form, the bipolar pulses are spectrally shaped by the filters with the transfer function $H(f)$. Subsequently, both in-phase and quadrature components are shifted from the baseband to the destination frequency range, using two modulators with cosinusoidal and sinusoidal carriers.

The received signal $\widehat{s}(t)$ is shifted back to the baseband, using a pair of in-phase and quadrature demodulators. This way the in-phase and quadrature baseband components are extracted. Next, both components are filtered by the receive filters $H(f)$ which perform *matched filtering*[4] at the same time. The filter outputs are sampled with the

[4]Assume that a binary stream is transmitted in form of the pulses $\pm h(t)$ lasting for T seconds through the channel introducing an additive white Gaussian noise. It can be proved that the optimal receiver

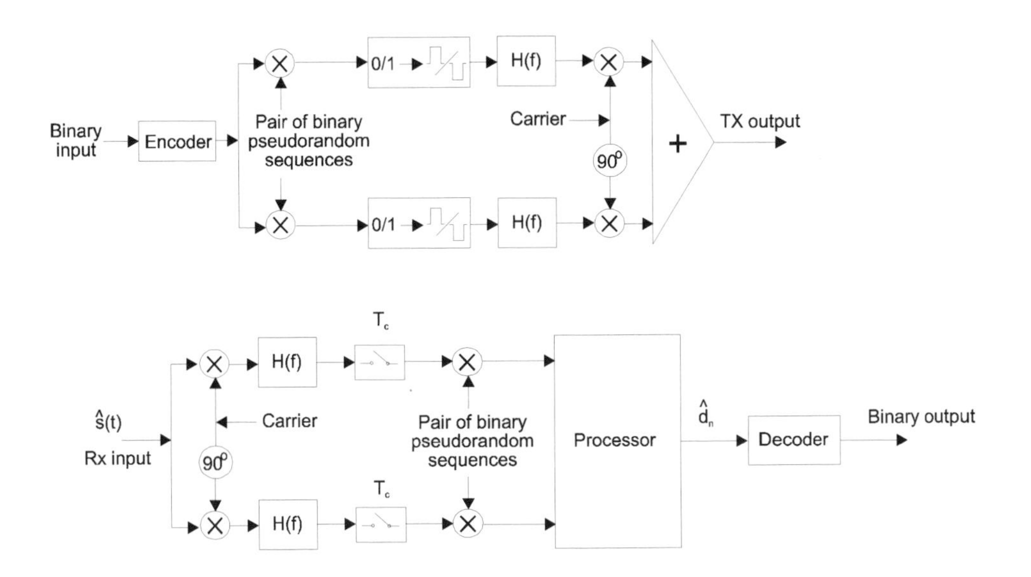

Figure 10.5 Basic transmitter and receiver structure for DS/CDMA

period T_c and multiplied by the same pair of synchronized pseudorandom sequences as those applied in the transmitter. The received samples are subsequently processed in the block denoted in Figure 10.5 as a processor. Its operation is described later in this chapter. The result of this processing is an estimate of a data sequence $\widehat{d_n}$. After deinterleaving and decoding the data sequence $\widehat{d_n}$ we receive a binary data sequence which is the estimate of the user data.

The contents of the processor block depends on the channel which is used for CDMA transmission. If the channel can be modeled as a constant gain with an additive white Gaussian noise, the processor contains a circuit which integrates the signal in the time interval of a user data symbol in which the spreading sequences of different users are mutually orthogonal. This integrator and the device performing multiplication by a pair of pseudorandom sequences form a correlator (see Figure 1.39 for comparison).

If the channel introduces the multipath, the situation is more complicated. Due to the high chip rate of the pseudorandom sequence used in the transmitter, the multipath signal components arrive at the receiver with such large mutual delays relative to the chip period that the particular path signals can be individually extracted through the correlation with the pseudorandom signal generated in the receiver and synchronized with that particular path signal. In this manner we receive signals from different paths

for this signal which ensures the minimum probability of error, is a circuit which correlates the received signal (the transmitted pulse plus noise) with the reference signal $h(t)$ over T seconds. Equivalently, the same result is achieved if the pulse corrupted by noise is filtered by a filter with the impulse response $h(T-t)$ and the filter output is sampled at the moment T. Such a filter is called the *filter matched to* $h(t)$.

carrying the same data symbol; however, their attenuations, phase shifts and delays are different. This phenomenon can be used constructively in spread spectrum systems and in CDMA systems in particular. We can apply a specific type of a receiver diversity - a *path diversity*, which is implemented using a well-known RAKE receiver. Let us analyze this type of the receiver in detail.

10.5 RAKE RECEIVER

Let the baseband signal at the output of the demodulator be described by the equation

$$r(t) = h(\tau; t) * s(t) + \nu(t) \tag{10.6}$$

where all functions are complex-valued. The function $h(\tau; t)$ characterizes the time varying multipath channel (including all filters), $s(t)$ is the baseband equivalent transmitted signal and $\nu(t)$ represents the additive noise. Let us note that the time variable τ reflects the time running from the moment of channel excitation, whereas the time variable t describes slow changes of the channel impulse response in time. For simplicity, consider the bipolar transmission in the form

$$s(t) = a_i u(t) \quad \text{for } 0 \leq t < T \tag{10.7}$$

where T is the duration of an information data symbol and $u(t)$ is the baseband modulated signal. One of the possible forms of $u(t)$ is a bipolar spreading sequence. In our derivation we follow [8].

Let the bandwidth of the signal $u(t)$ be limited to $W/2$ Hz. Thus, on the basis of the sampling theorem, this signal can be represented in the following form

$$u(t) = \sum_{n=-\infty}^{\infty} u\left(\frac{n}{W}\right) \frac{\sin\left[\pi W(t - n/W)\right]}{\pi W(t - n/W)} \tag{10.8}$$

so its spectrum is

$$U(f) = \begin{cases} \dfrac{1}{W} \sum\limits_{n=-\infty}^{\infty} u\left(\frac{n}{W}\right) \exp\left(-j2\pi f n/W\right) & \text{for } |f| \leq W/2 \\ 0 & \text{otherwise} \end{cases} \tag{10.9}$$

The signal (without noise) received at the output of the time-varying channel with the transfer function $H(f; t)$ when $u(t)$ is its excitation, is described by the formula

$$r(t) = \int_{-\infty}^{\infty} H(f; t) U(f) \exp\left(j2\pi f t\right) \mathrm{d}f \tag{10.10}$$

Using (10.9) in (10.10) we receive

$$r(t) = \frac{1}{W} \sum_{n=-\infty}^{\infty} u\left(\frac{n}{W}\right) \int_{-\infty}^{\infty} H(f;t) \exp(-j2\pi f(t - n/W)\mathrm{d}f =$$

$$\frac{1}{W} \sum_{n=-\infty}^{\infty} u\left(\frac{n}{W}\right) h\left(t - \frac{n}{W};t\right) \tag{10.11}$$

where $h(\tau;t)$ is the impulse response of the time varying channel $H(f;t)$. If we interchange the variables in (10.11), we get

$$r(t) = \frac{1}{W} \sum_{n=-\infty}^{\infty} u\left(t - \frac{n}{W}\right) h\left(\frac{n}{W};t\right) \tag{10.12}$$

Defining

$$h_n(t) = \frac{1}{W} h\left(\frac{n}{W};t\right) \tag{10.13}$$

we receive

$$r(t) = \sum_{n=-\infty}^{\infty} h_n(t) u\left(t - \frac{n}{W}\right) \tag{10.14}$$

It is clear from (10.14) that assuming the definition (10.13) the channel impulse response has the form

$$h(\tau;t) = \sum_{n=-\infty}^{\infty} h_n(t) \delta\left(t - \frac{n}{W}\right) \tag{10.15}$$

Let the effective duration of the channel impulse response be T_m seconds. Thus, the channel impulse response can be sufficiently well represented by $L = \lfloor T_m W \rfloor + 1$ coefficients $h_n(t)$ $(n = 0, \ldots, L-1)$, that is

$$h(\tau;t) = \sum_{n=0}^{L-1} h_n(t) \delta\left(t - \frac{n}{W}\right) \tag{10.16}$$

Let us return now to the analysis of the channel output when the signal (10.7) is transmitted. Let the pulse $u(t)$ last much longer than the channel impulse response. It means that the responses of the channel to the subsequent signals $u(t)$ practically do not overlap. The received signal is described by the formula

$$r(t) = \sum_{n=0}^{L-1} a_i h_n(t) u\left(t - \frac{n}{W}\right) + \nu(t) \quad \text{for } 0 \le t < T \tag{10.17}$$

Assuming that the coefficients of the channel impulse response are known, the optimum receiver is a correlator, with the following signal as the reference signal

$$q(t) = \sum_{n=0}^{L-1} h_n(t) u\left(t - \frac{n}{W}\right) \tag{10.18}$$

The correlator performs the operation

$$\widetilde{a}_i = \int_0^T r(t) q^*(t) \mathrm{d}t = \sum_{n=0}^{L-1} \int_0^T h_n^*(t) r(t) u^*\left(t - \frac{n}{W}\right) \mathrm{d}t \tag{10.19}$$

Usually the change of the impulse response is negligible within the data symbol period T. Therefore, the functioning of the receiver can be approximately described by the formula

$$\widetilde{a}_i = \sum_{n=0}^{L-1} h_n^*(t) \int_0^T r(t) u^*\left(t - \frac{n}{W}\right) \mathrm{d}t \tag{10.20}$$

Inspection of expression (10.20) allows us to draw the scheme of the receiver shown in Figure 10.6. The conjugated version of the pulse $u(t)$ propagates along the tapped delay line. The tap spacing is $1/W$ and the tap coefficients are h_n^* $(n = 0, \ldots, L-1)$. The tap signals weighted by the tap coefficients are correlated with the received signal $r(t)$. The outputs of the correlators are processed (usually simply added together). On the basis of the result of this processing the data decision is drawn.

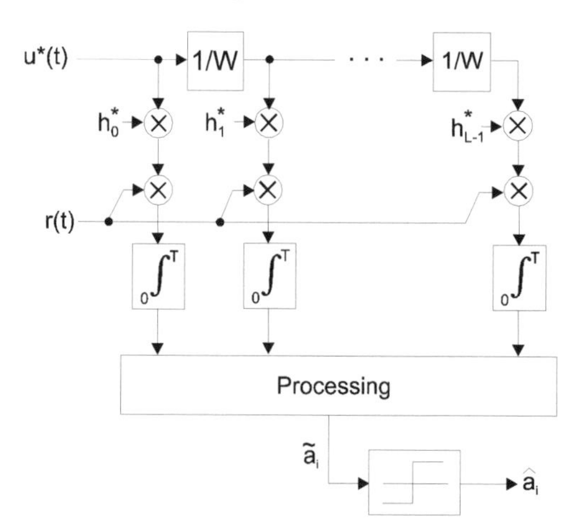

Figure 10.6 Scheme of the optimum receiver for a multipath channel if the pulse length is much longer than the channel impulse response

In spread spectrum systems the pulse $u(t)$ is simply a sequence of pseudorandom binary symbols (chips) and sampling with frequency W can be approximately replaced by sampling with the chip rate $1/T_c$. From the implementation point of view such operation is equivalent to giving the received signal to the input of the tapped delay line and correlating the tap signals with the conjugated pseudonoise sequence $u^*(t)$; however, to preserve the equivalence, the order of the weighting coefficients has to be reversed. This version of the receiver is shown in Figure 10.7. It is known as the RAKE receiver. Path correlators are often called RAKE "fingers". Each "finger" collects the signal from one path and all of them are optimally combined by being weighted by h_i^* ($i = 0, \dots, L-1$) and summed. In this respect this receiver performs path diversity with optimal combining.

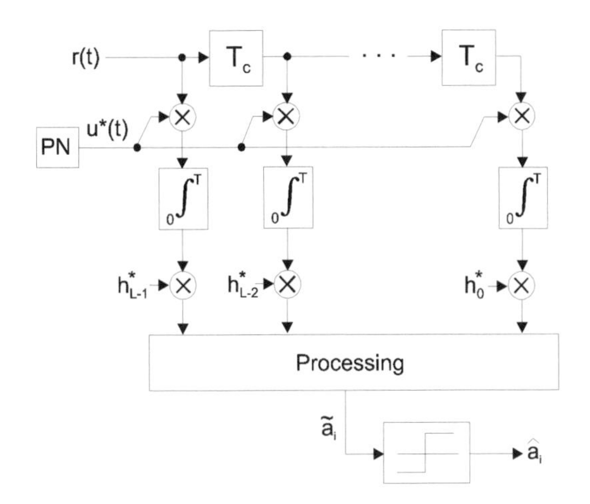

Figure 10.7 Basic structure of the RAKE receiver

In practice, due to the implementation reasons, at the chip rate of a few Mchip/s the number of RAKE fingers is often limited to 3–4. Those taps are selected for which the power of the signal path is the largest. In order to find them, a special correlator looks for the strongest paths, sequentially measuring the signal power on the output of each tap. The presented RAKE structure is optimal and equivalent to the filter matched to the spreading sequence convolved with the channel impulse response. Other suboptimal structures differ in the processing of the tap correlator outputs and are applied if the estimation of the tap weights is difficult.

10.6 JOINT DETECTION OF CDMA SIGNALS

So far we have analyzed the reception of a spread spectrum signal by a single user. Due to the multipath channel the signal arriving at the receiver is a sum of echoes delayed by multiples of chip period T_c. Thus, the zero off-peak autocorrelation function of the spreading signals is important for appropriate operation of the receiver. In that case RAKE is the optimal receiver.

The situation becomes more complicated if we analyze a CDMA system in which many users transmit their signals at the same time and in the same band. Thus, the cross-correlation properties of the spreading sequences applied by various users play a crucial role in the reception quality of CDMA receivers. In real systems the cross-correlation values are not exactly equal to zero, so the users interfere with each other. This interference is the basic limitation factor in CDMA systems. Although the RAKE receiver is no longer optimal in this case, it is often applied. In the reception of a signal of a specified user the RAKE receiver does not take into account the signals received from other users. These signals, after appropriate processing, could be eliminated or used constructively in the detection process.

Let us analyze a simple baseband model of the CDMA synchronous transmission from K mobile stations to the base station. The analyzed model is illustrated in Figure 10.8. It will allow us to consider the receiver structures which take into account the signals from all CDMA users. Such receivers are called *joint CDMA detectors.* The following considerations on joint CDMA detectors are based on [9] and [10].

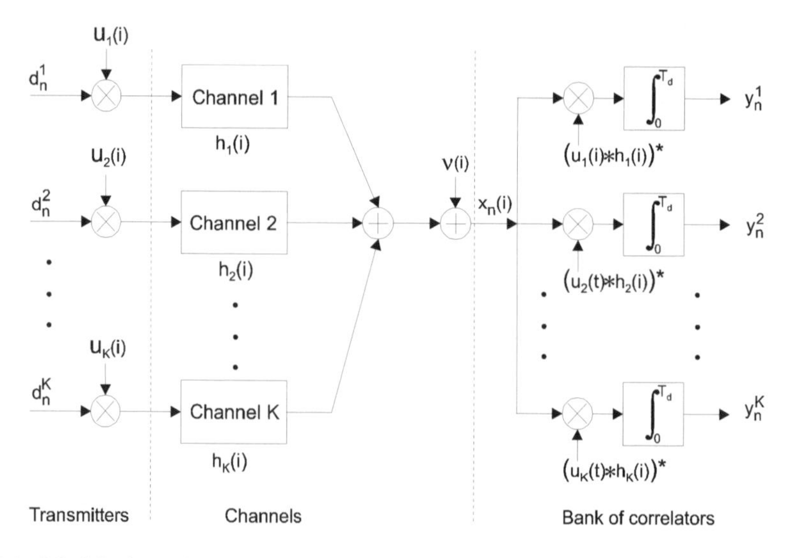

Figure 10.8 Model of synchronous CDMA transmission from mobile stations to the base station

Data symbols d_n^k $(k = 1, \ldots, K)$, where n denotes the information data symbol time index and k is the user number, are spread by the sequence $u_k(i)$ and transmitted through the channel with the impulse response $h_k(i)$. The sum of the signals generated by all users and of the additive noise $\nu(i)$ enters the base station receiver. The signal of each user is extracted by the correlator applying the reference signal $(u_k(i) * h_k(i))^*$. Let us note that this correlator is equivalent to the RAKE receiver. Decisions can be made on the basis of the correlator outputs y_n^k; however, due to the partial correlation of the user signals caused by the multipath channel and imperfect cross-correlation of the spreading sequences, the samples y_n^k contain information not only on the desired kth user, but also on the remaining users as well.

We will consider the basic structures which can be applied in joint detection of all user signals. Under the assumptions valid for the RAKE receiver, stated in the previous section, we can consider an equivalent system model, in which the spreading sequences are the convolution of the spreading sequence applied in the transmitter and the kth channel impulse response

$$g_k(i) = u_k(i) * h_k(i) \tag{10.21}$$

The reference sequence applied in the kth correlator is $g_k^*(i)$. Thus, we can treat the channels as those which introduce gains A_k $(k = 1, \ldots, K)$ only. Denoting the data symbols and the outputs of the correlators in the vector form as $\mathbf{d}_n = \left[d_n^1, \ldots, d_n^K\right]^T$ and $\mathbf{y}_n = \left[y_n^1, \ldots, y_n^K\right]^T$ we can describe the operation of the system with the formula

$$\mathbf{y}_n = \mathbf{R}_g \mathbf{A} \mathbf{d}_n + \mathbf{z}_n \tag{10.22}$$

where $\mathbf{R}_g$ is the cross-correlation matrix of the spreading sequences $g_k(i)$, $(k = 1, \ldots, K)$, $\mathbf{A}$ is the diagonal matrix of the gains introduced by the channels and $\mathbf{z}_n$ is the noise vector observed at the outputs of the correlators. On the basis of the vector $\mathbf{y}_n$, the base station detector derives the estimate $\widehat{\mathbf{d}}_n$ of the input data vector $\mathbf{d}_n$. The most obvious way of finding the data estimates $\widehat{\mathbf{d}}_n = \text{dec}(\widetilde{\mathbf{d}}_n)$ is to left-multiply both sides of (10.22) by the matrix $\mathbf{R}_g^{-1}$. As a result we have

$$\widetilde{\mathbf{d}}_n = \mathbf{R}_g^{-1} \mathbf{y}_n = \mathbf{A} \mathbf{d}_n + \mathbf{R}_g^{-1} \mathbf{z}_n \tag{10.23}$$

The detector which performs (10.23) is called a *decorrelating detector* [11]. In the ideal case it zeros interchannel interference. It is often called *Zero-Forcing* (ZF) detector. The consequence of its operation is noise amplification, which is expressed by the second term of (10.23). The decorrelating detector is a linear circuit whose implementation requires perfect knowledge of the cross-correlation matrix $\mathbf{R}_g$. The scheme of the basic part of the detector is shown in Figure 10.9.

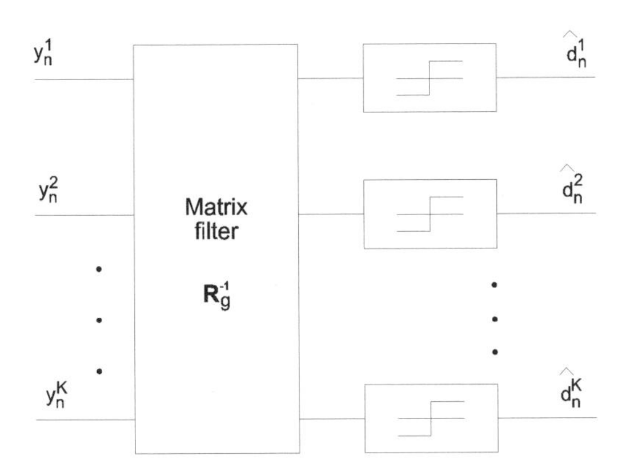

Figure 10.9 Structure of the decorrelating detector

The *Minimum Mean-Squared Error* (MMSE) detector is another type of linear detectors. Here the correlator output vector $\mathbf{y}_n$ is not multiplied by the inverse of the autocorrelation matrix $\mathbf{R}_g^{-1}$, instead, it is multiplied by the matrix $\mathbf{Q}_{\text{MSE}}$, which ensures minimization of the mean-squared error

$$\mathcal{E}_n = E\left[(\mathbf{d}_n - \mathbf{Q}_{\text{MSE}}\mathbf{y}_n)^H (\mathbf{d}_n - \mathbf{Q}_{\text{MSE}}\mathbf{y}_n)\right] \tag{10.24}$$

In the above formula $(.)^H$ denotes vector transposition and complex conjugation. One can show that minimization of $\mathcal{E}_n$ leads to the following form of the matrix $\mathbf{Q}_{\text{MSE}}$ denoted as $\mathbf{Q}_{\text{MMSE}}$

$$\mathbf{Q}_{\text{MMSE}} = \left[\mathbf{R}_g + (N_0/2)\mathbf{A}^{-2}\right]^{-1} \tag{10.25}$$

where $N_0/2$ is the power density spectrum of the additive white noise at the receiver input. Because the MMSE detector takes this noise into account, it generally provides lower probability of error than the decorrelating detector; however, it is sensitive to the estimation accuracy of channel gains.

The optimum joint detector operating according to the *Maximum Likelihood* (ML) criterion at the assumption that the noise samples from the channels are mutually statistically independent, finds such a data vector $\widehat{\mathbf{d}}_n$ for which the following expression is minimized

$$\left\|\mathbf{y} - \mathbf{R}_g\widehat{\mathbf{d}}_n\right\| \tag{10.26}$$

where $\|\mathbf{a}\|$ denotes the quadratic norm of vector $\mathbf{a}$. Minimization of (10.26) requires searching for the data vector $\widehat{\mathbf{d}}_n$ among M^K possible vectors (M is the size of the data alphabet). Although this search can be performed using the Viterbi algorithm, in many cases such a receiver is computationally too complex and cannot be considered for implementation. Therefore, suboptimal solutions have to be used. Among them, subtractive interference cancellation detectors are the most important. They can be divided into the following types

- the successive interference cancellation (SIC) detector (see Figure 10.10),
- the parallel interference cancellation (PIC) detector, and
- the zero-forcing decision-feedback (ZF-DF) detector (Figure 10.11).

In the SIC detector, the strongest user signal is found and the conventional detector is used to find the decision upon the carried data symbol. Subsequently, on the basis of this data symbol, the spreading sequence and channel characteristics, the signal estimate is regenerated and subtracted from the total received signal. This way, subsequent user signals are detected and eliminated starting from the strongest and ending with the weakest one.

The PIC detector estimates and subtracts the interference originating from all users in parallel. Usually such a detector has a multistage structure shown in Figure 10.10. In the first stage, tentative decisions are derived using conventional detectors or a

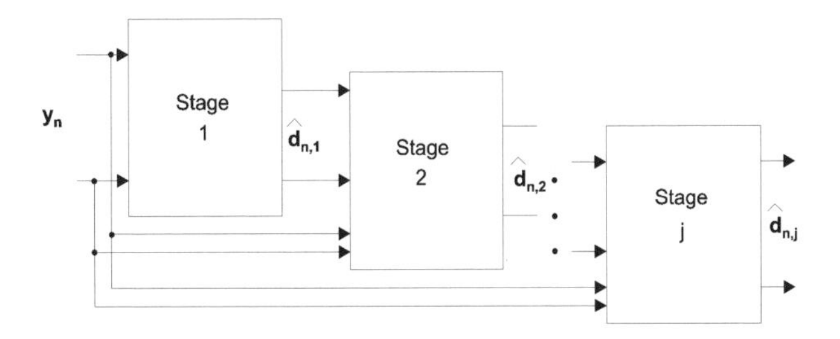

Figure 10.10 Multistage structure used in the successive cancellation detector

decorrelating detector. On the basis of these decisions, interference from all users can be regenerated and cancelled in the next stage. As a result, better quality decisions upon transmitted data are generated. This operation can be repeated in the next stages, gradually improving the performance of the detector.

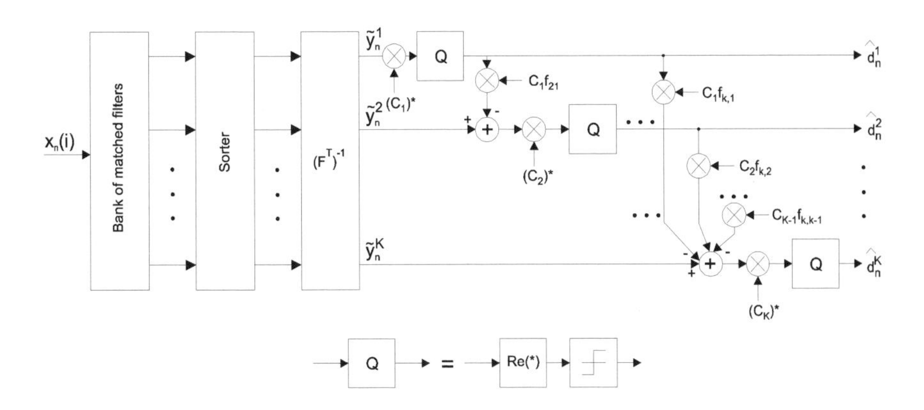

Figure 10.11 Zero-forcing decision-feedback detector

The zero-forcing decision-feedback detector combines two operations. The first operation relies on a bank of *Whitened Matched Filter* (WMF). These filters are based on the Cholesky decomposition of the correlation matrix $\mathbf{R}_g = \mathbf{F}^T\mathbf{F}$ where $\mathbf{F}$ is a lower triangular matrix. In the second operation successive decisions are made and decision feedback from stronger users is applied for weaker users.

All the detectors described above are computationally complex. Each of them requires the estimation of the correlation matrix, calculation of its inverse or matrix factorization. Despite computational complexity, however, joint detectors will be applied in several third generation CDMA mobile communication systems.

10.7 BASIC PROPERTIES OF A CDMA MOBILE SYSTEM

After presenting a survey on spreading sequences applied in CDMA systems and relevant receiver structures, let us concentrate on the basic features of a typical CDMA cellular system, which make it attractive from the implementation point of view.

The first attribute which differentiates a CDMA system from FDMA or TDMA systems is the value of the frequency-reuse factor. In FDMA and TDMA it was 1/7 or 4/12, while in CDMA, it is equal to one. It means that the whole spectrum can be reused in each cell. That value of the frequency reuse factor has a crucial effect on the system capacity and is possible due to different almost orthogonal spreading sequences applied in the neighboring cells. Spreading sequences are not ideally orthogonal due to the way they are generated and to the channel multipath. Therefore, interference among users arises. Since the number of users is usually large, the interference appears as an additive noise which can be tolerated up to a certain level for which the user data detection quality is still ensured. It is clear that the level of interference arising from a user is strictly associated with the power level of the signal generated by him. Therefore, the signal power should be kept at the lowest level which ensures sufficient reception quality.

The next attribute of CDMA systems follows from the considerations on the signal level. In CDMA systems a precise power control of user signals is a necessity. If the power control of the mobile station were not performed, the signal transmitted from the cell border would have much lower level at the base station than the signal reaching the base station from the mobile station located near it. It is the well-known *near-far effect.* The signal level from different mobile stations has to be the same with the accuracy of the order of 2 dB, regardless of the distance of the mobile stations from the base station. Otherwise the level of non-zero cross-correlation of spreading sequences used by the desired user and the interferer may be comparable with the level of the correlation peak of the desired user signal. This in turn may substantially deteriorate the signal detection.

Coarse power control is a procedure in which a signal is transmitted by the mobile station at the level which is inversely proportional to the level of the signal received from the base station. Precise power control is realized in a closed loop. The base station measures the power level of the signal received from the given mobile station and sends the appropriate command to the latter asking for increase or decrease in the level of the transmitted signal. The dynamic range of the power control is 80–100 dB at the maintained accuracy of 2.5 dB. It is necessary to keep these values to obtain the desired CDMA system capacity. Thanks to precise power control, it is possible to transmit a signal at the level about 20–30 dB lower than a mean level of the mobile station signal in an analog system. This is a very advantageous feature of a CDMA system which ensures lower interference with other users, lengthens battery life and decreases the radiation level of a mobile station.

Thanks to the RAKE receiver used in a mobile station and application of the same frequency bands in neighboring cells, the mobile station can receive signals simultaneously from two base stations. This case can occur at the border of two cells. Before the mobile station switches from the current base station to a new one, it can temporarily be connected to both of them. Such a dynamic process is called*soft handover*. It causes the handover to be made more smoothly and the reception quality at the cell border to be higher because the mobile station uses two base stations at the same time. This effect can be considered as a path diversity reception. It influences the overall system capacity. Figure 10.12 illustrates this situation.

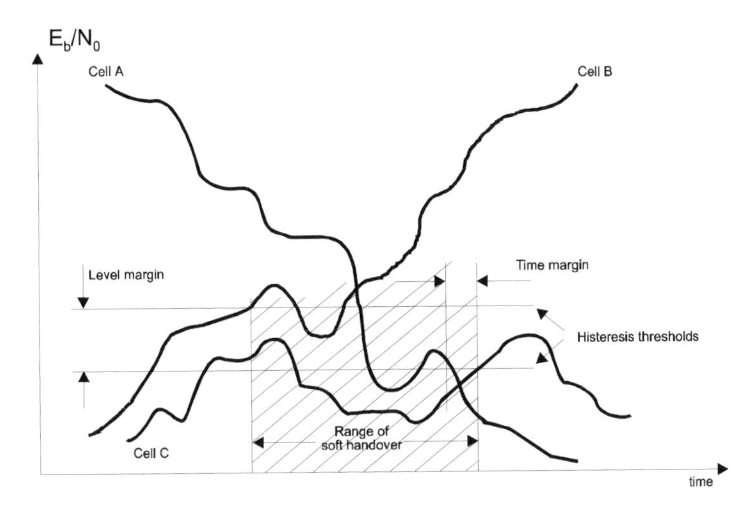

Figure 10.12 Illustration of the soft handover in the CDMA cellular system

The main limitation of the CDMA system capacity is the interference generated by the users with respect to each other. Similar to GSM, the interference level can be decreased by exploiting the speech activity detection. As we know, typical speech activity takes up about 35 percent of time. By measuring the user speech activity it is possible to digitally represent the speech signal by a variable rate data stream which is matched to the speech activity of the user. Thus, it is possible to control the power of transmitted signal depending on the current data rate and set it at such a level which is required to ensure a tolerable level of the error rate.

The next means of improving the system quality and its capacity is the application of strong error correction coding and interleaving. As we know, using the error correction coding is sensible if we achieve a positive coding gain, by which we understand a decrease of the required SNR at the acceptable error rate. In the case of positive coding gain achieved in the CDMA system, the power level of the transmitted signal can be decreased, the interference level is decreased too, which in consequence results in the increase of the system capacity.

A. Viterbi [1] presented an approximate formula to determine the number of users in a cell which can be active in the uplink connection. The formula has the form

$$N_u = \frac{(W/R)\, G_\nu G_A}{(E_b/I_0)\,(1+f)}\,[(1-\eta)F] \quad \text{Erl/cell} \tag{10.27}$$

where the variables have the following meaning:

- W/R – the spreading gain factor (W is the system bandwidth, R is the information data rate),
- E_b/N_0 – the required value of the ratio of the signal energy per bit to noise density, resulting from thermal noise and from the users located inside and outside the cell,
- f – the power ratio of interference from users outside the cell to the interference from users inside the cell,
- G_ν – the gain achieved due to variable speech activity,
- G_A – the gain achieved due to sectorized antennae
- $[(1-\eta)F]$ – the loss caused by imperfect power control and variable traffic intensity.

The application of the methods used to improve the system capacity and already described above, such as frequency reuse, the RAKE receiver and error correction coding, results in decreasing the denominator in (10.27). Currently achievable values of the parameters in (10.27) are: $G_\nu = 2.5$, $G_A = 3$, $E_b/N_0 = 5$ (7 dB), $f = 0.55$, $[(1-\eta)F] = 0.75$. As a result, the number of users which can be served in the cell is around $N_u \simeq 0.73W/R$, which is about five times higher than the number of users in a traditional FDMA system [1].

10.8 CONCLUSIONS

In this chapter we analyzed the basic features of the mobile communication system based on CDMA access scheme. We considered the desired properties of the spreading sequences. Then we derived the optimum receiver for a single user of spread spectrum signal. Subsequently, we presented basic types of multiuser detectors. Finally, we analyzed the characteristic features of a CDMA system and we concentrated on main factors which influence the system capacity. Finally, we quoted the formula for the approximate number of CDMA users [1].

REFERENCES

1. A. Viterbi, "The Orthogonal–Random Waveform Dichotomy for Digital Mobile Personal Communications", *IEEE Personal Communications*, First Quarter 1994, pp. 18-24

2. J. S. Lee, L. E. Miller, *CDMA Systems Engineering Handbook*, Artech House Publishers, Boston, 1998

3. J. D. Gibson (ed.), *The Mobile Communications Handbook*, CRC Press in cooperation with IEEE Press, 1996

4. D. Gerakoulis, E. Geraniotis, *CDMA Access and Switching for Terrestrial and Satellite Networks*, John Wiley & Sons, Ltd., Chichester, 2001

5. A. Finger, *Pseudorandom-Signalverarbeitung*, B.G.Teubner, Stuttgart, 1997

6. S. W. Golomb, *Shift Register Sequences*, Holden-Day, San Francisco, 1967

7. R. Gold, "Maximal Recursive Sequences with 3-valued Recursive Cross-Correlation Functions", *IEEE Trans. Inform. Theory*, IT-14, 1966, pp. 154-156

8. J. G. Proakis, *Digital Communications*, McGraw-Hill, Englewood Cliffs, 1995

9. A. Duell-Hallen, J. Holtzman, Z. Zvonar, "Multiuser Detection for CDMA Systems", *IEEE Personal Communications*, April 1995, pp. 46-58

10. S. Moshavi, "Multi-User Detection for DS-CDMA Communications", *IEEE Communications Magazine*, October 1996, pp. 124-136

11. R. Lupas, S. Verdu, "Linear Multi-User Detectors for Synchronous Code-Division Multiple-Access Channels", *IEEE Trans. Inform. Theory*, 35, 1989, pp. 123-136

11

Description of IS-95 system

11.1 INTRODUCTION

In Chapter 10 we showed how a CDMA system operates. In this chapter we will describe a particular implementation of the CDMA system, i.e. the IS-95 system known also as *cdmaOne*. Its air interface is described in the IS-95 standard [1], [2], [3]. In the early nineties, it became a reference system due to the innovative solutions applied. It has been used to test many new ideas and it has frequently served as a pattern for comparisons.

11.2 FREQUENCY RANGES

The IS-95 system operates in two bands. The *Frequency Division Duplex* (FDD) method is applied in both of them. The first band, Band Class 0, had been previously occupied by the AMPS and these frequency ranges which are now used by IS-95 are no longer used by the AMPS. Generally, the IS-95 standard allows for mobile stations to operate both in AMPS and IS-95 (CDMA) mode. The downlink is realized in the frequency range 824–849 MHz, and uplink in the frequency interval 869–894 MHz. There is 45 MHz difference between both bands. IS-95 can be also deployed in the PCS (*Personal Communication System*) 1800-MHz band [2] denoted as Band Class 1. The frequency ranges are 1930–1990 MHz for downlink and 1850–1910 MHz for uplink. Tables 11.1 and 11.2 present the assigned frequency bands in detail, showing their division between two possible systems or, in the second frequency range, among six blocks.

Table 11.1 Frequency ranges of Band Class 0

System	**Transmit frequency band [MHz]**	
	Mobile station	**Base station**
A	824.025–835.005	869.025–880.005
	844.995–846.495	889.995–891.495
B	835.005–844.995	880.005–889.995
	846.495–848.985	891.495–893.985

Table 11.2 Frequency ranges of Band Class 1

Block	**Transmit frequency band [MHz]**	
	Mobile station	**Base station**
A	1850–1865	1930–1945
D	1865–1870	1945–1950
B	1870–1885	1950–1965
E	1885–1890	1965–1970
F	1890–1895	1970–1975
C	1895–1910	1975–1990

11.3 DOWNLINK TRANSMISSION

In the IS-95 standard downlink and uplink channels are called *forward* and *reverse* channels, respectively. The rules for transmission in downlink and uplink are different, therefore we describe them separately. Let us start with the downlink.

In transmission from a base station to mobile stations all transmitters are located in the same place, so they operate in full synchronism. In the downlink transmission the following physical channels exist

- a pilot channel,
- a synchronization channel,
- paging channels,
- traffic channels.

Direct Sequence Spread Spectrum (DS-SS) transmission at the rate 1.2288 Mchip/s is applied on all channels. Two pseudonoise sequences of length 2^{15}, often denoted as PNI and PNQ, are used separately in the in-phase and quadrature branches. All base station transmitters generate the same PN sequence pair; however, the sequences applied in different cells are appropriately shifted in time with respect to the reference clock. This time shift identifies a given cell.

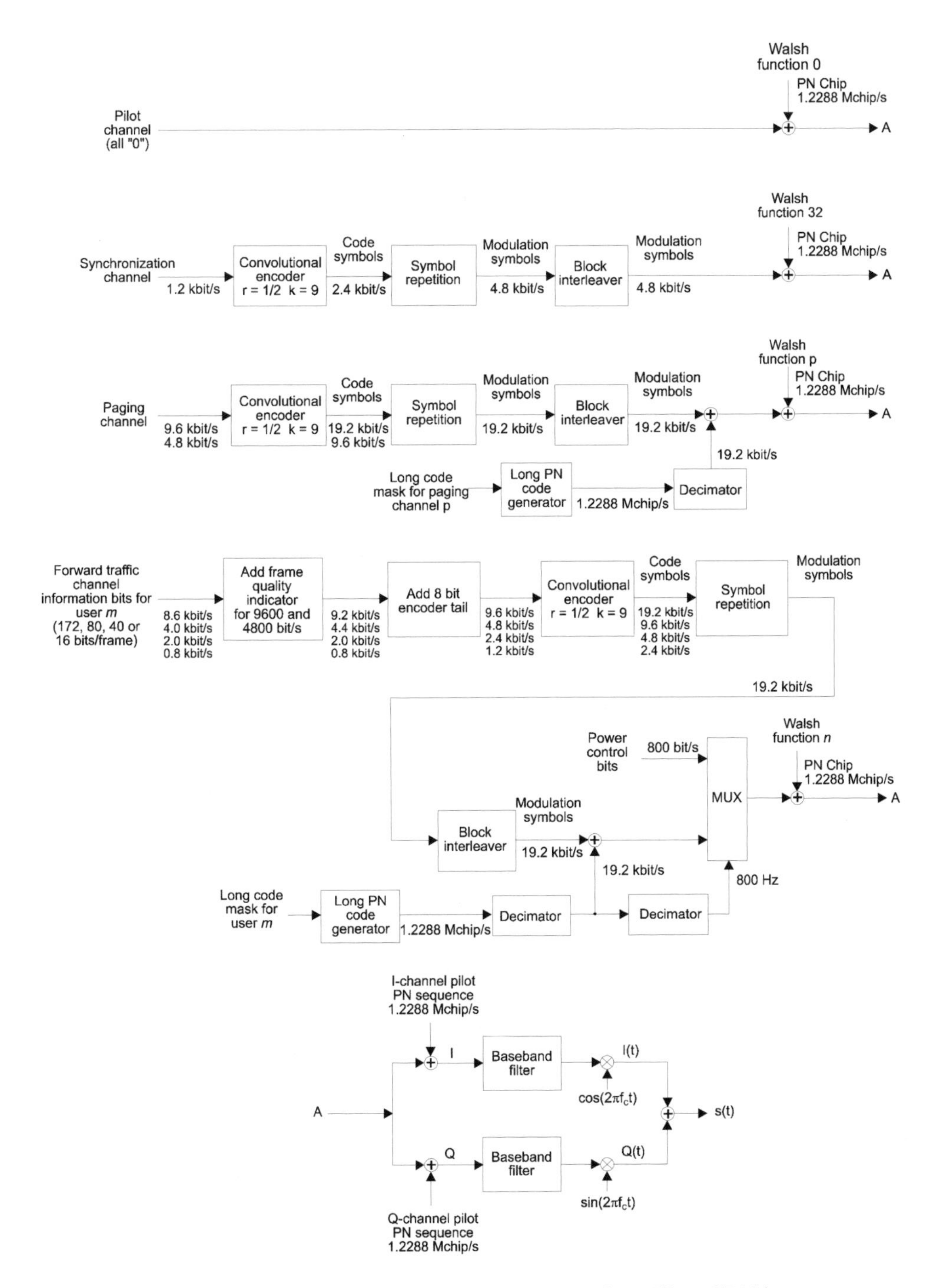

Figure 11.1 Scheme of the base station transmitter of the IS-95 CDMA system

In the case of sectorized cells these remarks refer to the sectors. The PNI and PNQ sequences are generated by LFSRs determined by the following polynomials

$$g_I(x) = x^{15} + x^{13} + x^9 + x^8 + x^7 + x^5 + 1$$
$$g_Q(x) = x^{15} + x^{12} + x^{11} + x^{10} + x^6 + x^5 + x^4 + x^3 + 1 \qquad (11.1)$$

As we remember, the LFSR can be applied to generate a pseudonoise sequence of length $2^N - 1$, ($N = 15$ in our case), therefore, in order to obtain a period of this sequence equal to 2^{15}, a single zero is introduced after the sequence of 14 consecutive zeros occurring once in the PN sequence. A pair of sequences generated by the LFSRs determined by the above polynomials is called a *short code.*

Figure 11.1 presents the scheme of a base station transmitter. In each cell/sector up to 64 channels can be used. This number results from 64 Walsh functions applied to ensure mutual orthogonality of the transmitted channel signals. So far, two kinds of sequences have been introduced. Applying a particular Walsh function is equivalent to using a patricular channel in a cell, whereas the time shift of the applied PNI and PNQ sequences with respect to the common clock marks the cell. For these reasons the Walsh sequences are frequently called *channelizing codes*, whereas PNI and PNQ sequences are *spreading sequences.*

The pilot channel is associated with the Walsh function W_0 (equal to logical zero) and the synchronization channel is associated with the function W_{32}. Sixty-two channels can be used by paging and traffic channels. Up to seven paging channels can be established, which apply the Walsh functions W_1 to W_7. Unused paging channels can be applied as traffic channels.

Data fed to the input of the synchronization, paging and traffic channels are the subject of convolutional coding, repetition and block interleaving. Two rate sets have been standardized in IS-95A [1]. In Rate Set 1 data rates equal to 1200, 2400, 4800 and 9600 bit/s are applied. In Rate Set 2 the data rates are 1800, 3600, 7200 and 14400 bit/s. Each mobile station has to support Rate Set 1 and can also realize Rate Set 2. For Rate Set 1, independently of the input stream data rate (equal to 1200, 2400, 4800 or 9600 bit/s), the data rate at the input and output of the block interleaver is 19.2 kbit/s. This is a result of applying the convolutional code of the coding rate $R = 1/2$ and the constraint length $L = 9$ (see the encoder scheme in Figure 11.2) as well as bit repetition applied for data rates lower than 9600 bit/s. The synchronization channel is an exception because its data rate after symbol repetition and interleaving is 4800 bit/s. Data from the paging and traffic channels are modulo-2 added to the output sequence of a pseudorandom generator which produces a data sequence of length $2^{42} - 1$ (called a *long code*) at the rate of 19200 bit/s, applying the mask assigned to the individual user or to the paging channel. The task of this operation is to mark an individual user and to ensure his/her transmission privacy. Each bit of the resulting 19.2 kbit/s data is represented by the whole period of a Walsh sequence of length equal to 64 symbols in the simple or logically negated form depending on the value of the data bit. The Walsh sequence functions as a spreading sequence. As a result, we receive spread spectrum signals at the rate of $19200 \cdot 64 = 1.2288$ Mbit/s. Subsequently, for each channel these signals are modulo-2 added to two pseudonoise sequences PNI and PNQ and the resulting dibit determines the phase of the QPSK modulator or, equivalently, it determines the amplitudes of the in-phase and quadrature component of the QPSK constellation (see Figure 11.3). After being shaped by the baseband filters, the in-phase and quadrature pulses are shifted to the destination band by the quadrature modulators. The signals from each channel are combined, resulting in the composite signal transmitted by the base station.

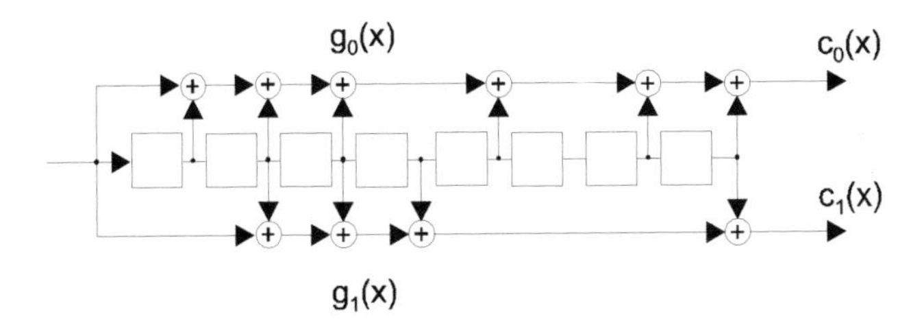

Figure 11.2 Scheme of a convolutional encoder applied in the downlink of IS-95

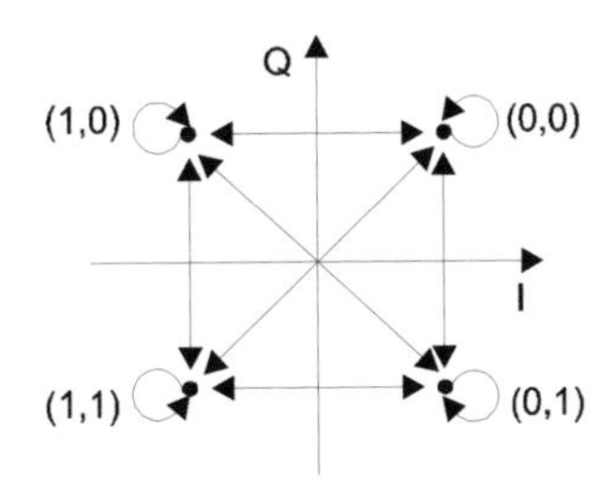

Figure 11.3 QPSK signal constellation used in IS-95 downlink

In the case of traffic channels, some of the bits characterizing the encoded and interleaved speech binary stream are replaced in the multiplexer by the power control bits which are sent to the mobile station using this channel. This data stream is a realization of a *power control subchannel.* The mean rate of the uncoded power control symbols is 800 bit/s. A "0" denotes the order to the mobile station to increase its mean power level by 1 dB, whereas "1" implies decreasing the mean power level by 1 dB. The position of power control bits within a data frame is determined by the decimated bits of the long-code generator. Thanks to this, power control bits appear in the data stream in pseudorandom positions.

If one of the data rates from Rate Set 2 is used on a traffic channel, the transmitter block sequence differs from that which is shown in Figure 11.1. This part of the scheme which is concerned with traffic channels should be replaced by the scheme shown in Figure 11.4. For Rate Set 2, the user data in 20-ms frame constitute 267, 125, 55 or 21 bits; therefore, the input rates are 13.35, 6.25, 2.75 or 1.05 kbit/s. For each 20-ms frame one reserved bit or flag bit, frame quality bits and an 8-bit encoder tail are added. The resulting data rates grow to 14400, 7200, 3600 or 1800 bit/s, respectively.

The application of a convolutional encoder of the constraint length $L = 9$ and the coding rate $R = 1/2$ results in doubling the data rates. For the data rates lower than 28.8 kbit/s, bit repetition is applied to achieve the highest data rate of 28.8 kbit/s. In order to match this data rate with the constant rate of 19.2 kbit/s in other channel types and in the channels applying Rate Set 1, 2 bits out of every 6 input bits are punctured. Other blocks in the traffic channel block chain are the same as for Rate Set 1, therefore they will not be described again.

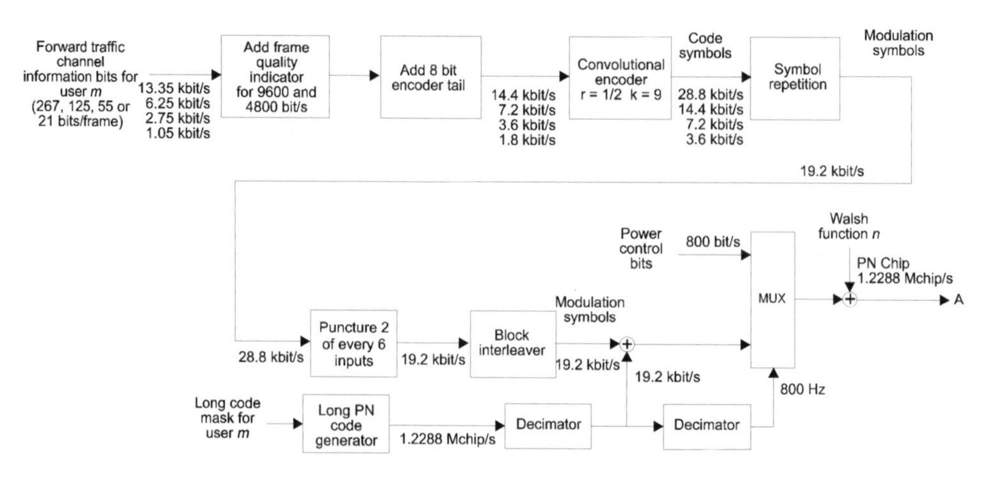

Figure 11.4 Fragment of the traffic channel structure applied in Rate Set 2

Data carried on the traffic channel are typically the result of QCELP (*Qualcomm Code Excited Linear Prediction*) coding[1] if speech is transmitted or they originate from a data terminal if data transmission takes place. The QCELP encoder generates a data stream at one of four possible rates which result from the user speech activity. The encoder operates on 20-ms frames. At the highest rate, 160 bits plus an 11-bit parity block are produced in each frame. The resulting data rate is 8.55 kbit/s. In the case of three lower data rates, there are 80, 40 and 16 bits in a frame, respectively, therefore the data rates are 4000, 2000 and 800 bit/s and the channel data rates are 9600, 4800, 2400 and 1200 bit/s. In the case of lower data rates each bit is repeated so many times, that the final data rate is 9600 bit/s. The repeated bits are transmitted with a lower power. This way interchannel interference is decreased which has a great impact on the system capacity.

It is worth noting that two more speech encoders have been defined. They are the *Enhanced Variable Rate Codec* (EVRC) and the *Algebraic Code Excited Linear Prediction* (ACELP) codec. The first one generates a 8000 bit/s data stream, whereas the second one produces a 13-kbit/s data stream.

The signal transmitted by the base station is received in the mobile station and is decoded four times according to four possible data rates at the speech encoder output.

[1]See the section on speech coding in Chapter 1.

The sequence which results in the lowest number of errors in the decoding process is selected as the output binary sequence.

The interleaving applied in IS-95 has a depth equal to 20 ms. In decoding of the convolutional code, performed in the mobile station, the soft-decision Viterbi algorithm is applied. It results in the coding gain of about 4.5 dB at the error rate of 10^{-3}.

In the downlink (forward direction) the receiver placed in a mobile station performs synchronous reception thanks to a pilot channel generated by the base station. The power of a pilot channel is 4 to 6 dB higher than the power of the traffic channel. The pilot signal allows estimation of the multipath channel. A three-tap RAKE receiver selecting the strongest signal paths is used. Other mobile receiver blocks are parallel to those contained in the base station transmitter and they will not be described here.

The operation of all base stations is synchronized with respect to the system clock by the *Global Positioning System* (GPS). This allows application of the same spreading sequences with appropriately selected time shifts. These time shifts are selected in increments of 64 chips providing 511 possible offsets with respect to the reference sequence. This allows identification of a particular cell uniquely.

Let us describe the functions performed by particular types of channels in the downlink.

A *pilot channel* provides the phase and reference timing for synchronous demodulation performed by mobile stations. As we have said before, the pilot channel transmits the unmodulated spreading signal which can be used by a mobile station for synchronization and channel/sector identification. The pilot signal power measurement performed in the mobile station is used in the power control loop. Measurements of pilot signals generated in the neighboring cells are applied in the handover procedure.

The *synchronization channel* broadcasts the data which allow for fast and reliable synchronization at the system level. The rate of the transmitted data stream is equal to 1200 bit/s. The stream carries the *Sync Channel Message* and padding bits. The Sync Channel Message contains the system and network identification, the offset of the cell/sector short spreading code, the long code state at the time specified in time system parameter, the parameters which allow determination of the current system time and local time with respect to the reference clock and the data rate of paging channels. Data transmission on the synchronization channel is organized in frames and superframes. A frame consists of a Start of Message (SOM) bit and 31 bits of data. Three frames constitute a 96-bit long superfame, which lasts 80 ms. The Sync Channel Message can be longer than a single frame, so padding bits are added to fill the data bits of a superframe. The Sync Channel Message has an 8-bit header determining the message length, the message bits of a minimum of 2 bits and a maximum of 1146 bits and 30 bits of CRC.

After synchronization acquisition performed using the pilot and synchronization channels, the mobile station starts to monitor the paging channel. The paging channel can be assigned up to 7 Walsh sequences (W_1–W_7). It can operate at the data rate of 4800 or 9600 bit/s. Several messages can be sent on a paging channel, which have a unified structure. A message contains an 8-bit message-length header, message data of length between 2 and 1146 bits, and 30 bits of CRC. The messages are divided into 47- or 95-bit parts which, after adding a one-bit flag, are subsequently transmitted in the form of *Paging Channel Half-Frames*. Eight half-frames form an 80 ms *paging channel slot*. A message can use up to 2048 slots. A single paging channel operating at 9600 bit/s can serve up to 180 paging processes. A mobile station monitors the paging channel assigned to it by the base station. It can do it in a slotted or unslotted mode. In the slotted mode the mobile station listens to the paging channel during the assigned page slots only. During the rest of the time the receiver is switched off and the energy is saved.

A few message types can be sent on the paging channel. They are:

- messages defining the system configuration, such as:
  - *System Parameter Message*, which provides configuration of paging channels, parameters needed for tuning to the pilot sequence and the number of paging channels used in the cell. While "entering" the system, the mobile station listens to the first paging channel where it can find this information.
  - *Access Parameters Message*, which indicates the parameters which have to be used by mobile stations during transmission to the base station on an *Access Channel*.
  - *Neighbor List Message*, which presents parameters of the neighboring cells, in particular the spreading sequence offset used in them.
  - *CDMA Channel List Message*, which gives the list of CDMA carriers
- paging messages – paging takes place if the base station receives the order to set a connection with a given mobile station. The paging signal is usually generated by a few base stations.
- order messages, which constitute a large class of messages ensuring proper system operation. They are used during mobile station registration, during confirmation of other operations, etc.

Some other important messages are:

- *Channel Assignment Message* – used to tell the mobile station to tune to a specified channel frequency.
- *Data Burst Message* – used to send data from a base station to the mobile station.
- *Authentication Challenge Message* – used to transmit data, which are necessary during the procedure of mobile station authentication.

In IS-95 *Traffic Channels* work in two rate sets. The Rate Set 1 is always supported, whereas the Rate Set 2 is optional. Sixty-one traffic channels can be realized with a single paging channel used in a cell/sector; however, applying seven paging channels decreases the number of traffic channels to 55.

Traffic channels are used to transmit user speech or data (*primary traffic*), data (*secondary traffic*) and signaling information. Due to the fact that a single traffic channel carries data at the rate equal at most to 9600 bit/s (RS1) or 14400 bit/s (RS2),

Table 11.3 Parameters of the forward traffic frames

Data rate bit/s	**Frame length** L bits	**Reserved/ flag** bit	**Data bits** D bits	**Frame-quality ind.** F bits	**Tail** T bits
9600	192	0	172	12	8
4800	96	0	80	8	8
2400	48	0	40	0	8
1200	24	0	16	0	8
14400	288	1	267	12	8
7200	144	1	125	10	8
3600	72	1	55	8	8
1800	36	1	21	6	8

the initial version of the IS-95 system underwent considerable modifications to achieve higher data rates [3]. It made it possible to use more than one channel in a link. Therefore the traffic channels applied in a link are called a *fundamental code channel* and *supplemental code channels.* Up to 7 supplemental code channels can be assigned to a link. The supplemental code channels always operate at the highest data rate within the rate set. Therefore, if Rate Set 2 is applied, the maximum data rate is $8 \times 14.4 = 115.2$ kbit/s.

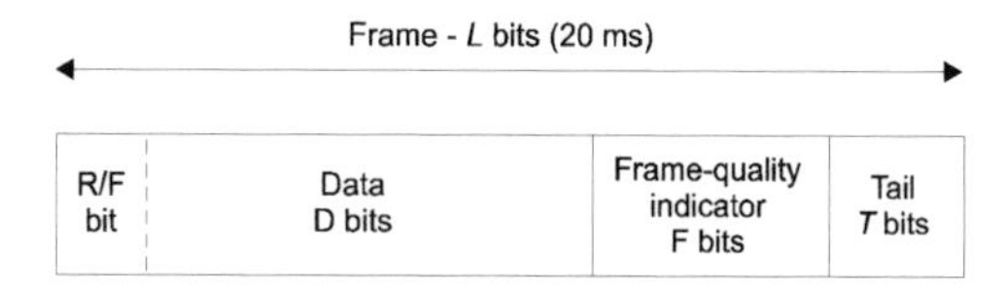

Figure 11.5 Forward traffic channel frame structure (R/F – reserved/forward flag bit)

The user data are transmitted in the form of 20-ms frames. The structure of a frame is shown in Figure 11.5. The lengths of particular fields depend on the data rate and are given in Table 11.3.

The frame-quality indicator field contains CRC bits calculated on the basis of the information bits in the frame. Data bits in a frame can be made up of primary, secondary or signaling traffic bits separately or a composition of them.

There is a long list of signaling messages transmitted by the base station on a traffic channel. The signaling messages are sent in the process of authentication, handover, power control, updating system parameters by sending information on neighboring cells, alerting, registration, etc. The signaling messages are similar in form to the messages sent on the paging channel. A message consists of an 8-bit message-length header, message data (so called *message body*) of a minimum of 16 bits and a maximum of 1160 bits and a 16-bit CRC block. Padding bits are added to fill the whole frame.

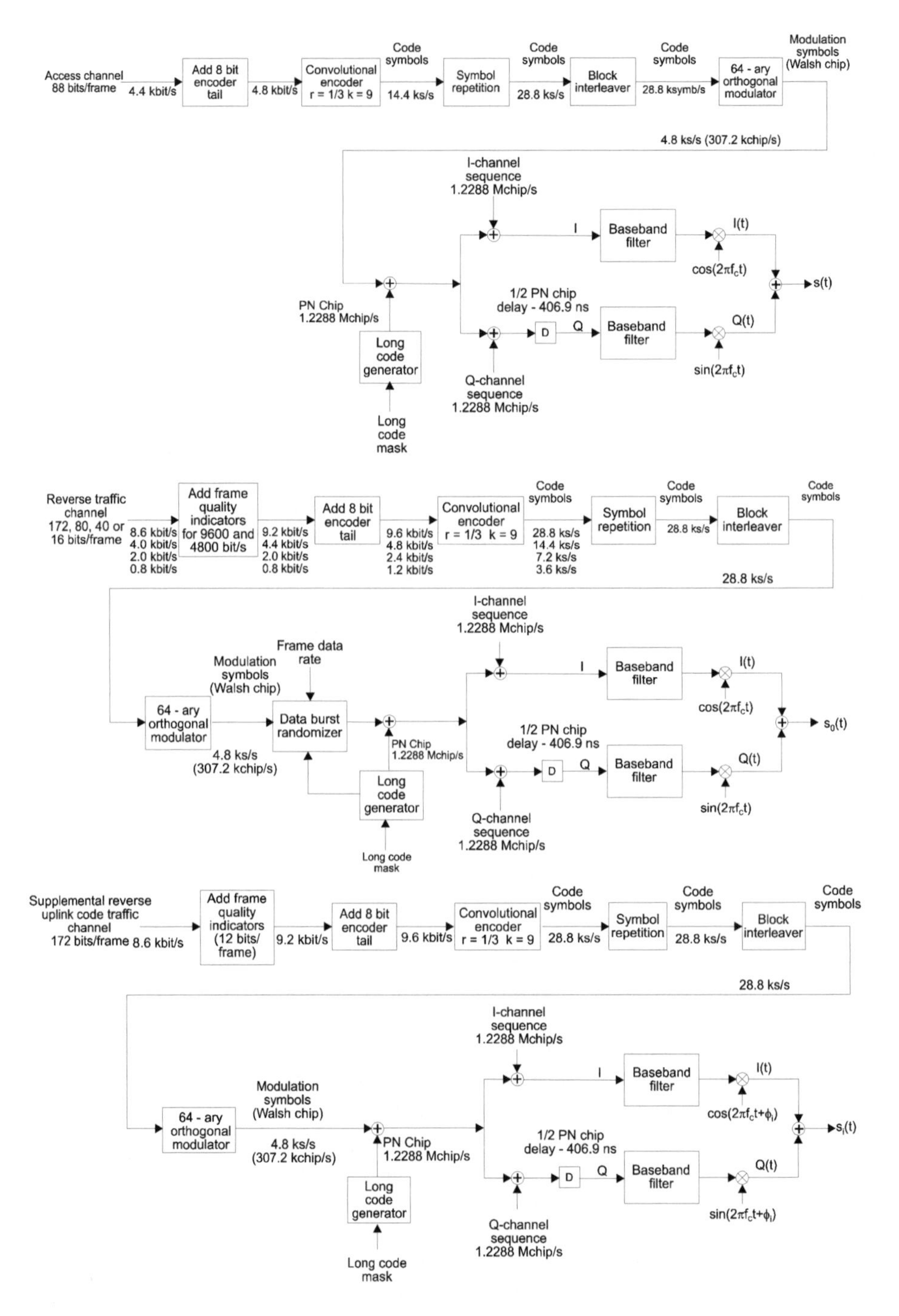

Figure 11.6 Mobile station IS-95 transmitter

11.4 UPLINK TRANSMISSION

Transmission from a mobile station to the base station differs considerably from that in the opposite direction. Figure 11.6 presents the scheme of a mobile station transmitter. There are parallel branches realizing *Access Channel* and uplink traffic channels. Let us note that as in the downlink (*forward*) transmission, besides a fundamental code traffic channel, up to seven supplemental code channels can be realized at the maximum data rate. The sequence of functional blocks for a single supplemental channel is shown in Figure 11.6. Let us note that the carrier phase for the ith channel is shifted by the angle ϕ_i with respect to the reference phase.

The same short spreading sequence 32768 chips long is applied in the uplink (*reverse*), and in the downlink transmission. All mobile stations in the cell apply the same time shift of the spreading sequence. The signals from different mobile stations are differentiated thanks to the application of a long pseudorandom sequence generated at the rate of 1.2288 Mchip/s using an individual mask which is unique for a given mobile station. The modulo-2 summing of the data sequence and the spreading sequence additionally improves transmission privacy.

A binary sequence characterizing the digital signal is encoded using a convolutional code of the coding rate $R = 1/3$ and the constraint length $L = 9$. The coded sequence is a subject of block interleaving in the time interval of 20 ms. The resulting output sequence generated at the rate of $R = 28.8$ kbit/s, is grouped into 6-bit blocks which constitute the address of one of 64 mutually orthogonal Walsh sequences. As a result, a stream of Walsh sequences at the rate of $R = 307.2$ kbit/s is achieved. Subsequently, the data sequence is fed to the data burst randomizer. The operation of this block depends on the current data rate generated by the voice encoder. If the user activity is low and the speech encoder generates data at the rate lower than 9600 bit/s, the data bits are repeated and the burst randomizer gates off some of them in a pseudorandom fashion. This way the power of the generated signal is minimized. The resulting data sequence is spread using PNI and PNQ short spreading sequences. After a binary-to-bipolar conversion and pulse shaping, it is used in the quadrature branches of an OQPSK modulator (see Chapter 1). OQPSK modulation allows for more efficient use of the mobile station power amplifier.

A similar scheme to that in Figure 11.6 can be drawn for Rate Set 2 uplink transmission; however, instead of the $R = 1/3$ code the $R = 1/2$ convolutional code is applied, so the data rates are 14.4, 7.2, 3.6 and 1.8 kbit/s.

Besides regular primary traffic (voice data stream), secondary traffic (data stream) or signaling can be sent in a 20-ms frame. Multiplexing of several streams can take place in two different ways. In the *blank and burst* mode, signaling data replace primary data. In the *dim and burst* mode, primary and signaling/secondary traffic are placed in the same frames.

There exists a long list of messages, which can be carried by an uplink (reverse) traffic channel. The messages are sent in a signaling part of a typical frame. These messages are associated with power measurement reports and power control, handover, authentication and several other orders.

In the *Access Channel* data are transmitted at 4800 bit/s. The access channel is used by a mobile station to initiate a call, to update the location of the mobile station and to respond to a paging issued by a base station. Each access channel is associated with one paging channel, so up to seven access channels can be established. Most of the functional blocks applied to realize an access channel are the same as for a traffic channel so they will not be described here.

The access channel message structure is similar to the structure of messages sent on the synchronization channel. It has an 8-bit message-length header, a data block of minimum 2 bits and maximum 842 bits and a CRC block of 30 bits. The messages are placed in 20-ms frames. Because the data rate on the access channel is 4800 bit/s, 96 bits are contained in a single frame. A frame is composed of 88 data bits and 8 encoder tail bits or it can contain an access channel preamble consisting of 96 zeros. An access channel preamble and a sequence of access channel frames form an *access channel slot.* Transmission within a slot begins with a short random delay to distribute transmission start times of several mobile stations using different access channels. When a mobile station uses the access channel for first time, it sends a sequence of probe messages at increasing power level until the appropriate power level for this mobile is determined. Apart from the case when two mobile stations use the same access channel and the same time shift of the pseudorandom sequence, the base station is able to receive signals from a few mobile stations. However, in order to limit the system load caused by access channels, the base station limits the number of simultaneous users of access channels. Controling the access to access channels is performed by sending the access parameters messages on the paging channel.

One of the most important control procedures performed in an IS-95 system is the mobile station registration. In this procedure the mobile station informs the base station on its location and status. There are various kinds of registrations:

- registration at power-up or at the moment of changing from an analog AMPS or alternative system to the digital CDMA system,
- registration at power-down, which indicates that the mobile station is no longer active,
- periodic registration due to timer expiration,
- distance-based registration performed when the mobile station has moved further than a certain distance from the place where it registered previously,
- zone-based registration performed after entering a new system zone (equivalent to a location area),
- registration caused by the change of some of the mobile station parameters,
- registration requested by the base station,
- registration ordered by the base station after successful use of the access channel by the mobile station.

Other important procedures are authentication and message encryption. A mobile station maintains an enciphering key and a set of shared secret data. The mobile station adds an 18-bit authentication signature to origination, page response, registration and data burst messages sent on the access channel. The shared secret data can be updated in response to an order from the base station. Enciphering user data can be switched on after the appropriate command (*Privacy Mode Command*) is issued by the mobile switching center.

11.5 POWER CONTROL

As we have already mentioned, power control is crucial for operation of a CDMA system. Various power control mechanisms are applied in the IS-95 system.

In the open loop power control, a mobile station measures the power received from the base station by attempting to set the level of the received signal in the *automatic gain control* (AGC) block. The transmit power of the mobile station is then determined from the following equation [4]:

$$\text{mean output power [dBm]} = -(\text{mean input power})\ \text{[dBm]} + \text{offset power} + \text{parameters} \tag{11.2}$$

The offset power level depends on the band in which the system operates and is equal to −73 dB for 800-MHz band and −76 dB for 1800-MHz band. The parameters used in (11.2) are transmitted on the synchronization channel and depend on the cell size, receiver sensitivity and the effective radiated power.

Let us note that the above method of power control is not very precise because frequency division is applied in the duplex transmission and up- and downlinks are separated by 45 MHz in the 800 MHz band and by 80 MHz in the 1800 MHz band. Therefore, the power level received by the mobile station only partially characterizes the uplink path loss. In particular, power control in the open loop is not able to compensate for short-term fading in the uplink. In order to cope with this problem the base station measures the received power from the mobile station and sends appropriate power control commands. Each 20-ms frame is divided into 16 power control groups. Each group is equivalent to 6 bits of the uplink sequence or one Walsh symbol used in the uplink orthogonal modulation. The signal-to-interference ratio is measured in each power control group, then it is compared with the reference SIR value and a decision is made if the power of the mobile station should be increased or decreased. As we have already mentioned, the power control bits are sent by replacing the transmitted bits in pseudorandom frame locations indicated by the decimated long pseudorandom generator. The dynamic range of the closed loop power control is ±24 dB. If the open loop is taken into account, then the joint dynamic range is ±32 dB for the 800-MHz band and ±40 dB for the 1800-MHz band. The final goal of the system physical layer is to ensure the required transmission quality measured by the tolerable level of the error rate. Therefore, the reference SIR is adjusted to meet this goal in a given cell.

Power control within a small dynamic range is also performed in the downlink. A base station gradually decreases the transmitted power in the 15- to 20-ms periods, until the frame error ratio measured by the mobile station exceeds an acceptable limit.

11.6 SIMPLIFIED CONNECTION SET-UP

Let us consider a mobile originated connection set-up. Let us assume that the mobile station is registered and its power is switched on.

First, the user dials the called user number and presses the "send" key. The mobile station sends an origination message to the base station using an access channel. The base station assigns the mobile station to the traffic channel using a paging channel and sends the called user number to the mobile switching center. After the MSC has completed the connection set-up, the mobile station is assigned to the reverse traffic channel and starts to transmit on it. As a result, the forward traffic channel and the mask of the long code used for ensuring data privacy are assigned to the connection.

The call flow for a mobile terminated call is the following. When a mobile system receives the call, it starts to set up a connection. The mobile station is searched by sending paging messages on the paging channels in the location area, i.e. in the cells or sectors in which the mobile station is probably located. After reception of a page the mobile station responds using the access channel and is assigned to a traffic channel. After full initialization of the traffic channel the base station sends special information causing the mobile station to ring. When the user answers the call, the base station is informed about his/her answer.

11.7 ENHANCEMENTS OF IS-95B FOR HIGH-SPEED DATA

As we have already mentioned, high data rates up to 115.2 kbit/s can be achieved in the IS-95 system without changing its physical layer, if more Walsh sequences are assigned to a link. Thanks to that, new services can be offered, such as database access, file transfer and electronic mail [5]. Most of these services generate bursty traffic. Therefore, IS-95B enhancements have been introduced to allow for high-speed data packet transmission. In packet transmission traffic in the forward and reverse link is asymmetric. In the forward direction the fundamental code channel and up to seven supplemental code channels are assigned to the link for the duration of a burst. In the reverse direction each supplemental channel is assigned a different pseudonoise sequence mask corresponding to a different pseudonoise sequence shift. Each mask is derived from the fundamental sequence mask. Each assigned channel is used at full data rate. Power control for all supplemental code channels is based on the fundamental code channel.

All this makes it possible to define the *high-speed packet data service option*. The service is established between the *Interworking Function* (IWF) and the mobile terminal (see Figure 11.7). During the negotiation procedure, the mobile terminal specifies its high-speed data capabilities understood as the number of parallel channels which can

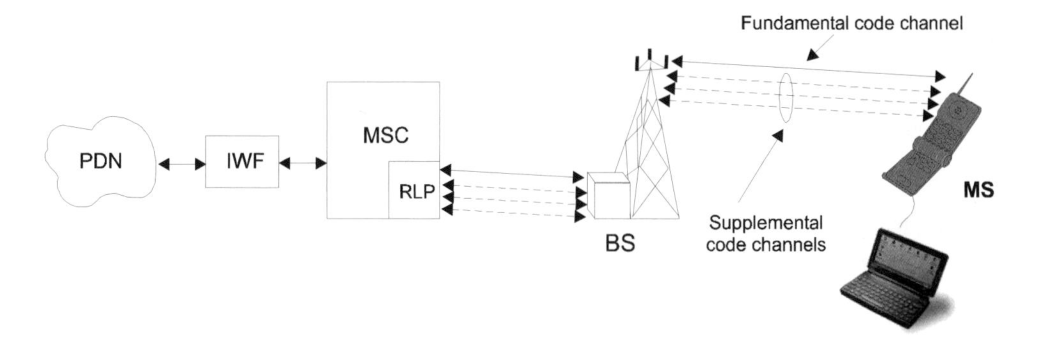

Figure 11.7 System configuration for high-speed packet data transmission (IWF – Interworking Function, PDN – Packet Data Network)

be applied in the up- and downlink. On the other hand, the base station specifies the maximum number of Walsh sequences which can be assigned to the negotiated up- and downlink. During packet data service realization, a mobile terminal can remain in two states:

- *Active state* – in which a traffic channel is assigned to the mobile station and the link and Point-to-Point Protocol (PPP) between the mobile terminal and the IWF is established,
- *Dormant state* – in which no RF or system resources are assigned to the connection, but the user's registration for packet data service and the PPP are maintained.

The mobile station realizing a packet data service remains in the active state for a certain period of time after sending a data burst in expectation of sending another burst. If the timer expires while waiting for the next data burst, the mobile terminal goes into the dormant state.

Let us briefly consider an uplink burst scenario [5]. Let the mobile terminal remain in the dormant state. If new data appear for transmission, the mobile terminal goes into active state. If the amount of data exceeds a predefined threshold, the mobile station requests supplemental code channels. This request is sent on the fundamental code channel. In response, the base station sends a message on the fundamental code channel in which it specifies the burst length, the number of supplemental code channels to be used and the starting moment of the burst. The mobile station can ask for an extension of the assigned burst before it ends. If there is no more data to send, the mobile station issues a message requesting zero supplemental code channels. In response, the radio resources are released and can be used by other users. A similar procedure can be described for the opposite direction. This time the IWF initiates the supplemental code assignment.

* * *

The architecture and the operation of the IS-95 system have been briefly sketched in this chapter. Details on realization can be found in the standards [1], [2] and [3]. It is worth noting that the system description of IS-95 conforms to the TR45/46 reference model [6], so almost all system entities have the same names as those applied in the GSM system description. Readers who are more interested in the operation of the IS-95 system and general rules of CDMA systems, are advised to study [7] whereas those who are interested in the system details may read [8].

Despite system complexity, the CDMA became the main technique applied in the proposals for third generation systems. Many solutions applied in IS-95 and patented by their inventors created a basis for the systems known as cdma2000 and UMTS. However, we have to realize that due to the limited spectrum resources, spectrum expansion existing in CDMA cannot be always applied, in particular if the data rates have to be very high. Therefore, other techniques, e.g. OFDM (see Chapter 1), have to be employed.

REFERENCES

1. TIA/EIA IS-95A, Mobile Station–Base Station Compatibility Standard for Dual Mode Spread Spectrum Cellular System, 1995

2. ANSI J-STD-008, Personal Station–Base Station Compatibility Requirements for 1.8 to 2.0 GHz Code Division Multiple Access (CDMA) Personal Communications Systems, 1995

3. TIA/EIA IS-95B, Mobile Station–Base Station Compatibility Standard for Dual Mode Wideband Spread Spectrum Cellular System, 1998

4. T. Ojanperä, R. Prasad (eds), *WCDMA: Towards IP Mobility and Mobile Internet*, Artech House, Boston, 2001

5. D. N. Knisely, S. Kumar, S. Laha, S. Nanda, "Evolution of Wireless Data Services: IS-95 to cdma2000", *IEEE Communications Magazine*, October 1998, pp. 140-149

6. TIA TR-46, Reference Model, 1991

7. J. S. Lee, L. E. Miller, *CDMA Systems Engineering Handbook*, Artech House Publishers, Boston, 1998

8. V. K. Garg, *IS-95 CDMA and cdma2000 Cellular/PCS Systems Implementation*, Prentice-Hall PTR, Upper Saddle River, N. J., 2000

12 Trunking systems

Trunking systems are the type of mobile communication systems which fulfill the need for communication between mobile vehicles of specialized government and public services such as police, emergency services, fire departments, and transport companies. Usually such services and companies use private communication systems called *Private Mobile Radio* (PMR).[1] In a conventional PMR system, each radio channel is permanently allocated to a given group of users. Due to the lack of appropriate channel management, it happens that despite the existence of free channels the users assigned to a congested channel cannot access the communication network. Such systems are sometimes called conventional dispatch systems.

12.1 THE IDEA OF TRUNKING

In order to rationalize the use of frequency channel resources, the idea of *trunking*, similar to trunking in traditional wireline systems, has been introduced. It relies on the possible assignment of a specified number of channels to all system users. The channel is allocated to a particular user dynamically for the time of connection. In consequence, we improve the use of channel resources, we can improve call privacy and service offer.

Figure 12.1 presents the idea of trunking in comparison with a conventional dispatch system. Let us note that the number of users currently served in the trunking system is equal to the number of channels. The number of users waiting for a connection is much smaller than in a conventional dispatch system. Obviously, the trunking system

[1] In the literature the abbreviation PMR is sometimes explained as *Professional Mobile Radio* [1].

requires channel management and ability of a mobile station to quickly synthesize the selected channel frequency.

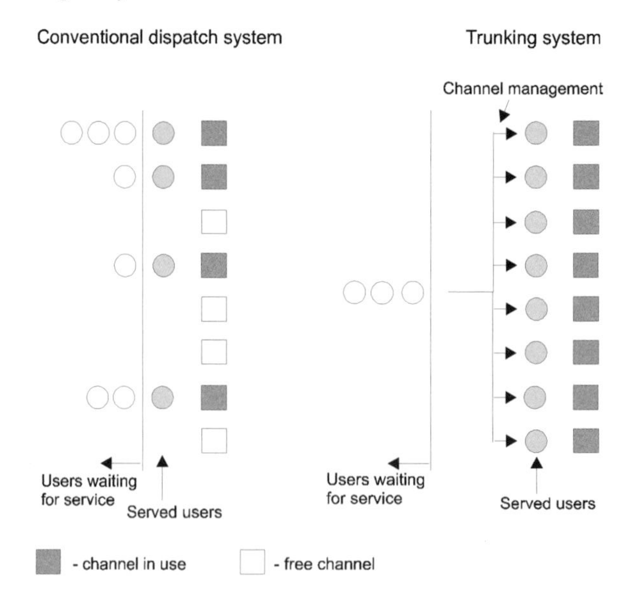

Figure 12.1 Channel assignment in a conventional dispatch system and in a trunking system

Let us note that thanks to the lack of a particular channel assignment to the given group of users, the system reliability is substantially improved. If a single channel is defective, it slightly decreases the whole system capacity without cutting off a whole group of users from access to the system, as would happen in a conventional dispatch system. At the moment of temporary loading of all channels, the trunking system does not reject the calls but queues the waiting users. The duration of a call can be limited by the system, which improves the effectiveness of connections.

Thanks to the fact that a channel is dynamically assigned to the user pair and during the connection another mobile station cannot use the same channel, in a certain sense the call is more private. In case of analog systems, further security improvements are rather difficult.

The peculiarity of trunking systems which distinguishes them from cellular systems at least partially results from the type of users. The characteristic feature of generated traffic is frequent connections of a dispatcher with mobile users (a fleet of vehicles). Sometimes many users are connected with the dispatcher simultaneously. On the other hand, connections with a PSTN are relatively rare. Trunking systems are fitted to specific types of calls or connections. These are:

- individual calls,
- group calls with connections to all users in a group,
- calls to a PSTN user (for selected users),

- alarm and alert calls,
- direct calls to public services,
- short digital messages,
- data transmission messages.

The characteristic feature of traditional trunking and dispatch systems is a *duo-simplex* mode applied in mobile stations. Base stations operate in duplex mode. During a call, the pair of users is assigned a pair of frequency channels (in up- and downlink). A mobile station is able to use only a link in one direction at a time. The base station receives the signal from a "talking" user on one frequency and transmits to the "listening" user on the other frequency. Due to this kind of use of channel resources, it is possible to realize the group calls without selection of individual numbers of the group members.

Trunking systems can be owned by a single company or can be public. The application of *Public Access Mobile Radio* (PAMR) is particularly advantageous from the point of view of spectrum efficiency and system organization. Participating companies have their own subnetwork within the PAMR network for their exclusive disposal. The companies do not need to have a licence for the use of the assigned frequencies. The network operator holds such a licence for all used frequencies.

12.2 MPT 1327 STANDARD

There are a few standards of analog trunking networks. Some standards are owned by companies (e.g. Motorola, Ericsson), others are public. In the second half of the eighties a whole family of standards has been established by the British Ministry of Post and Telecommunications. The most important is the signaling standard which describes the data exchange protocol between a base station and mobile stations and is described by the symbol MPT 1327 [2]. Many leading companies have offered trunking systems based on this protocol. Other standards associated with MPT 1327 are listed below.

- MPT 1317 specifies signaling.
- MPT 1318 is a memorandum on the trunking efficiency.
- MPT 1343 specifies the operations of the terminal equipment and defines the functions of system control and access to the traffic channel.
- MPT 1347 specifies the base station, describes the functions of the fixed network of the system and formulates the directives on the allocation of identity numbers.
- MPT 1352 specifies testing procedures to verify compatibility of mobile stations.

The MPT 1327 standard defines the random access to the system, so-called *Dynamic Framelength Slotted ALOHA*. The frames generated by mobile stations have variable length which is a multiple of time slots. If a frame is contained in a single time slot, then the next time slot can be applied by another mobile station in order to set up the connection. If the frame is longer than a time slot, then the mobile stations wishing to set up the connection select a free time slot randomly.

The connection set-up is supervised by a *Trunking System Controller* (TSC). It controls the frame length with respect to collisions between access attempts made by several mobile stations. Transmission of signaling data between the TSC and mobile stations takes place in the control channel. The standard specifies the digital transmission with MSK modulation at the data rate of 1200 bit/s. A control message is transmitted to mobile stations in the control channel with the period of 106.7 ms. The message alternately contains two control sequences of length equal to 64 bits. The first control sequence is the *Control Channel System Codeword* (CCSC) and the second one is the *Address Codeword* (AC) (see Figure 12.2).

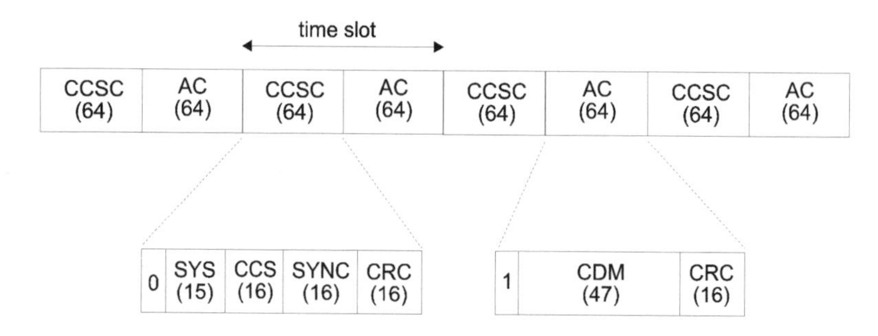

Figure 12.2 Structure of messages transmitted on the MPT 1327 control channel

The message starts with a CCSC word. It consists of a one-bit start flag equal to "0", 15 system identification bits (SYS - *System Identity*), 16-bit *Code Completion Sequence* (CCS), 16-bit preamble (SYNC) and 16-bit parity block CRC. This block contains the value $C4D7_H$, so the contents of the CCS word is so selected that the calculation of the CRC results exactly in the value of $C4D7_H$. The preamble is used in synchronization and contains the word $AAAA_H$.

The address codeword starts with the flag bit "1" followed by the *Codeword Message* (CDM) containing 47 information bits. The whole message is completed with a 16-bit parity block CRC.

The codeword message can have various meanings. It can be the information from the controller containing the number of a time slot in which the random access to the system can be realized. It can be the call sent by the system to mobile stations in expectation of the answer declaring the mobile stations to be ready to accept the connection, the order sent to the mobile station to switch from the control channel to the specified traffic channel or to switch from one traffic channel to another. It can also be a broadcast message indicating the system status. The codeword message can be also issued by a mobile station and can contain a connection request, or acknowledgement that the mobile station is ready to realize a call. The meaning of messages sent on the control channel is the following:

- ALOHA (ALH) – a message sent by the trunking system controller inviting initialization of a connection. The message ALH(n) indicates that the connection set-up is possible in the next n time slots.
- AHOY (AHY) – a message generated by the trunking system controller in expectation of the acknowledgement that a mobile station is ready to set up a connection,
- REQUEST (RQS) – a request for connection set-up issued by a mobile station,
- ACKNOWLEDGE (ACK) – an acknowledgement of RSQ or AHY message,
- GO TO TRAFFIC CHANNEL n (GTC) – a message requesting the mobile station to switch from the control channel or from the current traffic channel to the traffic channel n,
- BROADCAST (BCAST) – a broadcast message containing system information.

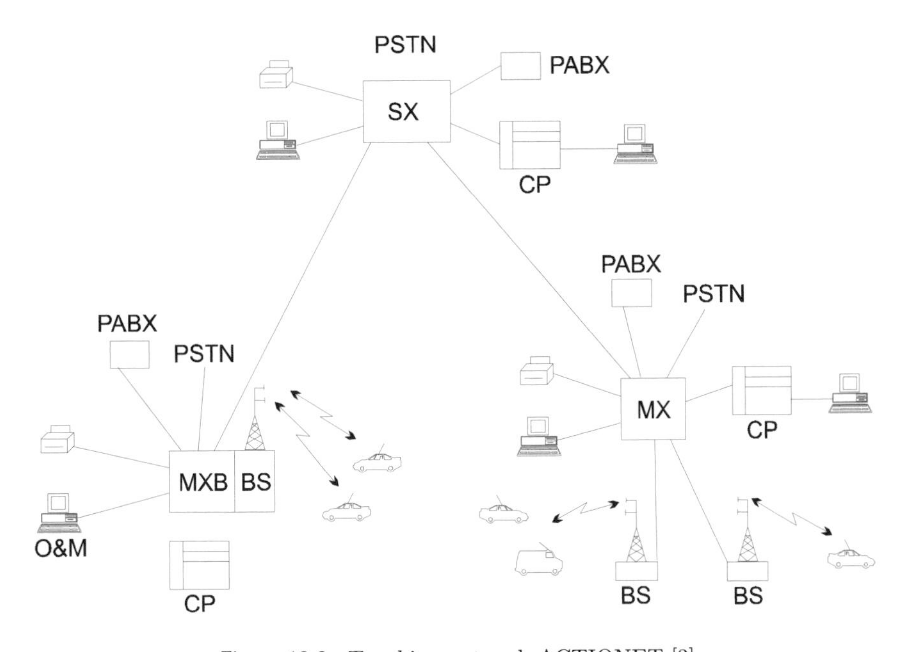

Figure 12.3 Trunking network ACTIONET [3]

An example of a trunking network functioning according to the MPT 1327 standard is ACTIONET by Nokia [3] (see Figure 12.3). This system can operate in a few different configurations. The simplest configuration consists of a number of mobile stations and a switching center (MBX) deployed jointly with the base station. This way a simple small private company network can be built.

The basic configuration features a mobile switching center (MX), one or a few base stations and a number of mobile stations. Larger configurations consist of a few mobile switching centers (MX) and base stations, a few control points (CP) or dispatch centers connected to a mobile switching center (MX). Such a network is usually connected with a PSTN or an internal PABX (*Private Branch Exchange*).

The main task of a mobile switching center MX in a local network is to provide automatic connection set-up between two mobile users, between a mobile and a fixed user of an PABX, and between a mobile and a fixed user of a PSTN exchange. The control point (CP) is a telephone console consisting of a keyboard, microtelephone and alphanumeric display.

If the trunking network covers a large area, an additional element integrating the local parts of the network is applied. It is a system exchange (SX). Its main blocks are doubled for high system reliability. In the case of a large network, the data update in the mobile station register can be necessary if some mobile stations leave their home MX service area. Very large networks consisting of a few independent traffic areas are also possible. It is assumed that mobile stations remain within their own traffic areas and traffic concentrates within these areas as well; however, there is a need for information exchange between traffic areas. Very large networks are equipped with additional blocks called *System Roaming Handler* (SRH) – the database of the current users' location, and *Call Information Handlers* (CIH) – the database of realized and attempted connections, used for tariffing purposes.

Typically, the links between particular mobile switching centers are implemented using a PSTN network. Each mobile switching center is equipped with the *Operation and Maintenance Center* (OMC). The OMC manages the system and equipment configuration and undertakes appropriate action in case of system failures or due to security reasons.

Generally, trunking networks based on the MPT 1327 protocol are characterized by a simple technology and service, and wide availability of equipment. They have, however, some disadvantages: poor spectrum efficiency, low frequency reuse factor, relatively low security of speech signals, high cost of the devices improving speech security, large network costs due to assignment of a single traffic channel to each carrier and low data transmission rates.

12.3 EDACS - AN EXAMPLE OF A PROPRIETARY TRUNKING SYSTEM STANDARD

The *Enhanced Digital Communication System* (EDACS) [4] has been developed by Ericsson. The company holds an exclusive right to build the network infrastructure and user equipment. The system was built to work mainly with security services, therefore it has some special features which are essential for these applications. It features a short access time (< 250 ms) and the possibility to transmit voice and data transmission on all channels. It improves the system reliability because the control channel can be located on any carrier. Because the system is applied in public security services, a special emergency call request initiated by pushing a single button is introduced. Thus, with the highest priority the mobile station is assigned the first free channel. The emergency call request is signalled to other mobile stations by a special sound (beep) and display of the identity of the mobile station sending the alarm.

The EDACS has narrowband and wideband versions. In the first one the channel separation is 12.5 kHz, whereas in the second one it is 25 kHz. EDACS can realize communications in four modes:

- analog voice transmission,
- encrypted digital voice transmission (in the wideband version of EDACS) (*Voice Guard*) at the data rate of 9600 bit/s,
- data transmission (4800 bit/s in the narrowband version or 9600 bit/s in the wideband version of the system),
- connection with a PSTN.

As we know, group calls are the basic way of communications in trunking systems. Group calls realized in EDACS have the following properties:

- blocking double transmission – a single user transmits to the group and the remaining users have to listen,
- possibility of searching the groups – the users can search many traffic groups,
- identity display – the identity of the currently transmitting mobile station is displayed on the screen of other mobile stations belonging to the same group,
- automatic addition to group – a mobile station is automatically added to its group after switching power on or after appearing in the system coverage area.

The system can operate in several configurations, depending on the size of the covered area. The basic configuration has a single base station. The configuration on level 1 supports full trunking, on level 2 the system is supplemented with electronic control of the dispatcher's console, on level 3 the data transmission is supported. The configuration on level 4 allows for wide area coverage and the simulcast principle is applied – several base stations transmit and receive simultaneously.

The wideband version of EDACS can operate in the frequency ranges 136–174, 403–515 and 806–870 MHz. The narrowband system operates in the range 894–941 MHz.

The EDACS trunking system, with its gradual extensions and service enhancements, has been intended by Ericsson for smooth transition to the second generation trunking system TETRA.

12.4 TETRA – A EUROPEAN STANDARD OF A TRUNKING SYSTEM

Due to the already listed disadvantages of analog trunking systems and their mutual incompatibility, in the early nineties the European Community undertook to define a new digital trunking system standard. As a result, a series of ETSI standards have been specified, which define *Terrestrial Trunked Radio* (TETRA).[2] In fact, two families of standards have been worked out. They are:

- Voice plus Data Standard (V+D),
- Packet Data Optimized Standard (PDO).

The V+D standard specifies the second generation digital trunking system, whereas the PDO standard family defines a packet data radio system.

TETRA has the following features:

- two- to four times higher spectral efficiency as compared with analog systems,
- better frequency reuse,
- much higher voice security due to the application of digital voice encoding with possible encryption,
- voice quality independent of the received signal power in a wide power level range,
- high rate of data transmission,
- possible cell sectorization.

12.4.1 Services offered in TETRA

The TETRA system offers a wide range of services in circuit- and packet-switched modes. The services are divided into two categories: teleservices and bearer services, similar to GSM. Within teleservices five different voice connections are offered:

- *individual call* – a point-to-point connection between two individual users,
- *group call* – a point-to-multipoint connection between a calling user and the group of users distinguished by their group number; transmission is performed in a half-duplex mode,
- *direct call* – a direct point-to-point connection between two users (mobile stations) without using a base station; at least one mobile station has a connection to the base station on a channel different from that used in the direct call,
- *acknowledged group call* – a point-to-multipoint connection between a user and a group of users distinguished by their group number; the called users acknowledge their presence in the connection,
- *broadcast call* – a point-to-multipoint connection in which the called users can only listen to the calling user.

[2] Formerly Trans-European Trunked Radio.

The following bearer services are offered in the V+D type of TETRA operation:

- 7.2–28.8 kbit/s circuit-switched, unprotected speech or data transmission,
- 4.8–19.2 kbit/s circuit-switched, minimally protected data transmission,
- 2.4–9.6 kbit/s circuit-switched, highly protected data transmission,
- connection-oriented point-to-point packet transmission,
- connectionless point-to-point packet transmission in a standard format,
- connectionless packet transmission in a special format (point-to-point, multipoint, broadcast).

The last three services are also offered in the PDO mode of TETRA.

Due to possible applications of TETRA (e.g. in the public security sector) additional services have been established, some of which are listed below:

- user's authentication, performed by the dispatcher,
- discreet eavesdropping of a conversation by an authorized user,
- priority call with interruption,
- setting up call request priorities,
- call holding, connect-to-waiting – the user can suspend the current call in order to take another call and later come back to the first call,
- identification of a calling user,
- short number addressing,
- dynamic creation and modification of user groups.

12.4.2 General architecture of TETRA

Figure 12.4 presents the TETRA reference standard and the interfaces placed between appropriate system blocks [8]. Figure 12.4 shows the types of stations used in the TETRA system, the control blocks and registers and possible connections with external networks.

A *Mobile Station* comprises a block called *Mobile Termination* which denotes the radio telephone and a block called *Terminal Equipment* which allows the user to transmit data.

A *Line Station* functions like a mobile station; however, it is connected to the switching and management infrastructure over the ISDN line. A line station can be used in a company network as a dispatcher's station.

The *Switching and Management Infrastructure* (SwMI) contains base stations, the main switching center (MSC), local switching centers (LSC) with location registers (LR) and the operations and maintenance center (OMC).

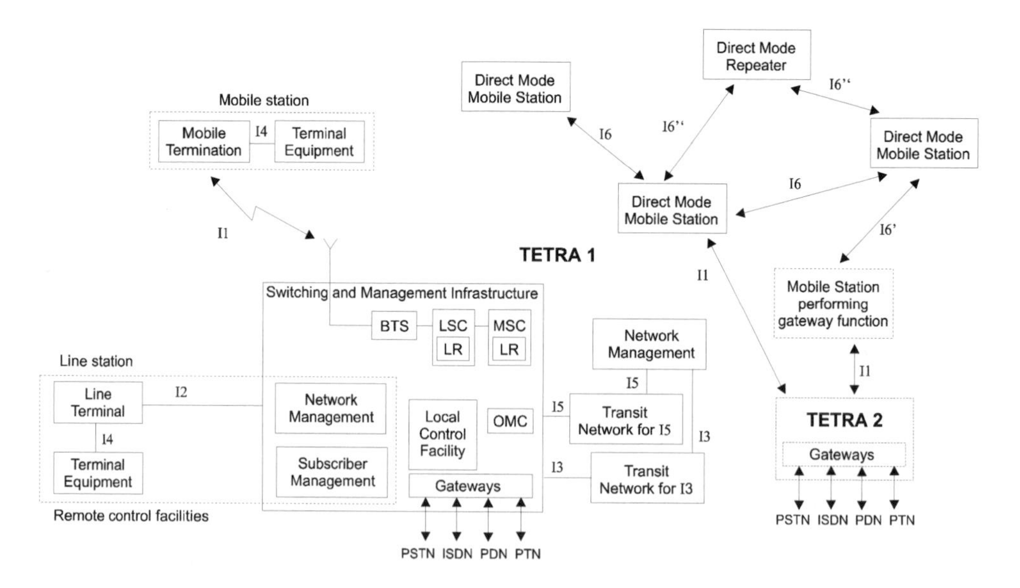

Figure 12.4 The TETRA reference standard with interfaces

Let us stress the possibility of establishing direct connections between mobile stations without using the network infrastructure. A number of interfaces between network elements exist. They are listed in Table 12.1.

Table 12.1 Types of interfaces applied in TETRA

Interface notation	Interface function	Interface notation	Interface function
I1	Radio air interface	I4	Terminal equipment interface
I2	Line station interface	I5	Network management interface
I3	Intersystem interface	I6	Direct mode radio interface

12.4.3 The TETRA physical layer

The following parameters of the TETRA physical layer have been standardized:

- frequency channel width equal to 25 kHz,
- four voice/data channels in the TDMA mode transmitted on each carrier with the interleaved control channel,
- direct mobile-to-mobile user calls possible within a limited distance between the mobile stations and with some technical limitations,

- frequency ranges in Europe:
  - uplink: 380–390 MHz, downlink: 390–400 MHz,
  - uplink: 410–420 MHz, downlink: 420–430 MHz,
  - uplink: 450–460 MHz, downlink: 460–470 MHz,
  - uplink: 870–888 MHz, downlink: 915–933 MHz,
- user terminals: handhelds, portable, vehicle-mounted or fixed with a possibility of a computer connection (data modem at the rate of 4800 bit/s),
- assignment of more than a single time slot to transmit data at rates up to 28800 bit/s (net) in a 25 kHz channel.

Let us briefly describe the TETRA physical layer [5]. We will analyse the following aspects:

- burst structure and multiplexing of logical channels,
- coding, sequence reordering, interleaving and scrambling,
- differential encoding and modulation.

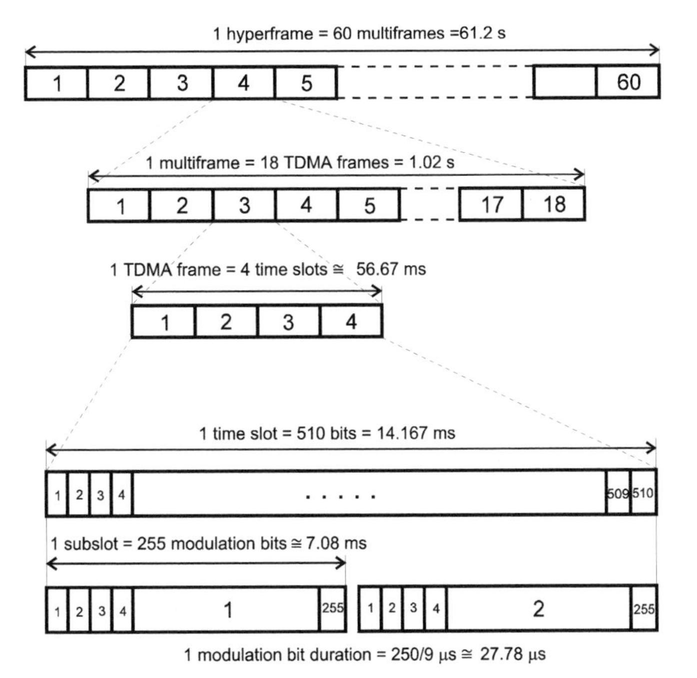

Figure 12.5 Time hierarchy of TETRA

As we have already mentioned, the system operates in the TDMA mode with four physical channels on each carrier separated from the neighboring carriers by 25 kHz. The basic time unit is a time slot. It lasts for 14.166 ms (85/6 ms). Within a time slot a binary sequence is transmitted at the rate 36 kbit/s. Thus, 510 bits, or equivalently, 255 four-level data symbols are placed in a time slot. Figure 12.5 presents the time hierarchy of TETRA.

The hyperframe is at the top of the hierarchy. It lasts for 61.2 s and it is divided into 60 multiframes. The duration of each multiframe is 1.02 s and each multiframe consists of 18 frames. The eighteenth frame is a control frame. Each frame has a duration of 56.67=170/3 ms and is divided into four time slots. The uplink slots can be further subdivided into two subslots. The bursts which carry information sequences are transmitted within a single time slot. The types of bursts will be described by the end of this chapter.

In TETRA three types of physical channels are defined. These are:

- *Control Physical Channel* (CP) which transmits exclusively control messages. One of the control physical channels is defined as the *Main Control Channel* (MCCH), whereas the others are called *Secondary Control Channels* (SCCH). The carrier transmitting the main control channel is called the main carrier. The MCCH is always transmitted in the first time slot on the main carrier.
- *Traffic Physical Channel* (TP) which mostly carries traffic channels.
- *Unallocated Physical Channel* (UP) which carries broadcast or empty messages.

Logical channels are assigned to the physical channels according to one of the following modes of operation:

- transmission modes:
  - *Downlink Continuous Transmission Mode* (D-CT),
  - *Downlink Carrier Timesharing Mode* (D-CTT),
  - *Downlink Main Control Channel Timesharing Transmission Mode* (D-MCCTT),
  - *Multiple Slot Transmission Mode* (MTS),
- control modes:
  - *Normal Control Mode* (NCM)
  - *Minimum Control Mode* (MCM).

The D-CT mode is mandatory for mobile stations. It means that a mobile station has to be able to co-operate with the base station if the latter works in this mode. The NCM mode is obligatory for all TETRA equipment. All mobile stations should be able to operate in the minimum control mode.

In the D-CT mode base stations apply continuous downlink bursts (see Figure 12.9 showing the burst structure used in TETRA). The transmission on the main carrier is continuous, although it can be discontinuous on other carriers.

In the D-CTT mode a carrier frequency can be shared by several cells. Each of the physical channels of this carrier (realized in four possible separate time slots) can be assigned to different cells. In this case a base station uses discontinuous downlink bursts.

In the D-MCCTT mode the main control channel is jointly used by different cells. Each frame of this channel is independently assigned to these cells.

The MTS mode is characterized by possible allocation of two to four physical channels to the same link. The aim of this allocation is to increase the data rate or to transmit mixed binary streams of voice and data.

The NCM mode requires the assignment of the main control channel (MCCH). A full range of control functions is ensured. In the NCM mode TETRA services are performed in a limited form. In this mode all physical channels of every carrier used by the system carry traffic channels.

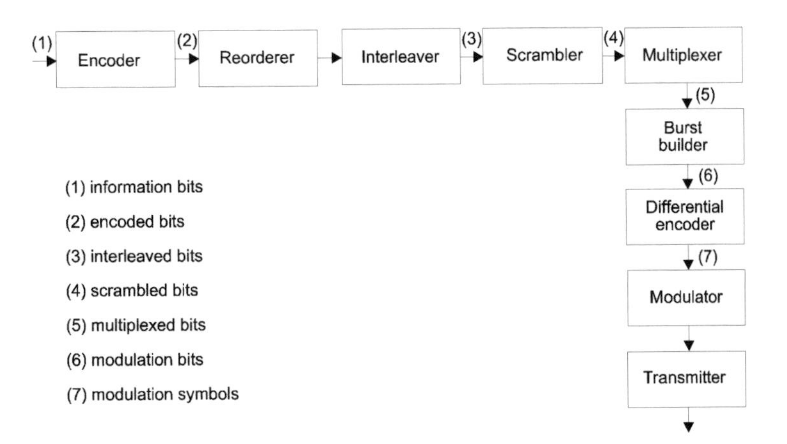

Figure 12.6 Block diagram of a TETRA transmitter [5]

Let us describe a block chain which prepares the signal for radio transmission. It is shown in Figure 12.6.

The bit stream received from the speech encoder or data terminal is subjected to channel encoding (block or convolutional), possible reordering, interleaving and scrambling. The type of the applied error correction code and interleaving depends on the type of the logical channel. The details of this type of processing are different for each logical channel type and can be found in Figure 12.7. The abbreviations in Figure 12.7 denote:

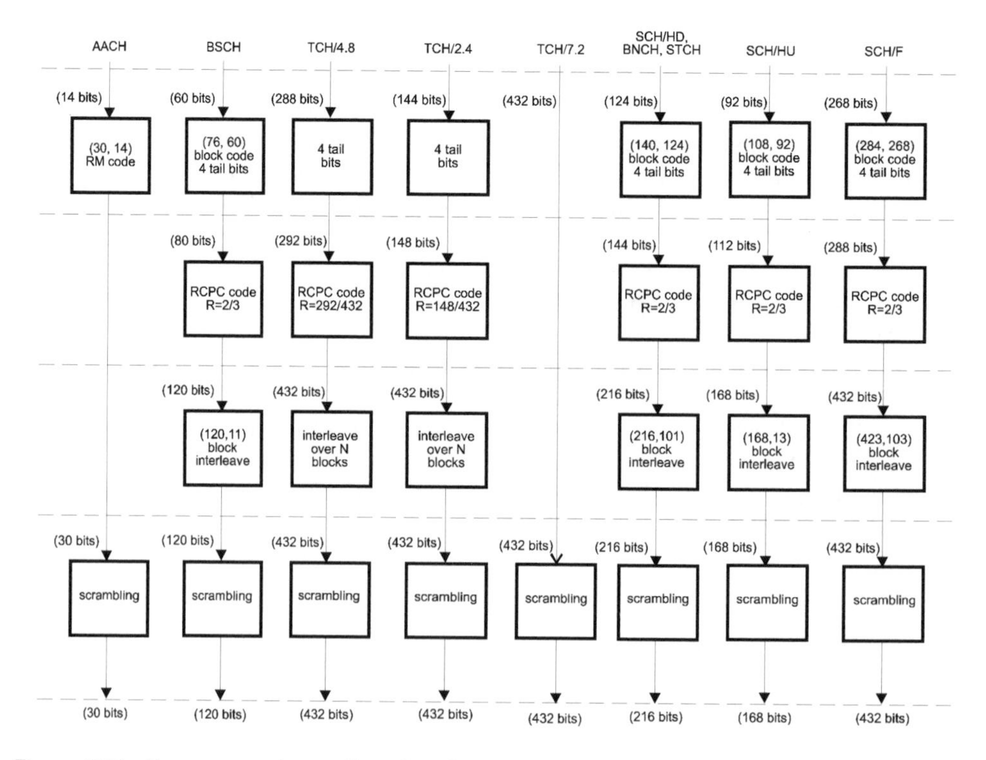

Figure 12.7 Error correction coding, interleaving and scrambling applied in several TETRA (V+D) logical channels

- shortened RM code – (30,14) Reed-Muller code,[3]
- RCPC code – *Rate-Compatible Punctured Convolutional Code* (see Chapter 1) realized in TETRA in two steps:
  - coding using the rate-1/4 16-state mother code
  - puncturing of selected bits in order to receive the desired coding rate.

The encoded and interleaved bits are multiplexed and fitted into the information fields of appropriate type of bursts used to realize a specified logical channel. The resulting binary data stream is differentially encoded and fed to the modulator.

As opposed to the rather complicated GMSK modulation applied in GSM, the TETRA system uses the so-called $\pi/4$-DQPSK modulation. It is a version of a differential quadrature phase shift keying. The binary rate is assumed to be 36 kbit/s.

[3]Reed-Muller codes (RM) are binary block codes which are equivalent to cyclic codes extended by a parity bit checking the parity of all code word bits. The (30,14) shortened RM code is derived from the (32,16) RM code by setting two information bits to zero and not transmitting them.

The binary stream is sliced into dibits which determine the phase shift with respect to the previous modulation period. These phase shifts can be $\pm\pi/4$ or $\pm 3\pi/4$. Figure 12.8 presents the signal trajectories for all possible dibit combinations when digital symbols are shaped by the baseband filter of the square-root raised cosine characteristic with the roll-off factor equal to 0.35. Despite the fact that the modulation does not have a constant envelope, its variability is limited and it never reaches zero. This is an advantage with respect to nonlinear distortions generated by the nonlinear power amplifier applied in mobile stations. $\pi/4$-DQPSK modulation has been already used in American TDMA cellular and PCS standards IS-54/136 [6]. The TETRA standard [5]

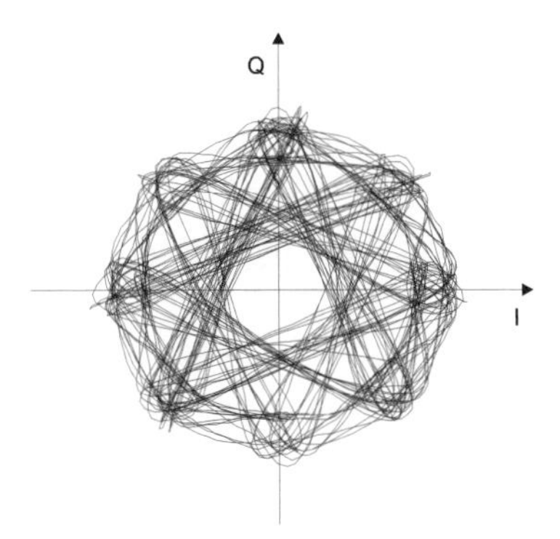

Figure 12.8 The in-phase and quadrature components of the $\pi/4 - DQPSK$ modulation with the square-root raised cosine filter applied in TETRA

also defines a set of channel frequencies for up- and downlink directions. In general, the frequencies of the nth downlink and uplink channels are expressed by the formulas

$$F_{up,n} = F_{up,\min} + 0.001G + 0.025(n - 0.5) \quad [\text{MHz}] \quad n = 1, 2, \dots, N$$
$$F_{dw,n} = F_{up,\min} + D \quad [\text{MHz}], \quad n = 1, 2, \dots, N$$

where $F_{up,n}$, $F_{dw,n}$ are the nth channel frequencies at the uplink and downlink transmission, respectively. G is the guard period given in kHz and D is the frequency separation between uplink and downlink bands.

The TETRA standard defines the classes of mobile and base stations, the level of intermodulation distortions and out-of-band unwanted emission. Ten classes of base station power levels have been set, starting from 40 W and ending with 0.6 W, which is equivalent to the power level per carrier between 46 and 28 dBm. In turn, the mobile stations belong to one of four power classes 30, 10, 3 and 1 W, which is equivalent to the nominal power level of 45, 40, 35 and 30 dBm.

As in the GSM system, a logical channel is a logical communication link between two or more link participants. Let us explain the types and tasks of the logical channels applied in TETRA.

The logical channels can be divided into two categories. They are:

- traffic channels carrying binary sequences representing voice or data in circuit switched mode, and

- control channels transmitting signalling messages and packet data.

There are a few types of traffic channels, such as:

- *Speech Traffic Channel* (TCH/S),
- circuit mode traffic channels at the net data rate 7.2 (TCH/7.2), 4.8 (TCH/4.8), and 2.4 (TCH/2.4) kbit/s.

Higher data rates than those listed above are possible by allocating more time slots to the same link. This way 9.6, 19.2 or 28.8 kbit/s can be achieved.

There are five categories of control channels.

- *Broadcast Control Channel* (BCCH) is a uni-directional channel for common use by all mobile stations. It transmits general information useful for all mobile stations. It is further subdivided into:
  - *Broadcast Network Channel* (BNCH) distributing network information to mobile stations,
  - *Broadcast Synchronization Channel* (BSCH) broadcasting information necessary for time and scrambling synchronization in mobile stations.
- *Linearization Channel* (LCH) is used for linearization[4] of the base station transmitters (BLCH – *Base Station Linearization Channel*) as well as mobile station transmitters (*Common Linearization Channel* – CLCH).
- *Signalling Channel* (SCH) is shared by all mobile stations; however, it may transmit messages to a single mobile station or to a group of them. In each base station at least one signalling channel is established. The signalling channel is divided into the following categories:
  - *Full Size Signalling Channel* (SCH/F) – a bidirectional channel transmitting full size messages,
  - *Half-size Downlink Signalling Channel* (SCH/HD) – used for transmission of half-size messages from base to mobile stations,
  - *Half-size Uplink Signalling Channel* (SCH/HU), same as above; however, used in the uplink direction.
- *Access Assignment Channel* (AACH) is a downlink channel, which, when transmitted on a physical channel, indicates the allocation of the uplink and downlink slots.
- *Stealing Channel* (STCH) is associated with the traffic channel. It uses part of the traffic channel capacity to transmit fast signalling information.

[4]Linearization is related to the power amplifier of a mobile station, which due to a limited source of energy and required power efficiency has to work in a nonlinear range of its characteristics. Due to non-constant envelope modulation used in TETRA, one can apply a predistorter, which distorts the signal before amplification in such a way that after passing through the amplifier the signal remains undistorted.

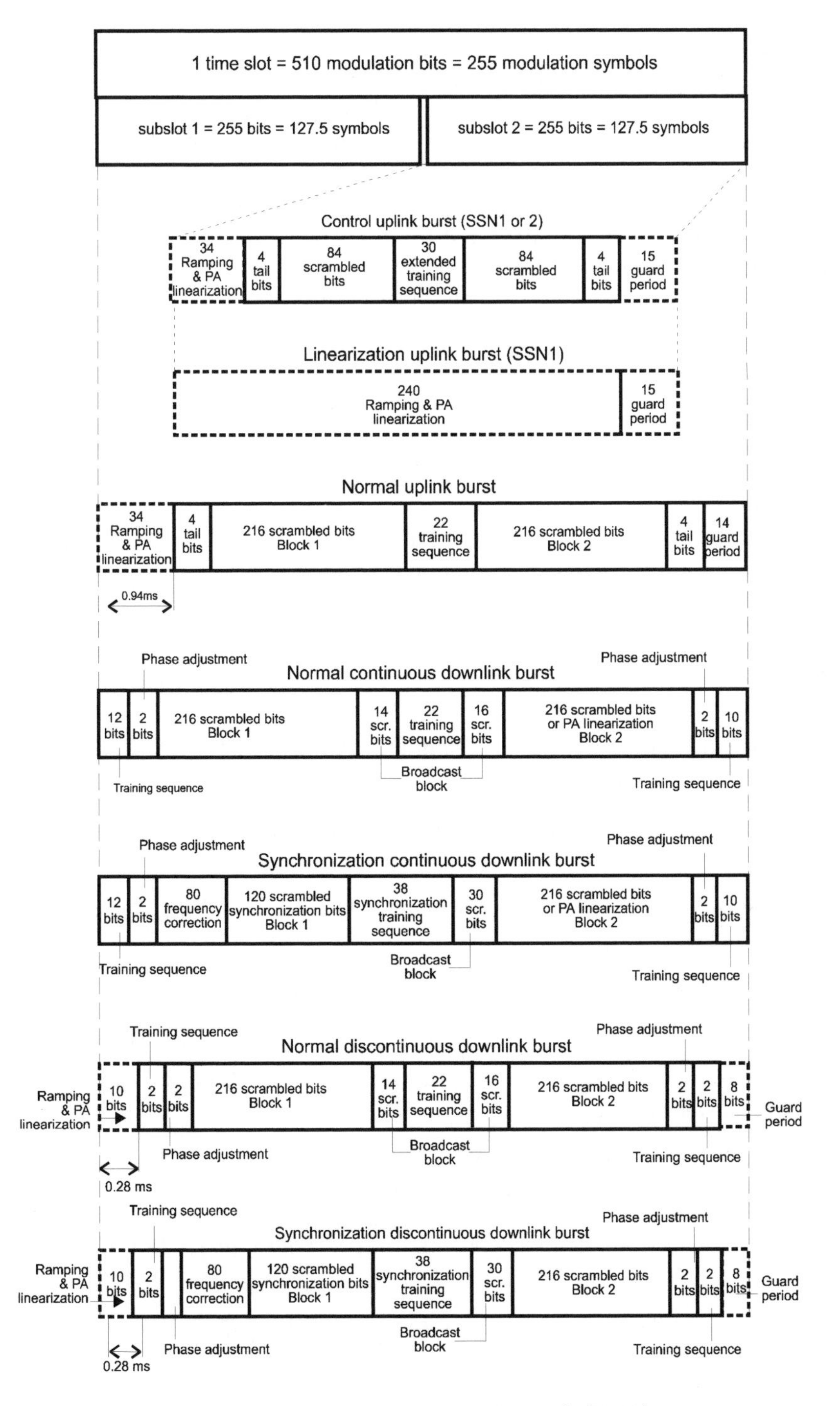

Figure 12.9 Burst structures in TETRA (V+D)

As we have already mentioned, transmission in the TETRA system is, as in GSM, performed using data bursts. A burst represents physical contents of the time slot or the time subslot. There are six types of bursts. They are shown in Figure 12.9. We clearly see a similarity of TETRA bursts to most of the GSM bursts described in Chapter 7. As in GSM, training sequences are placed in the middle of the burst. Due to a smaller number of time slots in a frame, the bursts are longer and they transmit higher numbers of bits as compared with GSM bursts. They also realize additional functions, in particular those associated with broadcasting. Let us note again a power amplifier linearization or frequency correction performed within the synchronization burst.

* * *

The information presented on the Terrestrial Trunked Radio (TETRA) dealt mostly with its services, types of connections and its physical layer. Higher layers of the system remained beyond the scope of this short survey. The interested reader is advised to study the TETRA standards or read the chapter devoted to TETRA in [7]. An overview of different trunked radio systems can be found in [8].

REFERENCES

1. J. G. Gibson (ed.), *The Mobile Communications Handbook*, CRC Press, Inc. and IEEE Press, Boca Raton, Fl., 1996

2. MPT 1327, A Signaling Standard for Trunked Private Land Mobile Radio System, 1988

3. E. Juszkiewicz, "ACTIONET - A Nokia Trunking System", *Przegląd Telekomunikacyjny* (in Polish), 1995, pp. 265-270

4. "EDACS – A Modern Radio Trunking System", Ericsson, 1996

5. Draft ETS 300 392-2, Radio Equipment and Systems (RES); Trans-European Trunked Radio (TETRA); Voice plus Data (V+D); Part 2: Air Interface (AI), Second Edition, December 1999

6. L. J. Harte, A. D. Smith, Ch. A. Jacobs, *IS-136 TDMA Technology, Economics, and Services*, Artech House, Boston, 1998

7. B. H. Walke, *Mobile Radio Networks: Networking and Protocols*, John Wiley & Sons, Ltd., Chichester, 1999

8. N. J. Boucher, *The Trunked Radio and Enhanced PMR Radio Handbook*, John Wiley & Sons, Inc., New York, 2000

13

Digital cordless telephony

Cordless telephony is worth consideration not only due to its meaning for the wireless access to a communication network but also because of some special applications, e.g. in a *Wireless Local Loop* (WLL). The latter topic will be the subject of Chapter 14.

Wireless telephony enables the user to realize a connection between a handset (mobile station) and a base station connected to a PSTN or to a *Private Branch Exchange* (PBX). The range of wireless phones is usually limited to a few hundred meters and it is assumed that a mobile user moves at pedestrian speed. There are several standards of wireless telephony, denoted by the abbreviations: CT1, CT2, CT3, DECT, PACS and PHS. CT*i* stands for *Cordless Telephony*, DECT denotes *Digital Enhanced Cordless Telecommunications*. PACS stands for *Personal Access Communications System*, and PHS denotes *Personal Handyphone System*. These two latter systems operate in America and Japan, respectively. They will be briefly described later.

Wireless phones are applied in the simplest form in many homes and offices. They differ in technical solutions, complexity level and cost. Many of them feature analog duplex voice transmission with FDD and frequency modulation. In practice, a mobile station operates with a single base station and the number of cordless phones in the range of a single base station is low. The information channel is not protected against eavesdropping. In the most primitive solutions an external phone can interfere with or can replace our own phone. In better solutions, at the moment of the connection set-up the mobile station or the base station searches the channel frequencies checking if they are not occupied by another transmission in close vicinity. The next improvement is the application of a password exchange between the mobile station and the base station in order to prevent a connection with an external phone.

In 1984 most of the European countries accepted the CT1 standard of analog cordless telephony; however, the CT1 system was unreliable and did not ensure call privacy.

Table 13.1 Basic parameters of CT2, DECT, PACS and PHS

System	CT2/CAI	DECT	PACS	PHS
Bandwidth [MHz]	4	20	U.S. PCS band	23
Frequency range [MHz]	864–868	1880–1900	U.S. PCS band	1895–1918
Carrier spacing [kHz]	100	1728	300	300
Access method	FDMA	FDMA/TDMA	TDMA/FDMA	TDMA/FDMA
Duplex method	TDD	TDD	FDD	TDD
Number of carriers	40	10	16 pairs/10 MHz	77
Channels/carrier	1	12	8/pair	4
Number of channels	40	132	-	308
Modulation	BFSK + gauss. filter 72 kbit/s	GMSK 1152 kbit/s	$\pi/4$-DQPSK 384 kbit/s	$\pi/4$-DQPSK 384 kbit/s
Handover	no	yes	yes	yes
Speech coding	ADPCM	ADPCM	ADPCM	ADPCM
Two-way call request	no	yes	yes	yes
Data rate [kbit/s]	32	32	32	32
Cell radius [m]	30–100	30–200		
Frame duration [ms]	2	10	2.5	5
Peak output power [mW]	10	250	200	80

This gave an impulse for introduction of digital speech encoding which, in consequence, substantially improved the technical abilities of cordless telephony. As a result, the CT2 standard was defined. It was supplemented by the *Common Air Interface* (CAI) standard formulated by the British Ministry of Post and Telecommunications. In 1989 this interface was described in the British standard MPT 1375 and in 1991 was standardized by ETSI. It allowed for cooperation of cordless phones and base stations from different vendors. Besides CT2, the CT3 standard was introduced by Ericsson in which some drawbacks of CT2 were avoided; however, it never gained importance because of the introduction of the DECT as the European standard of cordless telephony. Table 13.1 summarizes the basic parameters of CT2/CAI, DECT, PACS and PHS.

13.1 CT2 STANDARD

As we have already mentioned, CT2 was a British initiative, directly associated with the service called *Telepoint*. This service was introduced in the UK in 1989 and it did not gain much interest among subscribers; however, it became quite popular in Hong Kong and Singapore. The *Telepoint* systems were characterized by very small, inexpensive, low power (around 10 mW) mobile stations. They could be used in commercial and city centers, airports and railway stations in the radius of about 200 m from the base station. *Telepoint* was able to set up the uplink connection only. Moreover, it was not able to perform handover. In newer implementations of CT2 systems, two-way calling and intercell handover are possible [1]. Additionally, new features such as advanced location tracking have been introduced.

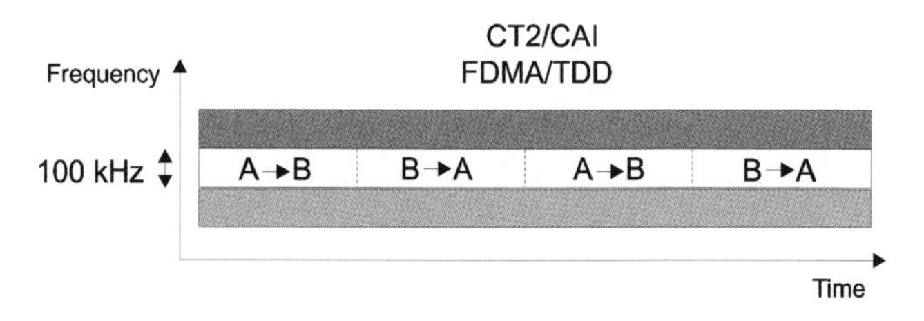

Figure 13.1 CT2 transmission mode

In the CT2 standard the 4-MHz band in the range between 864.1 and 868.1 MHz is used to realize 40 channels with carrier spacing equal to 100 kHz. During a connection a single carrier is assigned to a single user pair. User data streams are sent in the time division duplex (TDD) mode (see Figure 13.1). The binary data rate is 72 kbit/s. The frame structure applied in CT2 is shown in Figure 13.2.

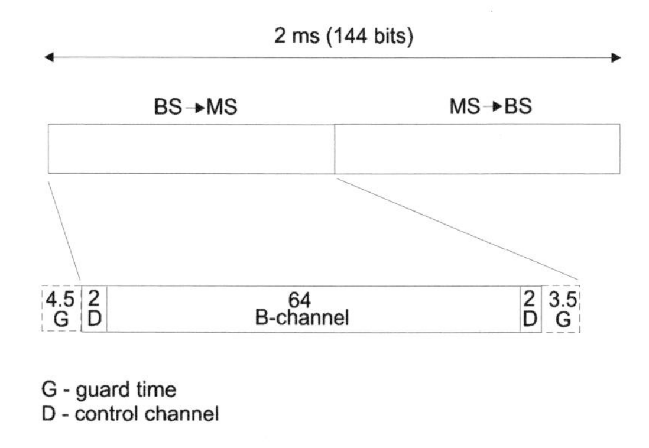

Figure 13.2 Frame and slot structure applied in CT2 (together with Multiplex 1.4 format)

The frame consists of a block transmitting the data stream from the base station to the mobile station and a block transmitting the data in the opposite direction. The frame lasts for 2 ms. Each data block (slot) transfers 64 bits of user information and 4 bits of control information divided into 2-bit portions. The neighboring blocks are separated by guard periods lasting for 3.5 and 4.5 bits, respectively. Summing all bits transmitted both in uplink and downlink blocks in the frame and adding the duration of guard periods, measured in bits, we receive 144 bits within a 2-ms frame. This value results in the 72 kbit/s data rate. The data stream modulates the carrier according to the BFSK (*Binary Frequency Shift Keying*) modulation with Gaussian shaping of frequency pulses. At the carrier spacing equal to 100 kHz, the spectral efficiency is 0.72 bit/s/Hz. A binary signal representing speech is received using the ADPCM encoder in accordance with ITU-T G.721 Recommendation (see Chapter 1). The speech quality is comparable to that achieved for 64 kbit/s PCM encoding.

Signaling in CT2 is organized in three layers. Layer 1 signaling is performed to initiate a two-way radio link, select a radio channel and configure the multiplexing of three types of channels used in different phases of the link. The *B-channel* is used for transmission of speech or data. The *D-channel* performs control and signaling purposes and the *SYNC channel* is used to ensure bit and burst synchronization. All these channels are properly multiplexed within the CT2 frames. The lower part of Figure 13.2 shows the so-called Multiplex 1.4. In Multiplex 1.2 only 2 bits of the D-channel are applied in a burst and the guard time on each burst end is lengthened by 1 bit. Multiplex 1 is used in the information (speech or data) transfer phase after setting up the connection and acquiring synchronization.

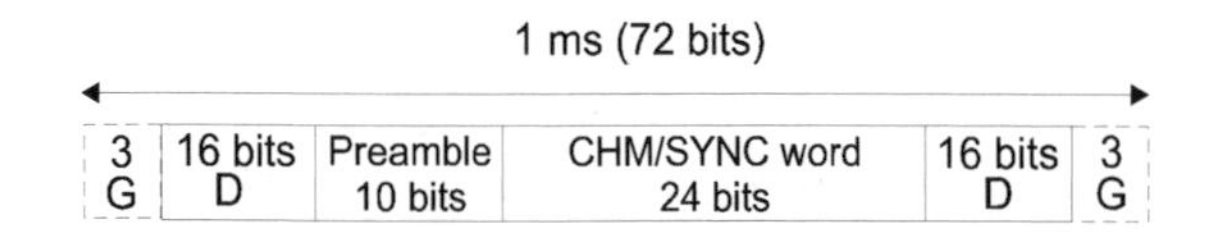

Figure 13.3 Organization of the CT2 burst in Multiplex 2

Multiplex 2 (see Figure 13.3) is applied during the connection set-up between the base station and a particular mobile handset. The base station uses the D-channel to send its identity to the called mobile handset and uses the SYNC channel to establish synchronization with it. The CHM word carries the so-called channel marker and is used to establish a channel. Setting up the channel is acknowledged by the SYNC word having special forms for the mobile handset and for the base station, respectively.

Multiplex 3 is used in the mobile-originated connection set-up. The mobile handset pages the base station for 10 ms by sending the preamble, its own identity and the identity of this base station the handset wishes to set a connection with. In the next period of 4 ms the handset listens and waits for its own identity and the SYNC message sent by the base station. The detection of these messages will allow the mobile handset to fully synchronize with the base station and to switch to the multiplex 2 format in the data exchange with the base station.

Layer 2 signaling is devoted to the control channel protocols for error detection and correction, message acknowledgement, link maintenance and link end-point identification [2]. Layer 2 packets are composed of a 64-bit address word and a sequence (from 0 to 5) of data code words, each 64 bits long. These packets are transmitted on the D-channel. If there is nothing to transmit, an idle D-channel is realized in which alternating zeros and ones are sent. Transmitting the packet messages is preceded by sending a synchronization pattern.

Finally, Layer 3 signaling is responsible for error-free transmission of telephony messages over the radio link. The messages refer to the mobile handset keypad, its display or PBX access.

The CT2 standard, although not the newest one, is still attractive in some special applications such as a wireless local loop. It can be successfully applied in residential applications, as well as in wireless PBX systems which ensure communications within company premises, warehouses, etc.

13.2 DECT SYSTEM

The DECT system is, like GSM, a result of the activity of the ETSI standardization committee. DECT has been described in a series of standards called *Common Interface*, which include: system overview [3], physical layer (PHL) [4], medium access control layer (MAC) [5], data link control (DLC) layer [6], network layer (NKW) [7], identities and addressing, security features, speech coding and transmission, public access profile and approval test specification (check other standards in the EN 300 175 and 176 series).

Like CT2, DECT realizes radio access to fixed networks in the areas of dense communication traffic. It consists of a set of small and inexpensive radio handsets, which are able to perform connections with base stations located in the vicinity (see Table 13.1). The DECT technology can be used in the following systems:

- residential systems,
- small business systems characterized by a single site and a single cell,
- large business systems characterized by multisite and multicell structures,
- public cordless access systems,
- wireless access to local area networks (WLAN),
- wireless local loops.

13.2.1 DECT architecture

The structure of a DECT system is based on microcells with the radius of a few hundred meters, in which the mobile handsets – *Portable Parts* (PP) communicate with the base stations – *Fixed Parts* (FP). The DECT system tolerates the movement of a mobile handset up to the speed of 20 km/h. Figure 13.4 presents the basic architecture of this system. Besides fixed and portable parts, the system contains the *DECT Fixed System* (DFS) which controls a number of FPs. The access to one of external networks is performed through the *Interworking Unit* (IWU). A simple *Data Base* (DB) cooperates with the DFS in order to manage the users' connections.

A particular DECT configuration depends on the size of the system and on its application. In residential systems there is a single base station (FP) and a few mobile handsets (PP) operating within the cell. Wireless private branch exchanges as well as Telepoint systems can work with a few base stations, as shown in Figure 13.4. Finally, large and/or distributed systems can have a few *Subsystem Control Units* (SCU) managing a number of base stations. The SCUs are connected with each other through the backbone ring network [11]. The DFS supervises the operation of the whole system.

Figure 13.5 presents a possible application of the DECT system in a *Wireless Private Branch Exchange* (WPABX) [8]. The WPABX contains not only a central processing

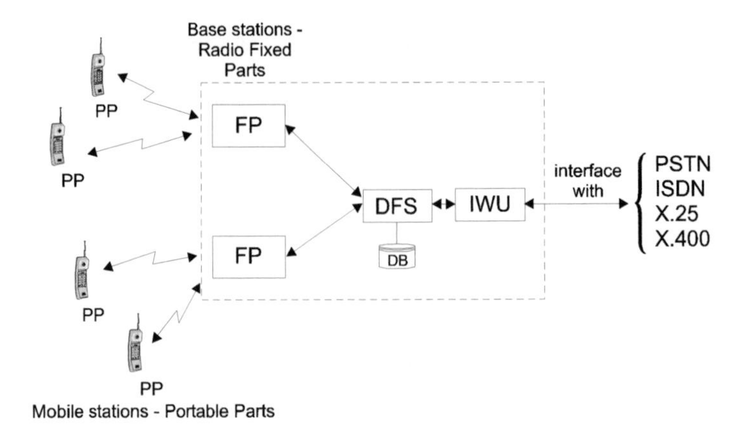

Figure 13.4 Typical configuration of a DECT system

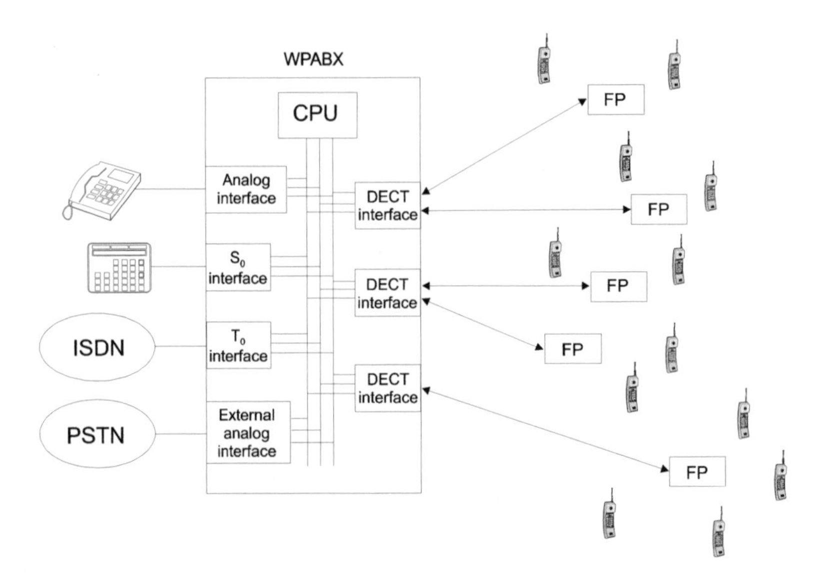

Figure 13.5 Typical configuration of the WPABX [8] (with permission of Alcatel Telecommunications Review)

unit but also a number of interfaces with networks and equipment other than DECT. Through the base stations (FPs) each DECT interface controls the set of mobile handsets (PPs) operating in the FP coverage areas.

13.2.2 DECT physical layer

Basic technical data of the DECT physical layer are presented in Table 13.1 and they will not be repeated here; however, let us note that DECT operates in TDMA/TDD

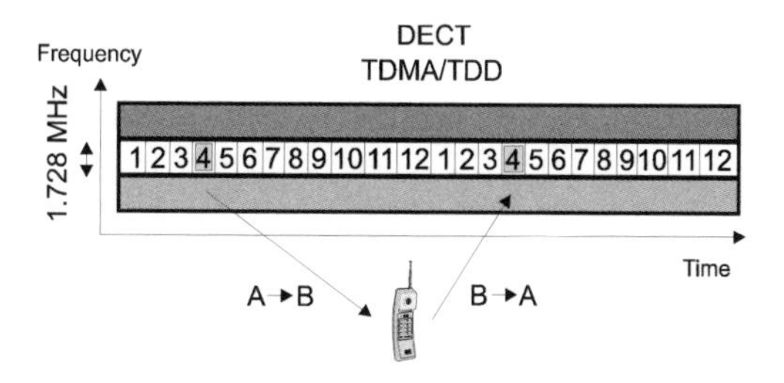

Figure 13.6 DECT transmission mode

mode (see Fig13.6). Twelve duplex channels are placed in a 10-ms frame divided into 24 time slots. The first twelve slots are used in the downlink, and the next twelve slots are applied in the uplink. Due to the applied time structure of packets placed in the slots, the data rate in DECT transmission is high and equal to 1152 kbit/s. The data stream in a single transmission channel realized in a particular time slot on the selected carrier is much slower. It is equal to 32 kbit/s in the traffic channel (so called B-field) and to 6.4 kbit/s in the A-field used for control and signaling. Figure 13.7 presents the structure of a typical DECT packet. Besides these packets, the DECT standard includes *short physical packets* which are 96-bit long, packets which are contained in a DECT half slot (180-bit long) and long packet lasting for 900 bits and requiring a double slot. A frame divided into 24 time slots constitutes one unit of the DECT multiframe. The multiframe consists of 16 frames and lasts for 160 ms.

One packet is transmitted in a time slot which lasts for 0.417 ms. This time is equivalent to 480 bits. A packet contains 420 bits, so the time needed for 60 bits is in fact a guard time. The guard time is used, as in GSM, to inhibit packet alignment inaccuracies within a time slot and to ramp on and off the handset's power amplifier. 420-bit packet starts with a 16-bit preamble and a 16-bit synchronization sequence. The next 388 bits are divided into a 64-bit control and signaling field (A-field) and a 320-bit information field (B-field). The packet ends with a 4-bit CRC block which checks the parity of the whole packet. Due to the crucial meaning of the control and signaling field for the system reliability, the A-field has an 8-bit header, a 40-bit data block, and a 16-bit CRC block. This block is a result of the applied BCH (63,48) error correction code supplemented with the overall parity bit. This code is able to detect up to 5 random errors, all error bursts up to the length of 16 bits and all error patterns with an odd number of errors. Depending on the phase of the current multiframe, the control and signaling block sends several messages associated with the MAC layer operations. They are:

- *Identities Information* – in which the base station sends its identity to the mobile handsets,

- *System Information and Multiframe Marker* which is sent once in a multiframe for acquisition of synchronization by mobile handsets, and contains information on the base station,

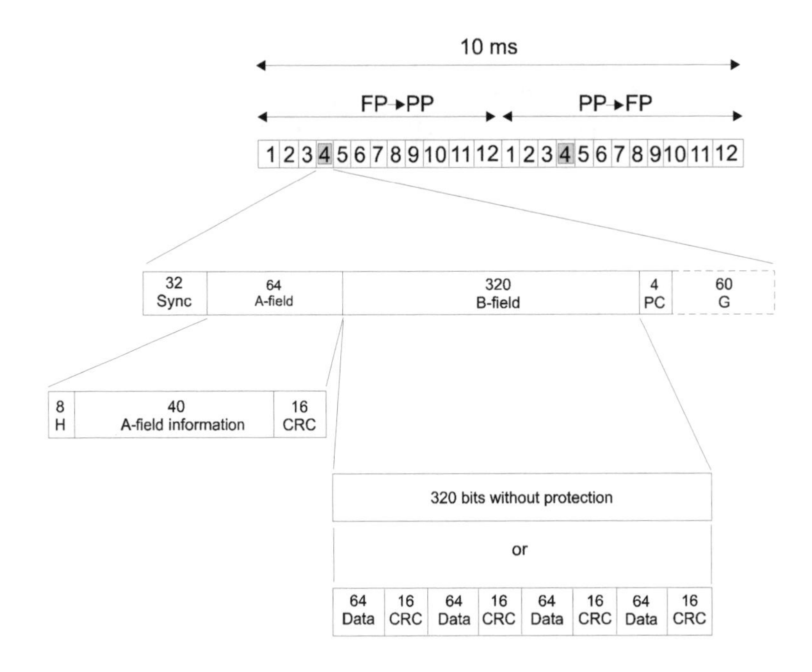

Figure 13.7 DECT packet structure: H – Header, PC – Parity Check, CRC – Cyclic Redundancy Check, G – Guard Time

- *Paging Information* in which the base station sends its broadcast service and some MAC layer information,
- *MAC Control Information* used for connection set-up, its maintenance and termination and handover requests,
- *Control Information of Higher Layers.*

The B-field can be protected by the forward error correction (FEC) code or can remain unprotected (see Figure 13.7). If FEC protection is used, user data are divided into four 64-bit blocks. Each of them is supplemented with a 16-bit CRC block.

Binary data are transmitted on the radio channel using GMSK modulation with $BT = 0.5$.[1] Such a signal is demodulated in a non-coherent receiver without an equalizer attempting to remove the influence of the channel multipath. Because the duration of a single bit at the data rate of 1152 kbit/s is 0.868 μs and a receiver without an ISI equalizer can tolerate the signal delay spread of about 10 percent of a symbol duration, therefore the delay spread cannot be larger than 200–300 ns. This value is related to about 100 m difference in the length of different signal paths. These conditions are fulfilled for many types of environment and for the cell radius of about 200 m. Otherwise

[1]Let us recall that B is a 3-dB bandwidth of a filter with a Gaussian characteristics, T is a single bit duration.

the locations of the base stations have to be carefully planned[2] and sectorized antennas can be introduced. Base station antennas should have the antenna gain of about 22 dBi [9]. The carrier-to-interference ratio C/I should be at least equal to 10 dB.

In DECT systems intra- and intercell handover is possible. It is assumed that the time interval between two handovers should be at least 3 seconds. It prevents unnecessary channel and cell changes, which otherwise could happen bacause of dynamic channel assignment implemented in DECT. The handover is performed at the above limitations if another signal than that from the current base station is received at higher power or another channel ensures higher link quality in the same cell [10].

13.2.3 DECT MAC layer

Medium Access Control layer is very important for the overall functioning of the DECT system. Its main tasks are following:

- system resources control,
- multiplexing of signaling channels,
- error protection.

The first task of the MAC layer consists in creating, maintaining and releasing radio bearers by allocating and releasing physical channels. The control of system resources is performed through realization of *Dynamic Channel Allocation* (DCA) and through the possible assignment of more than one time slot in a frame to a given connection.

As we remember from Chapter 5, *fixed statistical channel assignment* relies on such permanent allocation of the channels to the mobile communication system cells, which takes into account the expected traffic density in the particular cells. However, in the systems exploiting pico- and microcells temporary situations can occur in which the number of mobiles in a given cell exceeds the cell capacity determined by the number of allocated channels, while in the neighboring cells some channels remain unused. The dynamic channel allocation removes the above disadvantage. The channels are dynamically assigned to the cells, taking into account the needs of the users. However, this is not a completely free allocation because continuous monitoring of the interference level is necessary in order to determine if the distance between different cells using the same channel is not too short.

Each base station (FP) sends a marker called a *beacon signal* which means that at least one channel of the base station is active. In this channel system information and base station identity are broadcasted. If a mobile handset (PP) remains in the idle state, it tunes to the strongest base station and listens to the paging messages from all active radio channels. A mobile handset wishing to set up a connection searches the channels until it finds a beacon signal. In the same channel in the time slot in which uplink transmission takes place, the mobile handset sends a *set-up request* message in which

[2]It does not mean that channel frequencies have to be planned, as is done in a classic cellular telephony. In DECT systems the dynamic channel assignment is performed (see Chapter 5).

it asks for allocation of the carrier and of the time slot in which the signal exchange is currently proceeding. The base station (FP) responds in the same channel (in the downlink time slot). Due to the TDMA mode applied in DECT, the mobile handset has enough time for systematic search of other channels during the connection in order to find a channel featuring the required value of C/I. It eventually requests permission to switch to a new channel within the same cell or within a new one. As we see, due to the activity of a mobile handset in searching for a new channel, the *distributed* dynamic channel allocation is applied in the DECT system.

The second task of the MAC layer, i.e. the signaling channel multiplexing, relies on the control of transmission of the appropriate signaling messages in the A-field of the packets. The signaling channel multiplexing is related to the number of the current frame within the multiframe structure. The third task of the MAC layer – error protection – was presented while describing the DECT packet structure.

It is worth noting that it is possible to assign more than one slot to a single connection in the DECT system (*multislot assignment*). The number of time slots allocated in the uplink and downlink can be different. This allows application of the DECT standard in LANs.

13.2.4 DECT/GSM interworking

As we have already mentioned, the DECT systems are installed not only in a residential areas or in private companies but also in public places characterized by particularly high traffic density, such as airports, railway stations or city centers. However, it is also possible to apply DECT in co-operation with a cellular system, in particular with the GSM. It is possible if dual-mode GSM/DECT mobile stations are used. A dual-mode phone operates as a GSM phone in places where the GSM system is accessible or if the subscriber moves at a high speed. In the traffic spots covered by DECT in which the GSM system is not able to serve all subscribers, a dual-mode phone can switch to the DECT mode. In the area covered by GSM, the DECT-islands can be additionally placed. Figure 13.8 presents DECT/GSM interworking architecture. The DECT fixed system (DFS) is connected via the *Interworking Unit* (IWU) with the mobile switching center (MSC). The interface between the IWU and MSC is a GSM A-interface. The task of the IWU is to translate the DECT system mobility management protocol and the call control protocol into the appropriate GSM protocols. The reverse process takes place as well. The DECT DFS block is seen by the GSM MSC as a base station controller. The cooperation of both systems has been standardized by ETSI by defining the *GSM Interworking protocol*. In the joint DECT/GSM system data transmission and SMS services are provided.

13.2.5 Description of a typical DECT handset

In this paragraph we will present a block diagram of a typical DECT terminal based on the design shown in [8].

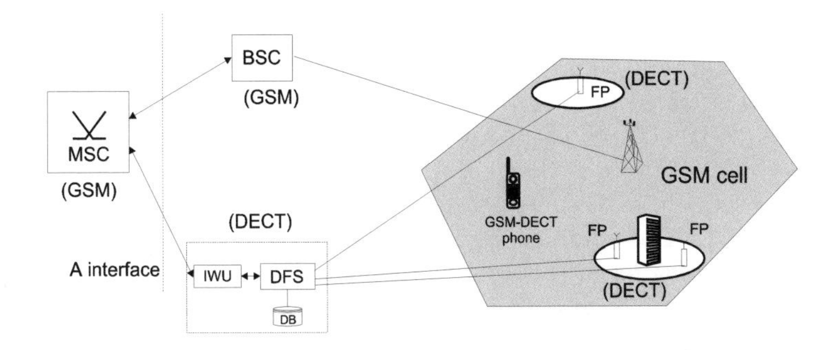

Figure 13.8 DECT/GSM interworking architecture

The transmit/receive section of the DECT handset can be divided into two parts: the RF part and the control and baseband DSP part. The block diagram of a mobile handset is shown in Figure 13.9.

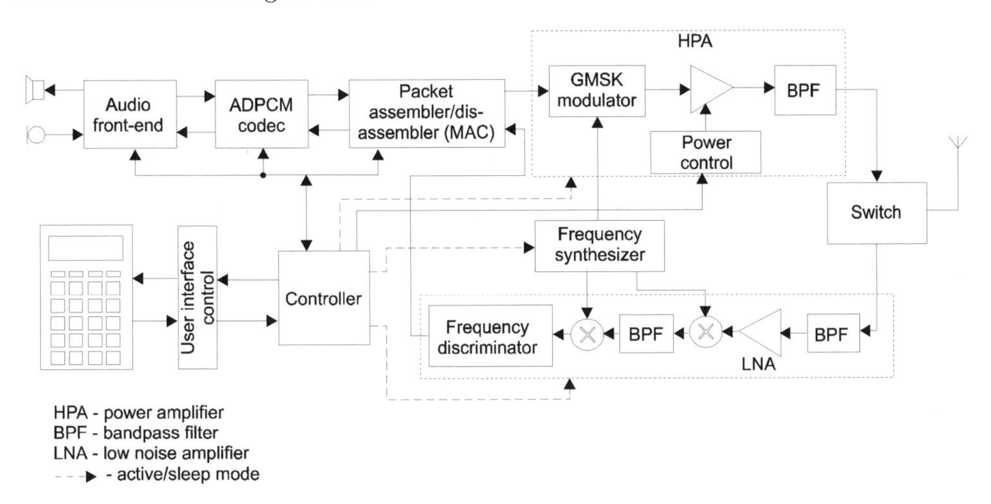

Figure 13.9 Block diagram of a typical DECT handset

In the transmit part the analog voice signal is converted to a digital form in a 13-bit linear A/D converter operating at the frequency of 8 kHz. Next, digital voice samples are encoded using an ADPCM encoder (see Chapter 1). This process results in the data stream of 32 kbit/s. The data bits are packed into 320-bit B-fields of the DECT bursts in the packet assembler block. The bursts are supplemented with the signaling and control messages carried in the A-field. The bursts formed in this way are transferred to the transmit RF part. The binary data stream is shaped by the Gaussian filter and controls the frequency modulator. In consequence, a GSMK signal is generated which is shifted to the final frequency range of 1880–1900 MHz. This signal is amplified and sent through the circulator or the active switch to the antenna.

The signal received by the antenna is directed to the receive section where it is filtered and then amplified in a low noise amplifier. Subsequently, it is down-converted to the first intermediate frequency, it is filtered and converted to the second intermediate frequency. The received signal is then processed in a noncoherent demodulator, mostly in a frequency discriminator.[3] The recovered TDMA burst is processed in a logical circuit, in which signaling and control signals, and the 320-bit block representing speech signal are extracted. The obtained signal is given to the ADPCM decoder, and after conversion to the analog form, it is sent to the handset speaker.

Transmit and receive sections are supervised by the controller which manages transmission protocols, signaling, controling the RF part and controling the user interface – the keypad, display and beeper. Let us note that there are signals which activate and deactivate transmitter and receiver blocks, control the transmitted power level and select the synthesized carrier frequency. Controling the sleep mode of particular handset blocks is an important issue with respect to the length of the time interval between subsequent battery loadings.

13.3 PERSONAL ACCESS COMMUNICATIONS SYSTEM (PACS)

The PACS (*Personal Access Communications System*) is a counterpart of DECT in America. The PACS standards are mostly based on the Bell Communcations Research wireless communications system (WACS). There are three versions of PACS:

- the version denoted mostly as PACS, which operates in a FDD mode in the licensed *Personal Communications System* (PCS) bands 1850–1910 and 1930–1990 MHz,
- the version known as PACS-UB, which operates in a TDD mode in the unlicensed band 1910-1930 MHz in the same way as the PACS system for the licensed band,
- the PACS-WUPE (*Wireless User Premises Equipment*) based on the Japanese PHS system (see the next paragraph).

The basic system parameters were listed in Table 13.1. PACS has similar advantages as DECT. It features small, inexpensive base stations (RP - *Radio Ports*) with a small coverage area, low-DSP-complexity and low-power small mobile handsets, the capability to provide telecommunication network access whose quality, privacy and reliability are the same as in wireline access. PACS has been optimized for services provided in indoor environment, and high density traffic environments. As a result, it is a cost-effective solution in high traffic areas. The simplicity of the applied radio protocol allows for easy integration of the PACS unit with any celullar mobile station used in America, such as AMPS, IS-95, IS-136 or PCS 1900.

Figure 13.10 presents a simplified PACS architecture. The PACS handsets (*Subscriber Units* - SU) communicate with the base stations called *Radio Ports* (RP) in the

[3]It is not the only demodulation method applicable in DECT; however, it is the least expensive one.

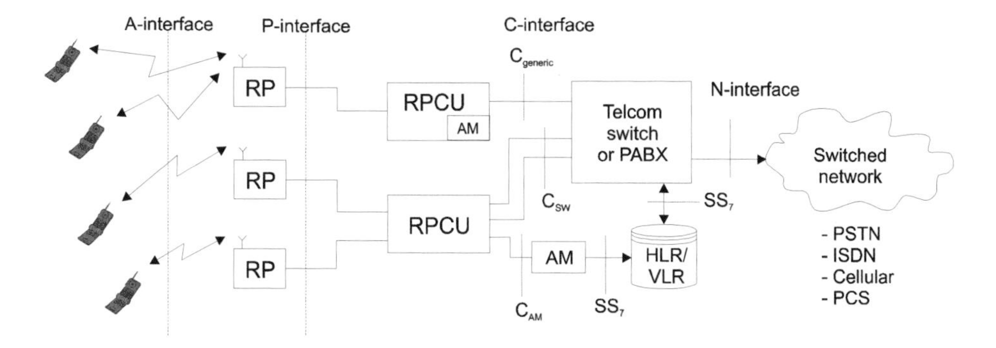

Figure 13.10 Simplified PACS architecture; RP – Radio Port, RPCU – Radio Port Control Unit, AM – Access Manager

TDMA/ FDD mode. RPs are connected to *Radio Port Control Units* (RPCUs). The link between an RP and RPCU is realized using a primary level of PCM hierarchy (T1 in America at the rate of 1.544 Mbit/s or E1 in the countries in which PCM-30/32 is applied), or a High Speed Digital Subscriber Line (HDSL). The radio ports which operate basically as radio modems are powered from the local exchange company supply voltage through HDSL. This way the need of a local power supply is eliminated. The radio ports (RPs) and PACS handsets (SUs) have been designed to be as simple as possible to decrease power consumption, among others. The peak transmit power of a PACS handset is 200 mW, whereas the average power is 25 mW. Most of the electronics and intelligence is contained in RPCUs. The *Access Manager* (AM) is a unit which performs network-related tasks such as: searching the remote databases due to visiting users, assisting at the connection set-up and call delivery, coordinating link transfer between different RPCUs in case of handover and management of the radio ports. The access manager can be a stand-alone unit, it can be attached to the RPCU or it can be associated with some other network elements.

The PACS radio transmission is organized in 2.5-ms frames consisting of 8 time slots (see Figure 13.11). The fifth slot of each frame is used by a *System Broadcast Channel* (SBC) in which the system information is transmitted at the rate of 16 kbit/s. On the physical SBC three logical channels are realized: the *Alerting Channel* (AC) used to alert the PACS handsets to incoming calls; the *System Information Channel* (SIC) applied to broadcast system information (identities, timers, protocol constants) and the *Priority Request Channel* (PRC) used by handsets to request emergency calls.

A single time slot has a duration of 120 bits so the data rate is 384 kbit/s. Out of 120 bits, 80 bits carry user information in the form of *Fast Channel* (FC). The remaining 40 overhead bits of the downlink burst are used for the following purposes:

- 14 bits of a synchronization pattern,
- 10 bits of a *Slow Channel* (SC) applied for an additional synchronization pattern, indication of word errors, signaling messages or subscriber data,

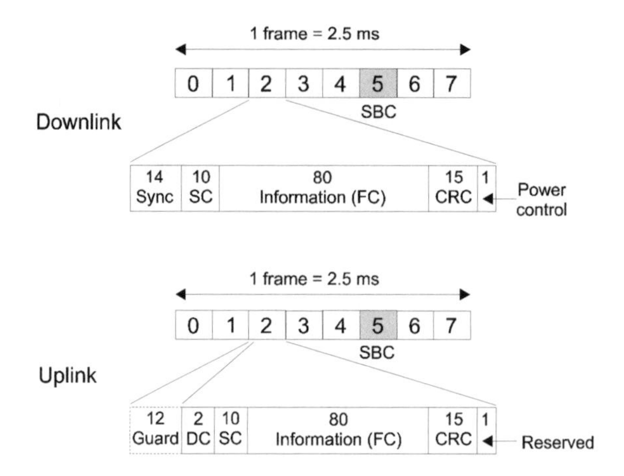

Figure 13.11 Downlink and uplink PACS frame structure (SBC - *System Broadcast Channel*)

- 15 CRC bits for error detection,
- 1 bit for power control of a handset.

Similarly, the burst in the uplink is organized in the following way:

- 12 bits of guard time,
- 2 bits needed for a reference symbol in differential encoding of $\pi/4$-QPSK modulation used in PACS,
- 10 bits of a slow channel,
- 80 bits of user payload,
- 15 CRC bits,
- 1 bit for future applications (reserved).

Let us note that the number of 80 bits per frame results in the data rate of 32 kbit/s for a single user. It is a sufficient data rate to realize 32 kbit/s ADPCM speech coding. The PACS system also provides subrate channels at the rates of 16 and 8 kbit/s. Two or more slots in a frame can also be aggregated to give throughputs higher than 32 kbit/s.

An interesting feature of PACS transmission is the diversity used both in handsets and base stations (radio ports). At the handset (SU) *preselection diversity* is applied. The handset performs diversity measurements which are possible due to continuous downlink transmission. As a result, prior to an incoming burst, it can choose the antenna which ensures higher performance. In the uplink, the base station full selection dual-receiver *selection diversity* is applied. Due to it the base station selects the signal from that receiver which gives a higher quality signal. Additionally, *switched transmitter diversity* is also applied in the handset and in the radio port. Both units inform each other if the previous burst contained any errors. If the errors have occurred, the transmitter selects the other diversity antenna for sending the next burst.

Besides the physical layer, the PACS standard defines a series of functions performed by the system which are necessary to realize the radio access and network functions.

The first group of functions is associated with the radio link maintenance and associated measurements. In order to keep the link quality during a call, the power control of the handsets and appropriate measurements are performed. Besides the radio signal strength, the SIR is estimated and the occurrence of bit errors in a slot is monitored. On the basis of these measurements several kinds of handover, called *Automatic Link Transfer* (ALT) in PACS, can be applied. The automatic link transfer can take place within the same frame by selecting a new time slot, or a new base station (RP). Depending on the unit which supervises the current RP and a new RP, the handover procedure can be more or less complicated. The simplest procedure is the link transfer between RPs managed by the same RPCU. The transfer between two RPs which belong to two different RPCUs and access managers is the most complicated procedure.

Another function which has to be performed by the system is associated with mobility management. After power on or after moving from one registration area into another, the handset registers with the visited location register (VLR). The VLR informs the user's HLR about his/her current locations. De-registration of the user can take two forms. In the first one, the handset is de-registered from the VLR if it has not registered again within a determined time period. The second form relies on periodical sending a polling message to the handset. The latter has to answer within the predefined time period, otherwise it is de-registered.

Other functions performed in PACS are associated with user authentication and enciphering data over the radio link. Service functions supporting call origination from the handsets, call delivery to them and some supplemental services are also realized in PACS.

PACS frequency planning is performed for base stations (radio ports) automatically without manual frequency planning. The applied automatic frequency assignment is called *Quasi-Static Autonomous Frequency Assignment* (QSAFA). This procedure is performed within the radio port control unit (RPCU) which manages a number of radio ports. The procedure is iterative and relies on measurements made by transceivers in the radio ports [12].

13.4 PERSONAL HANDYPHONE SYSTEM (PHS)

Personal Handyphone System (PHS) is the Japanese counterpart of DECT and PACS. Its basic parameters are presented in Table 13.1. The applications of PHS are similar to those of DECT, CT2 and PACS. Figure 13.12 presents different PHS configurations for home, office and public systems. Note that PHS is connected to the ISDN network; however, completely stand-alone PHS networks are also possible. The PHS system offers voice transmission with ADPCM 32 kbit/s coding, group 3 fax at 2.4–4.8 kbit/s and duplex modem transmission at 2.4 to 9.6 kbit/s. Additionally, there exists the capability of data transmission at 32 and 64 kbit/s.

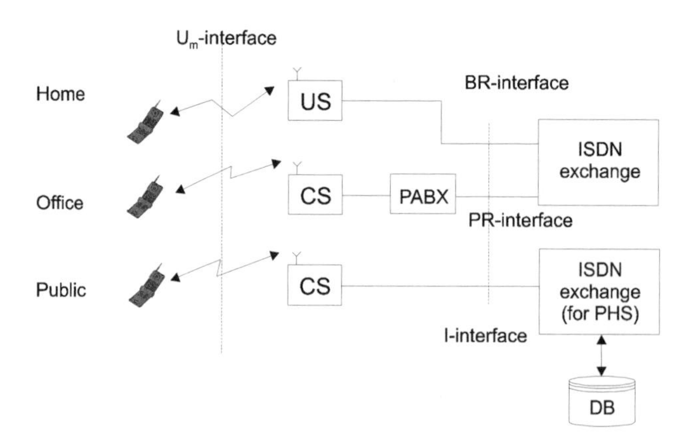

Figure 13.12 PHS applications (CS - Cell Station, PS - Personal Station, US - User Station, BR - basic rate, PR - primary rate, I - CS-to-digital network interface for PHS)

The allocated spectrum (1895-1918.1 MHz) is partitioned into 77 frequency channels, each of width equal to 300 kHz. Out of 77 frequencies, 40 frequencies are allocated to public services, whereas the remaining 37 carriers are devoted to home and office applications. The digital transmission is organized in the TDMA/TDD mode. Four duplex channels are carried in one frame. The first four slots are devoted to downlink transmission, whereas the remaining slots are used for uplink transmission. Figure 13.13 presents the frame organization together with the contents of the frame for traffic and control channels.

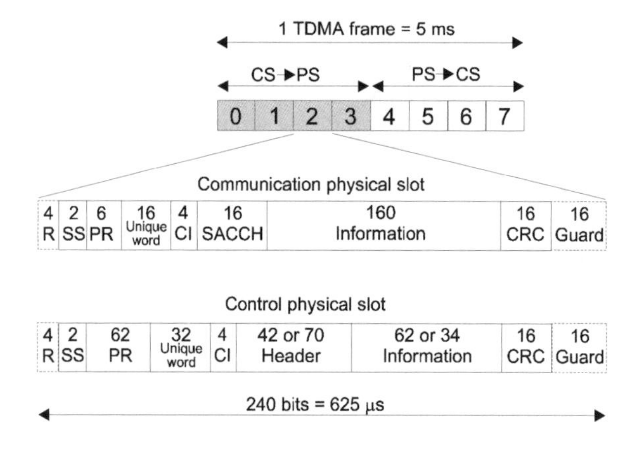

Figure 13.13 PHS time slot and frame formats

The slot duration is equivalent to 240 bits. Each slot starts with 4-bit ramp time (R) in which the handset (*Personal Station* - PS) or the base station (*Cell Station* - CS) switches on its transmitter. Next comes a 2-bit start symbol (SS) which serves as a reference symbol for the $\pi/4$-DQPSK demodulation. The preamble located in the next field is used for updating the slot synchronization. The *unique word* is a binary sequence known in advance, different for down- and uplink and control and communication physical slots. It is used to detect appropriate channels and to help, together with the CRC block, in error detection. The CI (*Channel Identifier*) field contains information on the type of the logical channel transmitted within the slot. The *information field* carries the user data sequence. In the case of control physical bursts this field is preceded by the header which contains the addresses of CS or CS and PS stations. The burst is ended with 16 CRC bits. The rest of the time slot is a guard time which lasts for 16 bits.

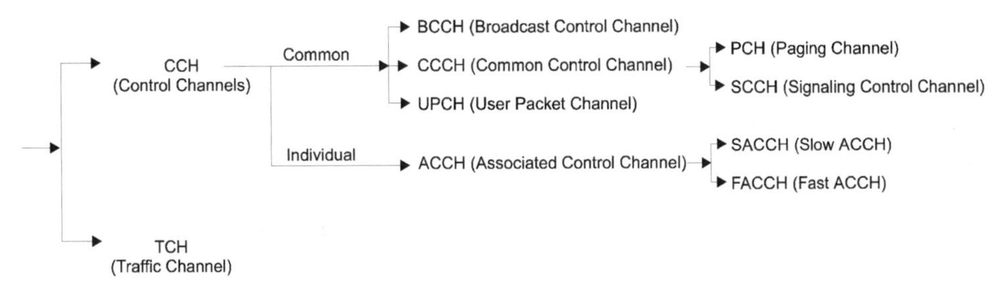

Figure 13.14 PHS logical channel structure

A number of different logical channels have been introduced in PHS. Their structure is shown in Figure 13.14. They are:

- *Traffic Channel* (TCH) – a bidirectional, point-to-point channel serving for transfer of user information,
- *Broadcast Control Channel* (BCCH) – a downlink channel broadcasting control information from CS carrying channel structure, system information and system restrictions,
- *Common Control Channel* (CCCH) – a control channel used for transmission of control information necessary for call set-up, divided into:
  - *Paging Channel* (PCH) – a downlink channel serving for sending paging information for PSs in the cells belonging to the same paging area,
  - *Signaling Control Channel* (SCCH) – a bidirectional, point-to-point channel transmitting information necessary for call connection,
- *User Packet Channel* (UPCH) – a bidirectional, point-to-multipoint channel used for control information and packet data (such as SMS if the PS does not have a call in progress),

- *Associated Control Channel* (ACCH) – a bidirectional channel, asociated with the TCH, which carries control information and packet data used for call connection; divided into slow (SACCH) and fast (FACCH) channels.

Sending broadcast and control information is organized in the form of the so-called downlink *Logical Control Channel* (LCCH) which has a superframe structure. In each nth frame the cell station (CS) transmits an LCCH slot. The superframe consists of m LCCH slots. Within the superframe the BCCH, PCH, SCCH and optionally UPCH channels are transmitted. The broadcast control channel is sent in the first LCCH slot of the superframe. For the uplink no superframe structure is determined. The personal stations start their transmission from sending SCCH slot to the cell station in accordance to the *slotted Aloha* principle.

As we have mentioned, the PHS system co-operates with digital networks (see Figure 13.12). This allows the user to take advantage of their signaling capabilities in several functions performed by the network which are related to user mobility, such as maintaining location information for all personal stations operating in the system, registration, authentication (supplementary service) and incoming call delivery.

* * *

Due to limited space, the description of CT2, DECT, PACS and PHS is very superficial. Much more information on DECT and PHS can be found in [11]. The highly motivated reader is advised to study the standards describing the PACS and PHS systems, e.g. [13], [14] and [15].

REFERENCES

1. I-ETS 300 131, Radio Equipment and Systems (RES); Common air interface specification to be used for interworking between cordless telephone apparatus in the frequency band 864.1 MHz to 868.1 MHz including public access services, November 1994, Second Edition

2. R. Pandya, *Mobile and Personal Communication Systems and Services*, IEEE Press, New York, 2000

3. EN 300 175-1, Digital Enhanced Cordless Telecommunications (DECT); Common Interface (CI); Part 1: Overview, V.1.4.2, June 1999

4. EN 300 175-2, Digital Enhanced Cordless Telecommunications (DECT); Common Interface (CI); Part 2: Physical Layer, V.1.4.2, June 1999

5. EN 300 175-3, Digital Enhanced Cordless Telecommunications (DECT); Common Interface (CI); Part 3: Medium Access Control (MAC) Layer, V.1.4.2, June 1999

6. EN 300 175-4, Digital Enhanced Cordless Telecommunications (DECT); Common Interface (CI); Part 4: Data Link Control (DLC) Layer, V.1.4.2, June 1999

7. EN 300 175-5, Digital Enhanced Cordless Telecommunications (DECT); Common Interface (CI); Part 5: Network (NKW) Layer, V.1.4.2, June 1999

8. V. Werbus, A. Veloso, A. Villanueva, "DECT – Cordless Functionality in New Generation Alcatel PABXs", *Electrical Communication*, 2nd Quarter 1993, pp. 51-57

9. K. David, T. Benkner, *Digitale Mobilfunksysteme*, B. G. Teubner, Stuttgart, 1996

10. S. Ghaheri Niri, R. Tafazolli, B. G. Evans, "Wide Area Mobility for DECT", *Proc. of GLOBECOM'96*, pp. 1119-1125

11. B. H. Walke, *Mobile Radio Networks: Networking and Protocols*, John Wiley & Sons, Ltd., Chichester, 1999

12. A. R. Noerpel, Y.-B. Lin, H. Sherry, PACS: Personal Access Communications System - A Tutorial, *IEEE Personal Communications*, June 1996, pp. 32-43

13. ANSI J-STD 014, Personal Access Communication Systems Air Interface Standard, New York, 1998

14. ANSI J-STD 014 supplement B, Personal Access Communication System Unlicensed (version B), New York, 1998

15. Research and Development Center for Radio Systems, Personal Handyphone System, RCR Standard, version 1, RCR STD 28, Tokyo, December 1993

14
Wireless local loops

14.1 INTRODUCTION

Deployment and exploitation of many new mobile communication systems, in particular digital systems, in which call privacy and security are technically feasible, have shown new possibilities of implementation of a local loop. Local loop[1] is the most expensive part of a public telephone network. In order to improve the access to telecommunication services, many countries, including some East European and developing countries, need to invest heavily in the telecommunication infrastructure. Connecting the subscribers with a fixed network, although not difficult from the engineering point of view, requires large funds and is time-consuming. Development of telecommunication infrastructure in low-populated areas can be economically unreasonable. In such areas a telecommunication operator cannot expect much profit because the traffic is very low.

Wireless Local Loop (WLL) technology [1], [2], [3] which has been introduced in recent years, solves a lot of problems associated with fast and inexpensive access to the telecommunication network. Because of a wireless character of communication, WLL does not incur large labor costs and time-consuming investments. The time needed for installment of a wireless local loop is very short. The telecommunication network can be quickly deployed in the desired area, even if the destination network will be wireline or if the network is only temporarily needed. The above advantages are particularly important in rural areas. Falling prices of the radio equipment resulting from the economy of scale in cellular systems give an additional impulse for the development of the WLL technologies.

[1]In a traditional PSTN a *local loop* (*subscriber loop*) is a pair of copper wires connecting a subscriber phone with a switching center.

The WLL technologies make the competition on the telecommunication market much easier. In many countries the telecommunication network is state-owned or owned by a single company. A local loop is the most difficult barrier in the way of breaking the monopoly of a telecommunication operator. Wireless local loop technologies are an important tool in demonopolizing the telecommunication market.

In this chapter we will overview several WLL technologies based on the PMP (*Point-to-Multipoint*) technique, cellular radio and cordless telephony. In this respect we will show new applications of the systems already described before.

14.2 PMP SYSTEMS

The PMP systems are not an example of mobile communication systems applied in the local loop implementation; however, we will describe them for completeness of the topic.

The PMP systems have been known for many years. Their basic feature is the transmission of digitized speech using PCM 64 kbit/s or ADPCM 32 kbit/s coding. Let us consider a typical architecture of such a system using ALCATEL 9800 shown in Figure 14.1 as an example [4].

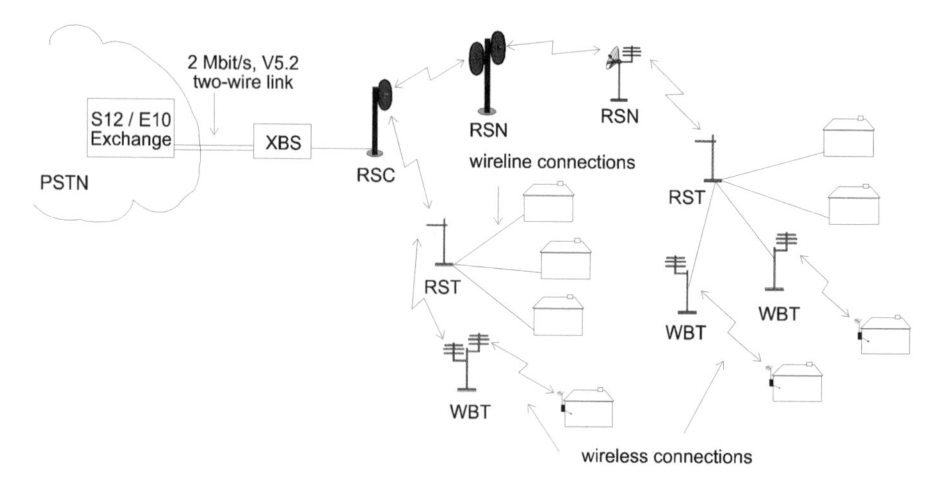

Figure 14.1 PMP system architecture on the example of Alcatel 9800

The system is connected with the PSTN through a central station (XBS). The central station is connected with the central radio station (RSC), which transmits and receives signals in the TDMA mode from the radio end stations (RST) or radio relay stations (RSN). If directional antennas are applied, the distance between line-of-sight antennas of the end station and the central station cannot be longer than 50 km. The relay stations are used if the distance between the central station and the end station exceeds this value or there are terrain obstacles which do not allow for mutual visibility in a single hop. The signals received by the end station are distributed to the subscribers using copper local loops with standard telephone sets connected to them. ALCATEL 9800 also supports wireless local loop connections applying the DECT technology. The length of a wireless local loop cannot exceed 5 km.

Generally, the PMP systems ensure the links with the PSTN with the quality close to that achieved in a fixed network. Let us note that thanks to PCM or ADPCM coding of the transmitted signal, binary signals are regenerated in the relay stations which operate in the same way to the repeaters applied in PCM networks. This way the quality deterioration associated with the number of hops is avoided. Thanks to a standard PCM or ADPCM coding of an acoustic signal, voiceband modems can be applied as in a conventional telephone channel. Binary streams representing PCM or ADPCM samples are transmitted without error protection, although the latter can be applied. Thanks to the digital form of a transmitted signal, it cannot be easily eavesdropped using typical analog equipment, as occurs in analog cellular radio. Due to coding and the frame structure which are compatible with PCM systems, it is possible to create the basic ISDN access[2] (2B+D) link or to set separate data transmission channels. Thanks to the PCM time structure and the desired compatibility with a fixed network, typical PMP systems ensure 30 or 60 PCM 64 kbit/s links. Thus, the PMP system with 30 PCM links is equivalent to 2.048 Mbit/s PCM 30/32 system. Assuming the probability of blocking equal to 0.01 and the mean intensity of traffic generated by a single subscriber equal to 0.05 Erl, the system can serve about 400 subscribers.

Transmission in the PMP systems is organized in packets. In the downlink packets are transferred in the TDMA mode. In the opposite direction the end stations transmit packets in a time division multiplex mode on a different carrier frequency. The length of packets sent by a given end station depends on the number of the subscribers connected to it and on the current traffic generated by them. Due to the time division multiplex mode, the end stations have to be synchronized. Very often the access method applied in the channel is *Demand Assigned Multiple Access* (DAMA).

The PMP systems use frequency ranges dedicated to the fixed access. They are usually allocated by the national institutions responsible for frequency management. Typically, the assigned bandwidth for a single link is about 2 MHz in each direction when the frequency division duplex method is applied. However, with respect to rural wireless local loops a frequency band around 500 MHz should be allocated to PMP systems. In the 1-3 GHz band the following frequency ranges are recommended by [5]:

- 1.350–1.375 GHz band paired with 1.492–1.517 GHz band for low capacity *Point-to-Point* and *Point-to-Multipoint* systems,
- 1.375–1.400 GHz band paired with 1.427–1.452 GHz band for low capacity PP and PMP systems,
- 2.025–2.110 GHz band paired with 2.200–2.290 GHz band for traditional multi-channel, multi-hop radio relay systems and for modern access radio applications,

[2]In the basic ISDN access a subscriber uses two 64kbit/s duplex channels (B-channels) for his/her applications and one 16 kbit/s duplex control channel (D-channel). This results in a standard 144 kbit/s transmission.

- 2.520–2.593 GHz band paired with 2.597–2.670 GHz band for PP and PMP systems for single and multi-hop applications.

There is an opinion that the PMP systems do not use the allocated spectrum very efficiently. Another drawback is high cost of directional antennas and the necessity to ensure that they are mutually visible.

14.3 APPLICATION OF CELLULAR TECHNOLOGY IN WLL

Massive production of cellular transmission equipment caused its price reduction. Therefore, it makes the cellular technology attractive for application in wireless local loops. There are examples of application of known cellular telephony standards such as NMT 450, TACS, AMPS, GSM and IS-95 in the design of WLL.

Application of the cellular technology in the WLL results in the simplification of the system design as compared with cellular telephony, because subscribers do not move. Handover and roaming in another network do not occur, so the network layer of the system is substantially simplified. However, the end station (equivalent to the mobile station in cellular telephony) has to be equipped with a standard interface which allows connection of a regular telephone set to it. The quality of the WLL and the offered services strongly depend on the applied technology. Generally, cellular systems used in the implementation of the WLL are particularly useful for realization of simple telephone services in large low-populated areas. The reason for that is a large radius of a cell (up to a few tens of kilometers) and limited cell capacity.

The WLL systems based on the analog cellular technology suffer from all limitations which occur in the mobile version of that technology. The quality of the voice signal is lower than in a conventional wireline telephony and it further decreases when the distance between the base station and the end station gets closer to the cell range limit. The voice signal remains practically unprotected. Call privacy can be ensured using an expensive analog vocoder. Data and Group III fax transmission in a voice channel are possible at the rate up to 4.8 kbit/s. Despite the drawbacks and limitations, however, several system are offered for low-populated areas. It is worth noting that the application of analog cellular technology in WLL allows further utilization of the already existing equipment and to get some investment return despite the fact that most of the users have already moved to much more modern digital systems.

It is possible to use GSM technology in WLL implementation. The above mentioned simplifications of a cellular system remain valid for GSM as well; however, the service offer is broader and link security is much higher than in analog systems. Application of GSM in WLL will be attractive if the tariffing of stationary calls is set on the competitive level as compared with the calls to/from mobile users.

The IS-95 system which was described in Chapter 11 has been modified for application in WLL as well. The WLL systems based on the IS-95 technology have been offered by Qualcomm (QCTel) and Motorola. The structure of such systems does not differ too much from IS-95. The QCTel system consists of the base station controller, base stations and the subscriber units. The subscriber units are telephone sets with an integrated radio part, stations equipped with the interface allowing connection of a standard telephone set to them, and coin-operated telephones. The speech signal is encoded using QCELP technique resulting in the 13.2 kbit/s data stream in each channel. Supplementary services such as Group III fax transmission and data transmission with the rate up to 14.4 kbit/s are also offered.

The base stations apply directional antennas and divide the covered area into sectors. The number of sectors is not higher than nine. In each sector up to 45 calls can be realized at the same time. Assuming the mean traffic intensity per subscriber equal to 0.1 Erl, the probability of blocking equal to 0.01 and setting the upper value of traffic intensity in a sector on the level of 35.6 Erl, 334 subscribers can be served in one sector. This results in 2700 subscribers who can be served in the coverage area of a single base station. The cell radius can reach up to 50 km.

Besides the WLL systems directly based on the IS-95 technology, there are systems using CDMA technique which are specially designed for realization of a WLL. Air Loop of Lucent Technologies [6] is the example of such a system. Figure 14.2 presents the architecture of the Air Loop system.

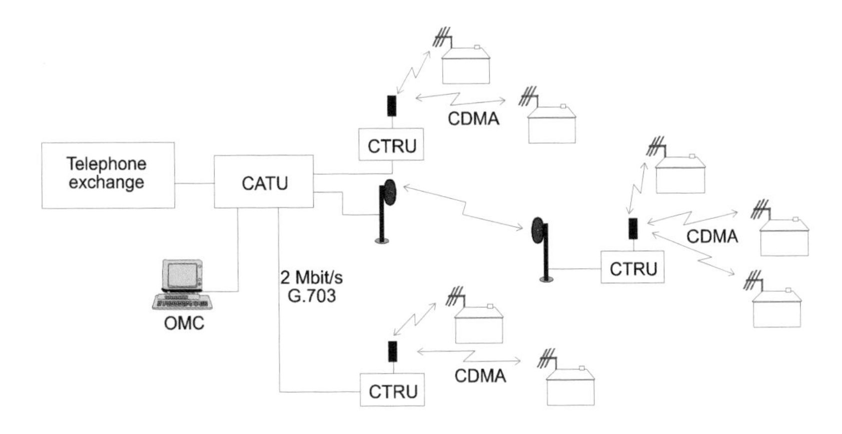

Figure 14.2 Air Loop system architecture

The telephone exchange is connected to the *Central Access and Transcoding Unit* (CATU). The CATU ensures analog or digital interface with the telephone exchange, it controls the call set-up, realizes data encryption and speech coding and processes data for control and monitoring purposes. It also realizes the interface with a set of *Central Transmitter/Receiver Units* (CTRUs) - the equivalents of base stations. One of the most important functions performed by CATU is speech transcoding from the PCM 64 kbit/s format to the ADPCM 32 kbit/s or LD-CELP[3] 16 kbit/s formats. The links between CATU and CTRUs can be realized in a wireline or wireless technique.

The CTRU transmits and receives signals to/from the end stations in the CDMA mode. The range of CTRU is up to 6 km without visibility of end stations and can

[3]LD-CELP is the acronym of *Low Delay Code Excited Linear Prediction*.

reach up to 15 km for line-of-sight transmission. The subscribers can also be connected to CTRU using a wireline link.

The end station in the Air Loop system is called *Network Interface Unit* (NIU) and consists of the *Subscriber Transmit/Receive Unit* (STRU) and *Intelligent Telephone Socket* (ITS). The STRU is integrated with the transmit/receive antenna. It is typically installed outdoors and connected to the ITS placed indoors. The ITS supports a basic rate ISDN (BRI) access (2B+D) or the access to two analog lines. There is a version of ITS which allows for the connection of eight analog lines. The configuration of the subscriber link in the form of an analog or ISDN link is controlled by software by the OMC. The analog interface can be connected to a standard telephone set, a Group III fax machine, a telephone modem with the data rate up to 19.2 kbit/s or with a public telephone set.

The Air Loop system operates using the CDMA principle in the frequency range between 3.6 and 4.0 GHz. The frequency interval between uplink and downlink is equal to 110 MHz. In a 10 MHz channel (uplink or downlink) typically 24 ISDN (2B+D) channels are placed. In another configuration one 80 kbit/s link and sixty 32 kbit/s links or one hundred and twenty 16 kbit/s links are located in a 10 MHz channel.

Assuming the mean traffic intensity per subscriber equal to 0.05 Erl, the probability of blocking equal to 0.01, and the omnidirectional base station antenna, the base station can serve up to 4060 subscribers. On the basis of the above description we can conclude that the Air Loop system is aimed at ensuring the ISDN or PSTN network access in medium-sized and medium-populated areas. A relatively large bandwidth requirement is a certain drawback of the Air Loop system. On the other hand, the system ensures call privacy thanks to applying speech coding and the spread spectrum technique.

The systems presented so far, which are based on cellular technology or its derivative, have a relatively large range and a limited capacity. It is reasonable to use cellular technology to implement WLLs in systems which are losing subscribers and in the cells whose capacity is too big for the real needs.

14.4 CORDLESS TELEPHONY IN THE REALIZATION OF WLL

Cordless telephony standards have been also applied in the realization of wireless local loops. There are solutions using CT2, DECT and PACS as the basis of the wireless local loop system. We will concentrate on the application of the DECT standard.

As we know, the DECT system has been optimized for urban and suburban environments with high traffic density [7]. The nominal range of the DECT base station is around a few hundred meters if the omnidirectional antenna is applied and the subscribers can move. However, if the end stations are stationary and directional antennas are used, the range of DECT base stations can be extended to 5 km. The range is limited by the structure of the TDMA frame and the propagation time which has to be contained in the guard time of a DECT burst, which is equivalent to the duration of 60 bits [8]. In the case of terrain obstacles or if the range extension is necessary, the DECT *Wireless Repeater Station* (WRS) [9] can be applied. In this manner the range is further extended by 5 km.

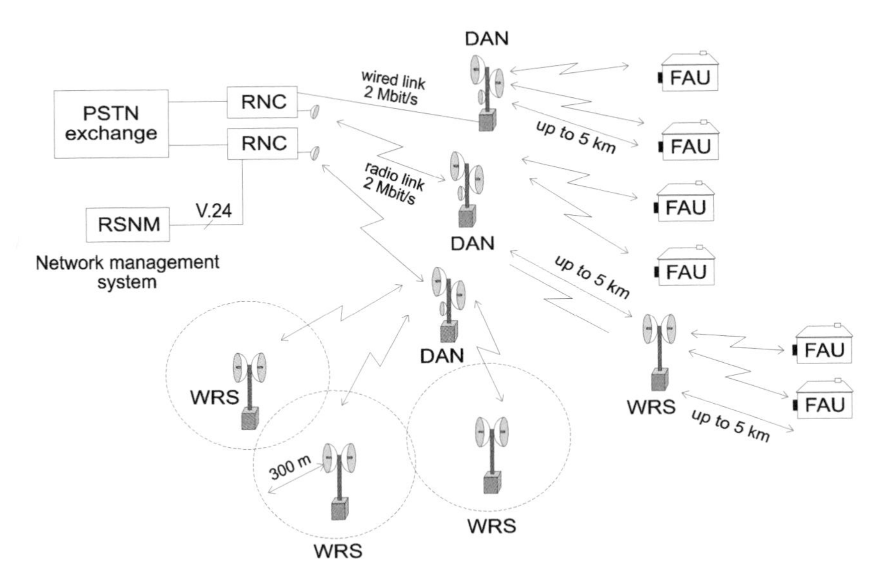

Figure 14.3 Architecture of the wireless loop system based on the DECT standard

Figure 14.3 presents the architecture of the WLL system based on the DECT standard. The *Radio Network Controller* (RNC) is connected with the set of base stations DANs (*DECT Access Nodes*) using 2 Mbit/s link realized by the radio relay systems or by wired connection. The DANs contain a set of directional antennas usually mounted in such a way that joint omnidirectional coverage is obtained. Subscribers are equipped with the *Fixed Access Units* (FAUs), which allow for connection of one or several telephone lines and which communicate with the base station (DAN) through a directional antenna.

The application of the WRS has been standardized by ETSI [9]. The WRSs can be applied in different ways. If several directional antennas are applied, they can be used to ensure coverage in the area which otherwise would not be accessible for the system. The WRS can also be equipped with a directional antenna for the communication with the base station (DAN) and the omnidirectional antenna for the communication with the end stations (FAUs) establishing a picocell environment. In a typical application the WRS is used with two directional antennas to extend the range of the system.

Owing to the general features of the DECT, the wireless local loops based on this standard are a particularly economic solution, if the WLLs are deployed in small, densely populated areas in which average or high traffic intensity is expected.

* * *

Our short survey of wireless local loop technologies shows that the WLL has become an attractive means of fast and economic realization of basic subscriber services in rural areas, in which the telecommunication infrastructure is poor or requires high investment costs, or in a densely populated urban environment in which modification of the existing telecommunication infrastructure can also be very expensive. The WLL technologies are also a tool of demonopolization of the telecommunication market. The selection of the appropriate WLL technology depends on many factors, in particular on the size of the area of the WLL system deployment, the expected density of the telecommunication traffic and the expected demand for more sophisticated services than a standard speech transmission.

REFERENCES

1. W. Webb, *Introduction to Wireless Local Loop*, Artech House Publishers, Boston, 1998

2. D. C. Cox, "Wireless Loops: What Are They?", *International Journal of Wireless Information Networks*, Vol. 3, No. 3, 1996, pp. 125-138

3. W. C. Y. Lee, "Spectrum and Technology of a Wireless Local Loop System", *IEEE Personal Communications*, February 1998, pp. 49-54

4. A. Adolski, "Fast and Inexpensive - ALCATEL 9800" (in Polish), *TELECOM Forum, Special Edition*, December 1996, pp. 8-9

5. ERC/Recommendation T/R 13-01, Preferred channel arrangements for digital terrestrial fixed systems operating in the range 1-3 GHz, Montreux, 1993

6. E. Juszkiewicz, "AT&T Air Loop - A High Capacity Access System", (in Polish), *Przegląd Telekomunikacyjny*, No. 8, 1995

7. S. Kandiyoor, P. van de Berg, S. Blomstergren, "DECT: Meeting Needs and Creating Opportunities for Public Network Operators", *Proc. of International Conference on Personal Wireless Communications*, New Delhi, February 1996, pp. 28-32

8. J. Henry, M. H. Kori, "DECT Based Rural Radio Local Loop for Developing Countries", *Proc. of International Conference on Personal Wireless Communications*, New Delhi, February 1996, pp. 44-46

9. ETS 300 700, Radio Equipment and Systems (RES); Digital European Cordless Telecommunications (DECT); Wireless Relay Station (WRS), March 1997

15

Satellite mobile communication systems

With the collaboration of Rafał Krenz, Hanna Bogucka and Wojciech Lasecki

15.1 INTRODUCTION

This chapter is devoted to personal satellite communications. This is a relatively new type of mobile communications. We will begin by placing personal communication systems against the background of land mobile communication systems. Then we will classify the types of satellite communications. The main part of the chapter is devoted to the description of the most important personal satellite communication systems. The author is aware of the unstable situation in the area of personal satellite communications, which has a serious impact on the final configuration of the systems and the actuality of their description in this book.

Considerations on future communication systems have led to the concept of *Universal Personal Telecommunications* (UPT). According to this concept, several communication networks such as fixed networks, land mobile and mobile satellite networks will cooperate with each other, creating a common integrated system which supports a wide selection of personal services. Each user is recognized by a single subscriber number independent of the network which he/she currently uses. Personal satellite communication systems play an important role in the UPT concept. The aim of these systems is to ensure the access to the telecommunication network anywhere on the earth, in particular in these places which are not covered by any other communication network such as PSTN or land mobile communication networks.

Table 15.1 First generation mobile satellite systems (DL - downlink, UL - uplink, APR - Automatic Positioning Reporting

Organization	Standard (introduction)	Services (coverage area)	Transmision rate (bit/s)	Applications
INMARSAT	A (1982)	speech trans., telex fax, data trans. (world)	analog FM, up to 9.6 kbit/s data transfer from term. up to 64 kbit/s	ships, oil platforms, transportable terminals
INMARSAT	B (1993)	speech trans., telex fax, data (X.25)	16 kbit/s (speech) 24 kbit/s (data)	gradual repl. of INMARSAT A
INMARSAT	C (1991)	telex and data trans. store & forward APR (world)	600 bit/s	small vessels, yachts, vehicles
INMARSAT	M (1992/93)	speech trans., fax, data trans. (world)	6.4 kbit/s (speech) up to 4.8 kbit/s	briefcase terminals, small boats
INMARSAT	Aero (1992)	speech trans., fax, data (world)		commercial and private airplanes
	AERO-C	data trans., store & forward	600 bit/s	
	AERO-L	two-direct. data exchange	600 bit/s	
	AERO-H	data trans., fax (G.3), speech trans.	10.5 / 4.8 kbit/s	
INMARSAT	D (1995)	paging	N.A.	messaging, remote control
Qualcomm	OmniTracs (1989)	two-way short mess., APR (North Am.)	5-15 kbit/s (DL) 55-165 kbit/s (UL)	long-haul transportation
ALCATEL Qualcomm	EutelTracs (1991)	two-way short mess., APR (Europe)	5-15 kbit/s (DL) 55-165 kbit/s (UL)	long-haul transportation

15.2 FIRST AND SECOND GENERATION OF MOBILE COMMUNICATION NETWORKS

The main task of the first and second generation mobile satellite systems [2], [25], [14] was to guarantee communication with ships and trucks travelling on long distance routes. The non-military and commercial service market has been dominated by INMARSAT (*International Maritime Telecommunication Satellite Organization*). The subsequent versions (A, B, C, M and aeronautical) of INMARSAT systems are presented in the first five rows in Table 15.1 [2].

As we see, the INMARSAT A, introduced in 1982, transmits voice using analog FM modulation, supports telex, facsimile and data transmission services from any place on the earth. The terminals are installed mainly on ships and oil platforms. INMARSAT B, which started to operate in 1993 replaces INMARSAT A and offers the same services in digital technique for similar groups of subscribers. The voice signals are transmitted at the rate of 16 kbit/s.

INMARSAT C has been in operation since 1991. Its main task is world-wide data transmission with the rates not exceeding 600 bit/s. The system offers store & forward telex and data services and *Automatic Positioning Reporting* (APR). The terminals are installed mostly on board small ships and vehicles.

The next version of the system, INMARSAT M, was introduced in 1992/93 and features more advanced speech coding at the rate of 6.4 kbit/s and data transmission at the rate of 2.4 kbit/s. The terminals are the size of a briefcase.

The aeronautical type of INMARSAT (1992) is devoted to communication with civilian aircraft.

In the late eighties, in the USA Qualcomm Inc. started to exploit the OmniTracs satellite system. A similar system, called EutelTracs, was introduced in cooperation with ALCATEL in 1991 in Europe [4]. Both systems are used to transmit short messages in uplink and downlink transmission directions and they support the APR services for mobile terminals. Both systems use the Ku band (12/14 GHz).[1]

All the systems mentioned so far use geostationary satellites which hang above the earth at the height of about 35780 km, on the equatorial orbit. The systems are characterized by large cost of terminals and offered services. The terminals and services are expensive, which limits the number of potential subscribers.

Second generation mobile satellite systems are characterized by substantially lower terminal cost, further progress in digitalization of signals and lower service prices. Most systems have limited coverage, e.g. ESA (*European Space Agency* - Europe) or DoCoMo (Japan). They usually operate in the L band. They mostly offer speech transmission and data and facsimile transmission at low rates. Like the first generation systems, they are used by a limited number of subscribers or they are specialized for maritime or aeronautical applications.

15.3 PERSONAL SATELLITE COMMUNICATIONS

As we have mentioned, the concept of universal personal communications assumes that a subscriber is accessible in any place on the earth through any network under a unique number. The personal satellite communication network is one of the networks which allow the subscriber located in any place of the globe to access a telecommunication network using a handheld terminal and to establish a link with at least one satellite. Figure 15.1 presents the configuration of a satellite system in the network of wireless telecommunication systems.

[1]The appendix at the end of the chapter lists the frequency ranges of particular bands.

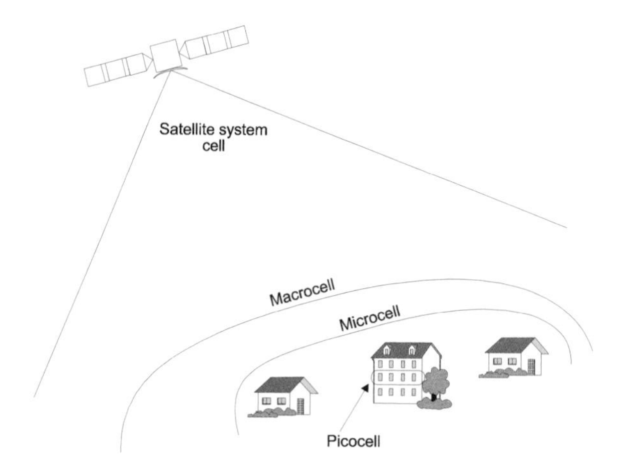

Figure 15.1 Location of the satellites systems among wireless systems

The systems operating in picocells (usually indoors) constitute the lowest layer of the mobile communication hierarchy. Wireless telephony working in microcells (e.g. city centers, airports, large supermarkets, etc.) is the next layer of the hierarchy. Classic cellular systems operating in macrocells process the traffic of moderate density, but allow for transmission to/from a fast moving terminal. The satellite system is superior to all the above systems. It can be used if there is no access to other telecommunication systems, e.g. in a sparsely populated area or if the serving system is temporarily overloaded. Therefore we can treat satellite systems as a supplement of cellular systems, but not as their replacement. The capacity of satellite communication systems is naturally much lower than the capacity of land mobile communication systems.

The handheld form of a mobile terminal implies serious consequences for the design of a whole satellite system. The power budget of the handheld is very tight due to its small energy source. One of the decisive factors in the commercial success of a handheld is the length of standby time after which the battery needs to be re-loaded. Due to health reasons a handheld kept during the connection close to the user's head cannot radiate too much power. An additional factor which can affect the commercial success of the system is the time delay in the transmission of human voice. A too long delay caused mainly by long propagation time to and from the satellite creates discomfort for the users. The length of the delay can also have a serious impact on the efficiency of block data exchange performed with the application of the ARQ method (see Chapter 1). Long propagation delay would require very large data buffers.

The above mentioned factors allow for differentiation of several types of satellite systems with respect to the location of their satellites. Taking into account the height of their orbits we divide the satellite systems into the following classes [3]:

- **LEO** (*Low Earth Orbit*) are the systems using satellites on low circular orbits. The satellites are nonstationary. Each of them is visible by the terminal antenna for a few minutes, several times during twenty-four hours. The orbit height is between

700 and 1500 km. Due to the low orbit radius, at least 40 satellites are necessary to ensure full earth coverage. The number of cells seen on the earth surface and moving due to the movement of satellites is equal to 3000 or more. The signal delay caused by the propagation to and from the satellite is no longer than 50 ms. The large number of cells results in a relatively large system capacity per spectrum unit. Unfortunately, the cost of the system per area unit is high. Low delay giving comfort in speech transmission is particularly advantageous. To launch LEO satellites into the appropriate orbits, rocket systems with moderate parameters are required; however, due to a high number of satellites, a high number of missions is necessary to place all of them in their orbits. The construction of the systems using LEO satellites requires a precise orbit design, which involves adjusting their positions in time. Due to short visibility periods of LEO satellites, the handover is relatively frequent. While the satellite travels over the horizon, its distance from the terrestrial terminal changes in a relatively large range, which implies providing complex automatic gain control in the satellite. The Doppler effect is also meaningful.

- **MEO** (*Medium Earth Orbit*), also called *Intermediate Circular Orbit* (ICO) are the systems with the satellites located at the height between 10000 and 15000 km over the earth surface. Due to a higher orbit, the satellite beams cover a larger area of the earth, so the required number of satellites is lower (between 10 and 15). The number of cells in MEO systems is about 800. Higher orbits result in higher propagation delay equal to about 150 ms. The negative consequence of fewer cells is the decrease of the frequency reuse if the FDMA or FDMA/TDMA are applied, which in turn leads to lower system capacity. The lower number of satellites implies lower system cost.
- **GEO** (*Geostationary Earth Orbit*) are the systems with geostationary satellites located in the equatorial plane in orbit with a height of about 35780 km. Only 3–4 satellites are needed for the coverage of the earth up to 75° of latitude. The propagation delay is longer than 300 ms. The number of cells depends on the number of beams radiated by a satellite; nevertheless, their number does not exceed 800. The capacity calculated per spectrum unit and the system cost of the GEO systems in comparison to the LEO systems are much lower. The GEO systems apply the well-tested space technology. The Doppler effect is negligible. On the other hand, in order to launch the GEO systems, expensive rocket systems are required. Due to the equatorial orbit, for the terminals located above 50° of latitude, the *elevation angle*[2] is so small that the signal is additionally attenuated by the higher layers of the atmosphere and the multipath effect occurs. Due to the orbit height, the GEO systems require high power radiation. The latter feature means that handhelds cannot be used in the GEO systems, therefore systems are eliminated from personal satellite communications.

Figure 15.2 illustrates the orbit location of three types of satellites considered above.

[2]The elevation angle is the angle between the line of horizon and the direction line to the satellite.

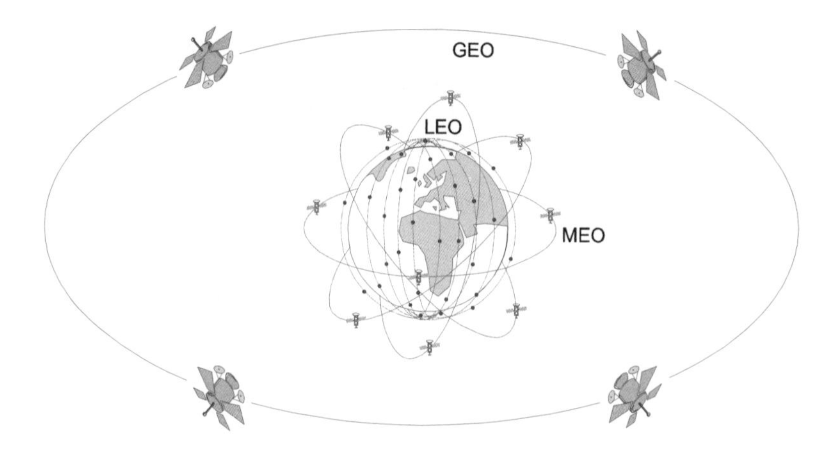

Figure 15.2 Several orbit constellations used in satellite mobile communication systems

Besides the division of the satellite systems with respect to their orbit height, we can also divide them according to the size of the satellites. From that point of view we categorize the satellite systems in the following way:

- large GEO satellites,
- large LEO/ICO satellites,
- small LEO satellites, further divided into:
  - LEO minisatellites weighing 100–750 kg,
  - LEO microsatellites weighing 50–100 kg.

The number of existing or planned satellite systems is quite high. In the next sections we will concentrate on personal satellite systems and representatives of future systems planned for multimedia and Internet access.

15.4 SERVICES OFFERED BY PERSONAL SATELLITE SYSTEMS

The list of basic services offered by personal satellite systems is the following:

- speech transmission,
- data transmission,
- paging,
- electronic mail.

Additional services offered by particular systems will be described when a given system is analyzed. In order to ensure commercial success of personal satellite communication, the system should offer:

- full global coverage,
- possibility of using inexpensive handheld terminals,
- low tariffs at the system introduction and low unit cost of access to the satellite and cellular systems,
- reliable connection to the PSTN,
- possibility of establishing direct connection between satellite system subscribers.

As we will see, not all of the planned or already introduced systems fulfill the above requirements. Generally, this segment of the mobile communication market is not as successful as land mobile cellular systems.

15.5 DESCRIPTION OF THE MOST IMPORTANT PERSONAL SATELLITE SYSTEMS

As we have already mentioned, the personal satellite market is undergoing constant changes and the systems which have not been introduced yet can still be modified, so the information contained below is to certain extent tentative.

15.5.1 Iridium

The Iridium system is one of two systems applying a combination of FDMA/TDMA. Its name comes from the element Iridium. One atom of Iridium has 77 orbiting electrons. At first, 77 satellites were to be used in the system. Later, after optimization of the constellation resulting in cost reduction, the number of 66 satellites appeared to be sufficient to ensure reliable communication.

The concept of Iridium [9], [13], [18], [17], [10] appeared first in 1987 and was presented to the public by the Motorola Corporation in 1990. The companies from many countries including Russia and China participate in the Iridium consortium. The Iridium satellites have been placed in six orbital planes inclined at 86° with respect to the equatorial plane. Thus, the satellites rotate almost between the geographic poles. In each orbital plane 11 evenly spaced satellites rotate in the circular orbit at the altitude of 780 km. The satellites located in the neighboring planes rotate in the same direction but are shifted in phase. Figure 15.3 presents the configuration of a few satellites and a part of their trajectories. The satellites which rotate in the first plane and the last (sixth) plane are the exception, because they rotate in opposite directions. The angular distance between orbital planes is equal to 31.6°, only the angular distance between the sixth and first plane is equal to 22°. The satellite constellation was selected according to the rule developed by Adams and Rider [1].

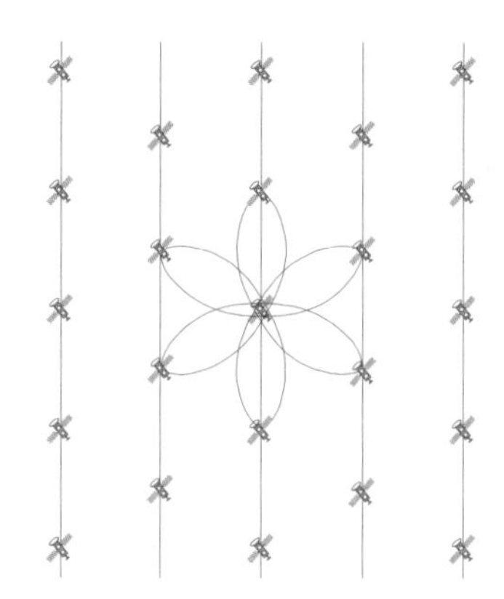

Figure 15.3 Communication between neighboring Iridium satellites [8]

Each satellite can communicate and route the traffic to two satellites from the same orbit: the preceding one and the following one, and two satellites from each neighboring orbits. The intersatellite links operate in the 23 GHz band and use four antennas. Therefore at any given moment, four intersatellite connections can be maintained by a given satellite, although there are six satellites located in its direct vicinity (see Figure 15.3).

The Iridium system was designed to perform communication with terrestrial networks using up to 250 gateways and two network control stations. Such a system provides sufficient redundancy to ensure reliability; however, 15 gateways are sufficient for the system operation. Each gateway has two tracking antennas communicating in the 20/30 GHz band with the currently active satellite and with the forthcoming satellite. The Iridium subscriber is connected with a PSTN through one of the earth gateways.

Intersatellite links allow for setting up a connection between two Iridium subscribers without using a terrestrial network. In a connection between an Iridium terminal and a terrestrial network subscriber, intersatellite links are used to minimize the route in the terrestrial network. The connection "goes down" to the terrestrial network through the gateway closest to the terrestrial subscriber. Figure 15.4 presents the general concept of the system with intersatellite links.

As we have already mentioned, the propagation delays in the Iridium system are relatively low. A single path delay between the satellite and a terrestrial terminal is between 2.6 and 8.2 ms. However, the joint delay introduced by the system is much higher, because the binary stream must be placed within the TDMA frame structure, and bacause all signal processing operations must be performed (e.g. speech encoding and decoding). The joint delay between the terrestrial terminal and a satellite reaches about 90 ms and is equal to about 9 ms in each intersatellite link.

Each satellite has three phase-array antennas[3] with 16 spot beams giving a total number of 48 spot beams creating cells on the earth surface. The cell diameter is about

[3]See Chapter 18 on adaptive antennas

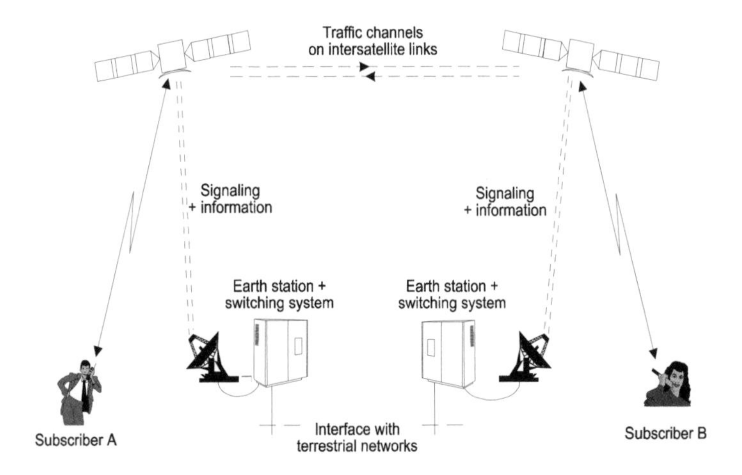

Figure 15.4 General concept of the Iridium system [11]

700 km. Among 66×48 = 3168 possible cells only 2150 cells are active. This is due to the fact that some of them are switched off in the polar areas in which the satellite orbits get closer to each other and cell overlapping is larger. The switching of the cells on and off requires a unified algorithm and a unified criterion. Each satellite has a look-up table which determines the moment of switching on and off at the change of 0.1° of latitude.

Because the Iridium system applies FDMA/TDMA as a multiple access method, the same channels can be used repeatedly. In Iridium a cell cluster consists of 12 cells. The satellite beams generated by the satellite antenna arrays are periodically concentrated on each cell for a prescribed time duration. At these moments, transmission from and to the subscribers located in that cell is performed in the TDD mode. Uplink and downlink transmission between a subscriber and a satellite is performed in the L band using QPSK modulation at the data rate of 50 kbit/s. The assigned frequency range for the uplink and downlink transmission is 1621.35–1626.5 MHz. The transmission is organized in 90-ms frames which consist of a paging and signaling channel, and four uplink channels followed by the associated downlink channels. The system bandwidth is divided into 124 carriers. Since there are 12 cells in a cluster and the satellite has 48 spot beams, each frequency is reused four time per satellite. Thus, theoretically 1984 connections could be established by a satellite; however, due to energy limitations 1100 connections can be realized at the same time.

The satellite connection with the gateways is performed in the K/Ka band, i.e. from 19.4 to 19.6 GHz in the uplink and from 29.1 to 29.3 GHz in the downlink direction. Intersatellite links are realized in the band between 23.18 and 23.31 GHz.

The basic service is speech transmission. For that purpose the speech encoder generates the data stream at the rate of 4.8 or 2.4 kbit/s. As in GSM, the user activity detection is performed to reduce the co-channel interference. Besides speech transmission, data transmission is offered at the rate of 2.4 kbit/s. Recently Iridium also offers direct access to the Internet at the rate of 10 kbit/s, independent of the PSTN. Other services such as paging, fax transmisison, SMS and position reporting are also listed. The assumed error rate is 10^{-2} for speech signal and 10^{-5} for user data. The user terminals are dual-mode. In a typical situation the terminal is connected to the land mobile network. If the latter is not accessible, the Iridium network is used.

In June 2001, Iridium is the only personal satellite communication system which has a truly global coverage. Other systems do not allow for transmission near the poles. Some systems are still in the deployment phase. This advantage of Iridium did not protect the system against financial problems which showed that a good and innovative technical solution is not enough to achieve commercial success. The services were very expensive, the phones were quite big and the investors had not foreseen the enormous progress of land mobile networks which took away a lot of potentional customers of Iridium. However, the system has recently been put back in operation by Iridium Satellite LLC. The prices of services have gone down and this time Iridium targets customers from US military, maritime, aviation, oil and gas, mining, construction, forestry, government, non-governmental organization/relief, and the yachting/leisure segment [6].

15.5.2 GLOBALSTAR

The GLOBALSTAR system [19], [15], [24], [12] was the initiative of Loral Communications. The system concept differs from that of Iridium in many aspects.

The reseach on the selection of LEO orbits showed that two approaches to setting their positions are possible:

- the orbits are located in the polar planes, which allows for full earth coverage,
- the orbits are located in the planes properly inclined with respect to the equatorial plane.

The first approach was applied in Iridium. The second approach has been used in GLOBALSTAR. The satellites will cover the area of the earth surface up to the determined latitude. The coverage of the GLOBALSTAR system has been set between $\pm 70°$ of latitude. The northern part of Greenland, Spitzbergen and the islands of Northern Canada remain outside the system coverage. Figure 15.5 presents the planned coverage map of GLOBALSTAR [19].

To ensure high transmission quality, the satellite elevation angle must be sufficiently large [19]. The value of this angle determines the transmission channel quality on the path between the satellite and a terrestrial terminal. For the elevation angle between 0° and 20°, the satellite channel has similar properties to the terrestrial mobile channel and suffers from the multipath phenomenon. For the elevation angle between 20° and 40° the multipath occurs due to diffuse reflections only. For higher elevation angles the channel quality is gradually improving. High quality transmission can be achieved above 70° of elevation.

In order to obtain good propagation properties, the orbits inclined at 52° have been selected. The satellites rotate in eight planes with the phase shift of 7.5° between each two successive planes. The rotation period is equal to 114 minutes. Figure 15.6 presents

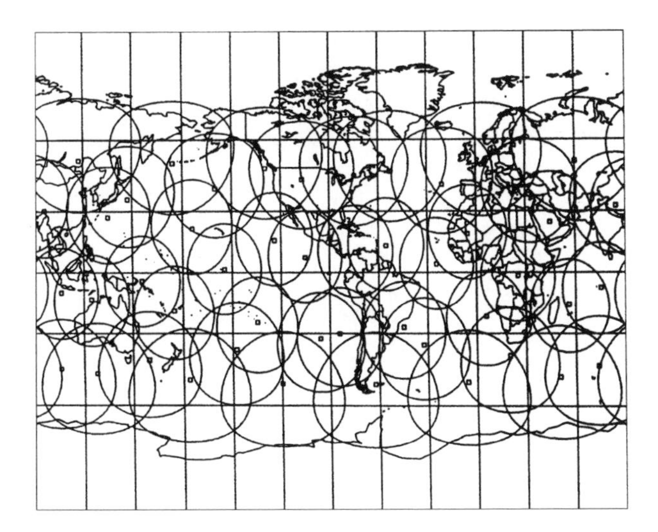

Figure 15.5 The planned coverage of GLOBALSTAR [19] (with permission of Alcatel Telecommunications Review)

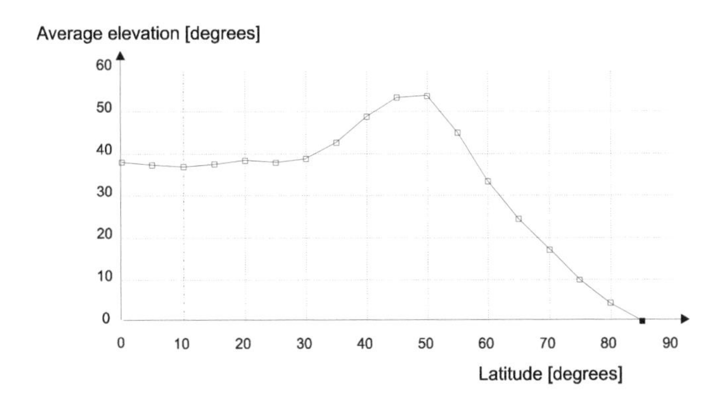

Figure 15.6 Average elevation of an active satellite as a function of latitude [19] (with permission of Alcatel Telecommunications Review)

the average elevation of an active satellite as a function of latitude. We see that up to the latitude equal to 60° the average elevation is quite high (not lower than 35°). A minimal elevation of the satellite seen by a terrestrial terminal is 10°. An important feature of the satellite constellation selected in GLOBLSTAR is simultaneous visibility of two satellites, which improves the system reliability.

The orbit altitude in GLOBALSTAR is equal to 1389 km. Due to higher altitude than in Iridium, a smaller number of satellites is necessary. Six satellites and one spare satellite rotate on each of eight orbit planes, so 48+8 satellites are located on the orbits. The satellite coverage area is divided into 16 cells. The cells are organized in a concentric manner with a single cell in the center, 6 cells in the first ring and 9 cells in the second ring. The minimum and maximum values of propagation delay are equal to 4.63 and 11.5 ms, respectively.

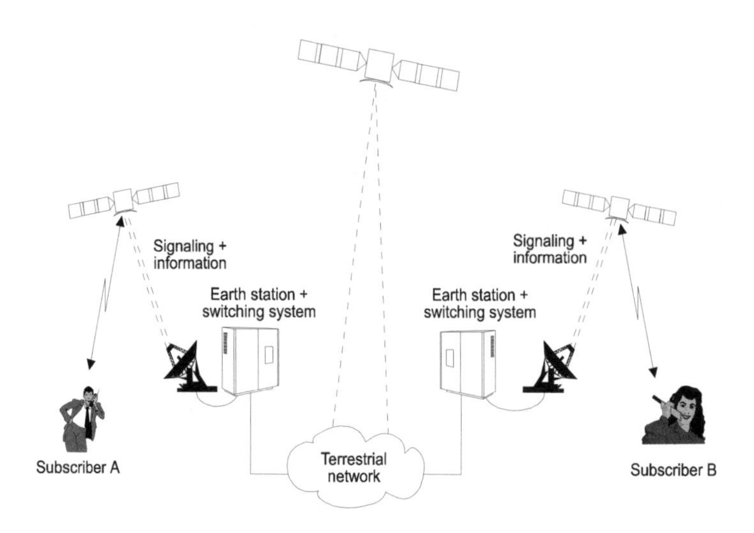

Figure 15.7 General concept of the GLOBALSTAR system [11]

GLOBALSTAR differs from Iridium in the general concept of the system as well. The GLOBALSTAR system works according to the *bent-pipe principle.* Bent-pipe is a signal relay scheme in which a signal from a terrestrial terminal is sent to a satellite, which relays the signal back to the Earth with minimal processing. This way the subscriber is connected via the satellite with the closest earth station. That station realizes the connection through a terrestrial network (see Figure 15.7). The satellite in the middle of the picture indicates that the connection between the GLOBALSTAR earth station and the station of another subscriber can be partially set using another satellite system. In consequence, the satellite part of the GLOBALSTAR system has to be transparent from the point of view of the remaining parts of this system. This transparency implies the necessity of creating the gateway infrastructure connecting GLOBALSTAR to public networks. The density of the gateways has been estimated to be one station per one milion km^2, which is equivalent to one station in a middle-sized European country. At such gateway density intersatellite links are not required. If gateways are built in each middle-sized country, the local network operators will not be omitted and thanks to the data bases contained in the gateways they will be able to monitor the traffic and establish the tariffs according to their own rules.

GLOBALSTAR applies the CDMA [7] in the transmission layer between a terminal and a satellite (and vice versa). GLOBALSTAR and a few other systems use the common band assigned to personal satellite communications at WARC'92[4]. Such a system must be robust against interference arising from other systems operating in the same band. The bands allocated to GLOBALSTAR and other potential CDMA personal satellite systems are 1610–1621.35 MHz for the uplink and 2483.5–2500 MHz for the downlink between a terrestrial terminal and a satellite. In the link between gateways and satellites the following bands are used: 5.091–5.25 GHz for the uplink and 6.875–7.055 GHz for the downlink.

The trasmission methods applied in GLOBALSTAR are very similar to those used in the cellular CDMA IS-95 system, therefore they will not be repeated here. We will only summarize the main transmission parameters.

The speech is digitally encoded at the rates 2.4, 4.8 or 9.6 kbit/s. Its allowable error rate is 10^{-3}. Data transmission rates range between 2.4 and 9.6 kbit/s and the maximum bit error rate is 10^{-6}. Besides voice and data transmission the GLOBALSTAR system offers the Short Messaging Service (SMS) and the facsimile service. The two latter services depend on the GLOBALSTAR service provider.

Let us consider a simplified procedure of the connection set-up in the GLOBALSTAR system [19]. After power on, the subscriber terminal attempts to find a terrestrial cellular network. If the attempt is successful, the terminal registers in the network and can use it for its own connections. If the attempt fails, the terminal tries to register in the satellite network. It looks for access channels and sends to the satellite a request for the CDMA channel assignment and registration. The mobile terminal also transmits its *International Mobile Subscriber Identity* (IMSI) to the *Satellite Station Controller* (SSC). In turn, the mobile terminal is located by one of the gateways within the satellite footprint. Next, the earth station performs subscriber's authentication, retrieving the data from the Home Location Register (HLR) in which users' records are stored. After successful authentication and location, the subscriber receives information concerning the registration with the SSC. After receiving this message, the mobile terminal synchronizes with the signaling channel of the gateway where it is registered. The gateway sends the mobile terminal registration message to its HLR. From that moment the mobile terminal is treated in the same way as a terminal of the land mobile network associated with the GLOBALSTAR system. We can describe this status as the integration of the GLOBALSTAR with the terrestrial network. The currently offered phones are two-mode or tri-mode. In Europe the users can apply the phones connecting with GSM 900 and GLOBALSTAR, whereas in the USA the phones are able to register with CDMA-800 (IS-95), AMPS-800 and GLOBALSTAR.

The configuration of the GLOBALSTAR network is based on the configuration of land mobile cellular networks. Thanks to a short delay introduced by the satellite system, the protocols and signaling systems used in the cellular networks can be applied without modifications. The interfaces are also similar to those in cellular systems. These interfaces are:

- the air interface between the mobile terminal, the satellite and the ground station,

[4]WARC - *World Administrative Conference*, now WRC - *World Radiocommunications Conference* is an international periodic conference at which common rules of using the spectrum are established.

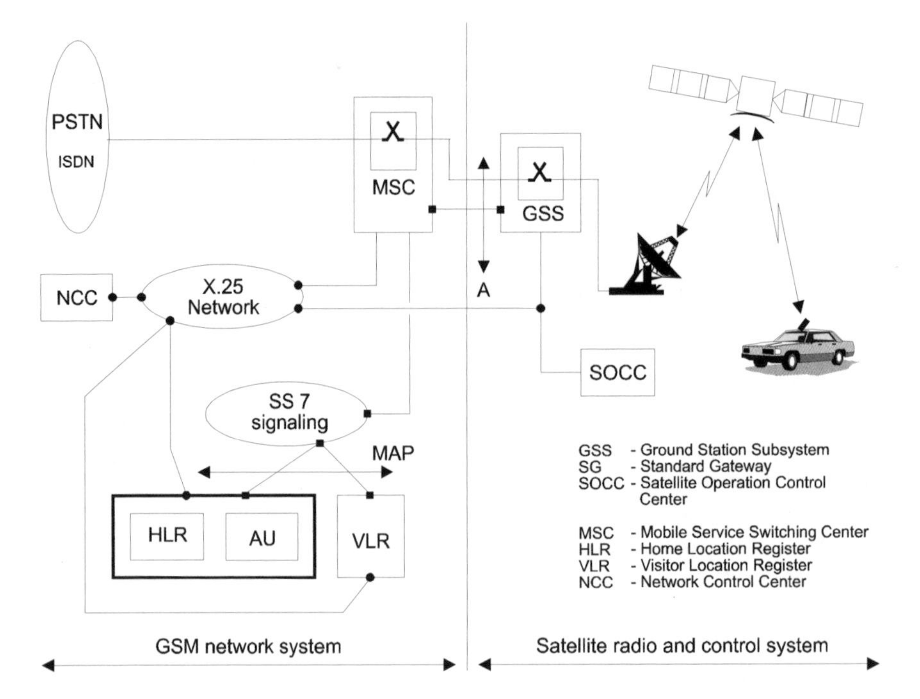

Figure 15.8 Simplified architecture of the SSC in GLOBALSTAR [19] (with permission of Alcatel Telecommunications Review)

- the interface between the SSC and the switching center (MSC) of the land mobile public cellular system,

- the interface associated with the data exchange protocol between the MSC and the data bases used in the management of mobile terminals.

Figure 15.8 presents a simplified architecture of the satellite station controller (SSC) connected with a part of a land mobile cellular network, e.g. GSM. We see that the appropriate ground station of the GLOBALSTAR system communicates with the MSC through the ground station subsystem. In order to perform the subscriber authentication, the MSC exchanges data with the HLR, VLR and AU registers. The route to PSTN is possible through the MSC. The PSTN can also serve as the network connecting with another SSC and the terminal registered with it.

In June 2001, the GLOBALSTAR system was already in operation, although it did not reach full coverage. Several vendors offered dual- and tri-mode mobile handsets, fixed phones and special phones for maritime applications.

15.5.3 ICO

The international consortium INMARSAT joined the race in the area of personal satellite communications by starting the so-called Project-21. In 1995 a special company called ICO Global Communications was founded. In 2001 the system was introduced by the New ICO company and in the summer of 2001 was in the deployment phase.

The New ICO system consists of a space segment and a dedicated ground network. The main task of the system is to provide global IP services, including connection to the Internet, data, voice and fax transmission [16]. The system will operate in three modes:

- circuit-switched mode based on the GSM standard
- packet-switched mode based on GPRS
- Internet protocol mode.

The space segment will consist of 10 active satellites and two spare satellites. They will be located in two mutually perpendicular orbits inclined at 45° to the equatorial plane. The satellites will operate in MEO/ICO orbits at an altitude of 10390 km. The satellites will be located in the orbit in such places that from every point on the Earth at least two satellites will be visible. Thanks to a relatively high altitude, the period of the orbit circulation is approximately equal to six hours and the average period of visibility of a single satellite is equal to about one hour. The minimum elevation angle for a mobile station is 10°. As a result, handover will be rare. The consequence of the MEO/ICO orbit selection is the propagation delay which is higher than for LEO systems; however, it does not exceed 48 ms for a single path.

The satellite system has the same bent-pipe architecture as GLOBALSTAR. Recall that the satellites act like mirrors reflecting signals from the earth. The assigned frequency band for the uplink is 1980–2010 MHz. The downlink will be placed in the 2170–2200 MHz band. The links from earth stations to a satellite and from a satellite to an earth station will use the 5.15–5.25 GHz and 6.975–7.075 GHz bands, respectively.

The satellites will communicate with terrestrial networks through the ICONET which is the dedicated terrestrial network. It is planned that ICONET will be based on the Internet Protocol. The ICONET will be built of 12 appropriately located *Satellite Access Nodes* (SANs - see Figure 15.9) connected by high-speed links. Six SANs will simultaneously act as *Telemetry, Tracking and Control* (TT&C) stations. They will control the positions of satellites and will be managed from the *Satellite Control Center* (SCC) located in England.

The main task of an earth station (SAN) is to provide an interface between the satellites and terrestrial networks and to perform routing through the ICO network. The earth stations are equipped with the following elements [16]:

- five antennas and the associated equipment to communicate with the satellites,
- packet-switched and circuit-switched equipment to perform routing through the ICONET and to interface with land mobile and wireline networks,

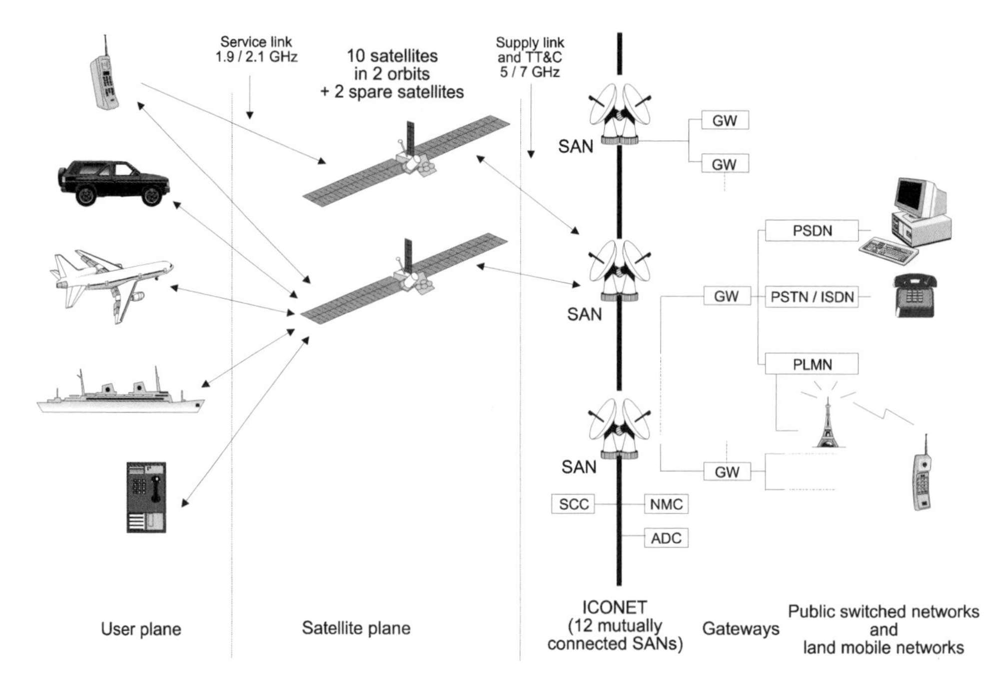

Figure 15.9 General architecture of the ICO system

- appropriate registers necessary for mobility and service access management,
- equipment necesssary to realize voice, facsimile and data messaging services,
- GPRS equipment which will direct traffic and store data and support several IP services.

It is expected that the future air interface of the New ICO system will support data rates up to 144 kbit/s and the communication protocols will be similar to those used in terrestrial networks. The New ICO system will directly cooperate with the GSM system without any special technological developments. It is expected that the New ICO system will be equipped with interworking functions which will enable the users of the New ICO system to roam in networks other than GSM. In the packet mode operation the New ICO system will support services similar to those offered by GPRS.

The general information on the New ICO system presented above differs in some aspects from what was available in the open literature a few years ago [22], [26]. Originally, the ICO system was mainly aimed at speech transmission and a limited rate data transmission with the maximum rate of 64 kbit/s at a connection with a fixed terminal. Currently its main task is IP networking.

15.6 FUTURE WIDEBAND ACCESS SYSTEMS

Two systems which will be described below are not personal satellite communication systems in the strict sense. Their main task, like that of New ICO, is to interface with Internet and broadband networks, although voice transmission is on the service list as well. The systems shortly discussed below are examples of broadband satellite transmission projects.

15.6.1 Teledesic

When the Teledesic company was founded in 1990 and the initial system design was completed (1994), the concept of Teledesic system was very futuristic. The aim of Teledesic is to build a global, broadband Internet-in-the-Sky™ network of fiber-like quality. In 1997 the Federal Communications Commission granted a licence to Teledesic and World Radiocommunications Conference designated the necessary spectrum. The start of the system operation is planned for 2005 [23].

Teledesic consists of a space segment and a ground segment containing terminals, network gateways and network operations and control centers. The space segment is in fact a satellite-based switch network whose main task is to provide reliable links between terminals.

The space segment will initially consist of 288 operating satellites (plus some spare satellites) rotating on 12 LEO orbits inclined at 40° to the equatorial plane. The altitude of the orbits will be equal to about 1375 km. In the earlier system versions the planned number of satellites was 840 and the inclination of their orbits was also different [8]. Due to broadband links which will be established between the terrestrial terminals and satellites, a large bandwidth is required. Therefore the communication between terminals and satellites will take place in the Ka band. The 18.8–19.3 GHz band is devoted to the downlink, while the 28.6–29.1 GHz band will be applied in the uplinks. In these high frequency ranges rain and line-of-sight obstacles seriously degrade wave propagation. Therefore the operating satellites have to be located at high elevation angles. The high elevation angle, the high system capacity required and the assumed satellite altitude result in a large number of satellites in the Teledesic constellation.

Teledesic will be optimized for service to fixed-site terminals; however, maritime and aviation applications will also be supported, so transportable terminals will also be operating. Typically, a terminal will support data transmission at rates up to 64 Mbit/s in the downlink and up to 2 Mbit/s in the uplink. Two-way data transmission at the rate of 64 Mbit/s is also planned for broadband terminals. The terminals will connect the end-user with other Teledesic users, or will interface with several networks operating with IP, ISDN, ATM and other protocols. Therefore, the Teledesic network will interface with Internet and will be used in multimedia communication, LAN interconnections, corporate intranets etc. [23].

The applied methods of multiple access follow the concept of the system as a packet network. Within a cell, the system resources are shared using *Multi-Frequency Time Division Multiple Access* (MF-TDMA) on the uplink and *Asynchronous Time Division Multiplexing Access* (ATDMA) on the downlink.

The Teledesic network will be equipped with network *Constellation Operations Control Centers* (COCC) and *Network Operations Control Centers* (NOCC). The COOCs manage the satellites, their deployment, replacement of fault satellites by the spares and de-orbiting. The NOCCs perform several network administration and control functions.

The Teledesic satellites contain switches on board. The Teledesic network is a packet-switched network. In this respect it is similar to an ATM network; however, the packets in Teledesic are longer than the 53-byte *ATM cells.*[5] Each packet consists of a header containing the destination address and sequence information, an error control field checking the validity of the header and a payload carrying user data. Transmission within the whole network is performed in the packet form. The change of the data format takes place in the terminals. Each packet is transmitted individually to the destination on the basis of adaptive packet routing algorithm which aims at minimizing the packet propagation delay and its variability. Let us recall that the LEO satellite network rotates with respect to the terrestrial terminals, which makes the network dynamic. The packets can arrive at the destination out of sequence along different paths, so they have to be reordered and buffered in order to ensure the appropriate quality. The switching between the satellite nodes will take place in space. Intersatellite links will be implemented between the satellites located in the same and adjacent orbits. The large number of satellites (and switches) and their interconnections will create a kind of mesh network which will be more tolerant to faults and will feature a natural adaptivity to traffic fluctuations and node congestion.

15.6.2 Skybridge

The Skybridge system is an initiative of Alcatel [5], [20], [21]. It will provide high speed access to the world's fiber optic backbone networks. In this respect the Skybridge system will attempt to fill the gap between fast development of fiber optical networks and the lack of possibilities of accessing them by many users despite the existence of the ADSL and cable modem technologies. The Skybridge designers estimate that their system will ensure broadband access to over 20 million users over the world. Skybridge will support multimedia applications over broadband Internet, video telephony, video conferencing, electronic commerce, telecommuting, distance learning, telemedicine, corporate networking, remote access to LAN, POTS (*Plain Old Telephony Service*) and live entertainment.

Skybridge will apply 80 LEO satellites plus several spares which will be located in the circular orbits in a so-called Walker constellation. The orbits with four satellites in each of them are contained in 20 planes inclined by 53° to the equatorial plane. The altitude of the orbits will be equal to 1469 km. For transmission between satellites and terrestrial terminals Skybridge will apply frequencies in the Ku band between 10 and 18 GHz. This band is also used by terrestrial microwave systems and GEO satellite systems; therefore, special means preventing interference between Skybridge and those systems have been planned. Among others, the Skybridge satellites cease

[5]Let us recall that in *Asynchronous Transfer Mode* (ATM) terminology an ATM cell is a 53-byte long packet constituting a basic transmission unit.

transmission to a given cell when an earth station recognizes that their transmission direction is close to the GEO systems' pointing direction. In consequence, for each gateway a non-operating zone is defined. It includes the satellite positions which would create interference with GEO systems. Thus, transmission from these directions will be realized by other earth stations.

The architecture of the Skybridge system can be divided into two segments:

- *Space segment* which includes the satellite constellation, the *Satellite Control Center* (SCC), the *Tracking, Telemetry and Command* (TT&C) ground stations and two mission control centers,

- *Telecommunication segment* which consists of gateways and Skybridge user terminals, which connect end users with broadband networks.

Each Skybridge satellite can create a number of spot beams of the diameter equal to 700 km. Thanks to the use of active antennas the beams are maintained pointed towards the terrestrial gateways located in a given spot beam for a prescribed period of time. Traffic transmitted by a subscriber terminal in a particular beam is directed towards the gateway in the same beam (and vice versa).

Up to 200 gateways are planned for the deployment necessary to ensure global coverage. The gateways will connect the Skybridge network with Internet servers, narrow- and broadband terrestrial networks or leased lines [21].

User terminals consist of the antenna equipment and the interface to the external multimedia equipment. Antennas used by residential users will be mounted on roofs and will have a diameter of 50 cm. Thanks to them a subscriber will be able to receive data at the rate of up to 20 Mbit/s and transmit data at up to 2 Mbit/s. A certain number of interfaces will connect the terminal to several PCs. The terminal will operate in the session-based data mode to handle Internet traffic.

Business (professional) terminals will apply 80 to 100 cm antennas and will transmit and receive data streams at the rates 3 to 5 times higher than residential user terminals. Thanks to a modular structure of professional terminals, they will receive up to 60 Mbit/s. Professional terminals will typically be connected to a Private Branch Exchange (PBX) or a Local Area Network (LAN). Two resource management modes will be supported: a *standard mode* in which $n \times 64$ kbit/s links for telephone and videotelephone traffic will be created, and *data mode* based on sessions for Internet or intranet traffic [20].

Skybridge network will apply ATM (*Asynchronous Transfer Mode*) transmission. As we know, in the ATM network data are transmitted in 53-byte long packets (cells). This unified way of transmission simplifies radio resource management, the required quality of service can be maintained and interworking with other networks is possible.

The capacity of the Skybridge system [20] is determined by several factors, i.e. the access to a wide frequency band, frequency reuse in different cells, multiple satellite visibility which can be used in high traffic areas, statistical multiplexing of different types of traffic and efficient use of satellite and terminal power. In each cell, the 750-MHz band can be used in the downlink and the 300 MHz band will be devoted to the uplink transmission. This asymmetry results from typical asymmetric traffic encountered in broadband multimedia networks. The Skybridge designers declare [20] that a cell will be able to support commercial traffic at up to 1 Gbit/s per satellite with up to 770 Mbit/s for professional traffic and up to 310 Mbit/s for residential traffic. Maximum traffic which can be handled by a gateway will be up to 3 Gbit/s due to the fact that up to three satellites are seen by the gateway. Generally, the whole system will be able to support up to 215 Gbit/s traffic.

* * *

In this chapter we concentrated on the most important satellite systems which mainly serve mobile users. We described three personal satellite communication systems and overviewed two projects aimed at broadband access which can revolutionize the access to the Internet and telecommunication networks in general. However, careful observation of deployment of several systems and the financial problems associated with this deployment show that the satellite segment of mobile communication systems is probably the most difficult and risky for investors. Let us hope that despite enormous progress of land mobile systems the satellite communications will become a meaningful part of integrated comunication systems in the future.

REFERENCES

1. W. S. Adams, L. Rider, "Circular Polar Constellations Providing Continuous Single or Multiple Coverage Above a Specified Latitude", *The Journal of Astronautical Sciences*, Vol. 35, No. 2, April-June, 1987

2. F. Ananasso, F. Delli Priscoli, "The Role of Satellites in Personal Communication Services", *IEEE J. Selected Areas in Commun.*, Vol. 13, No. 2, 1995, pp. 180-195

3. K. D. Carl, S. Ritterbusch, "GEO-MEO-LEO: satelittengestützte Systeme für PCS", *IK*, Berlin, No. 46, 1996, pp.45-50

4. J.-N. Colcy, R. Steinhäuser, "EUTELTRACS the European Experience on Mobile Satellite Services", *Proc. of Intern. Mobile Satellite Commun. Conference*, 1993, pp. 261-266

5. J. Couet, D. Maugars, D. Rouffet, "Satellites and Multimedia", *Alcatel Telecommunications Review*, Fourth Quarter 1999, pp. 250-257

6. K. Dawson, "Iridium Returns from the Dead", http://www.commweb.com, March 29, 2001

7. R. De Gaudenzi, T. Garde, F. Giannetti, M. Luise, "An Overview of CDMA Techniques for Mobile and Personal Satellite Communications", *Proc. of EMPS'94*, 1995, pp.78-104

8. P. P. Giusto, G. Quaglione, "Technical Alternatives for Satellite Mobile Networks", *Proc. of EMPS'94*, 1995, pp. 15-27

9. J. L. Grubb, "The Traveller's Dreams Come True", *IEEE Communications Magazine*, November 1991, pp. 48-51

10. J. E. Hatlelid, L. Casey, "The Iridium System: Personal Communications Anytime, Anyplace", *Proc. of Intern. Mobile Satellite Commun. Conference*, 1993, pp. 285-290

11. J. Huber, "Mobile/Personal Satcoms System Alternatives – Satellite and Network Aspects", *Proc. of EMPS'94*, 1995

12. J. B. Lagarde, D. Rouffet, M. Cohen, "GLOBALSTAR System: An Overview", *Proc. of EMPS'94*, 1995

13. R. Leopold, "Low-Earth Orbit Global Cellular Communications Network", *Proc. of IEEE Intern. Conference on Communications*, 1991, pp. 1108-1111

14. J. Lodge, "Mobile Satellite Communications Systems: Toward Global Personal Communications", *IEEE Communications Magazine*, November 1991, pp. 24-30

15. P. Monte, F. Way, S. Carter, "The GLOBALSTAR Air Interface: Modulation and Access", *Proc. of COST 227/231*, 1993, pp. 108-118

16. The New ICO System, http://www.ico.system/home.htm

17. S. R. Pratt, R. A. Raines, C. E. Fossa, M. A. Temple, "An Operational and Performance Overview of the IRIDIUM Low Earth Orbit Satellite System", *IEEE Communications Surveys*, http://www.comsoc.org/pubs/surveys, Second Quarter 1999, pp. 2-10

18. M. A. Pullman, K. M. Peterson, Y. Jan, "Meeting the Challenge of Applying Cellular Concept to LEO Satcom Systems", *Proc. of International Conference on Communications*, 1992, pp. 770-773

19. D. Rouffet, "GLOBALSTAR: a Transparent System", *Electrical Communication*, First Quarter 1993, pp.84-90

20. D. Rouffet, "Skybridge: System Description", *Alcatel Telecommunications Review*, Fourth Quarter 1998, pp. 269-275

21. P. Sourisse, "Skybridge: Global Multimedia Access", *Alcatel Telecommunications Review*, Third Quarter 1999, pp. 228-237

22. G. Symeonidis, P. McDougal, "Inmarsat and Mobile Satcoms in the 21st Century", *Proc. of Telecom'95*, Geneva 1995

23. Teledesic. Technology Overview, http://www.teledesic.com/tech/tech.htm

24. R. Wiedeman, A. Viterbi, "The GLOBALSTAR Mobile Satellite System for Worldwide Personal Communications", *Proc. of Intern. Mobile Satellite Commun. Conference*, 1993, pp. 291-296

25. P. Wood, "Mobile Satellite Services for Travellers", *IEEE Commun. Magazine*, November 1991, pp. 32-35

26. W. W. Wu, E. F. Miller, W. L. Pritchard, R. L. Pickholtz, "Mobile Satellite Communications", *Proc. of the IEEE*, Vol. 82, No. 9, September 1994

27. M. Amanowicz, "Satellite Personal Communication Systems", *Telecommunications Review*, (in Polish), No. 5-6, 1995, pp. 248-255

Appendix

Table A.1 Frequency ranges of electromagnetic waves used in communication systems [27]

Frequency range	Wavelength	Meaning
2 Hz–30 kHz	10^8–10^4	Very low frequency (VLF)
30 kHz–300 kHz	10^4– 10^3	Low frequency (LF)
300 kHz–3 MHz	10^3– 10^2	Medium frequency (MF)
3 MHz–30 MHz	10^2–10	High frequency (HF)
30 MHz–300 MHz	10–1	Very high frequency (VHF)
300 MHz - 3 GHz	1–10^{-1}	Ultra high frequency (UHF)
3 GHz–30 GHz	10^{-1}– 10^{-2}	Super high frequency (SHF)
30 GHz–300 GHz	10^{-2}– 10^{-3}	Extremely high frequency (EHF)

Table A.2 Subbands of SHF and EHF ranges [27]

Subband notation	Frequency range [GHz]
L	1–2
S	2–4
C	4–8
X	8–12
Ku	12–18
Ka	18–27
K	27–40
W	40–100

16

Wireless Local Area Networks

16.1 INTRODUCTION

In typical *Local Area Networks* (LANs) the network elements such as servers, terminals, printers and other peripherals are connected by a system of copper of optical fiber wires. Such a network consists of stationary nodes, terminals and the wireline infrastructure. Each network extension or reconfiguration requires additional wiring which is both time-consuming and costly. It is obvious that terminals cannot move.

Wireless LANs (WLANs) opens new possibilities for LAN users, which are mainly terminal mobility and easy reconfiguration. In general, wireless LANs have the following advantages [1]:

- *Flexibility* – WLAN nodes can communicate with each other within the network coverage area without major limitations in terminal locations. The terminals do not need to remain visible to each other. Walls and other typical obstacles in an indoor environment are mostly penetrated by electromagnetic waves if their frequency is not too high.

- *Simplified planning* – The network planning is related to the radio part; however, configuration of a network, in particular for *ad hoc* networks, is not necessary.

- *Possibility of a temporary network configuration* – Wireless communications open a possibility to construct a local network which is needed temporarily only (e.g. during large international exhibitions, sport contests, etc.).

- *Robustness to external conditions* – Due to the lack of connecting wires and the frequent use of WLAN terminals connected to laptops or other portable devices which can work without external power supply, the WLANs are not easily affected by extreme situations.

WLANs also have some disadvantages. Most of them are the result of using a radio channel as a signal propagation medium. The main disadvantages are the following [1]:

- *Lower transmission quality* as compared with wireline LANs – the error rate in a radio channel is of the order of 10^{-3}–10^{-4} or even worse. FEC or ARQ techniques are necessary to achieve higher quality. For comparison, the error rate for transmission over an optical fiber channel is at most 10^{-10}. Wireless channels are often time-varying and limited spectrum is at the disposal of a particular system. Spectral limitations and possible interference from other systems are a serious drawback of WLANs.

- *Local regulatory restrictions* – Several countries impose different spectral restrictions. This in turn imposes limitations on worldwide WLAN solutions.

- *Cost of wireless equipment* – it is still much higher than of the equivalent equipment for wireline networks.

- *Lower safety and security* – information transmitted on the radio channel can be intercepted much easier than in wireline LANs. An inappropriately used WLAN can be a source of interference for other sensitive devices such as medical equipment.

Wireless LANs rarely work independently of other networks. Typically, wireless transmission is used to access a wireline network. In other cases there is some kind of gateway between a WLAN and other networks, e.g. the Internet. Therefore, one of the features of WLANs which is highly desired is easy interoperation with other networks and WLAN transparency with respect to user applications.

16.2 TYPES OF WLANS

WLANs can be categorized according to several criteria such as the type of radio transmission, the frequency range applied (infrared or radio) or the multiple access. Here we will apply the criterion based on the network configuration. According to this criterion, we will divide WLANs into infrastructure-based wireless networks and *ad hoc* wireless networks.

Figure 16.1 presents the basic idea of *infrastructure-based wireless network*. The WLAN of this type features a wireline infrastructure which connects it to other wireline networks. Important elements of this type of network are network *Access Points* (APs) which interface the wireless terminals with the wireline network infrastructure. In most cases transmission is performed only between access points and wireless terminals, so two network terminals communicate via the appropriate access points. As we see, the APs work like base stations in a cellular system. The access point performs most of the transmission control procedures, leaving simple processing to wireless terminals. That

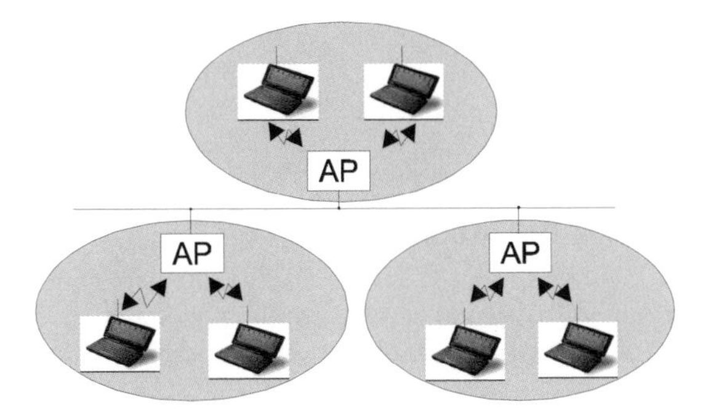

Figure 16.1 Example of the infrastructure-based network configuration

kind of network organization allows for easy access control and for collision avoidance. It can be realized if the access points manage the access to the radio channel.

The infrastructure-based wireless networks are not fully flexible because of the existence of a fixed part of the network. In turn, a second type of WLANs, called an *ad hoc* network, does not have this disadvantage, see Figure 16.2.

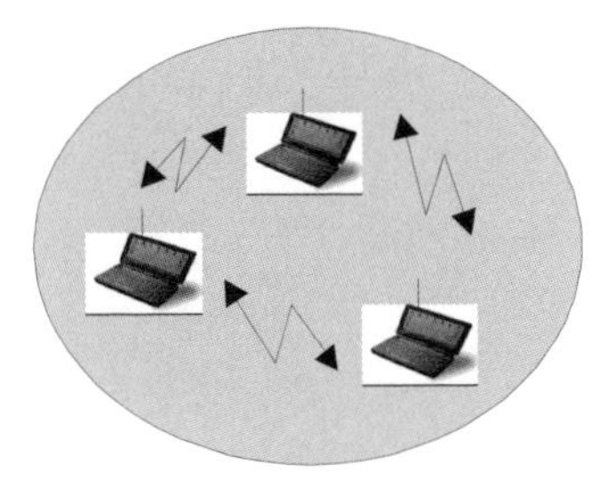

Figure 16.2 Structure of WLAN *ad hoc* network

Ad hoc wireless networks do not have any wired infrastructure. Wireless terminals not only connect the subscribers to the network but also function as network nodes. They can communicate with each other if they remain in the range of each other. The communication range can be extended if the wireless terminals are able to forward the message received from one station to another. The *ad hoc* network concept implies increased terminal complexity. A wireless terminal not only receives and transmits messages but also works as a network node. It has to compete for the network resources within multiple access procedures, it has to route the transmitted packets and set their priorities. In reality, a compromise between these two wireless network structures often occurs.

There is a selection of radio technologies, available data rates, ranges and possible applications of the WLANs. Below we will briefly describe the basic properties of the most important WLAN standards such as IEEE 802.11, HIPERLAN Type 1 and Type 2.

We will also consider the Bluetooth interface. More detailed information can be found in the books specially devoted to WLANs, such as [2] and [3] or in the appropriate chapters of [1] and [4].

16.3 HIDDEN STATION PROBLEM

Wireless transmission in WLANs can create some problems if the coverage areas of different parts of WLAN partially overlap. Let us consider an example shown in Figure 16.3a.

Station B remains in the range of stations A and C; however, the distance between stations A and C is so large that neither of them can detect whether the other one is transmitting. Let station A transmit to station B. Station C, using the CSMA/CD (*Carrier Sense Multiple Access with Collision Detection*) access mechanism, finds that the medium is free and starts to transmit to station B, too. In this manner a collision occurs. Both stations transmit to station B till the end of their packets without realizing that their packets cannot be received correctly. In this respect the waste is twofold. Firstly, a collision has occurred. Secondly, the time of a whole packet has been wasted. We say that station C is *hidden* for station A. A similar example can be considered for infrastructure-based WLANs (see Figure 16.3b). If two neighboring access points wish to transmit to a station located in the overlapped area and their transmission is not coordinated by the wired part of the system, a similar collision can occur.

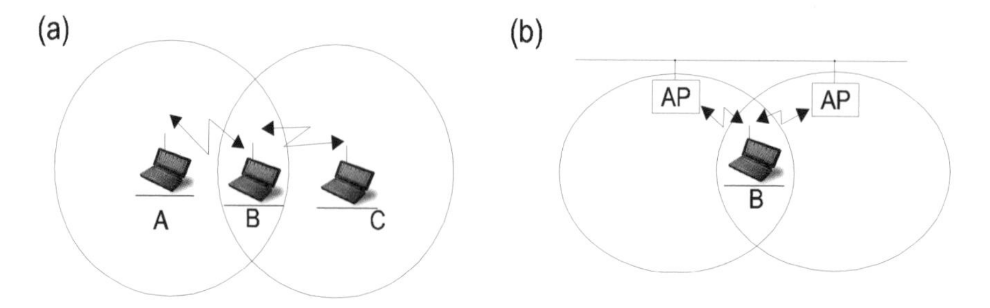

Figure 16.3 Illustration of the hidden terminal problem in *ad hoc* (a) and infrastructure-based WLANs (b)

The hidden station (node) problem can be partially avoided by ensuring the channel sensing range to be much greater than the receiving range (see Figure 16.4a). This method has been applied in HIPERLAN/1. Another solution is the application of a MACA (*Multiple Access with Collision Avoidance*) scheme shown in Figure 16.4b. In order to start transmission from station A to B, station A sends a *Request to Send* (RTS) packet first. The RTS packet contains the addresses of transmitting and receiving stations and the duration of the future transmission. Station B answers to station A with a *Clear to Send* (CTS) packet, which is also heard by station C. It contains the addresses of the sender and recipient and the duration of transmission, so station C

realizes that the medium will be occupied for the given period of time and it will not try to access the channel before that time. Collision can still occur if stations A and C send the RTS packets concurrently; however, these packets are much shorter, so the wasted time is much shorter, too. This method of overcoming the hidden station problem can be applied as an option in IEEE 802.11 WLANs (see Section 16.5).

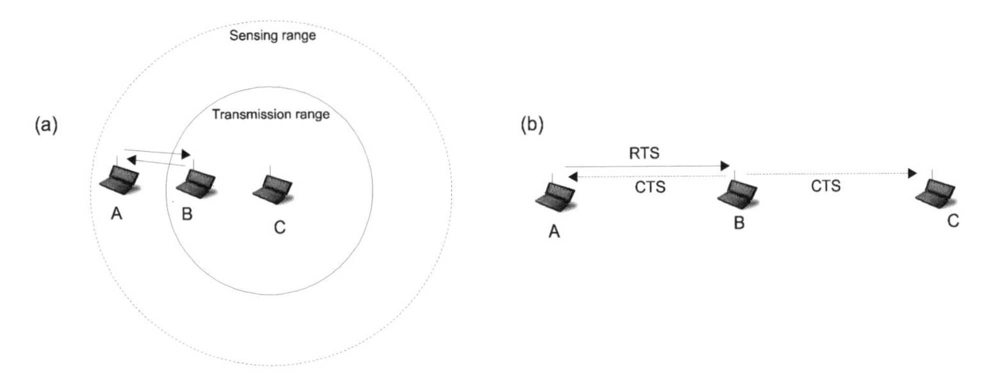

Figure 16.4 Avoiding a hidden station problem using different sensing and transmission ranges (a) and a MACA scheme (b)

The hidden station problem can be much easier avoided if spread spectrum techniques are used in the physical layer. It is sufficient that both transmitting stations send their pseudorandom signals (using direct sequence or frequency hopping methods) to the receiving station in different phases to achieve (almost) orthogonality of the signals from both stations, so one of them can be detected using the correlation method or the RAKE receiver.

16.4 HIPERLAN TYPE 1

The HIPERLAN Type 1 (*High Performance Local Area Network*) (HIPERLAN/1) standard was established by ETSI in 1996 [5]. It is the first of four ETSI standards dealing with wireless network access. The other are HIPERLAN/2 [6], HIPERACCESS (*Wireless ATM Remote Access*) [8] and HIPERLINK (*Wireless ATM Interconnect*). HIPERLAN/1 standard provides an ISO 8802 [9] compatible interface. The WLANs conforming to the HIPERLAN/1 standard can work in both network configurations described above.

HIPERLAN/1 network operates in the 5.15–5.3 GHz band, divided into five frequency channels. Table 16.1 lists the carrier frequencies of all channels. The HIPERLAN/1 terminals can move at the maximum speed of 1.4 m/s. Transmitted traffic can have an asynchronous or isochronous form. The terminal range is 50 m. The maximum data rate is about 23.5 Mbit/s. Figure 16.5 presents the reference model for HIPERLAN/1 networks. In turn, Figure 16.6 shows the HIPERLAN communication model.

Table 16.1 Carrier frequencies of HIPERLAN/1 channels

Channel No.	Carrier frequency [MHz]
0	5 176.4680
1	5 199.9974
2	5 223.5268
3	5 247.0562
4	5 270.5856

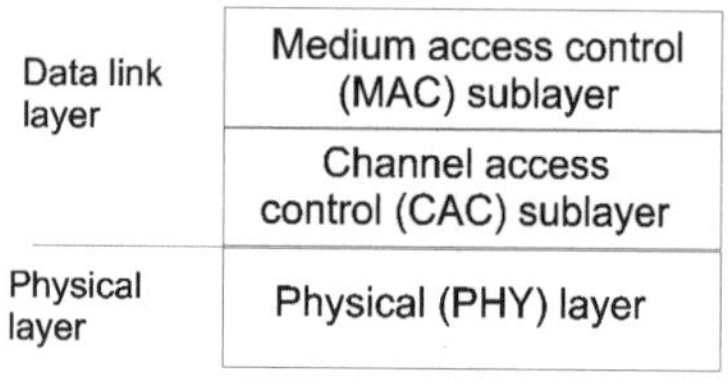

Data link layer	Medium access control (MAC) sublayer
	Channel access control (CAC) sublayer
Physical layer	Physical (PHY) layer

Figure 16.5 Reference layer model for HIPERLAN/1

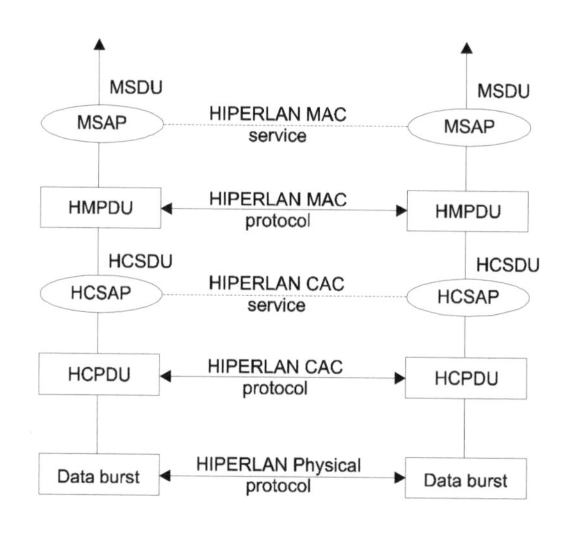

Figure 16.6 HIPERLAN communication model [5]

The MAC sublayer performs several functions which help to organize the operation of the HIPERLAN/1 network. The higher layers produce MAC service data units (MSDU), which enter the MAC layer through the MAC service access point. Due to the application of the HIPERLAN MAC protocol, HIPERLAN-MAC protocol data units (HMPDU) are created.

The *Channel Access Control* (CAC) sublayer contains a protocol which determines which nodes are allowed to transmit and specifies the access priorities. This layer offers a connectionless data transfer service to the MAC sublayer. Thus, the HMPDU constitutes a HIPERLAN-CAC service data unit, which enters the HIPERLAN-CAC layer through HIPERLAN-CAC service access point (HCSAP). Due to the CAC protocol, the HIPERLAN-CAC protocol data units (HCPDU) are formed and finally they constitute a payload of a physical data burst.

The physical layer protocol specifies transmission, reception and channel assessment techniques. The data bursts are transmitted between two stations (nodes) over the radio channel according to the HIPERLAN physical protocol.

There are a lot of similarities between the communication model philosophy and the layered structure typical for the OSI reference model shown in Chapter 1 in Figure 1.50.

16.4.1 HIPERLAN/1 MAC sublayer

The first function performed by the MAC sublayer is MAC address mapping, which allows differentiation of the terminal's own HIPERLAN network in the area where more networks can operate. The standard defines internal address structures. The address of a HIPERLAN station consists of two parts. The first part defines the network name and the second part determines the station identification. Within the address mapping function, IEEE-MAC addressing into HIPERLAN addressing is also performed.

The MAC sublayer also ensures communication security by defining the encryption-decryption algorithm. The algorithm requires an identical key and a common initialization vector for data encryption and decryption. The encryption and decryption are performed by modulo-2 addition of the sequence of user data to the sequence generated by the properly designed pseudorandom generator initiated by the key and the initialization vector. In order to achieve high security, the initialization vectors and the keys should be frequently changed.

The next function of the MAC sublayer is the addressing of MAC service access points (MSAP). According to the reference layer model, the HIPERLAN offers MAC services performed between MAC service access points (the "sender" and the "recipient"). The MSAPs are addressed using a 48-bit LAN-MAC address. Compatibility with the ISO MAC service definition is also ensured.

The MAC sublayer also manages data forwarding. Some HIPERLAN/1 terminals can operate as a relay for the packets sent between terminals which are beyond their common range. The forwarding can have unicast (point-to-point) or multicast/broadcast (point-to-multipoint) form. The relay stations maintain routing tables and a list of multipoint relays. The routing table maintained by a given relay station contains the addresses of the relay nodes which are the closest to that station on the way to a possible destination. In consequence, the transmitted packet carries the destination address and the next-hop relay address.

The last function performed by the MAC layer is power conservation understood as a power-saving mechanism in the stations which are supplied by battery power sources.

16.4.2 HIPERLAN/1 CAC layer

The most important part of the *Channel Access Control* (CAC) sublayer is the protocol called *Elimination Yield Non Pre-emptive Priority Multiple Access* (EY-NPMA). It is a version of the *Carrier-Sense Multiple Access* (CSMA) protocol with prioritization.

Let us imagine that several terminals wish to send a packet simultaneously. Only one of them can do it, otherwise a collision occurs. In the EY-NPMA protocol the process of obtaining the channel access is divided into three phases. Time is divided into channel access cycles. Each cycle starts with the channel access synchronization (CS). The synchronization is followed by the *prioritization phase.* The time of this phase is divided into five 168-bit slots starting from the slots of the highest priority ($p = 1$). If the terminal (HIPERLAN node) has the priority p, it senses the channel for the first $p - 1$ slots. If the channel remains idle, the node sends an access pattern (PA - *priority assertion*). If the node finds that the channel has already been occupied by the station with a higher priority, it does not send its access pattern and waits for the beginning of the next access cycle.

More than one station can have the same priority; therefore, the next phase of the access algorithm is necessary which is called the *contention phase.* It is further divided into the *elimination phase* and the *yield phase.* The elimination phase is divided into 0 to 12 slots lasting for 212 bits each. Each terminal which has not been eliminated in the prioritization phase sends an *elimination burst.* Its length is random, between 0 and 12 slots. The probability of continuation of the burst in the next slot is equal to 0.5, so the length n of the elimination burst has the probability distribution

$$P_E(n) = \begin{cases} 0.5^{n+1} & \text{for } 0 \leq n < 12 \\ 0.5^{12} & \text{for } n = 12 \end{cases} \tag{16.1}$$

After sending its elimination burst, each station senses the channel during a 256-bit long interval of *elimination survival verification* (ESV). A station gives up competing for the channel access, if during this interval the channel is occupied by another station, which is sending a longer burst. These stations which have survived this phase, i.e. all which have sent the longest bursts, take part in the yield phase. The yield phase lasts for up to ten 168-bit slots. Each station participating in this phase senses the channel during n slots ($0 \leq n \leq 9$). The probability that the given station senses the channel for n consecutive slots is

$$P_Y(n) = 1/10 \tag{16.2}$$

If the station has not detected any activity in the channel during the listening, it immediately starts to transmit its data burst, so the *transmission phase* begins. If the station has detected the signal of another station during channel sensing, it is eliminated from the channel access competition and waits till the beginning of the next access cycle.

Figure 16.7 illustrates the competition for the channel access between four stations denoted as A, B, C and D. It is assumed that the priority of stations A, B and C is $p = 4$. Station B survives all the phases of the EY-NPMA protocol and finally transmits its data burst. Let us note that despite quite a complex procedure, collision can still occur.

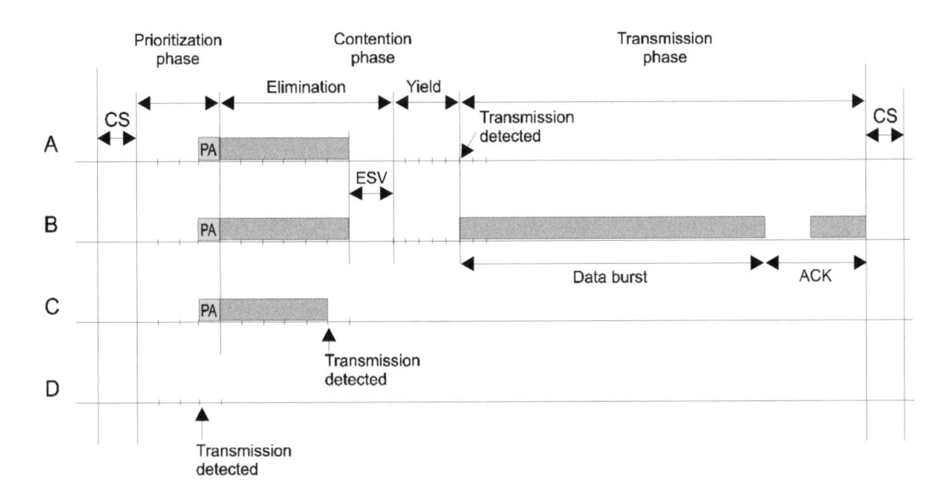

Figure 16.7 Illustration of EY-NPMA protocol

16.4.3 HIPERLAN/1 physical layer

Transmission on the radio channel is performed using a non-differential Gaussian Minimum Shift Keying (GMSK) with a normalized filter bandwidth-time product $BT = 0.3$ (see Chapter 1). Let us recall that B is a 3-dB signal bandwidth and T is a modulation period. Due to the channel multipath, the high data rate (23.5294 Mbit/s) and the application of a frequency pulse with a Gaussian shape, which lasts for several modulation periods, intersymbol interference occurs. This type of distortion is minimized by the application of an adaptive equalizer. Typically, in HIPERLAN/1 a decision-feedback equalizer (DFE)[1] is employed. Besides the high data rate applied in GMSK modulation, a low data rate is used mostly for control and signaling purposes. To perform these tasks, the FSK modulation at the rate of 1.4705875 Mbit/s is applied. In order to decrease the error rate, the HIPERLAN/1 standard specifies BCH (31,26) error-correction code used in the physical layer. This code is able to correct a single error, to detect two random errors, all error bursts not longer than 5 bits and most of the error bursts longer than 5 bits. Each data block is interleaved across sixteen codewords. This implies that the data block has $16 \times 26 = 416$ bits. After BCH encoding the block is 496-bits long. Figure 16.8 shows the data packet format applied in HIPERLAN/1.

Figure 16.8 presents one of the two types of bursts specified by HIPERLAN/1. They are the following:

[1] The decision-feedback equalizer is a type of equalization structure applied in the receiver. It uses a linear filter and a decision-feedback filter. The first one shapes the joint channel and filter characteristics, so that the intersymbol interference (ISI) at its output originates from the past data symbols only. Then, based on the past, already decided symbols, the decision-feedback filter synthesizes the ISI originating from these symbols and subtracts it from the sample being the subject of the current data decision.

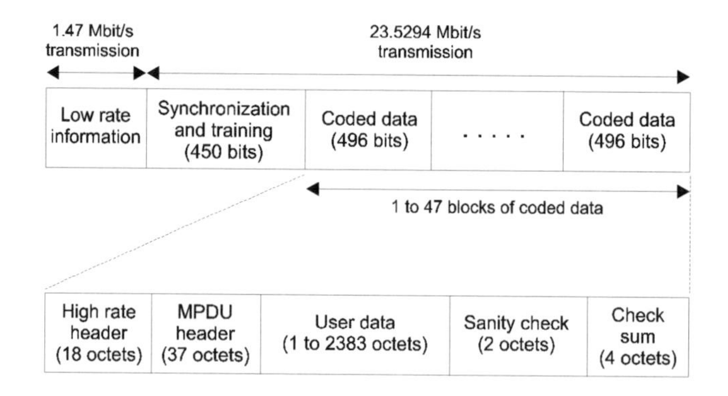

Figure 16.8 HIPERLAN/1 data packet format

- *LBR-HBR data burst* (shown in Figure 16.8), which consists of two parts: the *low bit-rate* (LBR) part and the *high bit-rate* (HBR) part
- *LBR data burst*, i.e. the burst consisting of a low bit-rate part transmitting the identification of a receiver.

The bursts are used to transmit user data, to acknowledge their correct reception, to participate in the channel access procedure, etc.

HIPERLAN/1 has been the initiative of the European Community and ETSI. However, other IEEE WLAN standards exist, which are worth consideration.

16.5 IEEE 802.11 WLAN STANDARDS

IEEE 802.11 WLAN standard results from an IEEE initiative to make a wireless extension of the existing LAN standards of the 802 series. The project started in 1990 and resulted in the publication of the standard in 1997. Figure 16.9 presents the IEEE 802 standard family showing the location of particular 802 standard versions in comparison with the first two OSI layers.

The standard describes the WLAN physical layer and the MAC layer. Three different transmission methods in the physical layer have been standardized. In two of them the *Industrial, Scientific and Medical* (ISM) band (i.e. 2.4000–2.4835 GHz) is used. It is available with some limitations all over the world. The third method applies infrared (IR) technology. All of them co-operate with the same MAC layer. The higher layers are common to other IEEE 802 networks, as Figure 16.9 shows.

Let us consider possible IEEE 802.11 architectures. The WLANs can be configured as *ad hoc* and structure-based networks. The *Basic Service Set* (BSS) is the fundamental building block of the IEEE 802.11 architectures [11]. It is a set of stations which can communicate with one another and remain under control of a single coordination function explained below. An *ad hoc* network containing an isolated set of stations remaining in the common range is a good example of an independent BSS.

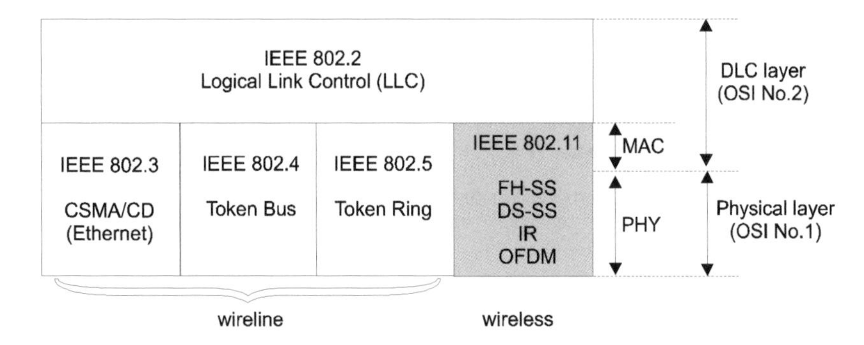

Figure 16.9 IEEE 802 standard family with two first OSI layers

Figure 16.10 presents the locations of the basic service sets (BSSs) in an infrastructure-based network. Let us note that the access points are connected by the *distribution system* (DS) and the latter can in turn be connected to the external IEEE 802.x networks through an interworking unit called a *portal.* Thanks to the integrated operation through the distribution system, the access points provide range extension and terminals from different BSSs can communicate with each other. This way an *Extended Service Set* (ESS) is formed. The distribution system works as a backbone network and can be built of any wireline or wireless networks such as IEEE 802.x, a metropolitan area network with fiber distributed data interface (FDDI) or another IEEE 802.11 network [11].

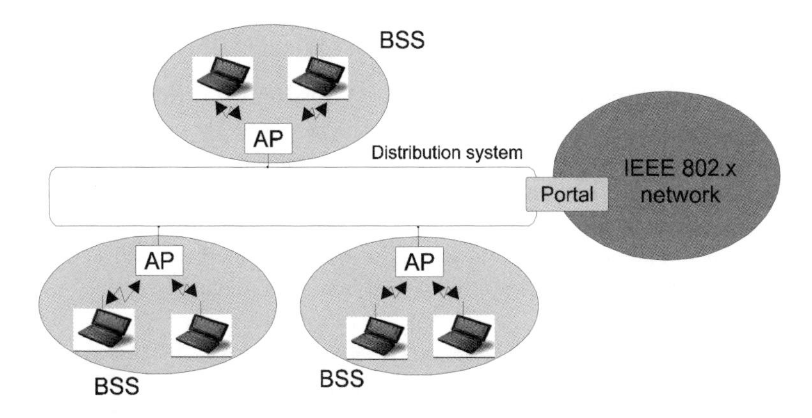

Figure 16.10 Architecture of the infrastructure-based IEEE 802.11 network

Before we consider the operation of the 802.11 WLANs, let us introduce the basic reference model [2]. Figure 16.11 shows its subsequent layers. The physical layer is divided into two sublayers. The lower one is called the *Physical Medium Dependent* (PMD) sublayer. It is related to the physical channel, and realizes transmission and reception including modulation and coding. The second sublayer is called the *Physical*

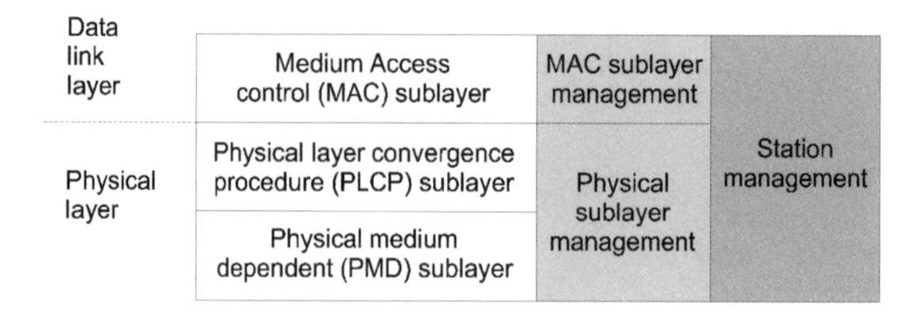

Figure 16.11 IEEE 802.11 reference model

Layer Convergence Procedure (PLCP) sublayer. Its main task is mapping the MAC sublayer protocol data units (MPDUs) into a packet format used by the PMD sublayer. Its second task is carrier sensing, whose result is transferred to the MAC layer. In the MAC layer the access mechanism based on CSMA is performed. Another task of the MAC layer is the fragmentation and encryption of data packets.

The physical sublayers are "supervised" by the physical layer management. This block maintains a physical layer management information base and supervises the adaptation of the physical layer to different link conditions. The MAC sublayer also has its management block. The MAC sublayer management is responsible for synchronization, power management and association and reassociation procedures. The station management defines how the physical layer and the MAC management sublayer cooperate with each other.

16.5.1 IEEE 802.11 physical layer

As we have mentioned, in the basic IEEE 802.11 version three different physical layers have been standardized. We will briefly describe each of them. We will discuss the modulation types and the packet format applied in each type of radio interface. Generally, packets sent over the channel consist of three parts: the PLCP preamble, the PLCP header and the MAC packet data unit (MPDU); however, each interface uses different preambles and headers. The basic data rate applied in each type of interface is 1 Mbit/s. The enhanced data rate of 2 Mbit/s is also allowable.

16.5.1.1 DS-SS physical layer In this type of physical layer in the air interface the *Direct Sequence Spread Spectrum* (DS-SS) technique (see Chapter 1) is applied. As we remember, in the DS-SS technique the user data are represented by a sequence of pulses (*chips*) of much higher rate than the original data bits. In this way the signal spectrum is spread along the frequency axis. In the IEEE 802.11 DS-SS interface the 11-chip Barker code[2] is applied as the spreading sequence. After spreading the data signal and using the result as the signal modulating the carrier at the rate 11 Msymbol/s according

[2]The Barker code used in the IEEE 802.11 interface is formed by the sequence +1, -1, +1, +1, -1, +1, +1, +1, -1, -1, -1.

to DBPSK or DQPSK scheme,[3] the signal bandwidth is roughly 22 MHz. In order to deploy BSSs which partially overlap or work in close vicinity of each other, their center frequencies must be separated by at least 30 MHz. The maximum transmit power is 1000 mW (EIRP) in the USA and 100 mW (EIRP) in Europe.

Figure 16.12 DS-SS PLCP packet structure

The DS-SS PLCP packet format is shown in Figure 16.12. The PLCP part is always transmitted at the rate of 1 Mbit/s. 1 or 2 Mbit/s data rates can be applied for the payload transmission. The fPLCP part of the packet has a number of fields with the following functions:

- *Synchronization* bits are used for synchronization, gain setting, frequency offset compensation and energy detection
- *Start Frame Delimiter* (SDF) is used for frame synchronization
- *Signal* bits carry information of the data rate (1 or 2 Mbit/s) used in the payload field
- *Service* bits are reserved for future use
- *Length* bits describe the length of the payload (MPDU) field
- *CRC* bits are used for header parity control.

The PLCP part of the packet is followed by the payload carrying MAC packet data unit (MPDU) of the length between 1 and 2048 octets.

In 1998 the IEEE 802.11 working group introduced a physical layer extension allowing transmission of data packets at the rate of 11 Mbit/s or at the fall-back rate of 5.5 Mbit/s, preserving the same chip rate and bandwidth of the DS-SS physical layer standardized so far. Instead of the 11-chip Barker code applied with DBPSK or DQPSK modulation, the *Complementary Code Keying* (CCK) was adopted in the payload field [13].

[3]Let us note that for *Differential Binary Phase Shift Keying* (DBPSK) the phase difference between two consecutive symbols carries information about a single bit, whereas in the case of *Differential Quadrature Phase Shift Keying* (DQPSK) each of four possible phase differences is mapped into two bits.

Complementary codes were invented by M. J. E. Golay [14]. A sequence x of length N is said to be complementary to another sequence y of the same length, if the following condition on the sum of the correlation functions is fulfilled [15]:

$$\sum_{k=0}^{N-1} (x_k x_{k+i} + y_k y_{k+i}) = \begin{cases} 2N & \text{for } i = 0 \\ 0 & \text{for } i \neq 0 \end{cases} \tag{16.3}$$

The CCK codewords used in the IEEE 802.11 physical layer extension are described by the expression

$$\mathbf{c} = \Big[e^{j(\varphi_1+\varphi_2+\varphi_3+\varphi_4)}, e^{j(\varphi_1+\varphi_3+\varphi_4)}, e^{j(\varphi_1+\varphi_2+\varphi_4)}, -e^{j(\varphi_1+\varphi_4)}, \\ e^{j(\varphi_1+\varphi_2+\varphi_3)}, e^{j(\varphi_1+\varphi_3)}, -e^{j(\varphi_1+\varphi_2)}, e^{j\varphi_1} \Big] \tag{16.4}$$

where φ_1, φ_2, φ_3 and φ_4 are the QPSK phases. The choice of all four QPSK phases indicates that each CCK codeword consisting of 8 complex chips carries information on 8 bits. Because the chip rate is equal to 11 Mchip/s (as in the basic version of the DS-SS physical layer), the final data rate is 11 Mbit/s. Let us note that the phase φ_1 is present in all the code elements in (16.4) and it is used for differential encoding; therefore, the receiver can apply differential decoding. For the fall-back data rate of 5.5 Mbit/s, 4 bits are mapped onto the CCK 8-chip codeword. Two bits determine one of four CCK code subsets and two remaining bits are used for rotation of the codeword by one of four possible value of the phase φ_1.

In the DS-SS version of the physical layer the radio interface ensures high data rates and high range; however, the DS-SS RF products cost more and use more power than the FH-SS technique used in IEEE 802.11 WLANs.

16.5.1.2 FH-SS physical layer The *Frequency Hopping Spread Spectrum* physical interface is characterized by high distortion immunity, high system capacity, low power use, medium range and low cost of the RF part [3]. The FS-SS system operates in the ISM band. For the purpose of this system 79 hopping frequencies with the raster of 1 MHz have been defined in the USA and Europe and 23 frequencies have been selected for Japan. As in the DS-SS radio interface, the transmit power is equal to 1 W (EIRP) in the USA and 100 mW (EIRP) in Europe. The data are transmitted using the Gaussian-shaped FSK modulation[4] on the carriers determined by the hopping pattern. For the basic rate of 1 Mbit/s the modulation is two-level, whereas for 2 Mbit/s four level GFSK is applied. Both applied GFSK modulations have the same root-mean-square deviation from the carrier frequency. The spectrum of the signal is fitted to 1 MHz bandwidth. The hopping has to occur at a specified rate. In the USA the lowest hopping rate is 2.5 hop/s. The access points determine the applied hopping rate. A mobile terminal detects the hopping rate during the process of association with a given access point. The hopping patterns are descibed by the 802.11 standard. They

[4]Gaussian-shaped FSK modulation is similar to GMSK modulation considered in Chapter 1. The main difference is the value of the applied modulation index h.

have been selected to minimize the use of the same frequency channel by different BSSs.

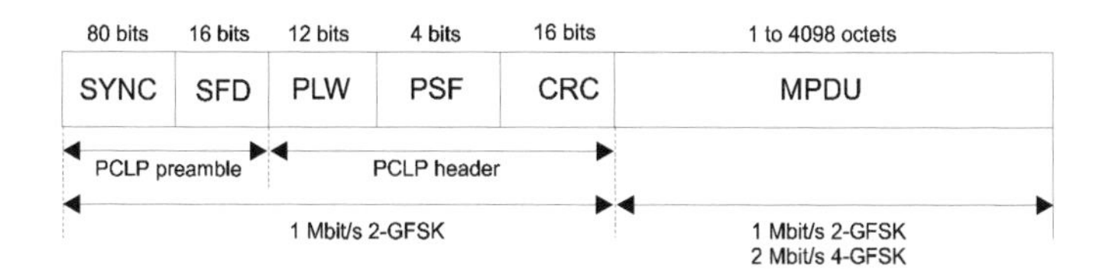

Figure 16.13 PLCP packet format for FH-SS physical layer

The FS-SS packet format differs from that used in the DS-SS radio interface. It is shown in Figure 16.13. Again, the packet consists of the PLCP preamble, PLCP header and the payload (MPDU). The fields of the preamble and the header are as follows:

- *Synchronization* is 80-bit long and is used for synchronization of the receivers and signal detection by the *Clear Channel Assessment* (CCA) process
- *Start Frame Delimiter* (SFD) indicates the start of the frame
- *PLCP_PDU Length Word* (PLW) shows the number of bytes in the payload, including the 32-bit CRC field at the end of the payload field
- *PLCP Signaling Field* (PSF) indicates the data rate (1 or 2 Mbit/s)
- *CRC* is the parity check of the header.

16.5.1.3 Infrared physical layer The third type of the physical layer standardizes digital transmission using infrared technology (IR). Digital signals are sent using infrared rays of the wavelength in the 850 to 950 nm range and *Pulse Position Modulation* (PPM). Two data rates equal to 1 and 2Mbit/s are standardized. For the lower data rate, transmitted bits are grouped in 4-bit blocks. The content of a block determines in which of the 16 slots the infrared pulse is transmitted. Thus, a 16-PPM is applied. For 2 Mbit/s the data stream is divided into 2-bit blocks and each of them determines in which of four possible slots the infrared pulse is radiated. Figure 16.14 illustrates the modulation process for 1 and 2 Mbit/s. The pulse duration is 250 ns for both data rates. Because the peak power of the signal is equal to 2 W, the mean power is equal to 125 mW at 1 Mbit/s and 250 mW at 2 Mbit/s.

The PCLP packet format is presented in Figure 16.15. We will explain only that field which is specific for this type of interface. This field is called *DC Level Adjustment* field. It contains a pattern which enables the receiving station to set the DC level of the signal.

The IR interface is the cheapest of all 802.11 physical interfaces and it does not need any frequency regulation. It is also resistant to eavesdropping. However, it has the lowest range among all 802.11 physical interfaces. It also has to operate indoors as ceilings are needed which could reflect the infrared signals.

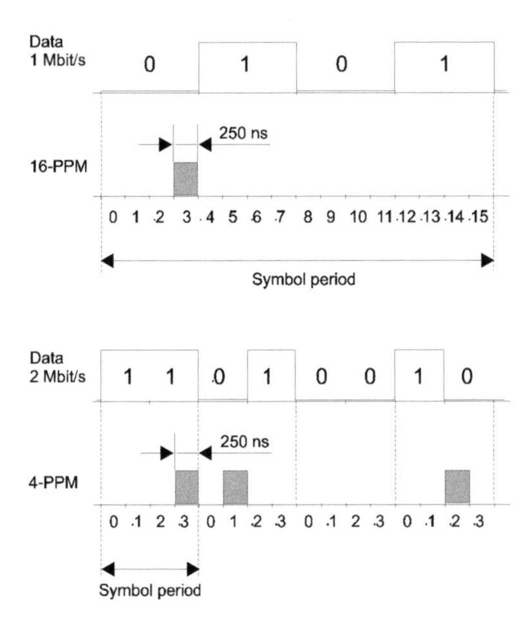

Figure 16.14 Illustration of *Pulse Position Modulation* (PPM) for 1 and 2 Mbit/s transmision in the IR physical layer

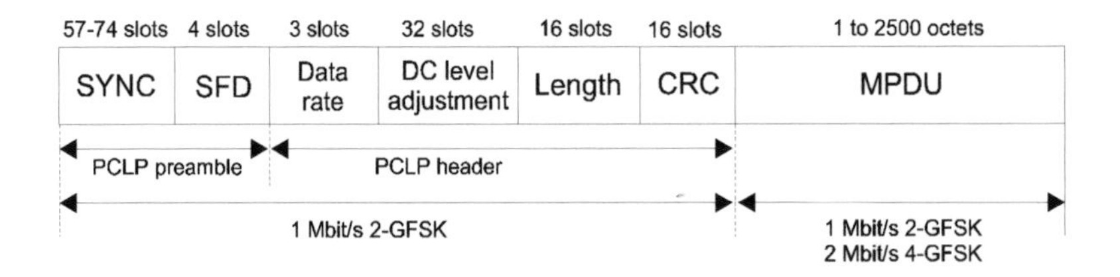

Figure 16.15 PLCP packet format in IR physical layer

16.5.2 IEEE 802.11 MAC sublayer

The MAC sublayer is located above the physical layer in the IEEE 802.11 reference model. Its main tasks are: channel allocation, protocol data unit (PDU) addressing, frame formatting, error checking, and fragmentation and reassembly of data blocks. Two basic operation modes of IEEE 802.11 WLAN are possible from the point of view of the MAC sublayer:

- a *contention mode* in which all WLAN terminals wishing to transmit a packet compete for the access to the channel,
- mixed mode in which the contention mode (used during the *contention period* (CP)) is periodically changed to the *contention-free mode* (realized during the *contention-free period* (CPF)).

The first mode is realized in *ad hoc* networks. The second one requires a *point coordinator* (PC) which controls the access to the channel through polling the terminals during contention-free periods. A BSS access point (AP) performs the function of the PC.

There are three types of MAC frames. *Management frames* are applied for timing and synchronization, authentication and terminal association and disassociation with the given access point. *Control frames* are applied in handshaking and acknowledgment procedures which take place mostly during contention periods. Finally, *data frames* are used to transmit user data and optionally can also contain polling and acknowledgment blocks during contention-free periods.

Frame control	Duration ID	Address 1	Address 2	Address 3	Sequence control	Address 4	Data	CRC

Figure 16.16 IEEE 802.11 MAC frame structure

Transmission in the MAC sublayer is organized in the form of MAC frames. Figure 16.16 shows a typical structure of the IEEE 802.11 MAC frame. The MAC frame is divided into the following fields:

- *Frame control* field, which determines the protocol version and the type of the frame (management, control or data frame). It also indicates whether the frame is fragmented and what the meaning of the address fields is (if the frame is directed to the distribution system (DS), arrives from it or whether the source and destination are mobile terminals or APs).

- *Duration ID* field, which determines the period of time during which the channel will be occupied. It is also used in the channel reservation mechanism (described below).

- *Adress fields 1 to 4*, which indicate the source and destination of the transmitted frame and are interpreted depending on the frame control bits determining the address meaning.

- *Sequence control* field, which contains a frame sequence number and is used to avoid frame duplication which would occur due to the acknowledgement mechanism.

- *Data* field, which carries data of the length up to 2312 bytes.

- *CRC* field, which is 32-bit long and is used for the ARQ (acknowledgement) procedure.

In the contention mode in which all stations compete for the channel, the so-called *Distributed Coordination Function* (DCF) is the basic channel access method. The DCF is based on CSMA (*Carrier Sense Multiple Access*) with collision avoidance. The carrier sensing is performed on two levels: at the physical layer interface (*physical carrier sensing*), and at the MAC sublayer (*virtual carrier sensing*). Physical carrier sensing detects other active WLAN terminals by measuring the signal power level and analyzing the received packets. In order to perform virtual carrier sensing, the duration ID field of a MAC frame carrying request to send (RTS), clear to send (CTS) or data is used. The terminals in the same BSS read the duration ID field and set their *Network Allocation Vector* (NAV).The NAV is a kind of timer. It determines the period of time in which the channel is busy. The channel can be sensed again after the NAV period increased by the so-called *Interframe Space* (IFS) time period. This idle period is used to set the priorities of the terminals. There are three types of IFS periods differing in length: short IFS (SIFS), point coordination function IFS (PIFS), and DCF-IFS (DIFS).

Figure 16.17 presents two typical situations in the contention mode, i.e. when RTS/CTS frames are not applied (Figure 16.17a) and when they are applied (Figure 16.17b). In the first case the source station, which previously obtained the right to transmit, sends a data packet. Other stations read the MAC frame and detect from the *Detection ID* field how long the channel will be busy. Therefore, they can set their NAV timers for the appropriate time period. The contention window starts after the additional DIFS period. The stations competing for the channel choose a randomized backoff time after which they sense the channel. The station which has selected the shortest backoff time finds the channel idle and starts to transmit data.

In the second situation, the source and destination terminals exchange the RTS and CTS packets first. The stations which are in the source terminal range set their NAV timer already after reading the RTS frame, which contains the Detection ID field as well. Other stations which are located in the range of the destination terminal set their NAV timer after detection of CTS. Some other stations start to count down time to the next possible channel sensing from the moment of start of the data packet. Let us note that the pair of source-destination terminals do not need to compete for the channel access till the acknowledgement of the data packet is received. If the packet is incorrectly received and has to be repeated, the source terminal has to contend for the channel again together with other stations which wish to send their packets. The *Contention Window* (CW) is set individually for a terminal up to a maximum value. If in an unsuccessful attempt to obtain channel access a specified backoff time was selected, in the next attempt the randomized backoff time is selected from the range limited by the previously selected value. In turn, if collisions occur, the contention window is doubled to increase the granularity of the backoff time selection and decrease the probability of selection of the same backoff time by more than one terminal.

A similar channel access procedure is performed if the data packet (MSDU - *MAC Service Data Unit*) is large and requires fragmentation. After gaining the channel access, the source terminal sends consecutive fragments of the MSDU acknowledged by the destination terminal. Both terminals do not need to compete for the channel access till the whole MSDU block has been transmitted.

The method of collision avoidance described above ensures fair access to all competing terminals; however, this mechanism does not guarantee a minimum delay of packet delivery to the terminals supporting time-bounded services [11].

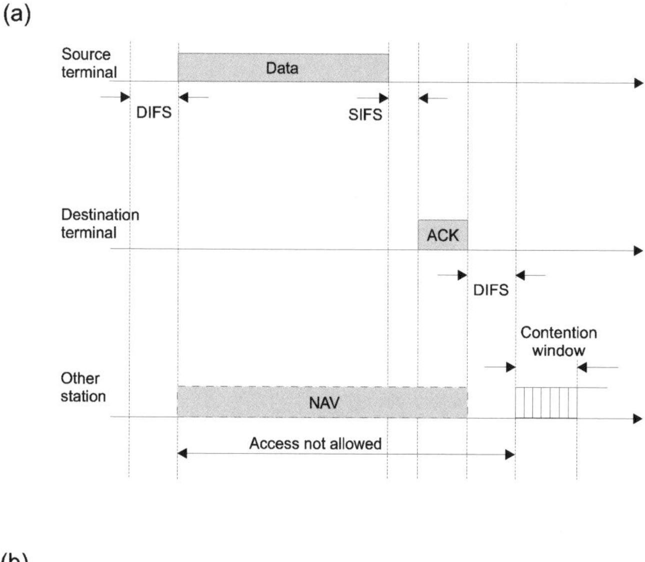

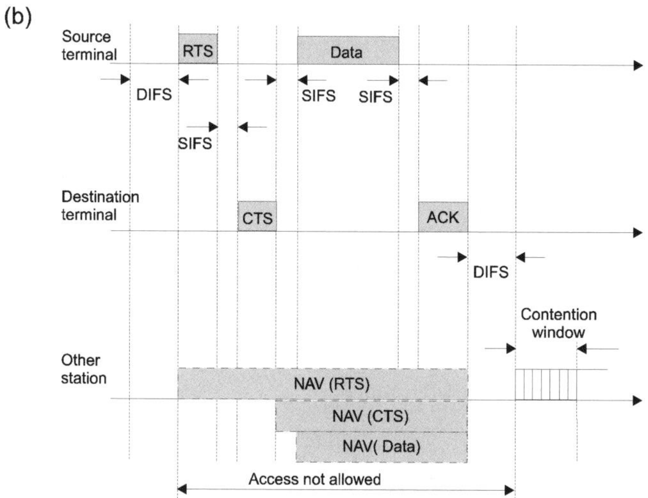

Figure 16.17 Transmission of MAC packet data unit without using RTS/CTS frames (a) and using CTS/RTS frames (b)

Besides the contention mode, in the infrastructure-based networks the terminals can periodically operate in the contention-free mode in which the point coordinator performs polling. Time is divided into contention-free repetition intervals which consist of contention-free periods (CFP) and contention periods. Let us consider the network operation during a CFP.

The point coordinator (PC) senses the channel and waits for the end of the current transmission. After detection the end of transmission, the PC waits for PIFS seconds

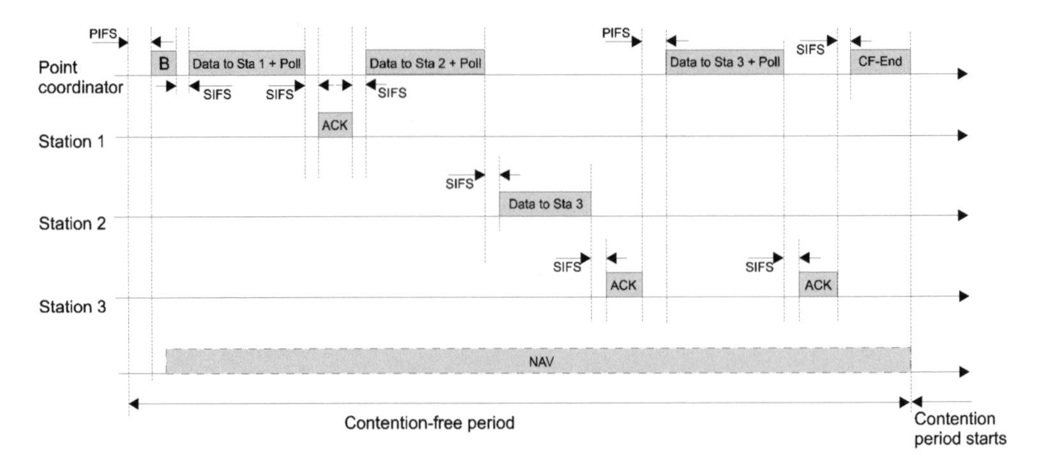

Figure 16.18 Example of the contention-free period

and starts transmission of the beacon frame. Let us note that the PIFS period is shorter than DIFS period, therefore the PC starts to transmit earlier than any other station could do. The beacon frame is used for synchronization and timing. After reception of this frame all terminals set their NAV timers for the whole contention-free period. SIFS seconds after the end of the beacon frame the PC sends one of three possible frames: CF-Poll, Data or Data+CF-Poll frame. Only the polled terminal can respond to the polling after SIFS period by sending an ACK frame or a Data+CF-ACK frame. The polled station can direct its frame to another station which answers using a CF-ACK frame, too. After PIFS seconds the point coordinator starts polling and sending data again. The contention-free period finishes with the CF-End frame. Figure 16.18 presents an example of a single contention-free period, in which the access point-to-terminal and station-to-station transmission take place.

16.6 IEEE 802.11 AND HIPERLAN STANDARDS FOR 5 GHZ BAND

Continuation of work on even higher data rates applied in WLANs led to the next IEEE 802.11 physical layer extension and the establishment of a new ETSI standard called HIPERLAN Type 2 [6], [7]. This was partly possible due to the fact that new bands in 5 GHz range were assigned to WLAN applications. Tables 16.2 and 16.3 present the frequency bands and power limits for US IEEE 802.11 and HIPERLAN/2 applications.

Two acronyms contained in Table 16.3 have the following meaning:

- DFS (*Dynamic Frequency Selection*) is a method of dynamic adaptive selection of carrier frequency, applied to avoid interference from other users,
- TPC (*Transmit Power Control*) denotes the RF power adjustment to ensure reliable communication between HIPERLAN access point (AP) and the most distant

Table 16.2 US 5.2 GHz Unlicensed National Information Infrastructure (UNII) band

Frequency band [GHz]	Maximum output power minimum of	
5.150 - 5.250	50 mW	4 dBm+10$\log_{10} B$
5.250 - 5.350	250 mW	11 dBm+10$\log_{10} B$
5.725 - 5.825	1000 mW	17 dBm+10$\log_{10} B$

Note: B is the -26-dB emission bandwidth in MHz

Table 16.3 HIPERLAN/2 frequency bands

Frequency band	RF power limit	Comments
5.150 - 5.350	200 mW mean EIRP	Indoor use only and implementation of DFS and TPC
5.470 - 5.725	1 W mean EIRP	Indoor and outdoor use and implementation of DFS and TPC

terminal located in the area managed by this AP, as well as between two HIPERLAN devices (in the uplink or direct link).

It is worth mentioning that the MAC layer of the IEEE 802.11 WLAN terminal working in the 5 GHz band is the same as that which was already described in the previous section; however, the layers above the HIPERLAN/2 physical layer are different and will be briefly described. The physical layer in both IEEE 802.11 and HIPERLAN/2 is very similar, therefore we will concentrate on HIPERLAN/2 only.

HIPERLAN/2 is designed to work in two configurations: business environment and home environment. The HIPERLAN/2 for business environment is an access network which consists typically of several access points (APs) connected by a core network. Each AP serves a number of mobile terminals associated with it. The coverage areas of neighboring APs may partially overlap. Roaming between the coverage areas managed by different APs is also possible.

The second type of the HIPERLAN/2 network is typical for home environment. In that environment an *ad hoc* network is typically created. Such a network can consist of a few subnetworks equivalent to cells in a cellular access system. Each subnetwork operates at a different frequency and has a central controller which is dynamically selected from HIPERLAN/2 terminals operating in this subnetwork.

HIPERLAN/2 can operate in the following modes:

- *Centralized mode*, in which access points are connected to the core network; communication between a mobile terminal and another device (e.g. another mobile terminal) is always performed via the access point to which the mobile terminal is associated.

- *Direct mode,* in which terminals exchange traffic directly between each other; however, this exchange is managed by a central controller (CC). The central controller can be connected to a core network and can operate in both centralized and direct modes.

16.6.1 HIPERLAN/2 physical layer

In the HIPERLAN/2 physical layer the following operations are performed. The stream of protocol data units (PDUs) received from the *Data Link Control* layer is first scrambled, then FEC encoded. Subsequently, the code words are interleaved and after partitioning the data stream into short data blocks the mapping onto data symbols is performed. The data symbols modulate the subcarriers of the *Orthogonal Frequency Division Multiplexing* (OFDM - see Chapter 1) signal. After forming a burst and shifting the signal into the RF band, the OFDM burst is transmitted over the radio channel.

Scrambling is performed by modulo-2 adding the PDU sequence to the output of the properly initialized linear feedback shift register (LFSR) determined by the polynomial $p(x) = x^7 + x^4 + 1$. N bits of the scrambled PDU sequence are supplemented with 6 tail bits. This block is a subject of FEC coding. The FEC coding consists of the rate $R = 1/2$ convolutional coding and two-level puncturing. The convolutional code has the constraint length $L = 7$ and is determined by the generator polynomials $g_1 = 133_{OCT}$ and $g_2 = 171_{OCT}$. The first level puncturing following the convolutional encoder is only applied to a part of the PDU train. The second level puncturing is performed on the output bits from the first level puncturing in cases of 9/16 and 3/4 coding rates. Next, the encoded bit stream is a subject of block interleaving.

As we remember, in OFDM the data stream is divided into a set of parallel lower rate data streams which modulate particular subcarriers. As a result, the signaling rate is much slower and the multipath effect can be contained in the first part of the OFDM data symbol which is not used in the detection process. Particular subcarriers are densely located on the frequency axis and their signals are mutually orthogonal in the prescribed time periods (see details in Chapter 1). Table 16.4 summarizes OFDM parameters of the HIPERLAN/2 signal.

By selection of the FEC coding gain and modulation formats applied in each subcarrier we can achieve a whole range of data rates. Table 16.5 presents achievable bit rates at the selected coding rates and modulation types.

The spectrum assigned to WLANs in the 5 GHz band is effectively used by placing the OFDM channels on the frequency axis with a 20 MHz carrier raster. The spectrum of a single OFDM channel is at least 16.25 MHz wide (see Table 16.4). The carrier frequencies in the lower band are 5.180, 5.200, ..., 5.320 GHz and the mean EIRP power of the generated signal is 23 dBm. In the upper band assigned to HIPERLAN/2 in Europe the carrier frequencies are 5.500, 5.520, 5.540, ..., 5.700 GHz. The OFDM spectrum is evenly distributed around a carrier frequency.

Let us recall that OFDM modulation is usually performed digitally by the IFFT (*Inverse Fast Fourier Transform*) algorithm. At the beginning of a received block a cyclic prefix is added. It is the replicated end-part of the IFFT output vector.

Table 16.4 The values of OFDM parameters applied in HIPERLAN/2

Parameters	Value	
Sampling rate $f_s = 1/T$	20 MHz	
Symbol part duration T_U	$64 * T = 3.2\,\mu s$	
Cyclic prefix duration T_{CP}	$16 * T = 0.8\ \mu s$ (mandatory)	$8 * T = 0.4\ \mu s$ (optional)
Symbol interval $T_S = T_U + T_{CP}$	$80 * T = 4.0\ \mu s$	$72 * T = 3.6\ \mu s$
Number of data subcarriers N_{SD}	48	
Number of pilot subcarriers N_{SP}	4	
Total number of subcarriers N_{ST}	52	
Subcarrier spacing $\Delta f = 1/T_U$	312.5 kHz	
Spacing between the two outmost subcarriers	$N_{ST} * \Delta f = 16.25$ MHz	

Table 16.5 HIPERLAN/2 data rates determined by the coding rates and modulations

Modulation	Coding rate	Nominal bit rate [Mbit/s]
BPSK	1/2	6
BPSK	3/4	9
QPSK	1/2	12
QPSK	3/4	18
16QAM	9/16	27
16QAM	3/4	36
64QAM	3/4	54 (optional)

The receiver blocks reflect those applied in the transmitter. After having been received from the antenna and after down-conversion, the signal is subsequently converted to the digital form. Next, the cyclic prefix is removed. The remaining block of samples is processed by the FFT algorithm, which performs block correlation with the reference subcarriers. The resulting vector is equalized and its elements are fed to the decision devices which find the most probable data symbols. After demapping the bit representation of each data symbol is obtained and the resulting bit stream is deinterleaved and FEC decoded. Finally, an N-bit long PDU train is recovered.

Transmission of digital data is organized in OFDM packets. Each packet starts with the preamble consisting of a certain number of short OFDM training symbols, followed by a few long training symbols. During transmission of regular information, four subcarriers are used as pilots, e.g. they transmit known data symbols. The pilots placed at the beginning of a packet are used for automatic gain control, for initial frequency correction and timing acquisition. The pilots used inside a packet are applied to adjust the phase offset and residual frequency offset. More information on pilots in HIPERLAN/2 can be found in [7]. Similar considerations on the preambles in the IEEE 802.11 OFDM packet are presented in [13] and [15].

16.6.2 HIPERLAN/2 data link control layer

The *Data Link Control* (DLC) is the layer situated on top of the physical layer. Its functions are divided into medium access control and radio link control.

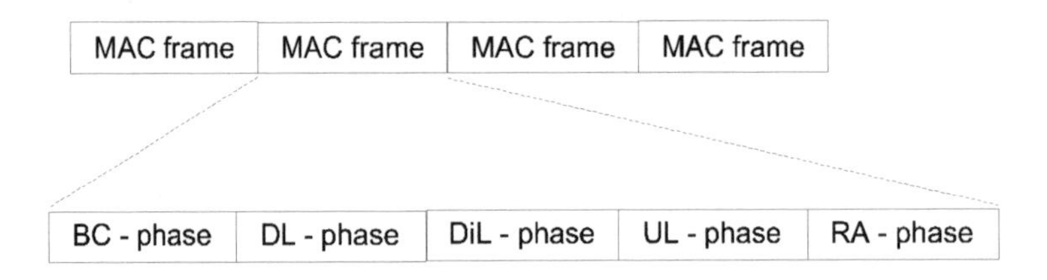

Figure 16.19 HIPERLAN/2 MAC frame format

In HIPERLAN/2 medium access control is based on the TDMA/TDD principle. From the point of view of MAC functions, time is divided into MAC frames which are 2 ms long. Each MAC frame consists of several phases shown in Figure 16.19. Their names and tasks are as follows:

- *Broadcast* (BC) phase – carries the broadcast control channel (BCCH) and frame control chanel (FCCH). The BCCH contains general information and status bits giving detailed broadcast information in the *Downlink* phase (DL). The FCCH indicates the structure of the current frame.
- *Downlink* phase – carries information directed from the AP or central controller (CC) to the specified mobile terminal or contains some broadcast information.
- *Uplink* phase – contains information sent by a mobile terminal to the AP or CC.
- *Direct Link* (DiL) phase – carries data directly between the involved mobile terminals under control of the AP or CC.
- *Random Access* (RA) phase – is used to transmit RCH (random access) channels which are applied for transmission of control information, assignment to the AP or CC and handover.

The lengths of the MAC frame phases are flexible. Within the DL, DiL and UL phases short and long protocol data units (PDU) are transmitted. The long PDUs are 54-byte long and contain user data and control information. The short PDUs are 9-byte long and contain only control information, carrying ARQ messages or information related to radio link control.

In the random access phase the RCH channels are realized by 9-byte packets transmitting radio link control messages and resource requests. The slotted Aloha is the random access method applied in the RA phase. Collisions are avoided thanks to using a binary backoff procedure controlled by the mobile terminals. The AP determines the number of RCH slots applied in the MAC frame.

The HIPERLAN/2 DLC layer also performs error control. Let us differentiate this type of error control from the FEC coding applied in the physical layer. Error control in the DLC layer relies on the selective repeat ARQ scheme (see Chapter 1).

Within the DLC layer several *Radio Link Control* (RLC) functions are performed. The first one is the *Association Control Function*, which realizes the following tasks: association of a MAC ID to a terminal and negotiation of the link capabilities, encryption key exchange (optional) and refresh, authentication, and beacon signaling in the AP/CC and disassociation. Another function is the *Radio Resource Control* which performs the following tasks: dynamic frequency selection, measurements performed by mobile terminals, reporting the measurements to the AP, frequency change perfomed by the AP and its associated mobile terminals, power saving procedure, transmit power control, handover and checking if the mobile terminal "is alive". Finally, *DLC Connection Control* (DCC) function deals with the connection set up, release or modification. The DCC function also allows a mobile terminal to join or leave a multicast group.

Finally, the *Convergence Layer* adapts the core network to the HIPERLAN/2 DLC layer [6].

* * *

As we have already mentioned, the MAC layer of the IEEE 802.11 WLAN based on the OFDM transmission is the same as the MAC layer cooperating with the remaining four physical layers in which a single carrier transmission is used. Therefore, we did not consider it twice. The minor differences betwen the physical layer of HIPERLAN/2 and IEEE 802.11 are listed in [13].

16.7 BLUETOOTH

Bluetooth [17], [18], [16] is a universal radio interface operating in the ISM band. It provides *ad hoc* wireless connectivity between portable electronic devices located in a short range from each other. The Bluetooth radio system is not a wireless LAN, although its philosophy is similar to it in many aspects. Therefore, we are describing Bluetooth in the chapter devoted to WLANs.

Bluetooth is named after Harald Blåtand, a Danish Viking king who lived in the early Middle Ages. Bluetooth is a result of a special interest group (SIG) formed by Ericsson in 1998 and promoted by Ericsson, Intel, Nokia and Toshiba, who formed a consortium. Many other leading communication and microelectronic companies joined the consortium. Bluetooth aspires to be a *de facto* standard for wireless radio interface used for communication between mobile phones, laptops, headsets, printers, projectors and many other devices of home or office use. Bluetooth is aimed to replace a web of cables connecting these devices by *ad hoc* wireless connections.

In order to perform this task, Bluetooth should fulfill the following requirements:

- The system should operate worldwide. From this point of view, the use of a worldwide available and licence-free frequency band is a prerequisite. Such a band is the ISM band, presented in the considerations on IEEE 802.11 WLANs. Let us recall that it ranges between 2.400 and 2.4835 GHz in the USA and Europe (with some exceptions in France and Spain) and from 2.471 to 2.497 GHz in Japan.

- The connection should support voice and data. It is an important requirement if headsets or multimedia devices are to be connected using this interface. This requirement determines the types of provided connections.
- The radio transceiver should be small, inexpensive and should operate at low power. This feature would enable to place the Bluetooth interface in a mobile phone, a headset, a personal digital assistant, or in any other battery-operated device.

The character of the connected devices and their mutual location indicates that a lot of *ad hoc* connections will coexist in the same area without being coordinated with one another. We can expect that a lot of small subnetworks with a limited number of terminals (devices) will operate in close vicinity. This fact and the available free band in which the system operates determine further technical requirements. First of all, the system should be immune to other systems using the same band and resistant to the interference from the connections made by other Bluetooth users. The lack of coordination between the connections indicates that spread spectrum technique has to be applied. The frequency hopping (FH) spread spectrum fulfills all the above mentioned requirements. It also makes it possible to build low-cost and low-power transceivers. In consequence, it has been selected as a transmission method in the Bluetooth interface.

In Bluetooth frequency hopping with time division duplex (FH/TDD) transmission is applied. Time is divided into 625-μs slots. The system band is divided into 79 one-MHz frequency channels. During each time slot the signal occupies one of these channels according to the selected hopping pattern. In consequence, there are 1600 hops per second. The direction of transmission alternates from slot to slot. In the USA and most European countries all 79 channels are used in the frequency hopping process. In France, Spain and Japan the available bandwidth is smaller, so only 23 hop frequencies are applied.

Each slot is used to send a packet. The packets are transmitted from the *master* to the *slave* or vice versa. A basic Bluetooth subnetwork is called a *piconet*. A piconet is established by the first station which initializes transmission to another device. Such a station becomes a *master* and performs this function till it switches off or moves outside the range of stations constituting the same piconet. The other stations in the piconet work as *slaves*. The overall number of stations creating a piconet cannot exceed eight. Transmission in a subnetwork takes place exclusively between the master and the slaves. The master coordinates the access to the channel on the basis of the polling principle, so the transmission is contention-free. The slave station whose address was contained in the master-to-slave packet sent in the preceding slot transmits to the master station. Piconets can partially overlap and a particular station can participate in a few of them at the same time or it can temporarily move to another piconet. A station can be a master in no more than one piconet. A configuration of overlapping piconets is called a *scatternet*.

A piconet which is established by a given station is determined by a Bluetooth FH channel. The FH channel is defined by the master station identity and its system clock. The identity determines the hop sequence, whereas the system clock indicates the hop phase. All slave stations belonging to the same piconet have to synchronize to the master station by adding an appropriate offset to their own clocks. Different piconets have different hop sequences. The sequences have been carefully selected to ensure their good statistical properties and immunity to mutual interference. At the same time their number should be large enough so they are not fully orthogonal. The lack of orthogonality is compensated by coding and ARQ techniques applied in data transmission. The sequence period lasts for about 23 hours. Thus, 32 consecutive hops span about 64 MHz of spectrum and all hop frequencies occur with equal probability [16]. By selection of a different station identity and clock phase the station is able to instantaneously switch to another piconet.

Table 16.6 The values of RF radio parameters applied in Bluetooth [18]

Parameter	Value
Modulation	GFSK ($h = 0.3$)
Peak data rate	1 Mbit/s
RF bandwidth	220 kHz (-3 dB), 1 MHz (-20dB)
RF carriers	79/23
Carrier spacing	1 MHz
Transmit power	$\leq$ 20 dBm

The packets placed in subsequent time slots are sent using Gaussian-shaped FSK modulation with the modulation index $h = 2\Delta fT = 0.3$. The basic transmission parameters are listed in Table 16.6. The packets have a fixed format. Each packet consists of 72-bit access code, 54-bit packet header and a payload. The payload can have 0 to 2745-bits. The packet format and the header fields are shown in Figure 16.20.

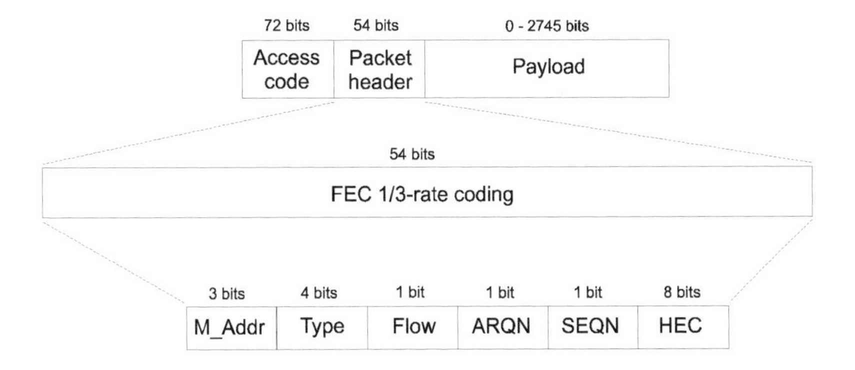

Figure 16.20 Bluetooth packet format and the header fields

The access code is unique for the FH channel (and the piconet). The stations listen to the access code of the received packet and check if the packet has been sent by a station belonging to the same piconet. If not, the rest of it is ignored. The access code is also used for synchronization and offset compensation.

A header is a packet field which follows the access code. It contains the MAC address of the slave which transmits the packet to the master or to whom the master sends a packet, the packet type code, the flow control bit, the ARQ and packet sequence bits and header error check (HEC) bits. The packet header is protected using the 1/3 repetition code.[5] The last field of the packet is usually a payload; however, there are packet types which do not contain the payload field. Typically packets do not exceed one time slot (625 μs); however, in some types of links multi-slot packets are used. They occupy three or five slots. A multi-slot packet starting at the kth time slot and lasting for $n = 3$ or 5 slots is transmitted on a single hop frequency f_k, but the next packet is transmitted using the hop frequency f_{k+n-1} resulting from the regular hop pattern.

One of the basic requirements for Bluetooth was to enable voice and data transmission. This determines the types of the links defined in the Bluetooth specification. They are:

- *Synchronous Connection-Oriented* (SCO) links supporting symmetrical, point-to-point, circuit-switched connections usually used for voice transmission. They are realized by reserving two consecutive slots at fixed slot intervals for forward and reverse transmission.

- *Asynchronous Connectionless* (ACL) links used for bursty data transmission and supporting packet-switched symmetrical and asymmetrical connections.

Depending on the type of the link, the applied coding rate, the payload length and the symmetry/asymmetry of a connection, several types of packets are defined. They are summarized in Table 16.7 for ACL links and in Table 16.8 for SCO links.

Table 16.7 Parameters of ACL packets [17]

Type	Payload header [bytes]	User payload [bytes]	FEC	CRC	Symmetric Max. rate [kbit/s]	Asymmetric max. rate [kbit/s]	
						Forward	Reverse
DM1	1	0 - 17	2/3	yes	108.8	108.8	108.8
DH1	1	0 - 27	No	yes	172.8	172.8	172.8
DM3	2	0 - 121	2/3	yes	258.1	387.2	54.4
DH3	2	0 - 183	no	yes	390.4	585.6	86.4
DM5	2	0 - 224	2/3	yes	286.7	477.8	36.3
DH5	2	0 - 339	no	yes	433.9	723.2	57.6
AUX1	1	0 - 29	no	no	185.6	185.6	185.6

[5] The code word of the 1/3-rate repetition code consists of a sequence of triplicated bits.

Table 16.8 Packet parameters in SCO connections [17]

Type	Payload header [bytes]	User payload [bytes]	FEC	CRC	Symmetric max. rate [kbit/s]
HV1	na	10	1/3	no	64.0
HV2	na	20	2/3	no	64.0
HV3	na	30	no	no	64.0
DV	1 D	10+(0 - 9) D	2/3 D	yes D	64.0 +57.6 D

Note: items followed by the letter D relate to data fields only

Let us note that the data rate of up to 723.2 kbit/s can be achieved in asymmetric links in which 5 consecutive slots are used to transmit long packets. In SCO links the basic data stream has the rate of 64 kbit/s and is mostly devoted to voice transmission. Bluetooth specifies two types of voice representation: a regular 64-kbit/s nonlinear PCM and a continuous variable slope delta (CVSD) modulation (see Chapter 1).

The applied coding rates are equal to 1/3 or 2/3. The 1/3 code is a repetition code already mentioned in the description of a packet header. The 2/3 code is a shortened Hamming code and is used both in SCO and ACL packets.

In ACL links the fast ARQ technique is applied. The station which transmits a packet is notified of the packet reception in the slot directly following the slot in which a packet in the opposite direction has been sent. The ARQN bit of the header carries an ACK or NAK message which is the response to the packet transmitted in the last slot.

A Bluetooth station remains in one of four modes. It can be in a connection mode, idle mode, park mode and sniff mode. In the connection mode transmission proceeds. In the idle mode the station wakes up for about 10 ms in a period of time ranging from 1.28 to 3.84 s and searches for its access code, which means a paging signal. If a piconet has been established, the slave station can switch into park mode in which it scans the piconet FH channel with a lower duty cycle than in the idle mode. In the sniff mode the station scans the channel even more infrequently. Except in the connection mode, in other modes a station saves power by being active only for a short fraction of time. This is particularly important for battery-operated devices.

Another important procedure in the Bluetooth system is establishing a connection. Let us assume that a station remains in an idle mode. It has to check if another station wishes to communicate with it. Therefore, it periodically wakes up and searches the channel for its access code. The searching relies on correlation of the received data with the station access code and lasts for about 10 ms. During the wake-up phase the station "visits" one of 32 hop frequencies which are repeated periodically. The 32-hop sequence is unique for each station. Let us consider the station which wishes to page a particular station remaining in an idle mode. The paging station must know the identity of the station it wishes to connect. Knowing the identity, it calculates the access code which is practically used as a paging message. The wake-up hopping sequence is also derived from the station identity. In consequence, the paging station transmits the access code. It is transmitted repeatedly on different wake-up hop frequencies. Let us note that during 10 ms in which the paged station wakes up and scans the FH channel the paging station visits 16 different hop frequencies[6] out of 32 used by the paged and paging station. The paging station cyclically sends the access code of the paged station during the sleep time of the station because it does not know the moments of its waking up. If there is no response from the paged station, the second half of 32 hop frequencies is cyclically visited and the paged station finally detects its access code. The maximum access delay is a doubled sleep time. If the paging station receives an answer from the paged station in the form of the access code, it sends the so-called HFS packet. It contains a real-time clock and identity information necessary to synchronize the hop sequences of both stations. This way a piconet is established, in which the paging station is a master and the paged station is a slave.

To establish a connection with a given station, its identity has to be known. If the identity is not known, the station which wishes to set a connection can broadcast an inquiry message asking all the stations in its range to send their addresses and clock information. The stations which received and decoded the inquiry message send their FHS packet, carrying their identity and clock information. In order to avoid collisions, a pseudorandom backoff mechanism is used in the FHS packet transmission.

Transmision in the Bluetooth system is protected against eavesdropping and unauthorized use. During the connection set-up, the authentication process of the participating stations is performed. The station initiating this process sends a 48-bit address to the recipient, which returns a 128-bit block. On the basis of this block, a 128-bit secret link key and the 48-bit address of the recipient the signed response (SRES) is generated in both stations. The signed response is transmitted by the initiating station to the recipient where it is compared with the SRES calculated locally. If both strings are identical, the authentication is successful.

The payload bits are additionally encrypted by being modulo-2 added to the data stream generated by the linear feedback shift register (LFSR), which is initialized in each time slot. More advanced security algorithms are applied at higher levels of transmission, which are beyond the Bluetooth system.

* * *

As we have mentioned, many important communication and electronic companies belong to the Bluetooth consortium. They pursue a smart policy of popularizing the Bluetooth standard. Since Bluetooth is a good solution to wireless connectivity of different devices in home and office environment, it may be expected to become more and more popular.

[6]During 10 ms 16 slots lasting for 625 μs occur.

REFERENCES

1. J. H. Schiller, *Mobile Communications*, Addison-Wesley, Reading, Mass., 2000

2. B. Bing, *High-Speed Wireless ATM and LANs*, Artech House, Boston, 2000

3. J. Geier, *Wireless LANs*, Macmillan Technical Publishing, 1999

4. B. H. Walke, *Mobile Radio Networks: Networking and Protocols*, Wiley & Sons, Ltd., Chichester, 1999

5. ETSI EN 300 652 V.1.2.1, Broadband Radio Access Networks (BRAN); High Performance Radio Local Area Network (HIPERLAN) Type 1; Functional Specification, 1998

6. ETSI TR 101 683 V.1.1.1, Broadband Radio Access Networks (BRAN); HIPERLAN Type 2; System Overview, 2000

7. ETSI TS 101 475 V.1.1.1, Broadband Radio Access Networks (BRAN); HIPERLAN Type 2; Physical (PHY) Layer, 2000

8. ETSI TR 101 177 V.1.1.1, Broadband Radio Access Networks (BRAN); Requirements and Architectures for Broadband Fixed Radio Access Networks (HIPERACCESS), 1998

9. ISO 8802, Information Processing Systems - Local Area Networks - Part 2: Logical Link Control. International Standard, 1990

10. IEEE 802.11, Information Technology – Telecommunications and Information Exchange Between Systems – Local and Metropolitan Area Networks – Specific Requirements, Part II: Wireless LAN Medium Access Control (MAC) and Physical Layer (PHY) Specifications, November 1997

11. B. P. Crow, I. Widjaja, J. G. Kim, P. T. Sakai, "IEEE 802.11 Wireless Local Area Networks", *IEEE Communications Magazine*, Vol. 35, No. 9, September 1997, pp. 116-126

12. R. T. Valadas, A. M. de Oliveira Duarte, A. C. Moreira, C. T. Lomba, "The Infrared Physical Layer of the IEEE 802.11 Standard for Wireless Local Area Networks", *IEEE Communications Magazine*, Vol. 36, No. 12, December 1998, pp. 107-112

13. R. van Nee, G. Awater, M. Morikura, H. Takanashi, M. Webster, K. Halford, "New High-Rate Wireless LAN Standards", *IEEE Communications Magazine*, Vol. 37, No. 12, December 1999, pp. 82-88

14. M. J. E. Golay, "Complementary Series", *IRE Trans. Inform. Theory*, April 1961, pp. 82-87

15. R. van Nee, R. Prasad, *OFDM for Wireless Multimedia Communications*, Artech House, London, 2000

16. J. C. Haartsen, "The Bluetooth Radio System", *IEEE Personal Communications*, February 2000, pp. 28-36

17. Specification of the Bluetooth System, Specification Volume 1, Bluetooth V.1.0 B, December 1999

18. J. Haartsen, "BLUETOOTH - The Universal Radio Interface for ad hoc Wireless Connectivity", *Ericsson Review*, No. 3, 1998, pp. 110-117

17

Third generation mobile communication systems

17.1 INTRODUCTION

In the previous chapters we presented examples of existing mobile communication systems. Among the second generation systems there are several cellular systems which have attracted millions of subscribers, such as GSM, IS-95, IS-136 and PDC (the last two only mentioned in this book), wireless telephony like DECT, PACS and PHS, data transmission systems like GPRS and EDGE or personal satellite communication systems (e.g. Iridium, Globalstar). A large variety of systems differing in applications requires different equipment. It is also worth noting that the existing systems operate in selected environments only. Thus, it would be desirable to create a universal system which could operate anywhere, anytime using a unified equipment.

The research on third generation systems was initiated long before the potential of the GSM and other second generation systems was exhausted.The aim was to create a global standard which would enable global roaming. The International Telecommunications Union (ITU) began to work on the third generation mobile communication system by defining the basic requirements. The system was initialy called *Future Public Land Mobile Telecommunication System* (FPLMTS) and now is know as *International Mobile Telecommunications* (IMT-2000). The basic requirements are as follows:

- Throughput rates up to
  - 2 Mbit/s indoors and for pedestrians
  - 384 kbit/s for terminals moving at no more than 120 km/h in urban areas
  - 144 kbit/s in rural areas and for fast moving vehicles
- Support of global mobility

- Independence of the IMT-2000 services from the applied radio interface technology. This allows use of different air interfaces. On the other hand, multimode terminals must be used.
- Seamless switching between fixed and wireless telecommunication services
- Support of circuit- and packet-switched services
- Support of multimedia and real-time services
- Implementation of the *Virtual Home Environment* (VHE). The user interface features typical for his/her home environment remain the same during roaming in different networks

The implementation of a system which would fulfill the above requirements was left to the following regional bodies:

- ETSI (*European Telecommunications Standard Institute*), which since 1995 has worked on proposals for UMTS (*Universal Mobile Telecommunication System*) WCDMA (*Wideband CDMA*) radio access,
- the T1P1 committee in the USA which has coordinated research on the evolution of the second generation systems used in the USA (IS-95, IS-136 and GSM 1900) which resulted in the proposal of a multicarrier CDMA based on IS-95,
- ARIB (*Association for Radio Industries and Businesses*) in Japan, which proposed an air interface very similar to the UMTS interface,
- TTA (*Telecommunications Technology Association*) in South Korea which proposed to use CDMA in the air interface proposal, too.

In 1998 the 3GPP (*3rd Generation Partnership Project*) group was established in order to define a common wideband CDMA standard. As a result the UMTS was defined. However, an alternative standard called *cdma2000* was also promoted by those partners who wished to extend the IS-95 system. This way the 3GPP2 (*3rd Generation Partnership Project 2*) was also established. In fact, three different IMT-2000 standards were agreed upon:

- UTRA (*UMTS Terrestrial Radio Access*) – wideband CDMA transmission with FDD and TDD modes and 5 MHz carrier spacing
- MC CDMA (*Multicarrier CDMA*)
- UWC136 (*Universal Wireless Communications*) – the standard based on the convergence of IS-136 and GSM EDGE. The UWC136 will be a natural extension of TDMA systems.

Figure 17.1 presents the evolution from the second generation systems to the third generation systems.

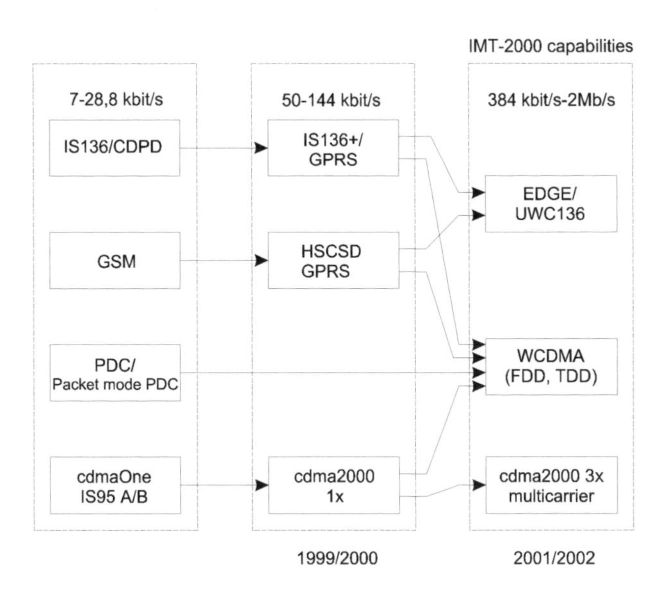

Figure 17.1 Evolution of the second generation systems resulting in the third generation systems

So far we have discussed the evolution of radio access standards. As we know, radio access is only a part of a whole communication system. Another part is the core network, which connects different elements of the radio access system with the fixed part of the system and with other networks such as PSTN, ISDN, the Internet and PDSN. The evolution has taken place in this respect, too. It has been decided that the UMTS and EDGE systems (evolving from GSM) will initially work with the GSM core network. On the other hand, the cdma2000 and EDGE (evolving from IS136) will initially work with the IS-41 core network. Both core networks will be equipped with interworking functions to enable roaming and other services. In a longer time perspective, all IMT-2000 networks will cooperate with the IP core network similar to the present core network of the GPRS system.

The implementation of third generation systems depenfd on the spectrum allocation made by the administrative bodies. Figure 17.2 presents spectrum allocation agreed upon during the World Radiocommunications Conference WRC-2000. Let us note that the spectrum for IMT-2000 has been allocated in almost all countries except the USA, where the third generation system will be allocated a new spectrum in the future, or the existing PCS system spectrum will change its application.

In the following we will briefly describe the most important third generation systems: UMTS and cdma2000. Currently, whole books are devoted exclusively to UMTS or cdma2000, so the reader searching for details is asked to study [1], [2] [3] and [4].

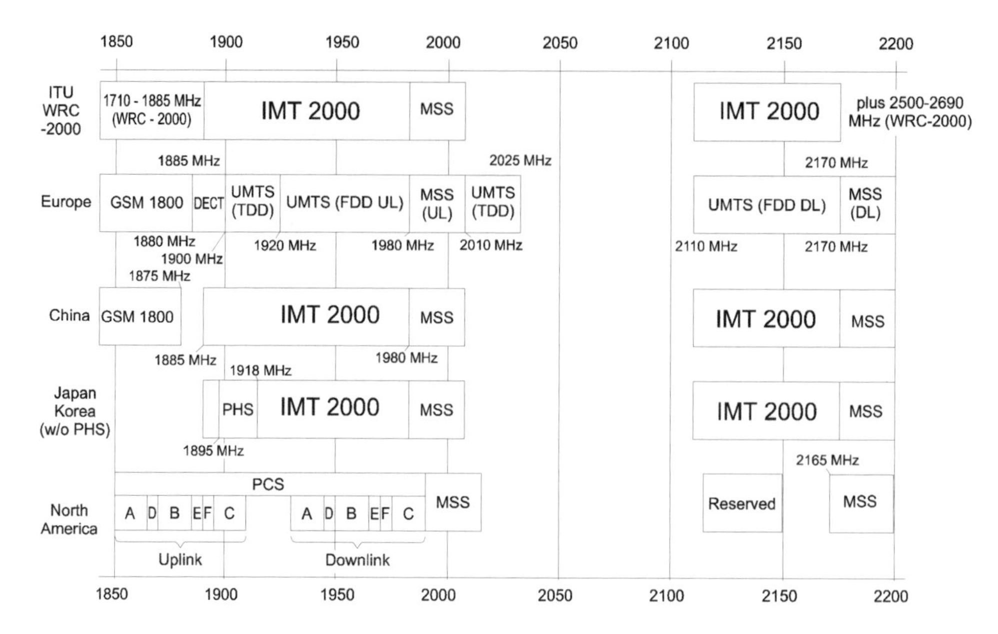

Figure 17.2 Spectrum allocation for IMT-2000 in different parts of the world agreed upon during WRC-2000 (MSS - mobile satellite systems)

17.2 THE CONCEPT OF UMTS

The basic purpose of introducing the UMTS is to support integrated digital wireless communications at the data rates up to 2 Mbit/s in the 2 GHz band. Table 17.1 presents the bands allocated to the UMTS.

Table 17.1 UMTS spectrum allocation

Frequency [MHz]	Bandwidth [MHz]	Destination
1900–1920	20	UMTS (terrestrial), TDD
1920–1980	60	UMTS (terrestrial) FDD, UL
1980–2010	30	UMTS (satellite) FDD, UL
2010–2025	15	UMTS (terrestrial) TDD
2110–2170	60	UMTS (terrestrial) FDD, DL
2170–2200	30	UMTS (satellite) FDD, DL

The UMTS requirements are similar to those stated for IMT-2000. They are:

- **Operation in various types of environment**. The terrestrial part of the UMTS will operate in several environments, from rural to indoor environment. There will be three basic types of cells fitted to these environments: picocells, microcells and macrocells with appropriate physical layers of the UMTS. The satellite segment of the UMTS will constitute a supplement of the land mobile system and will work in the areas of very low traffic density or under-developed infrastructure.

- **The choice of duplex transmission**. The bandwidth allocated by WRC'92 consists of two paired bands and two unpaired bands. This implies the necessity to apply frequency division duplex transmission in the paired bands and time division duplex transmission in the unpaired bands (see Tab. 17.1).

- **A wide service offer**. The UMTS should support a large selection of services, from voice transmission to fast data transmission. The traffic can be asymmetric. The system should be flexible enough to allow for the introduction of new services in the future. For this purpose the radio access system should support unidirectional links based on a few basic bearer services.

- **Cooperation with fixed networks**. The UMTS will be integrated with wireline wideband networks such as B-ISDN. *Intelligent Networks* (IN) technology will be applied.

Following the above requirements a list of services has been proposed and their quality has been defined. Table 17.2 (based on [5] with modifications) shows the UMTS service proposals. The UMTS offers voice, data and videotelephony transmission. Besides that, it can be treated as a wireless extension of integrated services digital networks. Therefore, the system supports basic ISDN access at the rate of 144 kbit/s (two B channels plus a D channel). Data transmission at the rate of 2 Mbit/s, possible in a limited range, allows transmission of compressed video.

Similar to the second generation systems, the UMTS will ensure a high level of security of transmitted data. It is important not only for the privacy of individual telephone calls, but also for support of tele-banking and e-commerce.

The UMTS applies a high quality *Adaptive Multi-Rate* (AMR) speech coding based on ACELP coding with discontinuous transmission and comfort noise insertion. It works at 8 different bit rates: 4.75, 5.15, 5.90, 6.70, 7.40, 7.95, 10.20 and 12.20 kbit/s. Three of them are compatible with speech coders applied in the existing second generation systems: 6.70 kbit/s – PDC EFR, 7.40 kbit/s – IS-641 (US TDMA/IS-136) and 12.20 kbit/s – GSM EFR. The data rate of the AMR coder depends on the network load, the service level specified by the network operator and the current SNR value.

17.3 UMTS RADIO ACCESS NETWORK ARCHITECTURE

As we have already mentioned, the UMTS takes advantage of the existing GSM and GPRS networks, which serve as a core network in the UMTS infrastructure. Figure 17.3 presents the UMTS radio access network architecture.

There are three main elements of this architecture:

- *User Equipment* (UE), which consists of

Table 17.2 Examples of services in UMTS

Type of service	Data rate [kbit/s]	Error rate	Allowable delay [ms]
Speech transmission	4.75–12.2	10^{-4}	40
Voiceband data	2.4–64	10^{-6}	200
Hi-Fi sound	940	10^{-5}	200
Videotelephony	64–144	10^{-7}	40–90
Short messages/paging	1.2–9.6	10^{-6}	100
E-mail	0–384	10^{-6}	many minutes
Facsimile (G4)	64	10^{-6}	100
Broadcast or multicast transmission	1.2–9.6	10^{-6}	100
Web browsing	16–64 (UL) 96–384 (DL)	10^{-6} 10^{-6}	seconds
Digital data without specified limitations	64–1920	10^{-6}	100
Access to data bases	2.4–768	10^{-6}	200
Teleshopping	2.4–768	10^{-7}	90
Electronic newspaper	2.4–2000	10^{-6}	200
Remote control	1.2–9.6	10^{-6}	100
Navigation and location	64	10^{-6}	100
Teleworking	32–64	10^{-6}	90

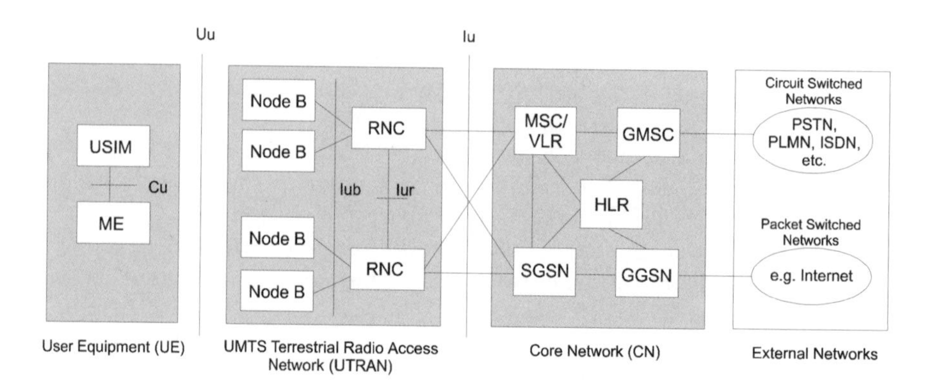

Figure 17.3 UMTS radio access network

- *Mobile Equipment* (ME), which is a radio terminal connecting a UMTS subscriber through the radio interface *Uu* with the fixed part of the UMTS system

- *UMTS Subscriber Identity Module* (USIM), which is a smart card similar to the SIM card used in the GSM system. The card contains the subscriber identity, authentication algorithms, authentication and encryption keys etc.

- *UMTS Terrestrial Radio Access Network* (UTRAN), which is a system of base stations and their controllers. It consists of two kinds of elements:
  - base stations called *Nodes B* (in accordance with the 3GPP), which perform physical layer processing such as channel coding, data interleaving, rate matching, modulation, etc. Briefly, a base station converts the data between the *Uu* radio interface and the *Iub* interface connecting a *Node B* with the *Radio Network Controller*.
  - *Radio Network Controllers* (RNCs), which control *Nodes B* connected to them and manage radio resources assigned to them. In this sense the RNC performs the data link layer processing and participates in the handover procedures. The RNC is considered a service acccess point of UTRAN for the core network. It is connected to a single MSC/VLR to route circuit-switched traffic and to a single SGSN to route packet-switched traffic.
- *Core Network* (CN) is shared with GSM and GPRS. It therefore contains typical elements both for circuit-switched and packet-switched systems, i.e.:
  - *Home Location Register* (HLR) performing functions similar to those in the GSM and GPRS systems,
  - *Mobile Switching Center/Visitor Location Register* (MSC/VLR) handling the circuit-switched traffic,
  - *Gateway MSC* (GMSC) connecting the UMTS with external *Circuit-Switched* (CS) networks,
  - *Serving GPRS Support Node* (SGSN) similar to that in GPRS and serving the packet-switched traffic,
  - *Gateway GPRS Support Node* (GGSN) connecting the UMTS with external *Packet Switched* (PS) networks.

Let us turn our attention to the interfaces shown in Figure 17.3. Most of the interfaces have been defined with such accuracy that elements of the UMTS can be produced by different manufacturers. Thus, the system can be attractive to many manufacturers and can gain more popularity. Below we list the UMTS interfaces and present their basic functions. They are as follows:

- *Cu Interface* connects the hardware part of the UMTS terminal with the USIM smart card. It conforms to a standard format of smart cards.
- *Uu Interface* is the radio interface between UMTS terminals and the base stations (*Nodes B*). It is precisely defined to allow for functioning of terminals of different brands. The novel contribution of the UMTS is this definition of the *Uu* interface.
- *Iub Interface* defines communication between base stations and an appropriate RNC.
- *Iur Interface* is the interface between different RNCs. Let us note that there is no equivalent of *Iur* interface in GSM; however, in the UMTS the *Iur* is necessary

to perform a soft handover with the participation of two base stations which are controlled by two different RNCs. This interface is also used if a connection from a base station is routed from the so-called *Drift RNC* to the *Serving RNC*, which finally, through the *Iu* interface, directs it to the core network.

- *Iu Interface* connects the UTRAN with the core network and is functionally similar to the *A* interface in GSM and the *Gb* interface in GPRS.

In the literature and the descriptions of UMTS standards one can find detailed considerations on different aspects of protocol layering. The interested reader is directed for example to [1]. Here we will only quote the general protocol model for UTRAN interfaces, which is directly related to above considerations on the UMTS radio access network and defined interfaces. This protocol model is shown in Figure 17.4.

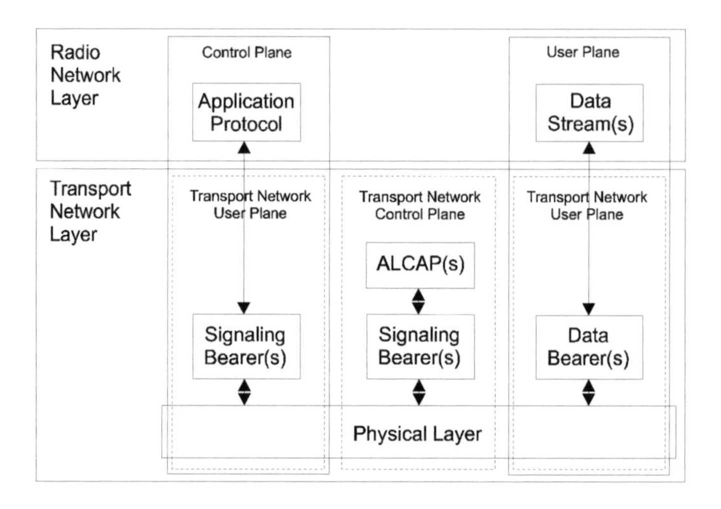

Figure 17.4 General protocol model for UTRAN interfaces

The protocol model consists of horizontal layers and vertical planes. Two main horizontal layers have been introduced:

- *Transport Network Layer*, in which data from the User Plane and Control Plane are mapped onto dedicated and shared *Transport Channels.* Transport channels, in turn are mapped onto *Physical Channels*

- *Radio Network Layer*, which is concerned with UTRAN-related issues, e.g. access to UTRAN via the *Iu* interface.

The following vertical planes have been defined across the Radio Network Layer and the Transport Network Layer:

- *Control Plane* used for UMTS control signaling, consisting of an *Application Protocol* specific for the appropriate interface and a *Signaling Bearer* used for the Application Protocol messages,

- *User Plane*, in which all user data such as encoded voice or packet data are transported. Within the user plane there are *Data Streams* and *Data Bearers* associated with them. Data streams are characterized by one or more frame protocols.

Within the Transport Network Layer the following vertical planes are defined:

- *Transport Network Control Plane* used for all control signaling within the Transport Layer. It consists of
  - *Access Link Control Application Part* (ALCAP) responsible for configuration of transport channels with respect to the given requirements and for combining the user and control data into dedicated and common channels, and
  - *Signaling Bearers* needed to perform the ALCAP.
- *Transport Network User Plane* which contain Signaling Bearers for the Application Protocol and data bearers in the User Plane.

17.4 UMTS AIR INTERFACE

The *UMTS Terrestrial Radio Access* (UTRA) interface was defined by the 3rd Generation Partnership Project (3GPP). It is often called WCDMA (*Wideband CDMA*) interface. WCDMA has two modes of operation differing in the kind of duplex transmission. As we have already mentioned, in the paired bands the UMTS operates in the FDD mode, whereas in the unpaired bands, the standard is the TDD mode. Both modes differ in their potential applications and details of their air interfaces. Table 17.3 presents the basic features these two modes.

Before we consider the WCDMA air interface, let us clarify its layered structure. Figure 17.5 presents the meaningful components of the air interface protocol architecture.

All the protocols can be placed in one of the three lowest OSI layers: the physical (PHY) layer, the Data Link Control (DLC) layer and the network layer.

The physical layer offers information transfer services in the form of transport channels. In the physical layer all the signal processing functions, channel coding, interleaving, modulation, spreading, synchronization, etc. are performed. One of them is mapping of transport channels onto physical channels.

The Data Link Control layer is divided into the following sublayers:

- *Medium Access Control* (MAC) sublayer
- *Radio Link Control* (RLC) sublayer
- *Packet Data Convergence Protocol* (PDPC) sublayer
- *Broadcast/Multicast Control* (BMC) sublayer.

Table 17.3 Basic parameters of WCDMA interfaces

	UTRA FDD	**UTRA TDD**
Multiple access method	CDMA	TDMA/CDMA
Duplex method	FDD	TDD
Channel bandwidth	5 MHz	
Chip rate	3.84 Mchip/s	
Frame length	10 ms	
Time slot structure	15 slots/frame	
Multirate method	Multicode, multislot and OVSF[a]	Multicode and OVSF[a]
Spreading (DL)	OVSF sequences for channel sep., truncated Gold seq. (2^{18}-1) for cell and user sep.	
Spreading (UL)	OVSF sequences, truncated Gold seq. (2^{25}) for user separation	
Spreading factor	4–512	1–16
Channel coding	Convolutional coding (R=1/2, 1/3, K=9), turbo coding (8-state PCCC, R=1/3), service specific coding	
Interleaving	Inter-frame interleaving (10, 20, 40 and 80 ms)	
Modulation	QPSK	
Pulse shaping	Square-root raised cosine with roll-off factor = 0.22	
Detection	Coherent, based on pilot symbols	Coherent, based on midamble
Burst types	Not applicable	Traffic, random access and synchronization bursts
Dedicated channel power control	Fast closed loop (rate - 1500 Hz)	UL: open loop (100 or 200 HZ) DL: closed loop ($\leq$ 800 Hz)
Intra-frequency handover	Soft handover	Hard handover
Inter-frequency handover	Hard handover	
Channel allocation	No DCA[b] required	Slow and fast DCA[b] possible
Intra-cell interference cancellation	Joint detection possible	Advanced receivers at base stations possible

[a] OVSF - *Orthogonal Variable Spreading Factor* – a type of spreading code used in UTRA, see the air interface description below.

[b] DCA - *Dynamic Channel Allocation*

The MAC sublayer performs data transfer services on logical channels, which are defined with respect to the type of information which is transferred on them.

The RLC sublayer performs ARQ algorithms, is responsible for segmentation and assembly of user data, controls the appropriate sequence of data blocks and ensures avoiding block duplication.

The PDPC sublayer provides transmission and reception of network Protocol Data Units (PDUs) in acknowledged or unacknowledged and transparent RLC modes. In turn, the BMC sublayer performs broadcast and multicast transmission service in transparent or unacknowledged mode.

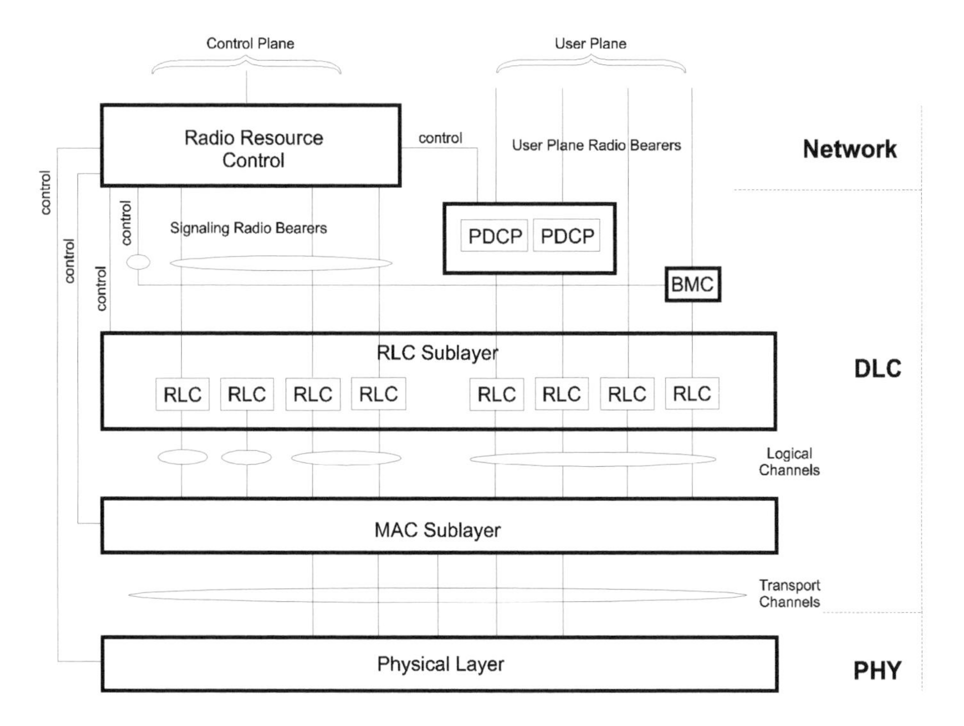

Figure 17.5 Air interface protocol architecture (after [6])

Finally, the lowest sublayer of the Network Layer shown in Figure 17.5 is the *Radio Resource Control* (RRC) sublayer. The RRC sublayer fulfills the following functions [7]: broadcasting of system information, radio resource handling, control of requested quality of service, measurement reporting and control.

Summarizing, to ensure that the higher sublayers perform particular functions, the MAC sublayer offers *logical channels* to those sublayers. In order to realize the MAC functions, *transport channels* remain at the disposal of the MAC sublayer. They are finally mapped onto *physical channels* in the physical layer.

Now let us consider all three types of channels. Let us start with logical channels. They are divided into two classes:

- *Control Channels* (CCH) used for the information transfer performed in the control plane
- *Traffic Channels* (TCH) used to carry information in the user plane.

The following types of logical control channels exist:

- *Broadcast Control Channel* (BCCH) used in the downlink to broadcast system control information,
- *Paging Control Channel* (PCCH) applied in the downlink to page the mobile station or wake it up if it is in the sleeping mode

- *Common Control Channel* (CCCH) applied in the downlink and uplink to transfer control information
- *Dedicated Control Channel* (DCCH) used in point-to-point transmission to transfer dedicated control information between the network and a mobile station during establishment of the RRC connection.

The logical traffic channels are divided into the following types:

- *Dedicated Traffic Channel* (DTCH) used in point-to-point transmission between a mobile station and the network to transfer user information; it can be established in the downlink and uplink,
- *Common Traffic Channel* (CTCH) applied in point-to-multipoint transmission, carrying information for a group of mobile stations.

The logical channels realized on the level of the RLC sublayer are mapped by the MAC sublayer onto the transport channels. Among transport channels there is a single *Dedicated Channel* (DCH) and six common channels.

The DCH is a point-to-point bidirectional channel carying both user data and higher level control data. It can be transmitted to the whole cell or a part of it if a beam-forming antenna is used. It can rapidly change its parameters (data rate, power level, etc.).

The common transport channels carry control data and small amounts of user data without establishing a dedicated connection with the user. There are the following common transport channels:

- *Broadcast Channel* (BCH), which sends system- and cell-specific information at a low data rate to reach all mobile stations in a cell; it carries part of the logical BCCH channel,
- *Forward Access Channel* (FACH), which operates in the downlink, sending control information containing another part of the BCCH channel and realizing the packet data link; there can be more than one FACH in a cell, of which at least one is transmitted at a low data rate and at high power,
- *Paging Channel* (PCH), which is a point-to-multipoint channel used to page a mobile station,
- *Downlink Shared Channel* (DSCH), which is an optional transport channel shared by several mobile stations; it provides dedicated user data and is associated with the Dedicated Channel (DCH),
- *Random Access Channel* (RACH), which is an uplink low rate channel. It has to be received by a base station when transmitted from any place in a cell and is used by a mobile station to set a connection or to send a small amount of data to the network,

- *Common Packet Channel* (CPCH), which is an optional uplink transport channel operating according to the contention principle and used for transmission of bursty data.

Each type of transport channel is associated with a *Transport Format* (TF) set. The TF determines possible mapping, encoding and interleaving of a given type of transport channel. The MAC layer procedure selects an appropriate transport format for a given transport frame. The features of the applied transport format are contained in a block called *Transport Format Indicator* (TFI), which usually accompanies the tranport blocks and indicates how the transport channel is realized.

After channel coding and interleaving, several transport channels can be multiplexed. This way the received data stream is assigned to the *physical data channel.* Accordingly, the multiplexed transport format indicators form a *Transport Combination Format Indicator* (TFCI) which is transmitted on a *physical control channel.* Figure 17.6 illustrates this process.

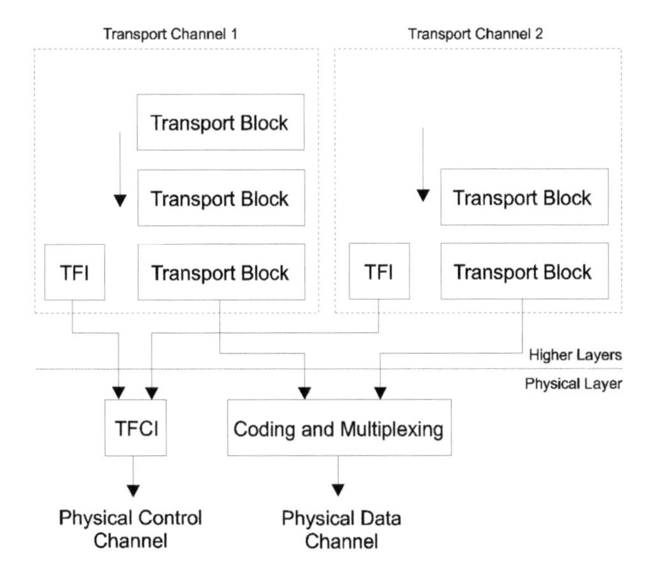

Figure 17.6 Example of mapping of the transport channels onto physical channels

Because transport channels can consist of a different number of transport blocks, the TFCI also informs which transport channel is active in a given frame. At the receiver, after decoding the TFCI, the data received in the physical data channel can be demultiplexed and decoded and the transport blocks of the appropriate transport channels can be extracted.

17.4.1 UTRA FDD mode

In the physical layer of the UMTS time is divided into 10-ms frames. The frames, in turn, are divided into 15 slots 666.67 μs long. In case of applying the FDD mode, this time division is not a result of using a multiple access method, because a CDMA scheme is applied; however, the slots are time units in which the transmission of the appropriate channel blocks takes place. Knowing that the basic chip rate is 3.84 Mchip/s, we find that each time slot lasts for 2560 chips.

In the UMTS FDD mode a physical channel is determined by the carrier frequency, the applied spreading sequence and the applied signal component (in the uplink, in-phase and quadrature components can carry different physical channels).

In the physical layer, two types of dedicated physical channels have been defined for uplink and downlink. They are:

- *Dedicated Physical Control Channel* (DPCCH)
- *Dedicated Physical Data Channel* (DPDCH).

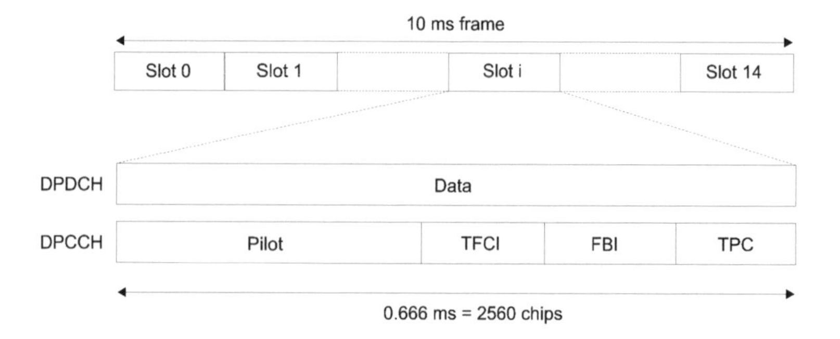

Figure 17.7 Uplink *Dedicated Physical Channel* structure

One dedicated physical control channel and up to six dedicated data channels are assigned to each connection. In the uplink the binary stream of the DPDCH is fed to the in-phase input of the transmitter, whereas the binary stream of DPCCH is fed to the quadrature input of the transmitter. If the number of data channels (DPDCH) is higher than one, then the odd-numbered channels are summed, weighted and transmitted using the in-phase component, whereas the even-numbered channels are summed, weighted and transmitted together with the control channel (DPCCH), using the quadrature component.

The DPDCH transmits user data and the DPCCH sends a pilot signal needed in the base station receiver for channel estimation, the TFCI block indicating the DPDCH format, the *Feedback Information* (FBI) for downlink transmit diversity and the *Transmit Power Control* (TPC) block for implementation of fast power control in the downlink (see Figure 17.7). Both data streams are spread using two different mutually orthogonal *channelization codes*. Thanks to them 4 to 256–ary spectrum spreading is achieved, depending on the information sequence data rate. The *Spreading Factor* $SF = 256/2^k$, $k = 0, 1, \dots, 6$, so the number of bits transmitted in one of fifteen slots of the 10-ms frame is 10×2^k. Because the data rates in both physical channels can be different, the mean power of the in-phase and quadrature inputs can be different, too, and the applied spreading matches the current data rate in the data channel, resulting in the desired bandwidth. The variability of the spreading factor is obtained thanks to the application of the *Orthogonal Variable Spreading Factor* (OVSF) codes [8].

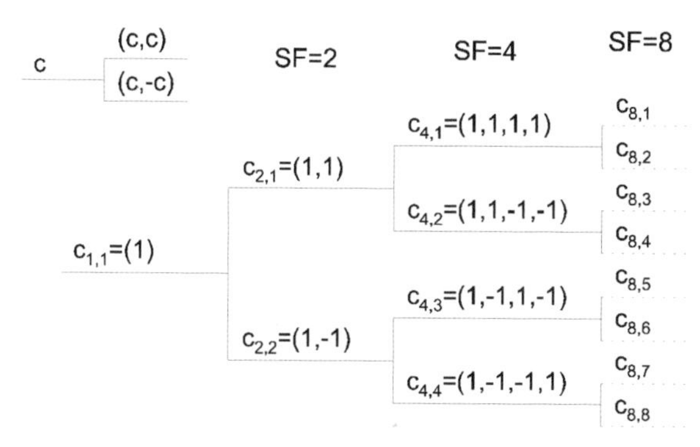

Figure 17.8 OVSF code tree

The OVSF codes are defined by a tree shown in Figure 17.8. Considering any node of the tree, we see that new code words are created by appending the preceding code word with itself (in the upper branch originating from the node) or with its negation (in the lower branch growing from this node). Mutual orthogonality of different code words is achieved by their appropriate selection from the code tree. A code word c_i is orthogonal to a code word c_j, if and only if the code word c_j is not associated with the branch leading from the branch associated with the code c_i to the root of the tree or is not located in the subtree below the code word c_i. For example, if bits of a particular data stream are spread using the code word $c_{8,5} = (1, -1, 1, -1, 1, -1, 1, -1)$ with the spreading factor $SF = 8$, then for another data stream requiring the spreading factor $SF = 4$ all the code words $c_{k,i}$ except $c_{4,3}$ can be applied.

The resulting pair of data streams spread by the OVSF code words can be interpreted as a complex signal with the real part being the in-phase component and the imaginary part being the quadrature component of the data signal. This complex signal undergoes a complex scrambling[1] operation. The scrambling sequence consists of two components interpreted as the real and imaginary parts of the complex scrambling signal. Let us note that the scrambling operation does not increase the bandwidth, i.e. the chip rate at the output of the scrambler is the same as at its input. The aim of scrambling is to differentiate cells. The reason for selection of a complex scrambling sequence is the possible unequal power of the in-phase and quadrature components carrying DPDCH and DPCCH or their sums, respectively. By applying a complex scrambling sequence the mean powers of both components become equal. Two types of complex scrambling

[1]A popular method of scrambling performed in the transmitter is modulo-2 summing of a binary information sequence with a given pseudorandom sequence. The pseudorandom sequence is selected to ensure good statistical properties of the transmitted sequence. Repeating the same operation in the receiver allows recovery of the original binary information sequence. In our case the modulo-2 addition is replaced by multiplication of scrambled signals by a scrambling bipolar signal.

sequences are standardized. The short sequence is 256-bit long. It is repeated at the frequency of 15000 Hz. Such sequence is applied when joint detection receivers (see Chapter 10) are used in the base stations. The long sequence is a pair of Gold sequences of the period $2^{25} - 1$ truncated to a 10-ms interval and applied when a regular RAKE receiver is used in the base station [7].

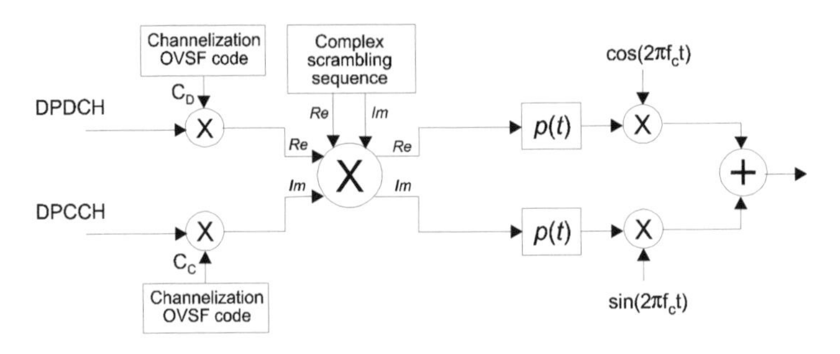

Figure 17.9 Generation of DPDCH and DPCCH signal in the uplink

The complex in-phase and quadrature pulse stream resulting from the complex scrambling operation is shaped by the filters with square-root raised cosine characteristics (roll-off factor $\alpha = 0.22$) and placed in the destination band by a pair of orthogonal modulators (see Figure 17.9).

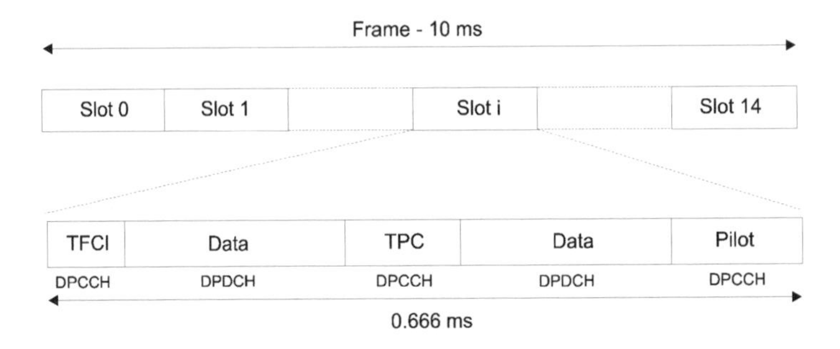

Figure 17.10 Frame structure in the downlink transmission of dedicated channels

The organization of the downlink transmission of dedicated channels is different from that applied in the uplink. The dedicated data and control channels are appropriately multiplexed, as shown in Figure 17.10. Next, they are serially demultiplexed into two parallel data streams which are the bases of the in-phase and quadrature components of the transmitted signal. The binary signals in each branch are spread using the same OVSF code word in both branches. The channelization code words used in the cell sector are selected from the same OVSF code tree. The spread signals in the in-phase and quadrature branch are treated as the real and imaginary parts of the complex signal, so such a signal is scrambled using a complex pseudorandom sequence which is 10 ms long. The complex pseudorandom sequence is created using two appropriately shifted truncated Gold sequences generated by the LFSRs of length 18. A mobile station that wishes to be synchronized with the base station has to find synchronism with this scrambling sequence; therefore, the number of different scrambling sequences is limited to 512. The sequences are divided into 16 groups with 32 sequences in each group. The scrambling sequences are assigned to a cell in the cell planning process. Figure 17.11 presents the process of generation of a WCDMA signal carrying dedicated channels in the downlink.

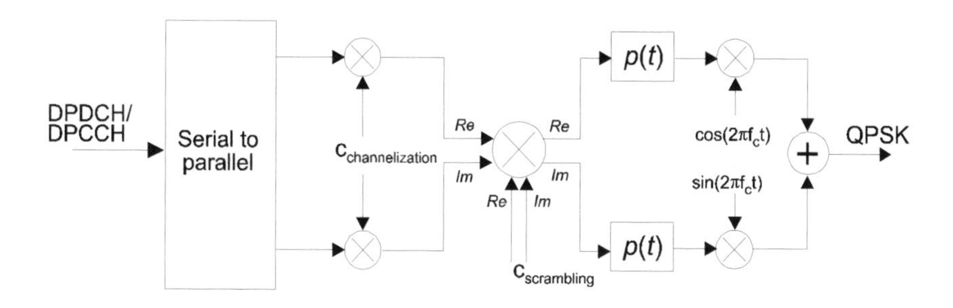

Figure 17.11 Generation of WCDMA signal in the downlink

Let us note that the downlink slot also transmits the pilot signal. The pilot signal ensures the coherent detection in a mobile station and makes it possible to use adaptive antennas in the downlink [7].

Besides dedicated channels, the following physical channels are also applied in the downlink:

- *Common Pilot Channel* (CPICH)
- *Synchronization Channel* (SCH)
- *Primary* and *Secondary Common Control Physical Channels* (P-CCPCH and S-CCPCH)
- *Acquisition Indication Channel* (AICH)
- *Paging Indicator Channel* (PICH).

In turn, in the uplink the *Physical Random Access Channel* (PRACH) and the *Physical Common Packet Channel* (PCPCH) are used. Let us briefly consider the above listed channels.

The *Common Pilot Channel* (CPICH) is transmitted by a base station in each cell to make the channel estimation possible. Therefore, the CPICH carries an unmodulated signal which is spread at the spreading factor equal to 256 and is scrambled with the cell-specific scrambling code. It has to be detected in the whole cell. Besides channel estimation, the CPICH channel is used in measurements in handover and cell selection operations.

The *Synchronization Channel* (SCH) is used in the cell search procedure performed by a mobile station. The SCH consists of two subchannels: the primary and secondary synchronization channel. The primary SCH uses a 256-chip long spreading sequence which is the same in all cells. The secondary SCH carries a sequence of fifteen 256-chip sequences (one sequence in each slot of a frame) which allows for frame and slot synchronization and determination of the group of scrambling codes used in the cell. Both synchronization channels are transmitted in parallel (using different sequences) within the first 256 bits of each slot. The remaining 2304 bits of each slot are occupied by the *Primary Common Control Physical Channel* (P-CCPCH).

The P-CCPCH carries the transport *Broadcast Channel* (BCH). It is transmitted continuously and has to be received by all mobile stations located in any place of the cell. Therefore, the spreading factor of the P-CCPCH channel is equal to 256 and the channel is transmitted with high power. A permanently allocated channelization code is also used. In order to ensure required transmission quality, the rate-1/2 convolutional code with interleaving over two frames (20 ms) is applied.

The *Secondary Common Control Physical Channel* (S-CCPCH) is used to transmit the following transport channels: *Forward Access Channel* (FACH) and *Paging Channel* (PCH). At least one S-CCPCH exists in a cell. If there is one S-CCPCH, it contains both transport channels. If there are more than one, both transport channels can use diferent physical channels. The spreading factor used in the S-CCPCH is fixed and determined by the maximum data rate applied. Again, the rate-1/2 convolutional code is applied in S-CCPCH. In the case of using FACH for data transmission, the rate-1/3 convolutional coding or turbo coding can be applied as well.

The *Physical Random Access Channel* (PRACH) and the *Acquisition Indication Channel* (AICH) are associated with each other, therefore we will describe them jointly.

The PRACH channel is used by a mobile station to access the network (by carrying the transport RACH channel). The random access is based on the slotted ALOHA principle. Two consecutive 10-ms frames are divided into 15 random access slots each 5120 chips long. First, the mobile station acquires timing and frame synchronization in the cell. Then, the mobile station detects the BCH channel in order to determine the random access slots available in a given cell, the scrambling codes and the signatures which can be used in the random access procedure. The mobile station also measures the received power and on the basis of this measurement it sets the power of the RACH preamble sent in the selected random access slot. The preamble is 4096 chip long and contains 256 repetitions of the selected signature. The mobile station periodically transmits the preamble in the available random access slots with gradually increased power till it detects the AICH preamble. When the AICH preamble is finally decoded, the 10 ms or 20 ms RACH message is transmitted.

A similar procedure is applied in data transmission using the *Physical Common Packet Channel* (PCPCH). The transport *Common Packet Channel* (CPCH) is realized using this physical channel. Again, a mobile station periodically sends 4096-chip preambles with gradually increasing power till it receives and detects the AICH preamble. Then it sends a CPCH CD (collision detection) preamble and after receiving the echo from the base station in the form of the *CD Indication Channel* (CDICH) it starts to transmit its packet. The message part lasts for $N \times 10$ ms, although it is restricted to a negotiated maximum length. Figure 17.12 presents the process of sending a packet using the PCPCH channel.

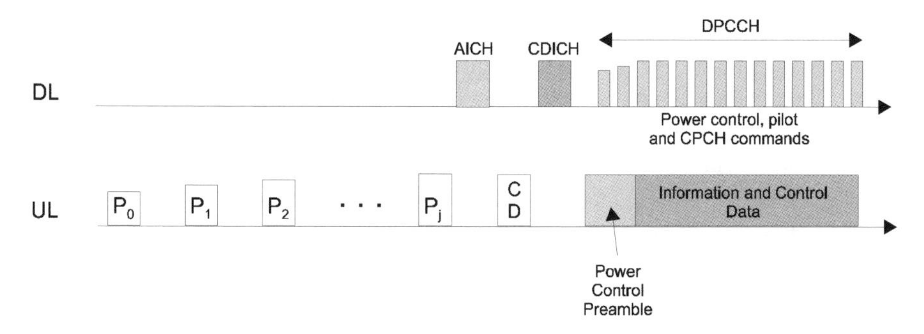

Figure 17.12 CPCH access procedure

One of the important procedures in the UTRA WCDMA is the *cell search*. After power on a mobile station has to find the closest base station. In the UMTS the base stations operate asynchronously and use different scrambling codes selected from the set of 512 sequences of 10 ms duration. In order to simplify and accelerate the cell search, the procedure is performed in the following steps:

- The mobile station searches for the primary synchronization channel (SCH) which is sent in the form of a 256-chip sequence common for all cells. This way the starting points of the slots are found. The search is performed using the filter matched to a known sequence. The mobile station receives the primary SCH signals from a few surrounding cells and selects the highest local maximum of the output signal of the matched filter, which corresponds to the closest (strongest) base station.

- For the detected starting points of the slots the mobile station attempts to acquire frame synchronization and tries to find a code group used in the secondary SCH. The synchronization is performed by the correlation of the received signal (the start of the correlation has been found in the previous step) with 64 possible secondary synchronization code words applied in the secondary SCH. All 15 possible slots have to be checked in order to find the slot No. 0, i.e. to find frame synchronization. The secondary synchronization code word determines a particular code group to which the scrambling code in a given cell is assigned.

- The scrambling code applied in the cell is found in the third step. For this purpose the mobile station correlates the received signal with all possible scrambling signals belonging to the given code group. After identification of the scrambling sequence, the mobile station is able to read the primary common control channel which holds system parameters sent on the broadcast channel (BCH).

Once a mobile station has been registered in the network it is allocated a paging group. The mobile station periodically reads the *Paging Indicator Channel* (PICH), looking for its paging indicator stating that in the secondary common control physical channel a paging message for the mobile station belonging to a given paging group is contained. If the mobile station has detected a paging indicator of its own paging group, it reads the PCH frame contained in the S-CCPCH in order to check if there is a paging message for it. The PICH channel uses a fixed data rate with the spreading factor $SF = 256$. Out of 300 bits transmitted within a 10-ms frame, 288 bits are used for paging indicators. There may be 18, 36, 72 or 144 different paging indicators, which are transmitted as the sequences of ones (if paging is indicated) or zeros.

As we know, power control has a crucial impact on the overall CDMA system capacity. In WCDMA FDD the open and closed loop power control is applied. The open loop power control is used in the RACH and CPCH transmission, as was described above. The accuracy of such an open control loop is not very high, mostly due to internal inaccuracies of a mobile station and due to measuring the signal arriving at the mobile station in a different band than the band of the signal whose level is controlled.

The closed loop power control is performed in each slot, so it is done 15000 times per second. As we remember, in the uplink dedicated physical control channel (DPCCH) block the transmit power control (TPC) field is placed in each slot, which indicates the change of power level by 1 dB or its multiple. In the case of soft handover, when the mobile station is assigned to two neighboring base stations, both of them send commands to the mobile station concerning the power control. In the mobile station the commands are appropriately weighted to work out a final decision on the power control.

The next procedure necessary for proper functioning of a cellular system is *handover*. In WCDMA there are several types handover:

- soft, softer and hard handover
- interfrequency handover
- handover between FDD and TDD modes
- handover between WCDMA and GSM.

In typical situations soft handover is performed, because cells usually operate on the same carrier frequency. A mobile station measures the power level of the common pilot channel (CPICH) and relative timing between cells. The mobile station entering the handover state receives the signal from all base stations participating in the connection. This is possible due to the application of the RAKE receiver whose "fingers" are synchronized with the scrambling and spreading sequences used in the current cell and the cell which will probably overtake the connection. The mobile station signals are received in the base stations participating in the handover procedure and are appropriately combined to perform *macrodiversity*. During the connection, the mobile station searches for a new base station from the list read from the BCH channel using a cell-search algorithm. Considering a new base station as a handover candidate implies sending by the mobile station a request to this base station to adjust the timing offset of the dedicated physical data and physical control channels (DPDCH and DPCCH) with respect to its primary common control physical channel (CCPCH). This way the frame timing differences between signals sent by different base stations and received at the mobile station can be minimized and the mobile station can receive the signal from the new base station as well.

Besides a soft handover, in WCDMA the so-called *softer handover* also occurs. In this state a mobile station is connected to two neighboring cell sectors served by the same base station, so the signals received in both sectors are already combined within the same the base station.

The main reason for *interfrequency handover* is the movement of a mobile station from one cell to another if both belong to different layers of the hierarchical cell structure e.g. pico-, micro- or a macrocell. A certain problem encountered in this case is performing appropriate measurements on other carrier frequencies than the one which is currently in use. There are two possible solutions to this problem. If a mobile station is able to apply space diversity, it uses a *dual receiver*. Then one receiver gets the signal from the current channel, whereas the second one is able to measure a signal on a new carrier frequency. To compensate the lost gain which in the normal case would be achieved due to the diversity reception, the signal sent on the current channel has to be additionally amplified by the base station. This is possible thanks to the closed loop power control.

The second solution of the measurement problem at the interfrequency handover is the application of the *compressed mode*, which is useful if a mobile station does not use a dual receiver. With a certain periodicity, the base station which normally sends its frames in 10-ms intervals transmits the frame contents in a shorter interval, i.e. in 5 ms, leaving the rest of the frame time for measurements performed by the mobile station at different frequencies. The shortening of the transmission time is performed by the application of code puncturing and changing the FEC rate. The signal power must be increased to compensate the loss of power following the application of the former two means. Figure 17.13 shows the principle of the compressed mode.

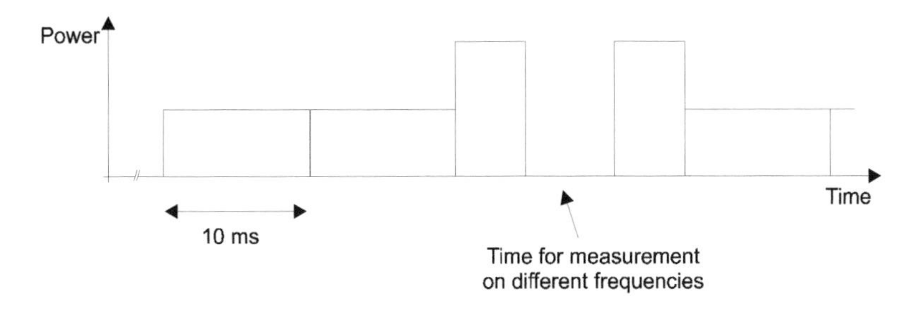

Figure 17.13 Principle of the compressed mode

Handover between FDD and TDD modes is possible if a mobile station can operate in a dual mode. A mobile station measures the power level of TDD cells, using the common control physical channels generated twice within a 10-ms frame of the TDD base stations. A similar situation occurs if a mobile station is able to operate in GSM - UTRA modes. This case is called *inter-system handover*. In the measurement periods a mobile station attempts to find a GSM frequency correction channel followed by the synchronization channel.

17.4.2 UTRA TDD mode

Let us consider the UTRA TDD mode. Its main parameters are presented in Table 17.3. Let us note that a lot of them are common for both FDD and TDD modes, which leads to strong similarities of both transmission types and simplification of mobile stations. As we remember, a part of the spectrum allocated to the UMTS is unpaired. In this part of the spectrum the FDD operation is excluded, so the TDD mode has been applied. Thanks to the time division duplex, time can be asymmetrically divided between two transmission directions. This division can be dynamically adjusted to the current kind of traffic. Multirate transmission can be easily implemented. Channel reciprocity is also an interesting feature of TDD transmission. Due to the fact that the same channel spectrum is used in both transmission directions, the measurements performed in one direction can be utilized in the opposite one, unless the channel is a fast time varying one.

Figure 17.14 presents the frame in the UTRA TDD mode. It lasts for 10 ms and is divided, as in the FDD mode, into 15 time slots. The physical channel is determined by the carrier frequency, the time slot within a frame and the applied spreading code. Several data rates can be achieved through allocation of an appropriate number of physical channels to the connection. This rule is also shown in Figure 17.14.

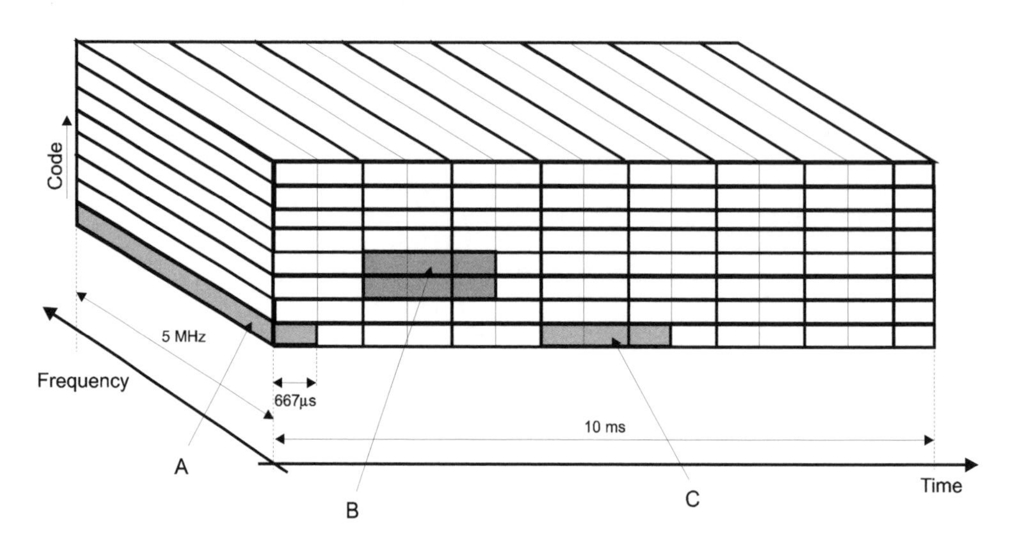

Figure 17.14 Example of the resource allocation for connections with different rates: A – a single channel (1 time slot +1 spreading code word), B – a connection using 3 time slots and 2 spreading code words, C – a connection applying 3 time slots and a single spreading code word.

The spreading codes used in the same slot are mutually orthogonal. They are selected from the OVSF code family. Particular channels are mutually synchronized. The frame consisting of 15 time slots is divided into two transmission directions. Several arrangements are possible [14]. The slot allocation can be symmetric with multiple switching between uplink and downlink within a frame. It can be asymmetric with multiple switching, or symmetric/asymmetric with a single switching within a frame. The main point is that at least one slot has to be allocated for uplink and at least one slot has to be allocated in the downlink in a 15-slot frame.

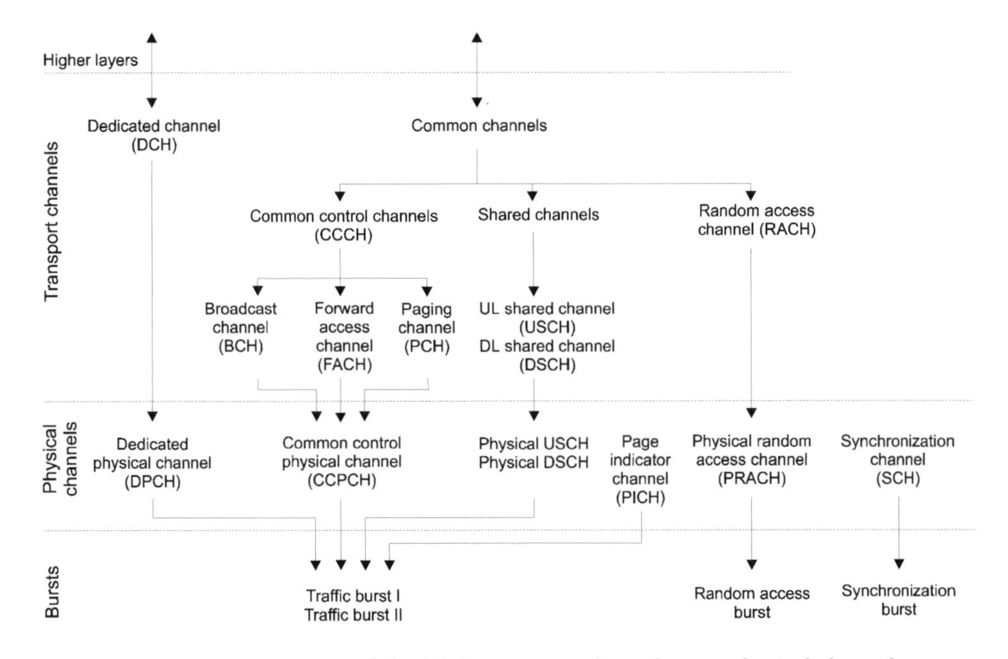

Figure 17.15 Mapping of the TDD transport channels onto physical channels

Transport and physical channels used in the UTRA TDD mode are similar to those applied in the FDD mode. The mapping of the transport channels onto physical channels is presented in Figure 17.15.

The length of each time slot is equivalent to 2560 chips. Transmission of physical channels is performed in the form of bursts. There are basically three types of bursts differing in the internal structure and the guard time length. Each burst contains two data fields, a midamble, and is ended with a guard period. Traffic bursts can also contain the *Transport Format Combination Indicator* (TFCI) and the *Transmission Power Control* (TPC) bit placed on both sides of the midamble. The transmission of TFCI and TPC is negotiated at the call set-up. The TPC field of a particular user is transmitted only once per frame. Figure 17.16 shows the structures of traffic and random access bursts.

Traffic burst type I is used in the uplink. Because of the application of a long midamble, up to 16 channel impulse responses can be estimated. The traffic burst type

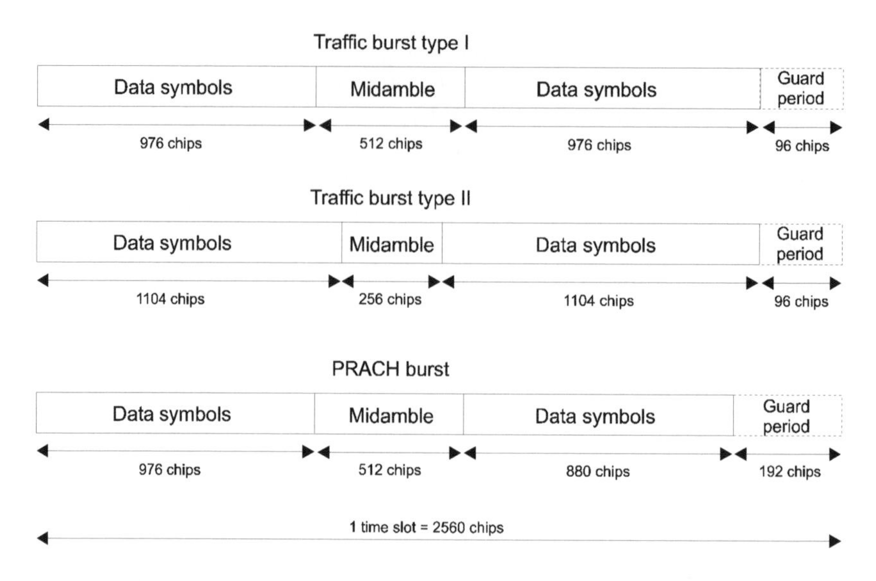

Figure 17.16 Traffic and random access bursts in TDD mode

II is basically used in the downlink and can be also used in the uplink if less than four users are allocated to the same time slot. The midambles used in the same cell are the cyclic shifted versions of the same basic code sequence. Different cells apply different basic code sequences.

Transmitted data is spread in the data symbol fields. Spreading is a two-step process performed by a channelization code and a complex scrambling code, in the same way as in the UTRA FDD mode. Dedicated physical channels use a spreading factor $SF = 16$ in the downlink. As we have mentioned, more than one physical channel using a different channelization code can be assigned to a high rate data link. In case of a single code used in a downlink physical channel, the spreading factor can be equal to 1. In the uplink, dedicated physical channels apply a spreading factor of the value between 1 and 16. Maximum two physical channels can be used by a mobile station per slot to increase the data rate. Simple calculations for the burst type II indicate that transmission with $SF = 16$ using a single code and a single time slot results in the 13.8 kbit/s raw data rate. On the other end, the application of 16 codes and 13 time slots gives the data rate equal to 2.87 Mbit/s. The same result would be achieved if instead of 16 codes with $SF = 16$, one code with $SF = 1$ were applied.

As in the FDD mode, in UTRA TDD the *Primary* and *Secondary Common Control Physical Channels* (P-CCPCH and S-CCPCH) are used. The P-CCPCH carries the BCH transport channel. For that purpose bursts type I with fixed spreading with $SF = 16$ are applied. The *Paging* (PCH) and *Fast Access Channel* (FACH) are carried by the S-CCPCH. Both types of bursts can be applied; however, the spreading factor remains fixed and is equal to 16.

The position of the P-CCPCH within a frame, i.e. the time slot number and the used spreading code, is written in the message carried by the *Synchronization Channel*

(SCH). A synchronization burst is shown in Figure 17.17. One or two synchronization bursts can be placed in a frame. In the first case the synchronization burst is emitted in the same slot of a frame as the P-CCPCH. Every time slot can be selected for this purpose. If two synchronization bursts are applied, then the *Synchronization Channel* (SCH) is allocated in the kth ($k = 0, 1, \dots, 6$) and $(k+8)$th slots of the frame, whereas the P-CCPCH is located in the kth time slot. Figure 17.17 illustrates the second case when $k = 0$. The SCH consists of a primary sequence and three secondary sequences, each of them 256-chip long. The sequences start at a specified time offset which is selected from 32 possible values. This prevents the capturing effect which would otherwise occur due to the mutual synchronization of base stations.

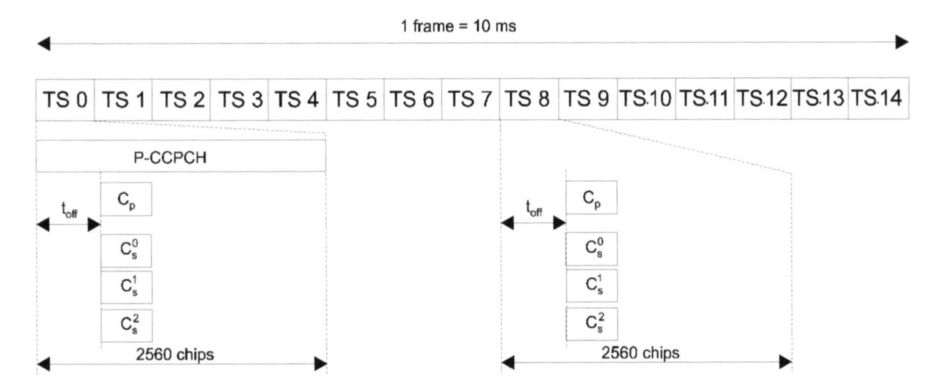

Figure 17.17 Position of the SCH in the TDD frame for $k = 0$ and two SCH bursts in the frame

A mobile station sends a request for access to the channel using the *Physical Random Access Channel* (PRACH). The request is realized by sending a random access burst, shown in Figure 17.16. Let us note that a longer guard time is applied in the burst. It allows operation at the time propagation differences resulting from the distance differences of the order of 7.5 km. Let us also note that a regular guard time applied in traffic bursts is equivalent to the duration of 96 chips which implies differences in propagation time equal to 25 μs. This in turn allows design of cells of the radius of 3.75 km, without applying the timing advance procedure.

A mobile station is paged by the message transmitted on the secondary CCPCH; however, first the *Page Indicator Channel* (PICH) has to be used, which is realized by replacing the S-CCPCH and carrying page indicators for appropriate groups of mobile stations. Generally, the paging mechanism is similar to that described for UTRA FDD and will not be described here.

Physical Uplink and *Downlink Shared Channels* (PUSCH and PDSCH) are used for setting and transmitting user-specific parameters such as power control, timing advance or directive antenna settings.

Due to the nature of the TDD mode, hard handover is applied. The network supplies a mobile station with a list of neighboring base stations whose strength should be measured. The mobile station performs the measurements in idle time slots. Similar handover types as in UTRA FDD occur in UTRA TDD. They are: TDD–TDD, TDD–FDD, WCDMA-TDD–GSM handovers.

In the design phase of the UMTS TDD system, several kinds of interference have to be taken into account. Detailed interference analysis [1] allows us to draw the following conclusions:

- TDD base stations managed by a single operator have to preserve frame synchronization; the frame synchronization between base stations managed by different operators is also desired.
- Allocation of asymmetric uplink and downlink traffic in the cell is not fully free, strong interference can potentially arise between transmission directions.
- *Dynamic channel allocation* (DCA) is a powerful tool used to avoid interference in the TDD band. Another possibility is the use of inter-frequency and inter-system handover.
- Special attention has to be paid to mutual influence of FDD and TDD systems, in particular those operating in lower TDD band and FDD uplink band.

Interference among users implies application of advanced receiver structures both in base and mobile stations. In base stations joint detection receivers (see Chapter 10) can be applied. It is a feasible solution due to the fact that the number of simultaneous users in a time slot is relatively low, so the computational complexity of such receivers is still acceptable, in particular when suboptimal solutions are used. In mobile stations, it is possible to apply single user detectors which combine adaptive intersymbol interference equalization with cancellation of multiple access interference (MAI).

* * *

TDMA transmission applied in the UTRA TDD mode has serious consequences for the cell coverage, since discontinuous transmission causes power reduction. Generally, in order to provide the same coverage area as in the FDD mode, more TDD base stations are needed. Therefore, the UMTS in the TDD mode can be applied as a system complementary to the FDD, especially for data transmission and asymmetrical links.

17.5 CDMA2000

As we have already mentioned, the 3rd Generation Partnership Project 2 (3GGP2) developed the air interface called *cdma2000* which has evolved from the second generation CDMA IS-95 air interface popular in America and South Korea.

Cdma2000 is one of the most important proposals for an IMT-2000 air interface. The first phase of cdma2000 called *cdma2000 1x* is an extension of the existing IS-95B standard. It allows doubling of the system capacity and increased data rates up to

614 kbit/s. The second phase called *cdma2000 1xEV* (Evolution) is a further enhancement of cdma2000 1x. It includes *High Data Rate* (HDR) technology providing data rates up to 2.4 Mbit/s. Initially, a dedicated carrier is devoted to high-speed packet data, whereas one or more additional carriers are used to realize voice connections. In later development of the 1xEV, packet data and voice transmission will be combined in the same carrier; however, packet services on a separate carrier can be possible [17]. Finally, in the third phase of cdma2000 development, called *cdma2000 3x*, three independent non-ovelapping CDMA channels are used, retaining backward compatibility with cdma2000 1x and IS-95B. Tripling the bandwidth and giving some additional freedom results in service enhancement and data rates up to 2 Mbit/s. Figure 17.18 shows the spectrum arrangement for cdma2000 1x and 3x. The cdma2000 is designed to operate in the following environments [4]:

- outdoor megacells (cell radius >35 km),
- outdoor macrocells (cell radius 1 - 35 km),
- indoor/outdoor microcells (cell radius $\leq$ 1 km),
- indoor/outdoor picocells (cell radius <50 m),
- wireless local loops.

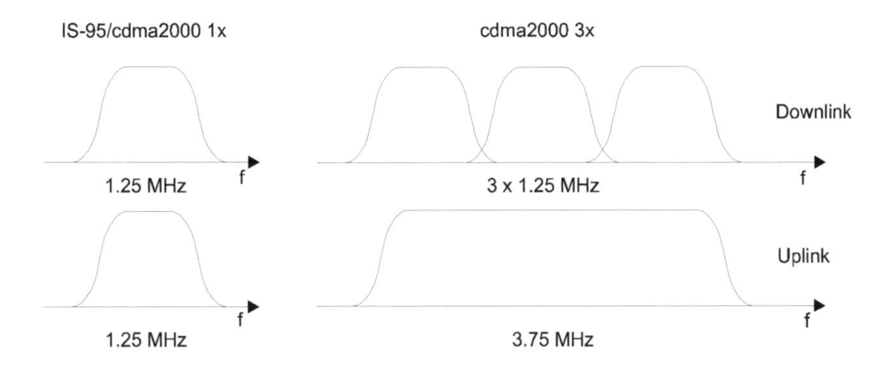

Figure 17.18 Spectrum arrangements for cdma2000 1x and 3x in downlink and uplink

Specific solutions for 3G systems in the USA partially result from the fact that no additional spectrum has been allocated to these type of systems. The PCS band (see Figure 17.2) currently partially used by cdmaOne (IS-95 based systems) and IS-136, has to be gradually reused by new 3G systems. Therefore backward compatibility is a highly desired feature of the 3G systems.

Let us concentrate on the basic features of cdma2000. Table 17.4, based on [3], presents the basic parameters of this system.

In cdma2000, as in IS-95, the downlink and uplink channels are called *forward* and *reverse* channels, respectively. Figure 17.19 presents the list of dedicated and common physical channels used in cdma2000.

Table 17.4 Basic parameters of cdma2000[3]

	cdma2000 1x	**cdma2000 3x**
Channel bandwidth	1.25 MHz	3×1.25 MHz
DL RF channel structure	Direct Spread (DS)	Multicarrier, DS on each carrier
UL RF channel structure	Direct Spread (DS)	Direct Spread (DS)
Chip rate	1.2288 Mchip/s (DL) 1.2288 Mchip/s (UL)	1.2288 Mchip/s per carrier (DL) 3.6864 Mchip/s (UL)
Frame length	20 ms / 5 ms option for signaling bursts	
Timing	Synchronous, derived from GPS	
Channel coding	Convolutional, turbo or no coding	
Modulation	QPSK	
Detection	UL: Coherent, pilot sequence multiplexed with power control bits DL: Coherent, common continuous pilot channel and auxiliary pilot	
Spreading factors	4 - 256	
Spreading in downlink	Variable length orthogonal Walsh sequences (channelization) m-sequence of length 2^{15}(the phase shift determines the cell)	
Spreading in uplink	Variable length orthogonal Walsh sequences (channelization) m-sequence of length 2^{41}-1 (the phase shift determines the user)	
Multirate	Variable spreading and multicode	
Handover	Soft handover, interfrequency handover	
Power control	Open loop and fast closed loop (800 Hz)	

As we see in Figure 17.19, the list of physical channels is very long. The functioning of many channels is very similar to the functioning of the physical channels in the previously described IS-95 system; therefore, we will consider them very briefly. Let us start with the forward link (equivalent to downlink in the UMTS).

The *Forward Pilot Channel* (F-PICH) is used by a mobile station to estimate the channel impulse response necessary in RAKE reception. It is also needed in cell acquisition and handover. It consists of a sequence of logical zeros, spread by the Walsh function No. 0. The actual pilot sequence is received due to the complex scrambling code determining the cell (see Table 17.4 and Chapter 11). The forward pilot signal is common for all mobiles in the cell, so the overhead due to it is not very significant.

The *Forward Sync Channel* (F-SYNC) is used by mobile stations to acquire system synchronization. There are two possible types of a synchronization channel [4]: the *shared* F-SYNC functioning in the IS-95B and cdma2000, which operates in the same area and *wideband* F-SYNC using the entire channel bandwidth and which can be applied in overlay (IS-95B and cdma2000) and non-overlay systems.

The *Forward Paging Channel* (F-PCH) is used to page mobile stations located in a cell and to send them several control messages to them, such as channel assignment, acknowledgments, etc. There may be more than one paging channel in a cell. The applied data rate is 9.6 or 4.8 kbit/s. The data is first convolutionally encoded ($R = 1/2$, $k = 9$), then repetition (if the input data rate is 4.8 kbit/s) and block interlaving are applied. The received signal is modulo-2 summed with the decimated long code which is characteristic for the paging channel. Finally, the signal is spread using an appropriate

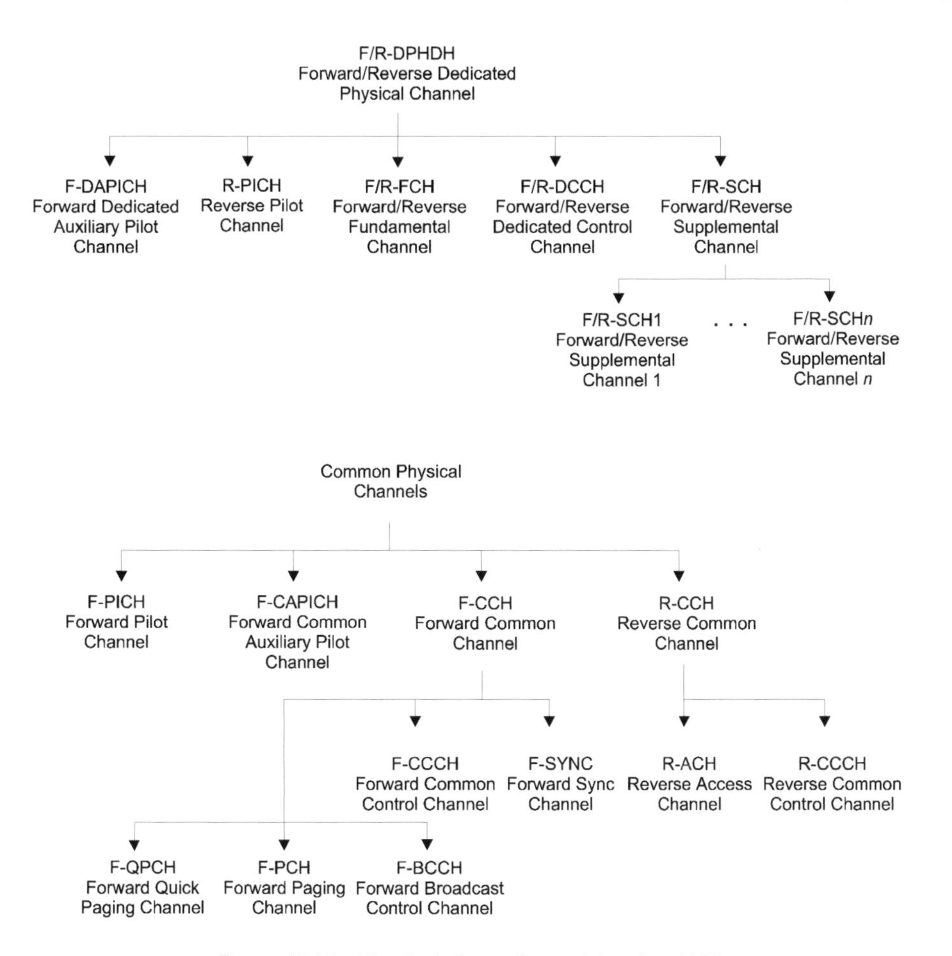

Figure 17.19 Physical channels used in cdma2000

Walsh function and is further processed by the output section. Shared and wideband paging channels are possible, similar to the sync channel.

The *Forward Common Control Channel* (F-CCCH) is used to carry MAC and network layer messages to mobile stations.

The *Forward Common Auxiliary Pilot Channel* (F-CAPICH) is applied to generate spot beams, using adaptive antennas. The F-CAPICH is used by mobile stations located in the generated spot beam. Similarly, the optional *Forward Dedicated Auxiliary Pilot Channel* is applied to target a particular mobile station.

The *Forward Broadcast Common Channel* (F-BCCH) is a type of paging channel which transmits overhead messages and SMS broadcast messages, so the paging channel does not have to transmit them. The number of the Walsh function used by the F-BCCH channel is transmitted on the sync channel.

The *Forward Quick Paging Channel* (F-QPCH) is used to page mobile stations operating in the slotted mode.

The *Forward Fundamental Channel* (F-FCH) is used to carry the downlink traffic. The 20 ms frame and the variable data rates are chosen from the data rate sets known from IS-95B: RS1 (1.5, 2.7, 4.8 and 9.6 kbit/s) and RS2 (1.8, 3.6, 7.2 and 14.4 kbit/s) (see Chapter 11). The existence of many configurations of coding, repetition and block interleaving results in a great number of available data rates (see [4] or cdma2000 standards [18]-[21] for details). All configurations yield the same number of 384 bits in a 20 ms frame, which is equivalent to 19.2 kbit/s for the RS1 rate set or 768 bits in a 20 ms frame equivalent to 38.4 kbit/s for RS2 rate set.

The *Forward Supplemental Channel* (F-SCH) is used to carry user information jointly with the fundamental channel at higher rates than can be achieved using the fundamental channel only. Convolutional coding is applied for lower data rates, whereas turbo coding is applied for higher data rates. More than one F-SCH can be allocated to the link at the same time. Several supplemental channels can have different requirements on error rates depending on applications. A wide range of data rates can be achieved, starting from 9.6 kbit/s and ending with 921.6 kbit/s for the RS2 data set applied in the multicarrier cdma2000 configuration. Is is worth noting that such data rates are practically achieved using multicode and multicarrier channel assignment. In transmission on supplemental channels the channelization Walsh functions can have different length (different spreading factor) depending on the input data rate. They are selected in such way that the bandwidth remains constant after spreading.

The *Forward Dedicated Control Channel* (F-DCCH) transmits point-to-point control data at the data rate of 9.6 kbit/s.

As an example let us consider the block chain for data transmitted on the forward supplemental channel with the input data selected from the data rate set RS2 [4] when $N = 3$ carriers as in cdma2000 3x are used (Figure 17.20).

The data block to be transmitted in a 20 ms frame is first appended with 16 CRC bits, then the encoder tail and reserve bits are added. Next the data block is convolutionally encoded by the code of the constraint length $k = 9$ and the coding rate $R = 1/4$. The resulting data are block interleaved and modulo-2 summed with the decimated output of the long code generator with the mask characteristic for the nth user. The data stream is demultiplexed into three branches generating CDMA signals on three carriers (f_1, f_2, and f_3). The binary streams are mapped into the in-phase and quadrature components and changed from the binary into bipolar form. Subsequently, both components are spread using the Walsh functions operating as the channelization codes and scrambled by the 2^{15}-long PN complex sequence. The phase of this sequence determines the cell. The in-phase and quadrature outputs of the scrambling process are spectrally shaped by the baseband filters and modulated using the appropriate carrier frequency. Let us note that besides the signals shown in Figure 17.20 other channels such as pilot, paging, fundamental channel, etc. are also transmitted.

Let us turn our attention to the uplink transmission. We will shortly describe the operation of most of the reverse physical channels shown in Figure 17.19. Reverse physical channels can be divided into dedicated channels which are a means of communication

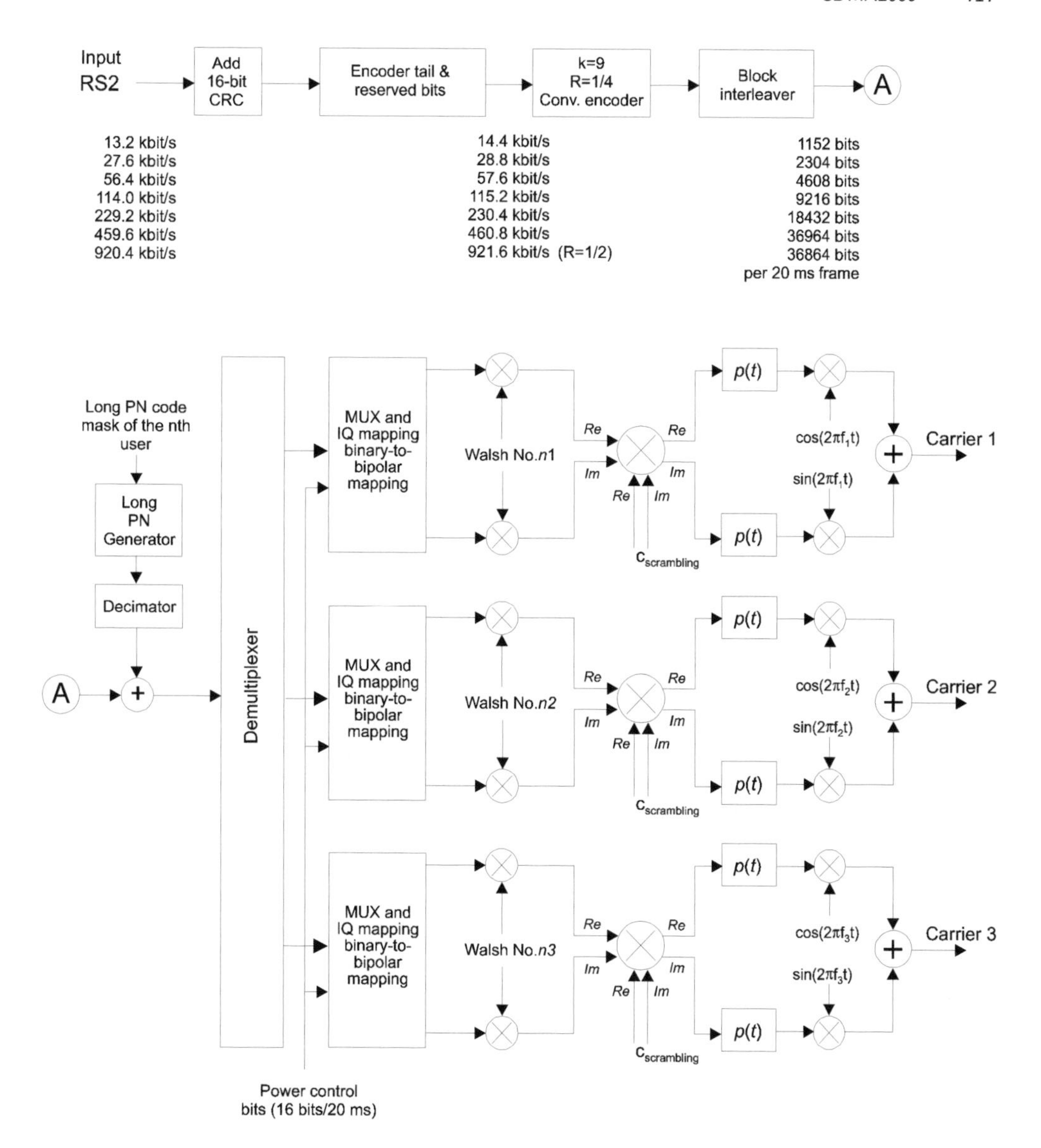

Figure 17.20 Transmission of data at the RS2 rates on the forward supplemental channel with multicarrier CDMA (cdma2000 3x)

between a particular mobile station and the base station, and common channels which are used to send information from multiple mobiles stations to the base station.

The *Reverse Access Channel* (R-ACH) is a multiple channel used by mobile stations to obtain access to the system resources. The slotted ALOHA principle is used in the R-ACH transmission. Let us note that due to the CDMA multiple access, more users can simultaneously attempt to get the access to the medium. More than one R-ACH channel can be used on one carrier frequency. The channels are then differentiated by the applied PN codes.

The *Reverse Common Control Channel* (R-CCCH) is applied to transmit MAC and network layer messages from a mobile station to the base station. The R-CCCH offers extended capabilities as compared with the R-ACH, which enable faster access in packet data transmission.

The *Reverse Pilot Channel* (R-PICH) consists of the pilot sequence originating from setting a fixed value at the channel input, multiplexed with the power control bits which are used in the closed loop power control. The R-PICH is applied in the base station for initial acquisition, time tracking, channel estimation and sequence synchronization used by the RAKE receiver and power control measurements [4].

The *Reverse Dedicated Control Channel* (R-DCCH) is associated with individual transmission from a mobile station to the base station.

The *Reverse Fundamental Channel* (R-FCH) is used to transmit user data. The data rates applied in R-FCH depend on the rate sets. They are equal to 1.5, 2.7, 4.8 and 9.6 kbit/s for RS3 and RS5 rate sets and are equal to 1.8, 3.6, 7.2 and 14.4 kbit/s for the rate sets denoted as RS4 and RS6.

The *Reverse Supplemental Channel* (R-SCH) is an additional channel to carry user data. It can be applied in two modes. In the first one the data rate does not exceed 14.4 kbit/s and the base station has to detect the real data rate without explicit information sent from the mobile station. In the second mode, higher data rates are possible but the data rate is known in advance.

The configuration of the fundamental, supplemental, pilot and dedicated channels differs from an analogous configuration in the downlink. The properly encoded and interleaved data streams sent on the fundamental, dedicated control and supplemental channels are assigned to the in-phase or quadrature components after application of the Walsh channelization codes. They are subsequently scrambled by the PN complex sequence modified by the user-specific long code. Finally, after pulse shaping, the in-phase and quadrature components are placed in the destination band by a pair of orthogonal modulators. Let us stress that the spreading performed by the channelization functions depends on the input data rates and the Walsh functions of different lengths can be used. This is a major difference between cdma2000 and IS-95B. Figure 17.21 presents the channel assignment to appropriate signal components, applied in the uplink (reverse) direction.

Several typical procedures, which have been previously described for UMTS, have to be performed in cdma2000 as well. The most important are: power control, handover, cell search and random access procedures.

Open loop and closed loop power control are applied in cdma2000. Typically, the level of transmitted signal is set on the basis of the measured level of the received signal. In cdma2000, as well as in the UMTS, the FDD mode is applied, so the accuracy of the open loop is limited. In the cdma2000 closed loop power control, the commands dealing with the power level modifications are transmitted in both directions 800 times per second. As a result, medium and fast fading can be compensated.

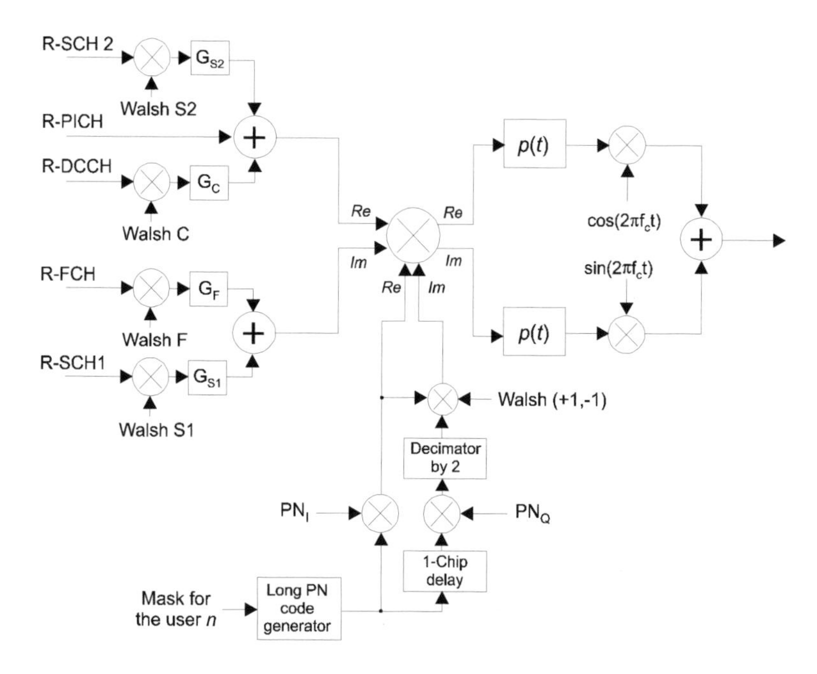

Figure 17.21 Assignment of pilot, dedicated control, fundamental and supplemental channels in the reverse link of cdma2000 (Walsh i denotes the Walsh channelizing function specific for the given type of channel)

The cell search procedure is performed similar to IS-95, because all the cells use the same complex PN sequence; however, in each cell the PN sequence is applied with a cell-specific phase shift. A mobile station searches for the pilot channel transmitting this PN sequence with appropriate phase shift selected from a finite number of shifts.

Cdma2000 applies the soft handover procedure. Intersystem handover between cdma2000 and IS-95 is also supported. The fundamental and supplemental channels are treated differently in the handover procedure. Generally, the number of base stations transmitting the supplemental channel to a mobile station in the handover phase is a subset of the base stations transmitting the fundamental channel [3]. The soft handover procedure tends to use as small a number of base stations as possible to minimize interference and maximize the system capacity.

* * *

In this section we presented a general overview of the physical layer of the cdma2000. The reader interested in the details of the cdma2000 operation is asked to study cdma2000 specifications [18]–[21].

17.6 APPENDIX - THE SOFTWARE RADIO CONCEPT

17.6.1 Introduction

As we noticed studying the previous chapters, the physical layers of the second and third generation systems have not been unified. Several incompatible air interfaces are applied. To obtain global roaming, a mobile station should be able to function with all types of air interfaces. This can be achieved by application of multimode terminals with duplicated or triplicated RF and DSP blocks. Another solution is the application of *Software Radio.*

In the concept of the *Software Radio* [22], [23], [24], transmitters and receivers of the base stations and mobile stations, implemented in a specialized hardware according to a particular standard, are replaced by a universal system. In this system the RF part is drawn to the minimum and the remaining parts consist of wideband A/D and D/A converters and a DSP processor which realizes the transmit/receive functions in software. The advantage of such a solution is not only the possibility to realize several standards of the air interface but also its universality which enables use of the applied system in the longer term and allows for gradual modifications reflecting the evolution of the standards. A short time from the start of the design to the development of a new product is another advantage. It allows for a faster response to the market needs.

The concept of transmit/receive blocks realized in software can be extended to the concept of *Mobile Software Telecommunications* [25]. The authors of this concept expect that many different standards will coexist and none of them will be able to realize all multimedia services. It is assumed that terminals will be intelligent, i.e. they will be software reconfigurable and will be able to communicate with different networks. Terminals will be reconfigurable according to different standards using a certain minimal software. It is expected that new JAVA-like programming languages will be developed which will allow for definition of new standards and services. Such a programming language will constitute a platform for realization of a communication session by defining the features of a physical connection and realized services. Therefore, at the connection set-up, its details will be defined such as: the speech encoder/decoder, applied modulation, data encryption, the features of realized services, associated protocols and the bandwidth requirements. Because most of the elements necessary to realize the communication session will be implemented in software, only a small part of the physical link realized in hardware needs to be standardized.

17.6.2 Minimum radio standard

In order to realize a communication session it is necessary to define a minimum radio interface which will be used to establish a basic connection and to set the remaining elements of the realized session and services. This interface has to enable network access and mobility management associated with the location of mobile terminals and their paging. In association with these tasks the *Network Access and Connectivity Channel* (NACCH) is defined. Figure 17.22 shows a possible placement of the NACCH modem in the mobile terminal architecture [25].

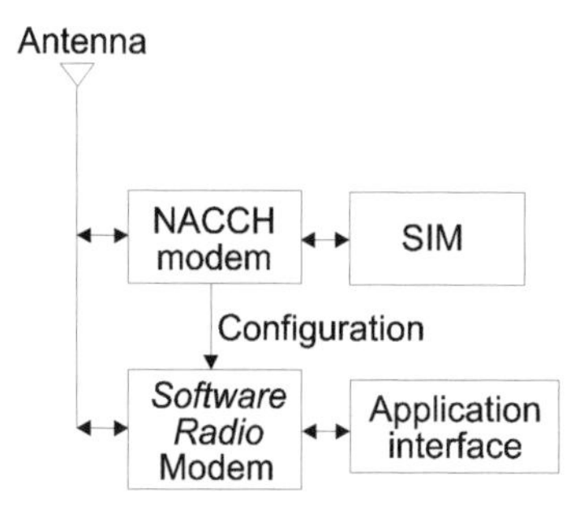

Figure 17.22 The concept of NACCH modem in the architecture of mobile terminal using *Software Radio* technology

Besides the mobility management, the NACCH modem should perform authentication and user registration together with establishing the traffic channels and demanded services. In order to do this, the NACCH modem should set the *Software Radio* modem configuration and program the traffic channels and services. Let us note that the terminal modem consists of two main parts – the NACCH modem and *Software Radio* (SR) modem. In the idle mode, only the NACCH modem is active. The above concept implies sending software subroutines necessary to program the SR modem by the air interface.

Several technical problems would arise in the implementation of this approach [28]. Transmission of the software subroutines by air must be practically error-free, otherwise the SR terminal will not be able to operate reliably. It cannot last too long, otherwise users will be disappointed. Therefore, transmission of the software subroutines by air requires an advance reliable procedure in which the software integrity is checked. On the other hand, this approach is more flexible than the second one – loading the SR software from a smart card.

Using a smart card storing the software enables reliable and fast programming of the SR modem; however, part of the flexibility is lost. This approach requires a whole network of distribution points in which new smart cards can be purchased. The intelligent card technology has to be further developed to enable a huge amount of SR software to be stored.

Another problem associated with the SR technology is the form of the software stored in network data bases or in the memory of a smart card. In order to enable competition among terminal vendors, the software cannot be hardware-dependent. It should be run on several DSP platforms. Therefore, it should have the form of a higher-level language instructions which would be compiled in the SR terminal. In consequence, an SR terminal should contain a compiler which would translate the higher-level language instruction sequence into a hardware-dependent program.

17.6.3 Basic elements of Software Radio architecture

Let us consider the basic scheme of a receiver applied in a mobile communication system (see Figure 17.23) [23].

The signal received in the antenna is filtered in a bandpass (BPF) filter, which extracts the band of a whole system. Subsequently, the signal is amplified in a low noise amplifier (LNA) and demodulated to the intermediate frequency band, using a programmable frequency synthesizer, a mixer and a bandpass filter. The number of intermediate steps is greater than one. After amplification in the automatic gain control (AGC) circuit, the signal is demodulated and shifted to the baseband, where the in-phase and quadrature components are derived. Both component signals are sampled and converted to the digital form. Subsequent processing is typically performed using an ASIC[2] DSP block or a DSP processor.

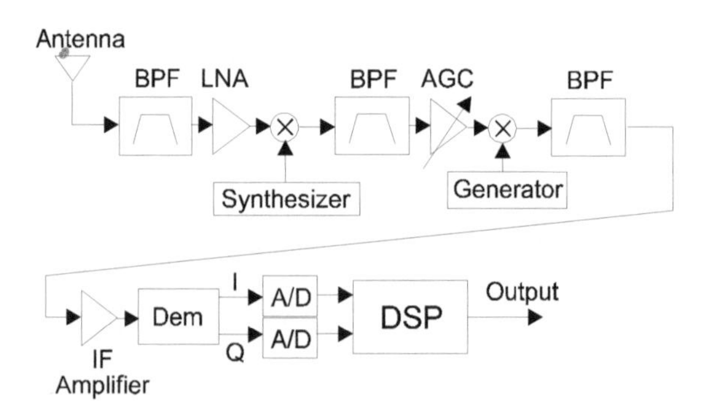

Figure 17.23 Architecture of a typical receiver of a mobile communication system

As we see, the receiver consists of many elements of which a large part is implemented in the analog form. Such a receiver is not a universal solution and is usually fitted to a single air interface. Instead, we can consider the scheme of an idealized receiver in which most of the functions are realized in software. Such a scheme is shown in Figure 17.24.

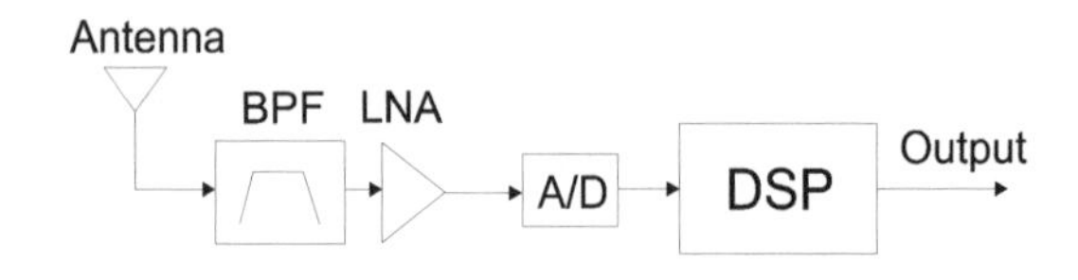

Figure 17.24 Scheme of the idealized software receiver

The analog-to-digital conversion is made in the radio frequency range and the remaining part of the receiver is implemented by a DSP block. This scheme is fully universal and well fitted to the *Software Radio* concept, although at the current state of the D/A technology it is not realizable at the required D/A conversion accuracy. However, it is possible to implement a digital programmable receiver with hardware support of

[2]ASIC – Application Specific Integrated Circuit

down-conversion. The *Programmable Down-Converter* (PDC) realizes down-conversion from the intermediate frequency to the baseband and performs additional filtration in the baseband. Figure 17.25 presents the scheme of the receiver with the PDC. We expect that fast progress in VLSI and A/D and D/A technologies will soon enable the realization of the SR terminal.

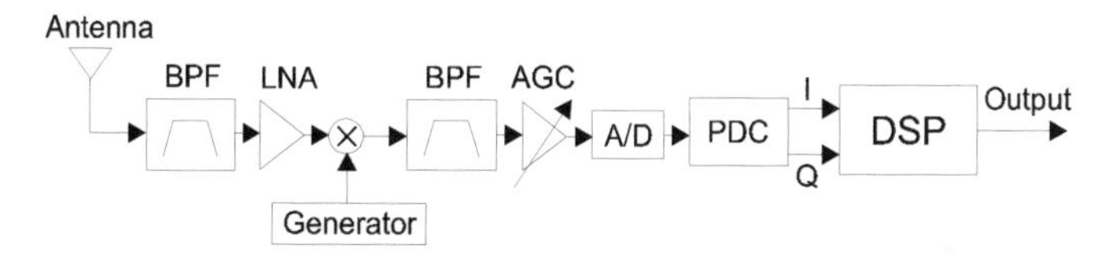

Figure 17.25 The scheme of a digital programmable receiver with the *Programmable Down-Converter* (PDC)

17.6.4 Software Radio in the realization of base stations

The application of the idea of the *Software Radio* in base stations is different from that in mobile stations [23]. The main task of a SR-implemented base station is the integration of the transmit/receive blocks operating on a single carrier in a unified, multifrequency transceiver for a whole base station. The main difficulty in the realization of such a transceiver is ensuring a sufficient A/D accuracy (14-bit conversion at the 50 MHz sampling frequency in case of the GSM system). Another challenge is the design of a highly linear high power amplifier (HPA) whose bandwidth covers a whole system band. In [26] the estimation of the required computational complexity of a GSM base station has been reported. It is expected that soon it will be possible to implement a GSM base station on a fast workstation or a server.

* * *

Software Radio is currently intensively researched by large scientific teams in Europe and the USA. The reader interested in this subject is advised to study the recently published books [29] and [30] and numerous papers published in proceedings of mobile communications conferences.

REFERENCES

1. H. Holma, A. Toskala (eds), *WCDMA for UMTS: Radio Access for Third Generation Mobile Communications*, John Wiley & Sons, Ltd., Chichester, 2000

2. J. P. Castro, *The UMTS Network and Radio Access Technology: Air Interface Techniques for Future Mobile Systems*, John Wiley & Sons, Ltd., Chichester, 2001

3. T. Ojanperä, R. Prasad (eds), *WCDMA: Towards IP Mobility and Mobile Internet*, Artech House, Boston, 2001

4. V. K. Garg, *IS-95 and cdma2000. Cellular/PCS Systems Implementation*, Prentice-Hall, Upper Saddle River, N.J., 2000

5. J. Rapeli, "UMTS: Targets, System Concept and Standardization in a Global Framework", *IEEE Personal Communications*, Vol. 2, No. 1, February 1995, pp. 20-28

6. ETSI TS 125.301, Universal Mobile Telecommunications System (UMTS); Radio Interface Protocol Architecture, (3G TS 25.301) V.3.4.0, March 2000

7. E. Dahlman, P. Beming, J. Knutsson, F. Ovesjö, M. Persson, Ch. Roobol, "WCDMA – The Radio Interface for Future Mobile Multimedia Communications", *IEEE Trans. Vehicular Technology*, Vol. 47, No. 4, November 1998, pp. 1105-1117

8. F. Adachi, M. Sawahashi, K. Okawa, "Tree-structured Generation of Orthogonal Spreading Codes with Different Lengths for Forward Link DS-CDMA Mobile", *Electronic Letters*, Vol. 33, No. 1, 1997, pp. 27-28

9. ETSI TS 125.211, Universal Mobile Telecommunications System (UMTS); Physical Channels and Mapping of Transport Channels onto Physical Channels (FDD), V.3.2.0, March 2000

10. ETSI TS 125.212, Universal Mobile Telecommunications System (UMTS); Multiplexing and Channel Coding (FDD), V.3.2.0, March 2000

11. ETSI TS 125.213, Universal Mobile Telecommunications System (UMTS); Spreading and Modulation (FDD), V. 3.2.0, March 2000

12. ETSI TS 125.214, Universal Mobile Telecommunications System (UMTS); Physical Layer Procedures (FDD), V. 3.2.0, March 2000

13. ETSI TS 125.221, Universal Mobile Telecommunications System (UMTS); Physical Channels and Mapping of Transport Channels onto Physical Channels (TDD), V.3.2.0, March 2000

14. ETSI TS 125.222, Universal Mobile Telecommunications System (UMTS); Multiplexing and Channel Coding (TDD), V.3.2.0, March 2000

15. ETSI TS 125.223, Universal Mobile Telecommunications System (UMTS); Spreading and Modulation (TDD), V. 3.2.0, March 2000

16. ETSI TS 125.224, Universal Mobile Telecommunications System (UMTS); Physical Layer Procedures (TDD), V. 3.2.0, March 2000

17. Qualcomm CDMA Technologies, The CDMA Difference, http://www.cdmatech.com/difference/faq/cdma_2000.html

18. 3GPP2 IS-2000.2, Physical Layer Standard for cdma2000 Spread Spectrum Systems

19. 3GPP2 IS-2000.3, Medium Access Control (MAC) Standard for cdma2000 Spread Spectrum Systems

20. 3GPP2 IS-2000.4, Signaling Layer 2 Standard for cdma2000 Spread Spectrum Systems

21. 3GPP2 IS-2000.5, Upper Layer (Layer 3) Signaling Standard for cdma2000 Spread Spectrum Systems.

22. Software Radios, Special Issue of *IEEE Communications Magazine*, May 1995

23. *Software Radio Workshop*, European Commission DG XIII-B, Brussels, May 29, 1997

24. Software Radio and Baseband Technology, Session 1 and 2, *Proc. of ACTS Mobile Summit*, October 7-10, 1997, Aalborg, Denmark

25. G. Fettweis, Ph. Charas, R. Steele, "Mobile Software Telecommunications", *Proc. of European Personal Mobile Communications Conference*, Bonn, 1997, pp. 321-325

26. Th. Turletti, D. Tennenhouse, "Estimating the Computational Requirements of a Software GSM Base Station", *Proc. of IEEE International Conference on Communications*, 1997, pp. 169-175

27. W. H. Tuttlebee, "Software Radio Technology: A European Perspective", *IEEE Communications Magazine*, February 1999, pp. 118-123

28. E. Buraccini, "SORT & SWRADIO Concept", *Proc. of ACTS Fourth Mobile Communications Summit*, 1999, pp. 587-593

29. J. Mitola III, *Software Radio Architecture: Object-Oriented Approaches to Wireless Systems Engineering*, John Wiley & Sons, Inc., New York, 2000

30. E. Del Re (ed.), *Software Radio. Technologies and Services*, Springer-Verlag, London, 2001

18

Application of smart antennas in cellular telephony

18.1 INTRODUCTION

Growing demand for cellular telephony services and wireless access to telecommunication networks results in a continuous need to increase the capacity of the existing systems. As we remember, the capacity of a cellular system, measured as the traffic density per area unit, depends on many factors. These factors are: the assigned bandwidth, the width of a channel required for a single carrier, the applied multiple access method, the kind of applied modulation, methods of reception and processing of the signal carrying information, the tolerable value of the signal-to-co-channel interference ratio, and the type of base station antennas, in particular the number of antenna sectors.

It seems that the technique of *smart antennas* is one of the most important factors which will allow a substantial increase in system capacity. Smart antennas consist of an array of antenna elements and digital signal processing hardware and software which perform beamforming and estimation of direction of arrival (DoA) [9]. Smart antennas can be applied both in the existing and in future systems. This technique has found application in the first generation analog systems such as NMT [1] and in the second generation GSM and DCS 1800 systems [2], [3] and the American TDMA IS-54 [4], [5]. The smart antenna technique finds application in the 3G UMTS system as well [6].

In this chapter we will present basic rules of operation of smart antennas and we will show their influence on the factors which have a crucial meaning for the system capacity. The topic of smart antennas is very wide. The motivated reader can find an interesting lecture on smart antennas and beamforming in the book [7] and an introductory chapter on adaptive antennas in the handbook [8].

18.2 BASIC PRINCIPLES OF SMART ANTENNA SYSTEMS

In a typical cellular system the communication between the base station and mobile stations located in a cell is performed using the omnidirectional or sector antenna. Typical sector antennas emit and receive the signals within the angle of 180, 120 or 60 degrees. Most of the energy emitted by the base station antenna is lost because at a given moment the mobile stations are located in specified places only. If the signal intended for a given mobile station was sent only in the direction of that station and the angle of emission was updated following the movement of the mobile station, a lot of energy would be saved. The energy savings can be expressed as a higher signal-to-noise ratio observed during transmission between the base station and the mobile station or as an increased range of the base station. The strong antenna directivity limits the multipath effect, too, because the most delayed paths, arising from the reflections from the obstructions located at large angles with respect to the desired direction, are avoided.

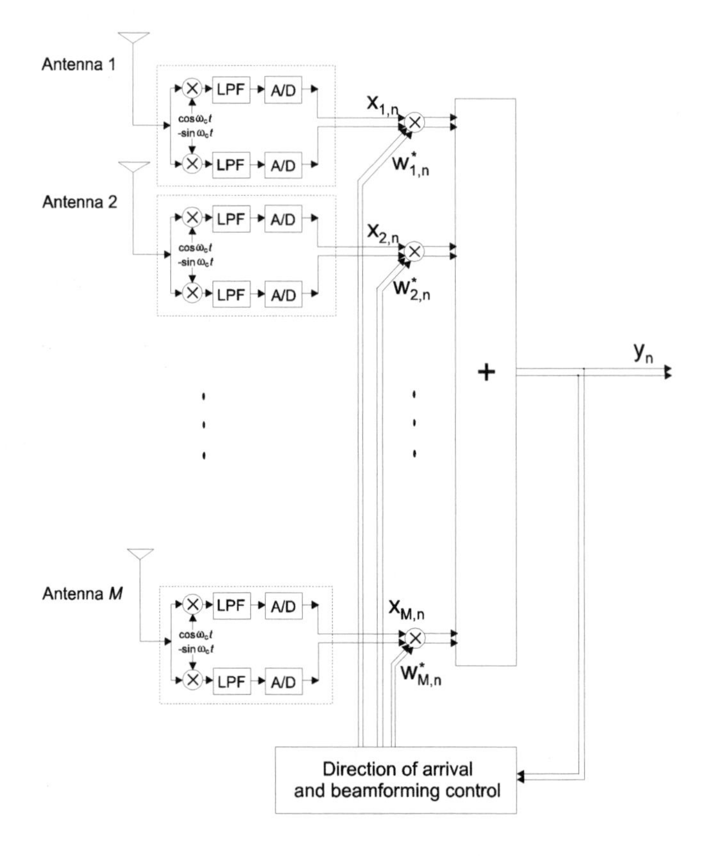

Figure 18.1 General scheme of the receiving part of the antenna array with the control system

Strongly directional antenna beam can be synthesized by the antenna array system shown in Figure 18.1. In this system the antenna array consisting of M linearly spaced antenna elements and a real time digital signal processing (DSP) controller are applied. The DSP controller shapes the space characteristics of the antenna array and at the same time it forms the shape and direction of the antenna beam. Other than uniformly spaced linear arrays are also applied.

In the receiving part of the system shown in Figure 18.1, the signals acquired by each antenna element are demodulated and their in-phase and quadrature components are extracted. As we know, the in-phase and quadrature components can be interpreted as the real part and imaginary part of the complex signal $x_{i,n}$, $(i = 1, \ldots, M)$, respectively. The signals $x_{i,n}$ are weighted by the weighting factors $w_{i,n}^*$ ($(.)^*$ denotes complex conjugate) and combined, giving the output signal y_n. The index n denotes subsequent time instants. Through appropriate selection of the coefficient set $w_{i,n}$ $(i = 1, \ldots, M)$ we are able to shape the space characteristics of the antenna array so as to maximize the power received from the desired mobile station or to minimize the influence of the interfering mobile stations.

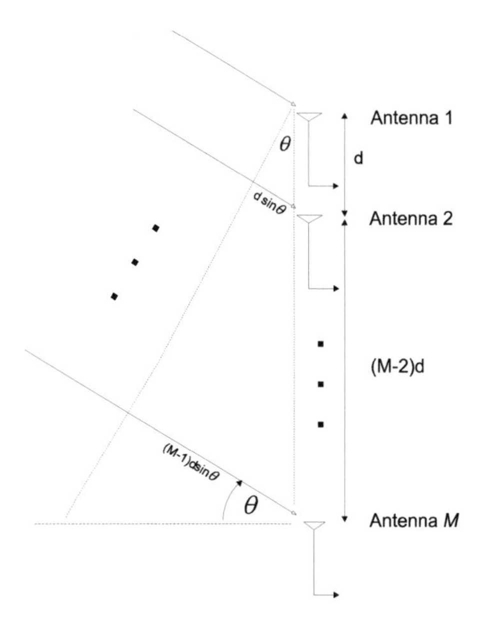

Figure 18.2 Signals reaching the antenna elements

Considering the reception of the signals generated by a sufficiently remote mobile station, we can assume that the waves carrying the desired signal reach M antenna elements along parallel paths (see Figure 18.2). If the antennas are located along a line (usually spaced by a half of the wavelength λ), we can notice that the difference in length between the reference path and the ith path is $d(i-1)\sin\theta$, $(i = 1, \ldots, M)$. In consequence, for the wavelength λ the difference in the path length is equivalent to the phase difference which is

$$\psi_i = \frac{2\pi d(i-1)}{\lambda} \sin\theta \tag{18.1}$$

Let us consider the case when the weight coefficients $w_{i,n}$ equalize the phase shifts between the signals received by antenna elements but do not change the signal amplitudes. The noise generated in the receivers connected to the particular antenna elements can be considered as mutually uncorrelated. Thus, the noise power on the combiner output increases M times. The desired signal received at each antenna element is phase-shifted due to the weight coefficients, and the components from each branch are positively combined. As a result, the signal amplitude increases M times so the power of the desired signal increases M^2 times. Therefore, the signal-to-noise ratio at the output of the signal combiner is M times higher than at the output of a single antenna receiver. In the decibel scale the obtained gain is

$$G = 10 \log M \quad [\text{dB}] \tag{18.2}$$

The application of $M = 8$ antenna elements theoretically results in the SNR gain of about 9 dB. In practice the gain is smaller, because the weight coefficients $w_{i,n}$ are used not only to correct the phase differences but also to lower the level of the sidelobes of the antenna array characteristics. This is done through the adjustment of the weights' magnitudes.

Consider another, a bit idealized, example showing the meaning of the antenna array [9]. The level of the received signal in function of the distance r from the base station can be estimated using the Hata propagation model [10] (compare Section 3.5.3). For typical GSM propagation conditions (carrier frequency range, the height of the base station and mobile station antennas) the Hata formula evolves to the form

$$L(r) = 100.1 + 33.3 \log r \quad [\text{dB}] \tag{18.3}$$

where $L(r)$ is the signal attenuation expressed in dB and r is the distance from the base station, measured in km. If for a single antenna the highest permissible attenuation level is achieved at the distance r_1 from the base station, then after application of the antenna array the same attenuation is achieved at the distance r_2, where

$$L(r_2) = L(r_1) + G \tag{18.4}$$

From (18.4) we conclude that $G = 33.3 \log \frac{r_2}{r_1}$, which results in the increase of the cell radius in the following proportion

$$\rho = \frac{r_2}{r_1} = 10^{G/33.3} = M^{0.3} \tag{18.5}$$

If we assume for simplicity that the base stations are distributed uniformly in the considered area S, then thanks to increasing the cell radius from r_1 (the area of a single cell is S_1) to r_2 (the cell area is now S_2) the number of necessary base stations decreases according to the reduction factor g given by the formula

$$g = \frac{S/S_2}{S/S_1} = \frac{S_1}{S_2} = \frac{r_1^2}{r_2^2} = \frac{1}{\rho^2} = M^{-0.6} \tag{18.6}$$

Figures 18.3 and 18.4 [9] illustrate formulas (18.5) and (18.6) drawn as a function of the number M of antenna elements. Space diversity with the maximum ratio combining or the selection combining using two antennas is very often applied in base stations. If we compare the gain obtained thanks to the antenna array with that obtained in the space diversity reception, the difference is obviously smaller than that shown in Figures 18.3 and 18.4.

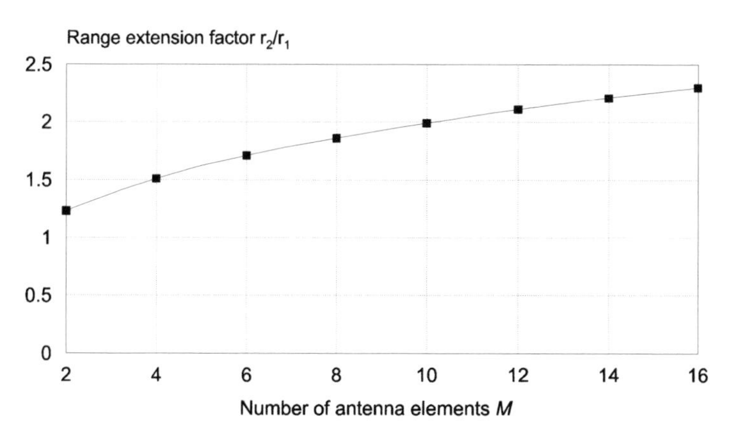

Figure 18.3 Range increase factor as a function of number of the antenna array elements M

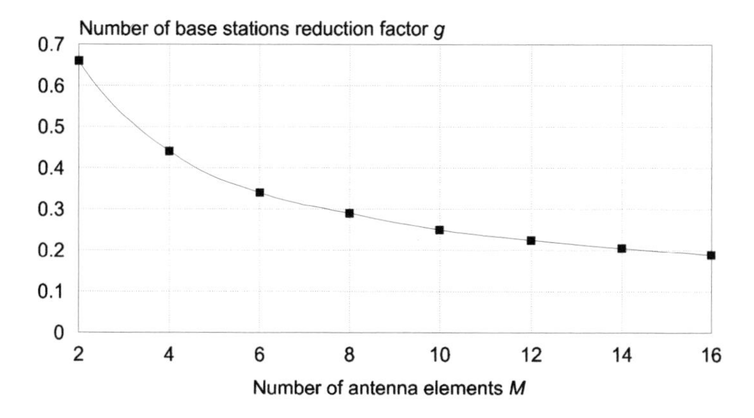

Figure 18.4 Reduction factor of the number of base stations as a function of the number of the matrix elements M

A substantial computational complexity of the receiver applying the antenna array results from the calculation of branch signal weights in real time. In the simple case a certain number of coefficient sets are calculated in advance and stored. Subsequently, the desired characteristic of the antenna array is synthesized by selection of the appropriate set of weights. This way we achieve the antenna array with a *switched lobe*. The task gets more complicated if the antenna array determines the *Direction of Arrival*

(DoA) of the user signal and has to track the change of the mobile station location. Such an antenna array is called a *dynamically phased array* [20]. Finally, if besides the tracking of the desired mobile station, the DoAs of interfering sources are to be determined, too, the antenna array attempts to maximize the signal-to-interference ratio by directing the main lobe to the desired station and placing the nulls of the array characteristics in the direction of the interferers. That kind of antenna array is called an *adaptive array.*

Let us analyze the antenna array consisting of M isotropic antenna elements linearly and uniformly distributed. Let L be the number of mutually uncorrelated mobile stations, treated as point signal sources and located sufficiently far away from the base station. We will treat the first antenna element as the reference. For the jth signal source the signal received at the reference element is

$$m_j(t) \exp\left[j2\pi f_0 t\right] \tag{18.7}$$

The delay of the arrival of the jth signal at the kth antenna element is given by the expression

$$\tau_k(\theta_j) = \frac{d}{c}(k-1)\sin\theta_j \tag{18.8}$$

so the signal received from the jth source at the kth receiver is

$$m_j(t) \exp\left[j2\pi f_0(t - \tau_k(\theta_j))\right] \tag{18.9}$$

In consequence, the signals from all the sources seen at the kth receiver are described by the formula

$$x_k(t) = \sum_{j=0}^{L-1} m_j(t) \exp\left[j2\pi f_0(t - \tau_k(\theta_j))\right] \tag{18.10}$$

so after weighting the signals from all receivers and combining them we receive the output signal in the form

$$y(t) = \sum_{k=1}^{M} w_{k,n}^* x_k(t) \tag{18.11}$$

or in the vector form

$$y(t) = \mathbf{w}_n^H \mathbf{x}(t) \tag{18.12}$$

where $(.)^H$ denotes transposition of the complex conjugated vector.

As we have already mentioned, several methods of the weights calculation are possible. In the most conventional one, in which the weights compensate the phase shifts only, their values are

$$w_{k,n} = \frac{1}{M} \exp\left[-j2\pi f_0 \tau_k(\theta_K)\right] \tag{18.13}$$

where the index K denotes the desired signal. In a more advanced method, the weights are selected to locate the zeros of the antenna array characteristics in these directions from which the undesired signals arrive. In order to consider this case, let us introduce the so-called *steering vector* $\mathbf{a}_j$. It is given by the formula

$$\mathbf{a}_j = \mathbf{a}_j(\theta_j) = \left[\exp\left(j2\pi f_0 \tau_1(\theta_j)\right), \ldots, \exp\left(j2\pi f_0 \tau_M(\theta_j)\right)\right]^T \tag{18.14}$$

and it is determined by the phase shifts caused in the antenna elements by the jth signal. Let θ_0 be the angle of arrival of the desired signal, whereas $\theta_1, \ldots, \theta_L$ are the angles of arrival of interfering signals. The vector of weights which sets the zeros in the characteristics in the directions indicated by the angles $\theta_1, \ldots, \theta_L$ and exposes the desired signal arriving at the angle θ_0 is determined from the following set of equations

$$\begin{aligned} \mathbf{w}_n^H \mathbf{a}_0 &= 1 \\ \mathbf{w}_n^H \mathbf{a}_j &= 0, \quad j = 1, \ldots, L-1 \end{aligned} \tag{18.15}$$

The same set of equations can be written in more elegant matrix form as

$$\mathbf{w}_n^H \mathbf{A} = \mathbf{e}^T \tag{18.16}$$

where $\mathbf{A}$ is the matrix with its columns being the steering vectors

$$\mathbf{A} = [\mathbf{a}_0, \mathbf{a}_1, \ldots, \mathbf{a}_{L-1}] \tag{18.17}$$

and $\mathbf{e} = [1, 0, \ldots, 0]^T$. Multiplying (18.16) on the right-hand side by $\mathbf{A}^H$, we receive

$$\mathbf{w}_n^H = \mathbf{e}^T \mathbf{A}^H \left(\mathbf{A}\mathbf{A}^H\right)^{-1} \tag{18.18}$$

As we see from (18.18), the calculation of weights requires the exact knowledge of the angles $\theta_1, \ldots, \theta_L$ for all interfering sources. Besides that, (18.18) does not take into account the noise. Instead, we can formulate the optimization problem for the weights in the following form

$$\min_{\mathbf{w}} P(\mathbf{w}) = \min_{\mathbf{w}} E\left[|y(t)|^2\right] = \min_{\mathbf{w}} \mathbf{w}^H E\left[\mathbf{x}\mathbf{x}^H\right]\mathbf{w} = \min_{w} \mathbf{w}^H \mathbf{R}_x \mathbf{w} \tag{18.19}$$

with the constraint $\mathbf{w}_n^H \mathbf{a}_0 = 1$. $P(\mathbf{w})$ is the mean power at the output of the combiner, $E[.]$ denotes expectation and $\mathbf{R}_x = E[\mathbf{x}\mathbf{x}^H]$ is the input signal correlation matrix. Formula (18.19) indicates that we minimize the mean power of the noise and interfering signals at the constant power of the desired signal. The solution of this optimization problem with the above constraint results in the following formula for the optimum weights

$$\mathbf{w}_{opt} = \frac{\mathbf{R}_x^{-1}\mathbf{a}_0}{\mathbf{a}_0^H \mathbf{R}_x^{-1} \mathbf{a}_0} \tag{18.20}$$

Calculating (18.20) requires the knowledge of the correlation matrix of the signals received at each antenna output.

A much easier method from the implementation point of view consists in finding the weights iteratively, minimizing the mean square error. This time the pattern signal d_n is required and the minimized function is determined by the expression

$$\mathcal{E}_{MSE} = E\left[\left|e_n\right|^2\right] = E\left[\left|d_n - \mathbf{w}_n^H \mathbf{x}_n\right|^2\right] = \\ = E\left[\left|d_n\right|^2\right] + \mathbf{w}^H \mathbf{R}_x \mathbf{w} - 2\mathbf{w}^H \mathbf{p} \tag{18.21}$$

where $\mathbf{p} = E\left[d_n^* \mathbf{x}_n\right]$ is the cross-correlation vector of the pattern signal and the vector of antenna elements' signals. The minimum of (18.21) equal to

$$\mathbf{w}_{MSE} = \mathbf{R}_x^{-1} \mathbf{p} \tag{18.22}$$

can be found iteratively using the well-known least mean squares (LMS) algorithm

$$\mathbf{w}_{n+1} = \mathbf{w}_n + \alpha e_n^* \mathbf{x}_n, \quad \text{where} \quad e_n = d_n - \mathbf{w}_n^H \mathbf{x}_n \tag{18.23}$$

The vector of weights $\mathbf{w}_{n+1}$ at the moment $n+1$ is received from the weights vector $\mathbf{w}_n$ at the nth moment by adding the updating terms depending on the current error e_n, the signals $\mathbf{x}_n$ at the output of each antenna branch and a small constant α which is called the step size of the algorithm. The value of α has a crucial influence on the convergence of the algorithm (18.23) and the final value of the mean square error. Figure 18.5 presents the characteristics of the 8-element antenna array with the weights which were received using (18.23) after 10000 iterations. Three interfering signals arrived at the angles equal to -45°, 20°, 60°. It was assumed that the angle of arrival of the desired signal was 0° and the SNR was equal to 10 dB.

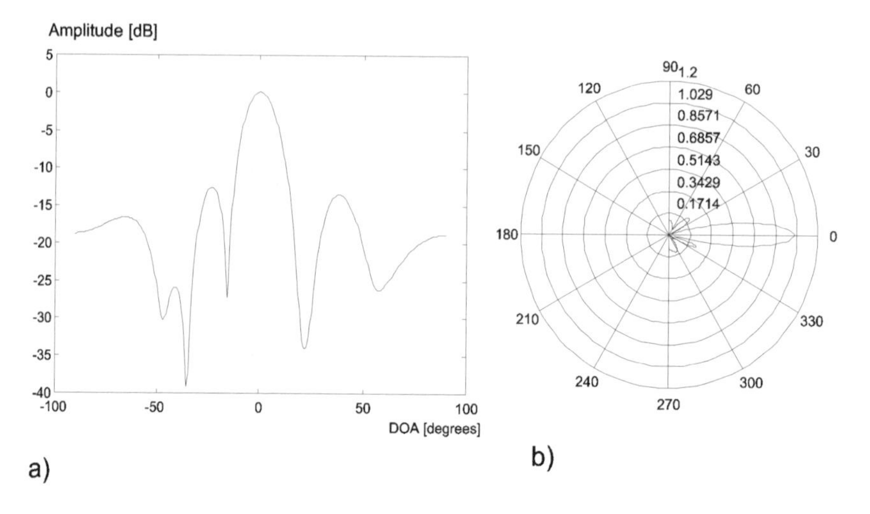

Figure 18.5 Two forms of characteristics of the 8-element antenna array achieved using the LMS algorithm (18.23)

So far we have been interested in the antenna array working in the receiver and we considered the weights adjustment performed for the uplink. The situation is more complicated in the downlink where the antenna array is applied in the transmitter. The base station has to form a beam in the direction of the desired mobile station to extract its signal from among the interfering signals. If the time division duplex transmission on the same carrier is used in communication between the mobile station and the base station, then, assuming that the channel varies slowly with respect to the frame length, the same weights can be used in the downlink as those calculated for the uplink. Here the channel reciprocity rule is applicable. However, if the frequency division duplex method is applied, transmission in the uplink and in the downlink takes place on different carriers. In this case the channel reciprocity rule is not applicable. Fortunately, the assumption of directional reciprocity is fully justified. It means that the antenna array in the transmitter should form a beam in the downlink in the same direction from which the signal has arrived in the uplink [20]. Thus, the estimation of *Direction of Arrival* (DoA) is necessary to estimate the angular directions of the mobile stations in the cell. This estimation is also useful in the handover procedure and in the determination of the concentration of mobile stations.

The DoA algorithms could be the topic of a separate chapter. The motivated reader is advised to study the tutorial paper [21]. Here we will present the most important methods without going into much detail.

The simplest method relies on the computation of the received signal spectrum as a function of the angle of arrival θ. Subsequently, the local maxima of the estimated spectrum are determined. Their arguments indicate the angular directions of the signal sources. Knowing that the received power P in the function of the vector of weights $\mathbf{w}$ can be estimated using the formula

$$P(\mathbf{w}) = \frac{1}{N}\sum_{i=1}^{N}|y_i|^2 = \frac{1}{N}\sum_{i=1}^{N}\mathbf{w}^H\mathbf{x}_i\mathbf{x}_i^H\mathbf{w} = \mathbf{w}^H\widehat{\mathbf{R}}_x\mathbf{w} \tag{18.24}$$

and setting the vector of weights $\mathbf{w}(\theta) = \mathbf{a}(\theta)/M$, we get the formula for the power received from direction θ as

$$P(\theta) = \frac{\mathbf{a}^H(\theta)\widehat{\mathbf{R}}_x\mathbf{a}(\theta)}{M^2}, \tag{18.25}$$

$$\text{where} \quad a(\theta) = \left[1, e^{j\varphi}, \ldots, e^{j(M-1)\varphi}\right], \quad \varphi = \frac{2\pi f_0}{c} d\sin\theta$$

The drawback of this method is its low resolution because the resulting power is not only contributed by the main lobe steered by the weights but also from the side lobes of the antenna array characteristics.

Several other methods of finding the DoA have been investigated. Among them the most important are MUSIC [22] and ESPRIT [23].

In its basic version, the MUSIC algorithm estimates in the first step the correlation matrix $\mathbf{R}_x$ of the received signals from M antenna elements. If L different signal sources are expected, then in the next step $M-L$ smallest eigenvalues of $\mathbf{R}_x$ are found. Next, the M by $M-L$ matrix $\mathbf{U}$ of the eigenvectors associated with the above eigenvalues is formed. Finally, the maxima of the following spectrum

$$P_{MUSIC}(\theta) = \frac{1}{|\mathbf{a}^H(\theta)\mathbf{U}|^2} \tag{18.26}$$

are searched for. They indicate the DoA of all signal sources.

Description of the ESPRIT algorithm requires longer derivation and it will not be presented here. Both algorithms attracted a lot of interest and many improvements have been proposed for them.

18.3 PHASES OF SMART ANTENNA DEPLOYMENT

Introduction of the smart antenna technique can be performed in three phases [9].

In **Phase 1** smart antennas are applied in the uplink only. Thanks to them the range of a base station increases. This is equivalent to a possible decrease in the power emitted by a mobile station. This last possibility is particularly advantageous due to health reasons. The other advantage, which we have already considered, is a possibility of deploying fewer base stations in a given area. This feature is particularly attractive for DCS 1800, PCS 1900 and other similar systems which have smaller cells than standard cellular telephony, and allows reduction in the cost of deployment and system exploitation in the areas with low traffic density.

In **Phase 2** smart antennas are applied in the base station in both up- and downlink. The basic aim of this development phase is to reduce the interference through beamforming, sending the signal only in the direction of the mobile station and tracking its location. Such technique is often called *Space Filtering for Interference Reduction* (SFIR). It allows reduction in the power emitted by the base station. Thanks to the reduction of the co-channel interference, the cell cluster size can be decreased. This in turn has an impact on the system capacity through increasing the frequency reuse factor. Figure 18.6 presents graphically the operation of the cellular system with adaptive antennas.

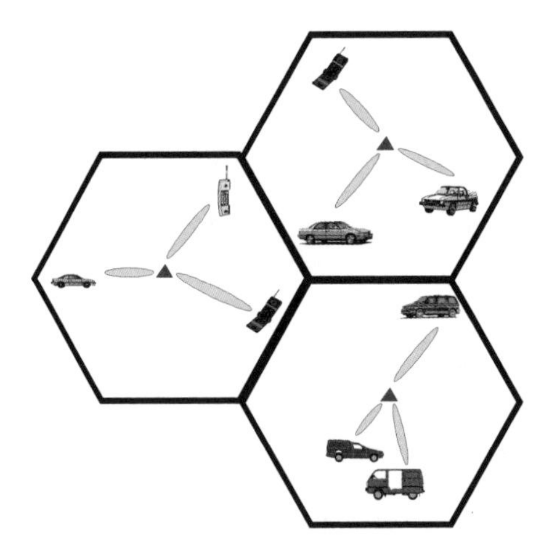

Figure 18.6 Operation of the cellular system using adaptive antennas for tracking locations of mobile stations

Finally, in **Phase 3** the known techniques of multiaccess such as FDMA, TDMA and CDMA are supplemented with *Space Division Multiple Access* (SDMA). Thanks to beamforming, the tracking of a mobile station and the co-channel interference reduction, more than one mobile station can use the same channel in a cell if the mobile stations are appropriately angularly distributed. This results in a substantial increase in the system capacity and flexibility.

18.4 INFLUENCE OF THE SMART ANTENNA TECHNOLOGY ON THE CELLULAR SYSTEM CAPACITY

One of the main reasons for the interest in the smart antenna technology in mobile systems is the expected system capacity increase measured as the traffic served per unit area. Another, more precise measure of the gain achieved thanks to this technique is the spectral efficiency η_s characterizing the traffic per unit area per Hz. This parameter is given by the formula

$$\eta_s = \frac{N_c G_c}{W_{sys} S} \quad [\mathrm{Erl/m^2/Hz}] \tag{18.27}$$

where N_c is the number of channels in a cell, G_c is the offered traffic load in a cell, S is the cell area and W_{sys} is the bandwidth used by the cellular system.

The values of the expected gains achieved thanks to the application of the smart antenna arrays depend on the ways of their operation and on the assumptions made in the system analysis. Due to the complex nature of the phenomena and the interdependence of many parameters, elaborate simulation models are usually built to estimate the gain achieved by the application of antenna arrays at some simplifying assumptions. These assumptions refer to the locations of mobile stations in a cell, the way the mobile stations move and the influence of the mobile stations and base stations from other cells.

In the previous paragraph we described three phases in the introduction of smart antenna technology in cellular systems. The gain estimations shown below are associated with the second and the third phase. In [13] the increase in the spectral efficiency of a system with SFIR in comparison to a conventional system has been estimated to be equal to $\sqrt{M}$. For the system using SDMA the gain has been found to be $\sqrt{KM}$, where K is not greater than 3 for $M = 8$ antenna elements. [14] reports the simulation results in which handover was taken into account. It was assumed that the signal-to-co-channel interference ratio decreases below the threshold level of $9+6 = 15$ dB for GSM. The value of 9 dB results from the GSM standard defining the signal-to-co-channel interference ratio necessary for reliable system operation. The additional 6 dB margin is results from the shadowing effect. Realistic antenna radiation patterns have been assumed. The detailed results have been summarized in Table 18.1, quoted according to [14]. The propagation model was assumed in which the received power decreases with the fourth power of the distance from the base station. Let us note the size of the cell cluster was equal to $N = 3$ or $N = 1$, which indicates potential use of the whole set of carrier frequencies in each cell. Let us compare these numbers with typical values of $N = 4$ at the application of $120°$–sector antennas in GSM or with $N = 7$ or $N = 12$ in the analog systems. When the SDMA technique is applied, the spectral efficiency increases considerably. Some losses are observed if the power control in the mobile stations is not perfect; however, they are not very meaningful.

Table 18.1 Increase of spectral efficiency η_s due to the application of intelligent antenna technique as compared with the systems using omnidirectional or sector antennas, N - the size of the cell cluster

		System with omnidirectional antenna	System with sector antennas
Sector antennas		2.3	1
SFIR	N = 3	2.3	1
SFIR	N = 1	7	3
SDMA	N = 3	4.8	2.8
SDMA	N = 1	9.8	5.4

The price which has to be paid for the capacity increase is not only growing complexity of the antenna system and the RF part of the transmitter but also the introduction of a complex digital signal processing block which controls the radiation patterns of the antenna array. Due to the number of M received signals, the front-end blocks of the receiver are also getting much more complicated, particularly if the maximum likelihood detection is performed in each branch [15]. Channel allocation rules in the systems using SDMA principle become more complicated too, because the allocation algorithm has additionally to take into account the angular distance between the mobile stations [16]. The frequency of intra-cell handovers increases, too.

Generally, the SDMA technique can be treated as a substantial enrichment of the existing multiple access methods such as FDMA, TDMA and CDMA. This has been confirmed by field measurements reported in the literature [17], [18] and [19].

REFERENCES

1. H. Andersson, M. Landing, A. Rydberg, T. Öberg, "An Adaptive Antenna for the NMT 900 Mobile Telephony System", *Proc. of IEEE Vehicular Technology Conference*, 1994

2. U. Forssén, J. Karlsson, B. Johannisson, M. Almgren, F. Lotse, F. Kronestedt, "Adaptive Antenna Arrays for GSM900/DCS1800", *Proc. of IEEE Vehicular Technology Conference*, 1994

3. M. Tangemann, U. Bigalk, C. Hoeck, M. Hother, "Sensivity Enhancements of GSM/ DCS1800 with Smart Antennas", *Proc. of the Second European Personal Mobile Communications Conference*, 1997

4. J. Winters, "Signal Acquisition and Tracking with Adaptive Arrays in the Digital Mobile Radio System IS-54 with Flat Fading", *IEEE Trans. on Vehicular Technology*, Vol. 42, No. 4, 1993

5. K. J. Molnar, G. E. Bottomley, "D-AMPS Performance in PCS Bands with Array Processing", *Proc. of IEEE Vehicular Technology Conference*, 1996

6. G. Tsoulos, M. Beach, S. C. Swales, "Adaptive Antennas for Third Generation DS-CDMA Cellular Systems", *Proc. of IEEE Vehicular Technology Conference*, 1995

7. J. Litva, T. K.-Y. Lo, *Digital Beamforming in Wireless Communications*, Artech House, Boston, 1996

8. S. R. Saunders, *Antennas and Propagation for Wireless Communication Systems*, J. Wiley & Sons, Ltd., Chichester, 1999

9. M. Tangemann, "Smart Antenna Technology for GSM/DCS1800", *Proc. of the Second Workshop on Personal Wireless Communications* (PWC), Frankfurt, Dec. 10-11, 1996

10. M. Hata, "Empirical Formula for Propagation Loss in Land Mobile Radio Services", *IEEE Trans. on Vehicular Technology*, Vol. 29, August 1980

11. H. Krim, M. Viberg, "Two Decades of Array Signal Processing Research", *IEEE Signal Processing Magazine*, July 1996

12. M. Haardt, J. A. Nossek, "Unitary ESPRIT: How to Obtain Increased Estimation Accuracy with a Reduced Computational Burden", *IEEE Trans. on Signal Processing*, Vol. 43, May 1995

13. M. Tangemann, C. Hoek, R. Rheinschmitt, "Introducing Adaptive Array Antenna Concepts in Mobile Communications Systems", *Proc. of RACE Mobile Telecommunications Workshop*, Amsterdam, May 17-19, 1994, Vol. 2

14. J. Fuhl, A. Kuchar, E. Bonek, "Capacity Increase in Cellular PCS by Smart Antennas", *Proc. of IEEE Vehicular Technology Conference*, 1997

15. G. E. Bottomley, K. Jamal, "Adaptive Arrays and MLSE Equalization", *Proc. of IEEE Vehicular Technology Conference*, 1995

16. N. Gerlich, M. Tangemann, "Towards a Channel Allocation Scheme for SDMA-based Mobile Communication Systems", *ITG Fachbericht* No. 135, *Mobile Kommunikation*, 26-28 Sept. 1995, Neu-Ulm, Germany

17. P. Mogensen, K. Pedersen, P. Leth-Espensen, B. Fleury, F. Frederiksen, K. Olesen, S. Larsen, "Preliminary Measurement Results from an Adaptive Antenna Array Testbed for GSM/UMTS", *Proc. of IEEE Vehicular Technology Conference*, 1997

18. S. Anderson, U. Forssén, J. Karlsson, "Ericsson/Mannesmann GSM field-trials with adaptive antennas", *Proc. of IEEE Vehicular Technology Conference*, 1997

19. P. Chevalier, F. Pipon, J.-J. Monot, C. Demeure, "Smart Antennas for the GSM System: Experimental Results for a Mobile Reception", *Proc. of IEEE Vehicular Technology Conference*, 1997

20. P. H. Lehne, M. Pettersen, "An Overview of Smart Antenna Technology for Mobile Communication Systems", *IEEE Communication Surveys and Tutorials*, http://www.comsoc.org/publications/surveys/, Vol. 2, No. 4, Fourth Quarter 1999

21. L. C. Godara, "Application of Antenna Arrays to Mobile Communications, Part II: Beam-Forming and Direction-of-Arrival Considerations", *Proceedings of IEEE*, Vol. 85, No. 8, August 1997, pp. 1195-1247

22. R. O. Schmidt, "Multiple Emitter Location and Signal Parameter Estimation", *IEEE Trans. Antennas Propagat.*, Vol. AP-34, 1986, pp. 276-280

23. R. Roy, T. Kailath, "ESPRIT - Estimation of Signal Parameters via Rotational Invariance Technique", *IEEE Trans. Acoust., Speech, Signal Processing*, vol ASSP-37, 1989, pp. 984-995

Index